Optical Wireless Communication Theory and Technology

Series Editor

Xizheng Ke, School of Automation and Information Engineering, Xi'an University of Technology, Xi'an, Shaanxi, China

The book series Optical Wireless Communication Theory and Technology aims to introduce the key technologies and applications adopted in optical wireless communication to researchers of communication engineering, optical engineering and other related majors. The individual book volumes in the series are thematic. The goal of each volume is to give readers a comprehensive overview of how the theory and technology in a certain optical wireless communication area can be known. As a collection, the series provides valuable resources to a wide audience in academia, the communication engineering research community and anyone else who are looking to expand their knowledge of optical communication.

Xizheng Ke · Jiali Wu

Coherent Optical Wireless Communication Principle and Application

Science Press
Beijing

Springer

Xizheng Ke
School of Automation and Information Engineering
Xi'an University of Technology
Xi'an, Shaanxi, China

Jiali Wu
School of Automation and Information Engineering
Xi'an University of Technology
Xi'an, Shaanxi, China

ISSN 2731-5967 ISSN 2731-5975 (electronic)
Optical Wireless Communication Theory and Technology
ISBN 978-981-19-4822-0 ISBN 978-981-19-4823-7 (eBook)
https://doi.org/10.1007/978-981-19-4823-7

Jointly published with Science Press
The print edition is not for sale in China mainland. Customers from China mainland please order the print book from: Science Press.

This Springer imprint is published by the registered company Springer Nature Singapore Pte Ltd.
The registered company address is: 152 Beach Road, #21-01/04 Gateway East, Singapore 189721, Singapore

Preface

The rapid development of artificial intelligence has resulted in higher requirements for the communication industry. The optical communication industry is focusing on larger channel capacities, longer transmission distances, higher data transmission rates, and better transmission quality. The intensity modulation/direct detection method adopted by the traditional optical communication technology cannot cope with this high-speed communication, owing to its limited single-channel bandwidth.

Coherent optical communication is favored by researchers because of its unique advantages of rich modulation modes, large communication capacity, and high receiving sensitivity. It can fundamentally improve the selectivity of the receiver and increase the relay-free transmission distance of optical signals. To achieve more accurate, effective, and reliable coherent optical communication, breakthroughs are needed in corresponding key technologies, such as carrier frequency stability, beam coupling, and the correction of the polarization direction and wavefront distortion of the signal light at the receiving end.

The book is divided into nine chapters, which provide an overview of optical wireless coherent detection, coherent optical communication, spatial optical-fiber coupling and beam control, beam polarization control technology, double-balanced detection, adaptive optics correction, wavefront sensorless correction, spatial coherent optical communication wavefront correction of LC-SLM-R, and the influence of the beam mode on the performance of coherent detection systems. The principles and key technologies of coherent optical wireless communication are also introduced. Array coupling reception and large aperture coupling are proposed for spatial optical coupling. To improve the mixing efficiency of coherent detection receivers, signal optical polarization control and adaptive optics technology are proposed.

This book is a result of the dedicated efforts of the author and his graduate students, from the Optoelectronic Technology Research Center of Xi'an University of Technology. The author and his graduate students have conducted a considerable amount of theoretical research and a number of experiments on the coherent optical wireless communication in recent years. Their efforts have contributed to the further development of optical wireless communication technology.

The related work of this book has been supported by the Shaanxi Provincial Key R&D Program (Key Industrial Innovation Chain) Project (2017ZDCXL-GY-06-01), the research plan project of the Shaanxi Provincial Department of Education (18JK0341) and the Xi'an science and technology plan project (2020KJRC0083).

This book is a preliminary summary of the research conducted by the author on coherent optical wireless communication. Due to the limited knowledge of the author, there may be some inevitable inconsistencies in the book. Readers are welcome to suggest corrections.

Xi'an, China
Spring 2022

Xizheng Ke
Jiali Wu

Introduction

This book focuses on optical wireless communication. It introduces beam coupling, polarization control, and wavefront correction of signal light in detail. Further, it discusses the basic principles and heterodyne detection conditions of coherent optical communication, space optical-fiber coupling by array coupling reception, and large aperture coupling. The use of simulated annealing, particle swarm optimization, and single particle swarm optimization algorithms for controlling the polarization state of signal light is discussed. The stochastic parallel gradient descent algorithm, adaptive optics correction, liquid-crystal spatial light modulation for signal light wavefront distortion correction, and the influence of beam modes on coherent detection performance are also discussed. This book provides both theoretical deductions and experimental results, which can enhance the reader's understanding of the theories presented herein.

This book is suitable for engineering and technical personnel, college and university teachers, graduate students and undergraduate students in their final year, who are engaged in the field of optical wireless communication.

Contents

Chapter 1
Optical Wirelss Coherent Detection: An Overview

Contents

1.1 Optical Wireless Coherent Communication

Free-space optical communication (FSOC), also known as optical wireless communication (WOC), is a communication technology that uses laser beams as information carriers to directly transmit data, audio, and video signals in space [1]. With the development of the high-capacity, long-distance communication technology, WOC has broad application prospects in the fields of inter-satellite communication and the last-kilometer access of local networks. Therefore, WOC has become a research hotspot in optical communication technology [2].

Generally, WOC adopts intensity modulation/direct detection (IM/DD) and subcarrier modulation (SCM)/DD. IM refers to the use of the intensity of light to carry information, including the on–off keying (OOK) mode and pulse-position modulation (PPM) mode. DD refers to direct detection, using a photoelectric detector, of

X. Ke and J. Wu, *Coherent Optical Wireless Communication Principle and Application*, Optical Wireless Communication Theory and Technology,
https://doi.org/10.1007/978-981-19-4823-7_1

the intensity of a laser converging on a photosensitive surface. The IM/DD mode has the advantages of its simple principle and low cost; however, it has the disadvantages of low frequency band utilization and poor reception sensitivity [3]. In SCM, the binary bit information is modulated to the carrier frequency. Thereafter, the modulated subcarrier is used to modulate the optical carrier [4]. This modulation technology can use the spectrum resources effectively through the wavelength division multiplexing technology and improve the capacity of transmission [5].

When compared with the DD technology, the coherent optical detection technology can improve the receiver sensitivity and signal-to-noise ratio (SNR), realize multiple modulation modes [6], ensure appropriate selectivity [7] and increased communication capacity [8], and improve the performance of the WOC system. The requirements of coherent light detection for light sources, modulation, transmission, and reception are significantly stricter than those for DD [9]. It is necessary to address the problem of light source stability to realize coherent optical detection [10]. The frequency stabilities of the transmitter and receiver light sources are significantly high in a coherent optical communication system. Furthermore, wavefront matching and polarization matching of the received signal light and local oscillator light are necessary to ensure that the receiver has high detection sensitivity [11]. However, the atmospheric environment significantly affects the detection sensitivity of coherent light, thereby limiting the popularization and application of the coherent light detection technology. To mitigate the effects of the atmospheric turbulence, developed countries such as the United States, Europe, and Japan have adopted some atmospheric-disturbance-compensation technologies in the design of WOC systems [12, 13]. Some research progress in this direction has been made also in China [14]. Notwithstanding these developments, mitigating the effects of the atmospheric turbulence and improving the received SNR and optical detection sensitivity remain open areas in the field of wireless laser communication. Therefore, in-depth research in these areas has potential theoretical and application values.

1.2 Optical Wireless Communication: Development Status

In the 1980s, some countries began to study the theory and engineering of the FSOC technology using various modulation methods. The United States, Japan, and some European countries have a leading position in the research on optical communication technology.

In the early 1980s, the MIT Lincoln Laboratory (MIT-LL) and National Aeronautics and Space Administration (NASA) jointly developed the NASA-ACCS communication satellite demonstration system [15]. The transmitter in this system adopts differential phase shift keying (DPSK) modulation and receives through optical heterodyne detection. The collaboration successfully launched a satellite loaded with the system in 1989. In 1995, the Jet Propulsion Laboratory (JPL), supported by NASA and managed by the California Institute of Technology, developed a laser communication demonstration system (LCDS) [16], which adopted OOK modulation for the signal light and continuous wave modulation for the beacon light. The data transmission rate of the system was 750 Mbit/s. In 1995, the US Ballistic Missile Defense Organization implemented the Space Technology Research Vehicle 2 (STRV-2) experimental plan. The communication unit of STRV-2 adopted the IM/DD mode semiconductor laser emission and avalanche photodiode reception [17]. Its designed data transmission rate was 1 Gbit/s. In 2000, Lucent and Astro Terra successfully realized a commercial WOC experimental system with four wavelengths near 1550 nm, wavelength-division multiplexing of 10 Gbit/s, and a transmission distance of 5 km [18]. In October 2013, NASA successfully realized a laser communication link between the lunar laser communication terminal and the ground station [19]. Its data transmission rate was 622 Mbit/s. The project adopted 16-PPM, superconducting nanowire single-photon detection and ground terminal array transceiver technology, which effectively mitigated the problems of the light intensity flicker and low receiving power density in the atmospheric turbulence channel. Photon reception with high sensitivity was realized. In the late 1980s, the European Space Agency (ESA) began to formulate the semiconductor laser inter-satellite link experiment (SILEX) [20] plan to establish laser communication links between satellites and develop the required subsystems and communication units. In 1994, British, Canadian, and Belgian companies jointly developed a small optical user terminal (SOUT) and very small optical user terminal (VSOUT), which realized an inter-satellite communication link using IM/DD [21, 22]. The communication terminal developed in the SILEX program in 2001 successfully realized the first inter-satellite laser link experiment [23]. A binary phase shift keying (BPSK)/coherent detection mode was adopted, and the communication rate was 50 Mbit/s. In 2005, Deutsches Zentrum für Luft- und Raumfahrt (DLR), Germany, developed a BPSK modulation/homodyne detection optical-communication terminal, called laser communication terminal (LCT). The data transmission rate in this communication experiment reached 5.625 Gbit/s. In 2006, Davis [25] reported that the sensitivity of polarization modulation was approximately 3 dB higher than that of OOK modulation. In 2009, Popoola [26] of Northumbria University reported that atmospheric turbulence significantly reduced the performance of the DD system, and SCM was better than OOK, which is a modulation method that effectively suppresses the effect of atmospheric turbulence. In November 2014, a second generation LCT [24] laser communication was established between the AlphaSat satellite and Sentry 1A [27]. BPSK modulation/homodyne detection was adopted in that study, and the test rate was 600 Mbit/s.

In 1995, the IM/avalanche photodiode (APD) DD communication terminal laser communication equipment developed by Comprehensive Communication Research Laboratory, Japan, successfully performed two-way laser communication with an American atmospheric observation satellite for 8 min [28]. In May 2006, a laser communication experiment was successfully realized between the Optical Inter-orbit Communication Engineering Test Satellite (OICETS) and the National Institute of Information and Communication Technology (NICT) ground station in Japan. The laser-utilizing communication equipment (LUCE) laser communication terminal onboard OICETS [29, 30] used an optical power of 100 MW and an 847-nm laser with left-handed circular polarization modulated by non-return to zero-code direct IM. Its data rate was 49.3 Mbit/s and the bit error rate (BER) was 10^{-6}.

In 2002, the Chengdu Institute of Optoelectronic Technology, Chinese Academy of Sciences, developed a WOC terminal with a transmission rate of 10 Mbit/s, working wavelength of 850 nm, transmission distance of 1–4 km, and transmission power of 3–30 mW [31]. In 2002, the University of Electronic Science and Technology developed a ground demonstration system for satellite optical communication verification [32]. Wuhan University is engaged in research on atmospheric laser communication and beacon light acquisition and tracking. It performed an atmospheric laser communication experiment in 2002 [33], adopted the IM /DD mode, completed the ground simulation test of an automatic tracking servo system of space optical communication in 2008 [34], realized a 4.6 km FSOC link, and completed 5 Gbit/s data transmission using dense wavelength division multiplexing in 2011 [35]. In 2009, Guangxi Guilin Laser Communication Research Institute developed a laser atmospheric communication system with a communication distance of 5 km, communication rate of 155 Mbit/s, and BER of less than 10^{-7} [36]. In October 2011, a satellite-to-ground communication terminal developed by the Harbin University of Technology was launched on the Ocean 2 satellite. It achieved a single-channel data transmission rate of 504 Mbit/s [37]. In September 2013, the Changchun University of Technology successfully completed an aircraft-to-aircraft long-distance laser communication test [38].

The Xi'an University of Technology has been studying the atmospheric laser communication system since 2000 and has made significant progress in optical design, coding, and modulation [39]. A wireless laser communication system terminal has been successfully developed. The system has common digital interfaces, including serial and parallel ports and Ethernet. The system can realize full-duplex communication between computers or terminal equipment and reliable transmission of data, audio, and video. Its transmission rate and distance are 20 Mbit/s and 1 km, respectively. At present, it has been applied to expressway video monitoring systems.

It is evident that the transmitter used in the FSOC technology has developed from the initial OOK modulation mode to other reliable modulation technologies such as SCM [40], PPM [41], multi-pulse position modulation [42], digital pulse position modulation [43], and BPSK modulation. Such reliable modulation technologies with high efficiencies and strong anti-interference abilities facilitate a high transmission rate and low BER communication and play an important role in improving the performance of communication systems. The receiver of FSOC also develops into a coherent detection owing to the low sensitivity of direct detection.

1.3 Research Status at Home and Abroad

In this section, we summarize the research status of inter-satellite coherent optical detection and allied technologies at home and abroad.

1.3.1 Inter-Satellite Coherent Optical Detection

1.3.1.1 Abroad

The United States, Europe, Japan, and other developed countries have performed extensive research on the coherent optical detection technology in the field of inter-satellite optical communication. Recently, they have entered the experimental stage of inter-satellite coherent optical communication systems.

(1) United States

In the mid 1980s, the MIT-LL performed a laser inter-satellite transmission experiment (LITE) and developed an optical heterodyne detection communication terminal. This experimental device is depicted in Fig. 1.1. The transmitter used in this device uses the DPSK modulation method. The experiment used a 30 mW semiconductor laser, 200 mm telescope system with a data rate of 220 Mbit/s, communication distance of 40,000 km, and modulation rate of up to 1 Gbit/s [44].

JPL successfully performed a 64-PPM coherent optical communication experiment in 2004. The experimental device is depicted in Fig. 1.2. In this experiment, two Nd-doped yttrium aluminum garnet lasers with a wavelength of 1064 nm were used as signal light and local oscillator light sources. The optical heterodyne detection method was used at the receiving end, with an intermediate frequency (IF) of 6 MHz [45].

(2) Germany

In Germany, Lange successfully realized 5.625 Gbit/s BPSK-modulated homodyne detection optical communication in October 2006 [46]. Two LCTs were used for ground-to-ground coherent optical communication experiments between Canary

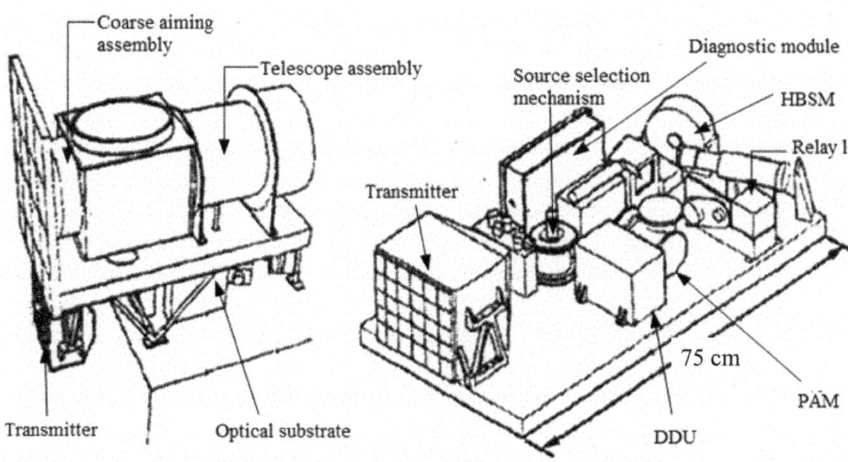

Fig. 1.1 Sketch of MIT Lincoln laboratory LITE [44]

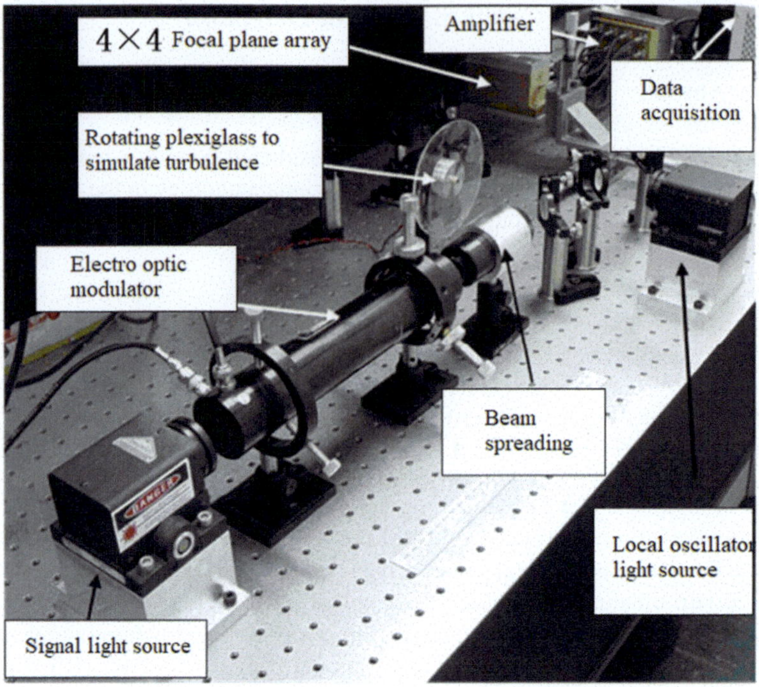

Fig. 1.2 Coherent light detection experimental device at JPL [45]

Islands, Spain, and La Palma and Tenerife islands 142 km away. The homodyne BPSK receiver used in the LCT is depicted in Fig. 1.3. The actual communication bit rate reached 5.625 Gbit/s without optical filtering, and the BER was between 10^{-6} and 10^{-4}.

In 2008, the near-field infrared experiment (NFIRE) of the Ministry of Defense successfully realized two-way broadband data transmission with a transmission rate of up to 5.6 Gbit/s. BPSK modulation/homodyne coherent optical detection was adopted, and the BER was better than 10^{-9}. This was the first time that a long-distance spatial coherent optical communication experiment was realized [47] in Germany. And, a laser link to realize satellite-to-ground bidirectional laser communication was successfully established between the terraSAR-X satellite and an optical ground station (OGS) in Germany. The link holding time was 8 min, and the data transmission rate was 5.6 Gbit/s. Experimental results indicated that the BER of the uplink was greater than 10^{-5}. The LCT homodyne coherent optical detection terminal was used at the transmitter and receiver [48]. An image of the LCT terminal, which comprises a coarse aiming device, telescope, and receiver, is depicted in Fig. 1.4.

In the same year, the infrared reconnaissance satellite NFIRE of the US Department of Defense established two-way laser communication with the OGS in Germany. The data transmission rate was 5.625 Gbit/s, and the link holding time

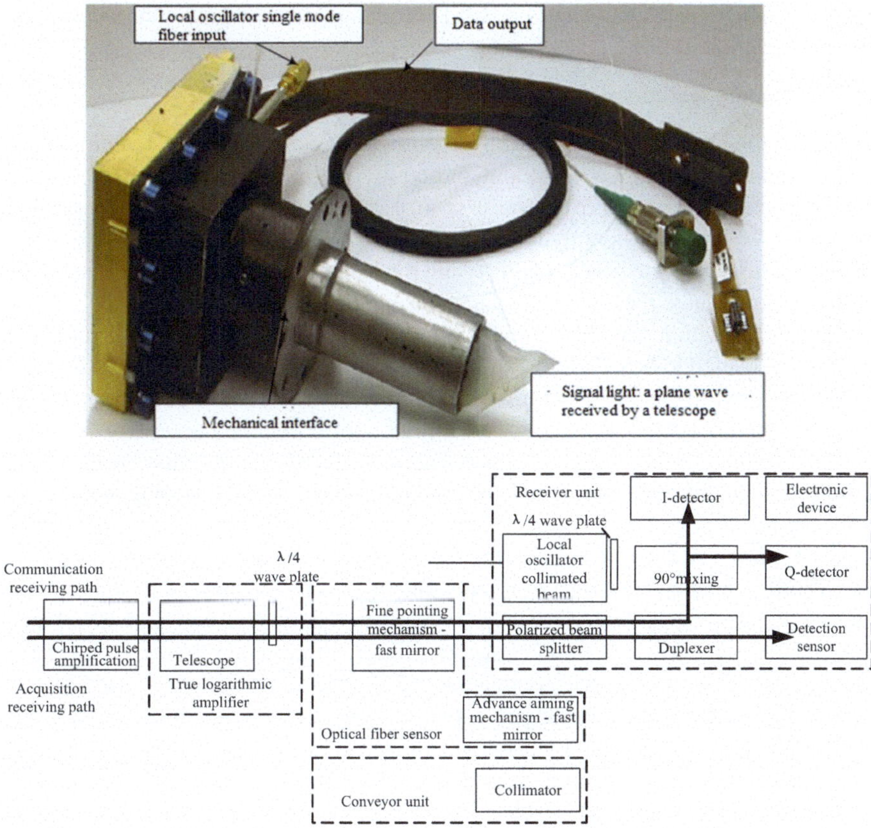

Fig. 1.3 LCT homodyne BPSK receiver [46]

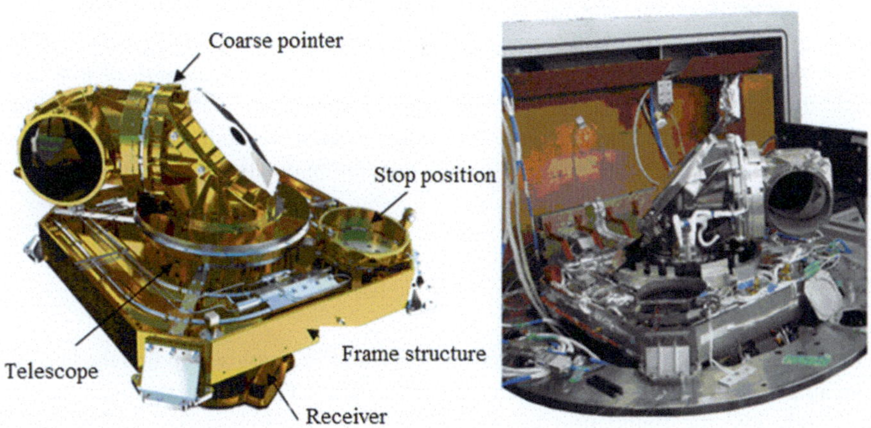

Fig. 1.4 LCT structure [48]

was 177 s. The experiment used the BPSK modulation/homodyne coherent optical detection technology [49].

(3) Japan and other countries

In 1997, the National Institute of Information and Communication Technology of Japan began to implement a development plan for the next-generation LEO system (NeLS). The objective of this plan was to realize satellite-to-ground and inter-satellite laser links. A DPSK/coherent detection system was used for this purpose. The communication data rate was 2.4 Gbit/s, and the laser wavelength was 1550 nm. The terminal demodulated the signal using a plane optical wave interferometer to recover the data stream with a detection limit of 56 photos/bit [50, 51]. The NeLS plan was originally expected to be officially put into commercial operation in 2010. Although the plan remains suspended as of now, it acts a reference for subsequent projects. In 2011, the University of Tokyo reported a digital optical heterodyne receiver suitable for intersatellite communication. It was based on BPSK modulation, with a data transmission rate of 12 Gbit/s. It may be noted that digital coherent optical detection is the core technology for next-generation intersatellite optical communication [52].

Based on various research breakthroughs-the short-range optical inter-satellite link (SROIL), intersatellite link front end, and optical cross links-of several collaborations, Contraves Space AG (a Swiss company) has designed and developed the OPTEL series laser communication terminals for commercial use to meet the requirements of various space applications [53]. The key technology used in this series is the high bit rate homodyne coherent optical communication technology. The OPTEL series meets the requirements of the miniaturization, high bit rate, and low energy consumption. These terminals use the 1064 nm laser BPSK modulation/homodyne detection. The OPTEL-25 communication laser uses a diode-pumped single-frequency single-mode tunable Nd-doped yttrium aluminum garnet laser as the main vibration amplification structure of the fiber laser amplifier. It uses emission wavelength of 1064 nm, 808 nm laser diode as the pumping source, 977 nm laser diode as the pumping source for the ytterbium-doped fiber amplifier, output optical power of 1.25 W, and laser diode as beacon laser. It has a wavelength of 808 nm, maximum output power of 7 W, and transmitting aperture of 135 mm. A variable-window charge coupled device (CCD) detector is used for acquisition and coarse tracking, and a micromechanical optical fiber scanning position detector is used for communication and fine tracking.

1.3.1.2 China

In the 1980s, the University of Electronic Science and Technology studied the heterodyne atmospheric laser communication system with a CO_2 laser as the light source. This study laid the foundation for the research on coherent optical communication systems [54]. In 2007, Wang [55] designed and simulated an optical heterodyne detection system for atmospheric laser subcarrier multiplexing. In 2007, Zhao [56] proposed a receiver scheme integrating heterodyne detection and optical power tracking, established a light-field distribution model on the detector surface, and analyzed the heterodyne efficiency of signal light and local oscillator light under the best matching and several incomplete matching. In 2008, Zhang [57] used experiments to prove that heterodyne signals can be obtained when the signal optical power in heterodyne detection is six orders of magnitude lower than the local oscillator optical power. Furthermore, they verified that the light source and spatial conditions for coherent detection were significantly higher than those for DD. In 2009, Wang [58] of the University of Electronic Science and Technology analyzed the optical coherence system based on BPSK modulation/homodyne detection, focusing on the design and implementation of the optical phase-locked loop technology. In 2010, Zhao [59] of the Changchun University of Technology studied the heterodyne asynchronous demodulation reception scheme and reported that the mixing efficiency decreases sharply with an increase in the mismatch angle when a spatial mismatch angle exists between the signal light and the local oscillator light. When the wave surface of the local oscillator Gaussian light bends, the mixing efficiency is maximized when the radius of curvature tends to infinity. In 2014, Luo [60] of the Beijing University of Posts and Telecommunications designed and built a spatial coherent optical communication experimental platform and performed indoor and outdoor spatial optical transmission experiments of 10 GBaud quadrature phase shift keying (QPSK) modulation. The indoor and outdoor distances were approximately 6 m and 600 m, respectively. Luo concluded that the BER was 10^{-6} at approximately −17.6 dBm outdoors and the indoor turbulence was weak, with its sensitivity being approximately 7 dB higher than that of outdoor.

There are two methods for coherent optical detection. In one method, the signal light and local oscillator light coherently generate differential frequency signals on a photosensitive surface. However, when part or all of the space laser communication link is in the atmosphere, atmospheric absorption or scattering causes serious attenuation of the intensity of the laser, and atmospheric turbulence causes the beam drift, light intensity flicker, angle-of-arrival fluctuation, and wavefront distortion. Consequently, the sensitivity of heterodyne detection is significantly reduced. The other method is based on coupling the space light into the optical fiber, and then making the space light coherent with the local oscillator light through a 90° mixer. However, the fiber coupling efficiency and Doppler shift in atmospheric turbulence are two main factors that affect the performance of space optical communication.

1.3.2 Coherent Optical Detection in Optical Fiber Communication

Coherent optical fiber communication technology was a research hotspot in the field of optical fiber communication in the 1980s. At that time, this technology mainly used the advantages of high coherent receiving sensitivity to increase the transmission distance of optical fiber communication. However, with the emergence of Er-doped fiber amplifiers, coherent optical communication lost favor during the 1990s and the IM/DD mode became the primary transmitting/receiving technology for optical fiber communication [61].

In 2003, Taylor was the first to demonstrate a 10-Gbit/s DPSK modulation homodyne digital coherent optical detection system [62]. The block diagram of this digital coherent receiver is depicted in Fig. 1.5. The laser linewidth of the signal light and local oscillator light generated by the system was 100 MHz. Taylor used the electrical domain equalization technology to compensate for the dispersion effect in the fiber. In 2006, a company in the United States launched an optical heterodyne detection receiver with a bandwidth of 10 Gbit/s, with the sensitivity reaching -30 dBm when the BER was 10^{-9} [63]. In 2007, the Nortel Network realized a 40-Gbit/s dual-polarization four-phase shift keying optical transmission technology [64]. The receiver in this scheme is a real-time homodyne detection system. The receiver uses a distributed feedback laser with a linewidth of 100 kHz. The signal light and local oscillator light are mixed through an asymmetric optical fiber coupler, and the IF signal is obtained on the photosensitive surface of the photoelectric detector. The signal is collected using an analog-to-digital converter (A/D), and the QPSK signal is demodulated using a digital signal processor (DSP). In 2008, the AT and NEC Laboratory collaborated for an experiment of a 640-km single-mode ordinary optical fiber transmission and coherent light detection of 8 $\times$ 114 Gbit/s polarization multiplexed 8-PSK signal with an interval of 25 GHz. At the receiving end, a 90° mixer was used to accomplish polarization and phase diversity reception, and an adjustable external cavity semiconductor laser was used as the local oscillator laser with a linewidth of 100 kHz. Automatic polarization control was achieved through electrical domain processing [65]. In 2010, Masato [66], in Japan, reported a transmission experiment of 140 Gbit/s with a transmission distance of 150 km by using polarization multiplexing and homodyne optical coherent detection. The local oscillator at the receiving end used a wavelength-adjustable local oscillator laser; the bandwidth was adjusted to 4 kHz; optical mixing was achieved using a 90° mixer; and the local oscillator phase locking was realized through the pilot optical signal and detected using a balance detector. In 2011, a successful homodyne detection experiment, with a 1550-nm, 5-Gbit/s 40-km bidirectional transmission optical network, was reported in Denmark [67]. In 2011, Alcatel-Lucent collaborated with Bell Laboratories to accomplish a 160-Gbit/s time division multiplexing QPSK coherent optical detection experiment over a transmission distance of 2400 km [68].

In 2007, Feng et al. studied digital optical coherent detection [69] and made a polarization diversity optical 90° mixer in a digital coherent receiver, as depicted in Fig. 1.5, using optical fiber devices. The difference between the four output losses was less than 0.3 dB, and the delay difference was less than 5 ps. They achieved in-phase quadrature (IQ) of optical mixer, digital dispersion compensation. The basic digital signal processing algorithm modules such as digital polarization demultiplexing and carrier phase recovery realize the 40-Gbit/s polarization multiplexed QPSK signal loop transmission of 1200 km, and the received BER is less than 10^{-3}. In 2008, Zeng [70] simulated an asynchronous demodulation coherent optical heterodyne balanced receiver in an optical fiber communication system. Their results indicate that the

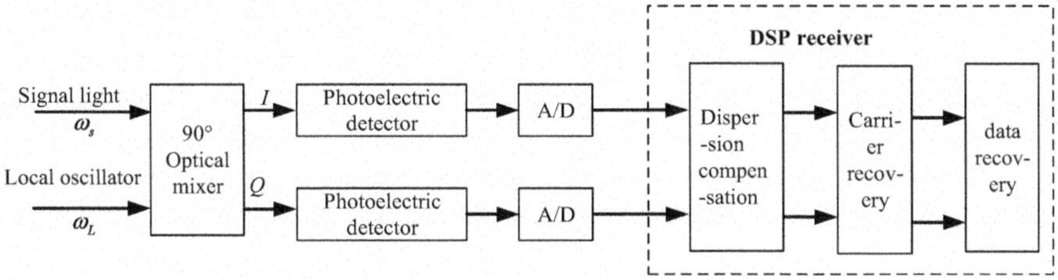

Fig. 1.5 Block diagram of digital coherent receiver [62]

quality factor Q value and receiver sensitivity of the coherent optical receiver decrease significantly with an increase in the linewidth of the light source laser. Increasing the IQ power to a certain extent can improve the sensitivity of the receiver: the higher the modulation rate, the lower the sensitivity of the receiver. In 2008, Zhu studied the heterodyne detection technology of a single-channel 2.5-Gbit/s signal-quadrature amplitude modulation and realized a 10-GHz IF signal. Zhu reported a sensitivity of -21.4 dBm at a BER of 10^{-8}. In 2011, Luo [71] studied a 40-Gbit/s coherent optical detection polarization multiplexing QPSK optical fiber communication transmission system. They used the back-propagation algorithm based on the Markov equation to achieve nonlinear compensation.

As depicted in Fig. 1.5, the digital coherent optical detection technology used in optical fiber communication systems comprises an optical 90° mixer, a balanced detector, a high-speed A/D, and a DSP processing algorithm. The optical 90° mixer and digital signal processing algorithm form its core. Digital optical coherent detection mainly includes the following digital processing algorithms: polarization control [72, 73], chromatic dispersion and polarization mode dispersion compensation [74–76], and carrier phase estimation algorithms [77].

In summary, there are many similarities and differences between the coherent optical detection system and the spatial coherent optical detection system used in optical fiber communication systems.

The similarities are as follows:

1. The coherent optical detection technology in optical fiber communication and in space is based on the principle of coherent optical detection, which uses signal light and local oscillator light to coherently generate differential frequency signals.
2. The homodyne detection requires optical phase-locked loop technology, and the heterodyne detection technology requires a narrow linewidth and high stability laser.
3. Both use multilevel digital phase shift keying (MPSK) demodulation technology or quadrature amplitude modulation and demodulation technology.

The differences are as follows:

1. The spatial coherent optical detection system suffers from the problem of beam alignment.

2. To improve the detection sensitivity of coherent optical at the receiving end, the optical fiber coherent optical detection system must control the polarization state and compensate for the dispersion and polarization mode dispersion. The spatial coherent optical detection system requires considering compensating for any interference such as wavefront distortion, random polarization change, and Doppler frequency shift caused by the atmospheric channel.

1.3.3 Free-Space Coherent Detection Communication System

1.3.3.1 Development Status Abroad

(1) United States

In the mid-1980s, the MIT-LL in collaboration with NASA, Aerospace Corporation, US Air Force, Missile Defense Agency, and JPL set up LITE using the FSK modulation mode and a modulation rate of up to 220 Mbit/s. The performance and space environment of each module of the system were tested. However, the project failed to perform the actual test and performed only a demonstration experiment on ground [78]. Figure 1.6 depicts the experimental device used in LITE.

In the late 1980s, NASA developed a new laser communication program. The objective of this program was to relay communication from the geostationary earth orbit (GEO) using an advanced communications technology satellite (ACTS) terminal laser to a ground station. ACTS laser communication experiments included DD and coherent detection experiments on ground and airborne counter terminals. Based on the experience of the LITE program, MIT-LL developed an APT system, an optical platform, and a heterodyne receiver. In the early 1990s, the JPL performed research on the inter-satellite coherent optical communication system and realized communication using a PSK modulation communication experiment demonstration of an inter-satellite coherent detection system with a code rate of 100 Mbit/s [79].

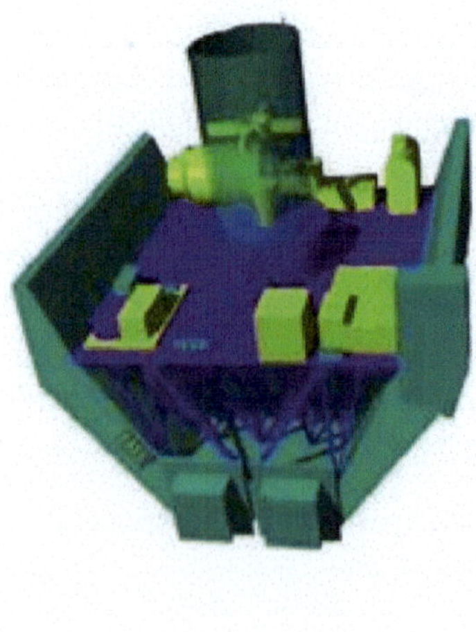

Fig. 1.6 Experimental device used in LITE [78]

Table 1.1 Parameters for laser communication system

Parameter	Value
Communication wavelength	0.8 μm
Detector response	0.517
Receiver optical efficiency	0.8
Transmitter optical efficiency	0.8
Bit error rate	10^{-6}
Inter-satellite transmission distance	3000 km
Transmission speed	2 Gbit/s

However, JPL shifted its focus to the IM/DD communication system until the end of the twentieth century. To improve the channel capacity of inter-satellite communication, the JPL laboratory shifted its research focus back to the coherent optical communication system. During this period [80], the LL developed a communication system with a communication rate of up to 1 Gbit/s. Simultaneously, the convolutional coding and decoding technology of the space channel was also being studied in depth [81]. In 1999 [82], for the low earth orbit (LEO) under the vibration condition of the inter-satellite platform, LL analyzed the SNR, BER, and other performance parameters of the coherent optical communication system under various modulation modes. The parameters of the laser communication system used in the test are listed in Table 1.1.

In 2008, NASA launched the Deep Space Optical Communication (DSOC) program, including Lunar Laser Communication Demonstration (LLCD). Its purpose was to verify the feasibility of laser communication between the Earth and Moon using small and light communication terminals. A schematic of this system is depicted in Fig. 1.7. The Laser Communication Relay Demonstration (LCRD) project can be considered as an extension of the LLCD program. The main objective of the LCRD program is to use a GEO satellite as a relay to realize continuous communication between two on-ground terminals. As depicted in Fig. 1.8, the LCRD program is an important step for NASA to improve its space laser communication capability.

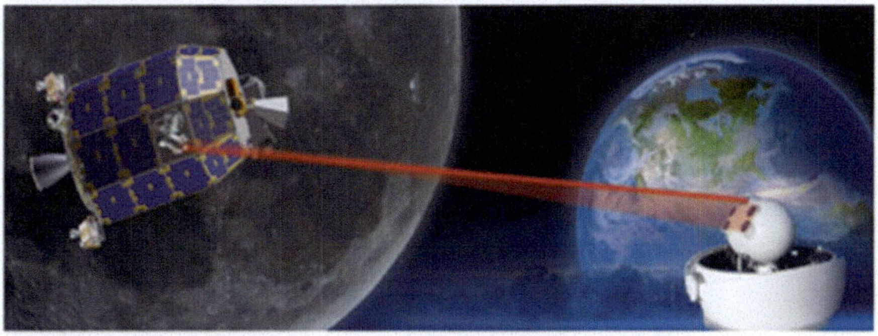

Fig. 1.7 Schematic of LLCD [86]

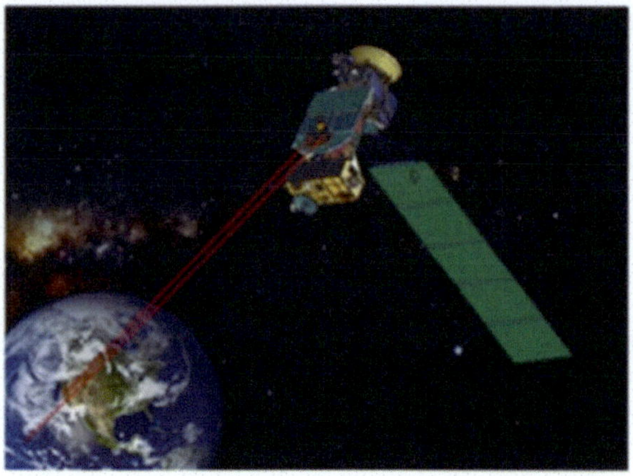

Fig. 1.8 Schematic of LCRD plan communication [86]

In 2013, as a prototype of the LCRD program, NASA announced that it had successfully established a duplex communication link between the Lunar Atmosphere and Dust Environment Explorer operating in the lunar orbit and the on-ground terminal for the first time. This project used pulse PPM, with the uplink and downlink communication rates being 20 Mbit/s and 622 Mbit/s, respectively [19, 83–85]. The DSOC program is part of the space exploration mission of NASA. In 2017, DSOC was expected to return data from Mars at a rate higher than 250 Mbit/s, with a communication distance of approximately 630 million kilometer, a terminal mass of 28 kg, and a power of 76 W. The system will continue to carry aircraft for deep space laser communication technology verification in 2023 [86]. Figure 1.9 depicts the schematic of the planned DSOC link.

In the mid-1980s, the MIT-LL used the semiconductor laser and heterodyne detection technology to successfully develop a coherent optical communication prototype [15, 38, 87–89]. This prototype was used for the LITE. The prototype comprised a

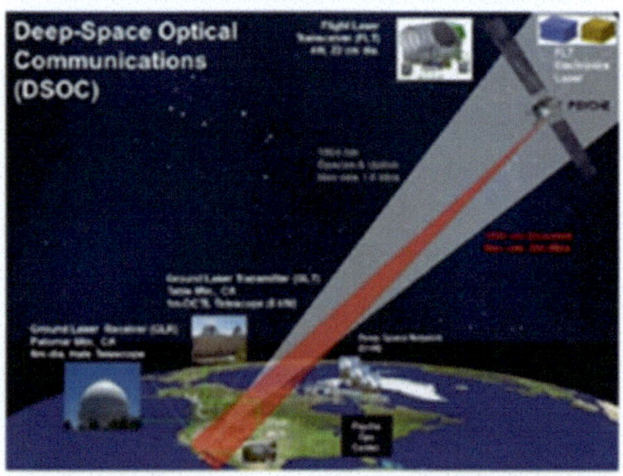

Fig. 1.9 Schematic of the planned DSOC link [86]

30-mW semiconductor laser transmitter, an Er-doped fiber amplifier with a wavelength of 1550 nm, and a telescope system of 200 mm diameter. It employed DPSK, with the modulation rate reaching up to 1 Gbit/s. The uplink communication rate was 220 Mbit/s, and the receiver adopted the heterodyne detection technology. However, no flight experiments were performed as part of this project.

In 2004, a 64-PPM coherent optical communication experiment was successfully performed by JPL. Both the signal light and the local oscillator light used a laser of 1064 nm wavelength. The output power of the signal laser was 2–3 mW, and the maximum output power of the local oscillator laser was 50 mW. There was a certain difference between the ground frequencies of the two lasers. The optical heterodyne detection method was used at the receiving end. A least square algorithm was used to track the differential frequency signal. The mixer output the frequency of a differential frequency signal of 6 MHz [90].

(2) Europe

Europe has also made some breakthroughs in the field of free-space coherent optical communication [91–99]. Some major research institutions in Europe are the ESA, DLR, and French military. ESA began research work in the field of space optical communication in 1977. In 1985, it started SILEX [100]. As part of SILEX, ESA launched Artemis into the GEO on November 20, 2001. For the first time, laser inter-satellite communication was realized between a satellite and the French SPOT-4 satellite in LEO. The communication setup is depicted in Fig. 1.10. The communication distance in this experiment was 45,000 km, and the BER was less than 10^{-6} [101]. The preferred communication scheme for SILEX was the homodyne coherent detection system with a communication wavelength of 1064 nm. Finally, the advanced DD scheme was used because the technical specifications of the lasers available at that time did not meet the requirements of the experiment.

Since 1989, ESA has performed in-depth research on coherent optical communication systems and other key technologies based on YAG solid-state lasers as part of SILEX. In 1996, the SROIL plan was officially launched in Switzerland. The plan is

Fig. 1.10 Schematic of inter-satellite communication between GEO and LEO in SILEX [91]

(a) Short range

(b) Medium range

Fig. 1.11 Images of SROIL planned communication terminals [102]

committed to developing a miniaturized and lightweight communication terminal for short-range communication between small LEO satellites. In 1998, the SROIL model of an optical demonstration terminal was successfully completed [102]. The transmitter used 1064 nm as the communication wavelength, the BPSK modulation mode and homodyne coherent detection scheme, transmitting antenna aperture is 350 mm, receiving antenna aperture is 40 mm, communication code rate is 1.5 Gbit/s, and BER is less than 10^{-6}. The terminal has a total mass of approximately 15 kg, power consumption of approximately 40 W, and dimensions of 30 cm × 20 cm × 50 cm. In addition, SROIL plans to develop medium- and long-range inter-satellite laser communication terminals. The terminal for medium-range communication keeps the light source, modulation mode, and detection mode unchanged, and can realize 6.5 Gbit/s rate communication on two LEO satellites 6000 km apart. The BER is less than 10^{-9}. The optical antenna aperture of the terminal is 100 mm, and its mass is 25 kg. Figure 1.11a, b depict the images of the short- and medium-range optical inter-satellite links.

In 1995, ESA placed a contract with Oerlikon-Contraves Space for developing an improved series of high-speed, miniaturized, lightweight, and low-power OPTEL terminals for commercial use [103]. Using the homodyne coherent detection mode, it developed inter-satellite communication terminals for short-, medium-, and long-range links. The corresponding models are OPTEL-02, OPTEL-25, and OPTEL-80. The terminals used in this improved OPTEL series use semiconductor lasers with a communication wavelength of 1064 nm as light sources. The transmitting terminal uses BPSK modulation, and the receiving terminal uses homodyne coherent detection.

Based on the LCT developed by SILEX, TESAT-Spacecom of Germany designed and developed a second-generation LCT, which uses the BPSK modulation mode and homodyne coherent detection. In 2005 [104], a 142-km BPSK-modulated homodyne coherent detection communication link with a transmission rate of 5.625 Gbit/s was established between islands La Palma and Tenerife in the Canary Islands of Spain, as depicted in Fig. 1.12. This experiment directly verified the feasibility of the homodyne BPSK scheme in a near-earth atmospheric turbulence channel and established an

Communication link between La Palma and Tenerife

Communication terminal in La Palma

Communication terminal in Tenerife

Fig. 1.12 Communication schematic of homodyne coherent detection system between two islands 142 km apart [104]

experimental base for coherent optical communication between satellites and the earth.

In 2008 [105], the LCT of the German satellite mission TerraSAR-X [106–109] was carried on the NFIRE of the United States. The homodyne BPSK communication mode was used in the LCT for the first time to realize inter-satellite two-way communication at the 5.625 Gbit/s rate. This experiment was the first in the world to establish an experimental link for coherent optical communication between two satellites 6000 km apart. Next, a similar coherent communication link between satellites and on-ground stations was established. In 2010, a 6.5-cm-diameter OGS was set up by ESA at Tenerife Island, and the TESAT coherent optical communication terminal of meters established a satellite ground communication link with the communication terminal of NFIRE at a rate of up to 5.635 Gbit/s. The ground communication terminal used an adaptive optics (AO) system to compensate for the distorted wavefront of the signal light, as depicted in Fig. 1.13.

Germany continues to improve the second-generation LCT [112], named LCT-TDP1 (see Fig. 1.14) to realize 1.8 Gbit/s data transmission over a communication distance of up to 45,000 km. In 2014, the Alphasat GEO satellite and Sentinel 1A

Fig. 1.13 Ground communication terminal of NFIRE [110]

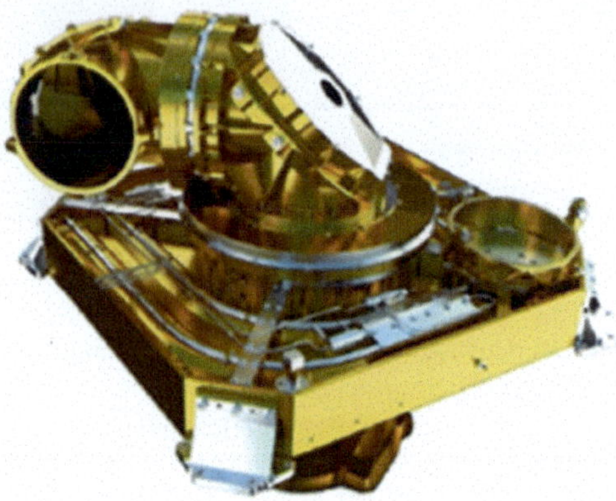

Fig. 1.14 Improved communication terminal LCT-TDP1 [112]

equipped with the LCT-TDP1 communication terminal LEO satellite successfully established an international coherent detection communication link and performed a data relay transmission experiment for the first time. To enhance the space data transmission rate and transmission capacity while maintaining the strategic independence of space communication, ESA launched the European Data Relay Satellite (EDRS) Project. In 2016, EDRS's first laser communication data relay payload, named EDRS-A, was successfully launched by the European Universal Satellite (Eutelsat). This launch marks the entry of space laser communication into the commercial application stage from the experimental verification stage. In 2018, the EDRS-C [113] satellite relay communication payload equipped with an LCT-TDP1 communication terminal was launched.

In Europe, many institutions are actively engaged in research on space optical coherent detection. These institutions have successfully completed the experimental verification of inter-satellite, satellite-to-ground, and near-earth communication links. They have made useful achievements in high-speed coherent optical communication and have entered the stage of engineering commercialization.

Since 1989, the Swiss Contraves Space Center agency has improved the key technology and system research on coherent optical communication while developing the SILEX [114] program. From 1996 to 1998, the Contraves Space Center agency successfully developed SROIL [53]. In the terminal simulation model, the light source of the transmitting section is a 1064-nm semiconductor-pumped Nd:YAG laser. The receiver uses the homodyne detection mode. Its communication code rate can reach 1.5 Gbit/s, and the BER is better than 10^{-6}. The ESA has also developed medium- and long-range inter-satellite optical communication terminals, which also use a 1064-nm semiconductor-pumped Nd:YAG laser and homodyne detection mode to establish communication links over a distance of 6000 km with a communication rate of 6.5 Gbit/s and a BER of less than 10^{-9}.

ESA and Oerlikon Contracts Space of the United Kingdom jointly designed and developed the OPTEL series laser communication terminals for commercial applications of the inter-satellite laser communication [53]. The objective of this series is to meet various application requirements in space. The key technology of the communication terminal is high-bit-rate homodyne detection technology at the receiving end. Its advantages include miniaturization, a high bit rate, and low energy consumption. The transmitting end adopts a 1064-nm excitation optical and BPSK modulation.

In 2016, building on the research in the earlier decades, the EDRS of ESA used BPSK modulation at the transmitter, with a communication rate of 1.8Gbit/s and a communication wavelength of 1064 nm. Similarly, NASA has been committed to the development of LCRD. The system uses a 1550-nm laser and DPSK modulation. Japan Aerospace Exploration Agency (JAXA) is also developing the Japanese Data Relay System. The communication terminal uses a 1550-nm laser and DPSK modulation, with a communication rate of 1.8Gbit/s [115–117].

In Germany, Lange successfully performed an experiment on spatially coherent optical communication using homodyne detection in October 2006 [118, 119]. To achieve reliable homodyne detection, a 1064-nm Nd:YAG laser with high-frequency stability and optical phase-locked technology, together with BPSK), was used in this experiment. The transmitting and receiving terminals were located at La Palma and Tenerife islands on Canary Islands, Spain, respectively. The communication distance between the LCTs was 142 km, the maximum transmission power was 1.2 W, and their height from the ground was 2350 m. A unidirectional spatially coherent optical communication with a transmission rate of 5.625 Gbit/s was successfully realized, and the BER was between 10^{-6} and 10^{-4}. Atmospheric attenuation was greater than 20 dB. The link continues to be in use even now.

In March 2008, Germany and the United States jointly launched the TerraSAR-X imaging satellite in Germany and NFIRE in the United States. The long-distance two-way broadband data transmission space coherent optical communication between these reconnaissance satellites was successfully realized. The transmission distance of the link was 5000 km. The lasers at the transmitting and receiving ends were 1064-nm Nd: YAG lasers. The transmitting end used BPSK modulation, with a data transmission rate of 5.6 Gbit/s and space transmission optical power of 0.7 W. The diameter of the aperture of the terminal telescope was 124 mm. The receiving end used the homodyne detection method of the optical phase-locked technology. The BER was less than 10^{-9} [120, 121]. When light propagates in turbulent medium, light intensity flicker and beam drift will affect the quality of the communication link.

In the same year, Germany successfully established a satellite-to-ground bidirectional laser communication link between the TerraSAR-X satellite and an OGS. The link transmission time was maintained for 8 min. The data transmission rate was 5.6 Gbit/s. LCTs were used at both the transmitter and receiver ends. The transmitting end used the BPSK modulation mode, whereas the receiving end used the homodyne detection terminal with optical phase-locked technology [122]. The communication wavelength was 1064 nm.

(3) Japan

Research on space coherent optical communication in Japan started late but developed rapidly. In 1994, Japan successfully accomplished the space optical communication plan of ETS-VI, which further verified the feasibility of the satellite-to-ground laser communication link. The objective of the OICETS project is to develop a miniaturized space communication terminal. In 2005, OICETS successfully launched a LUCE terminal, with a communication rate of 49.3 Mbit/s and communication distance of 600–1500 km [124, 125]. In May 2006, a two-way communication link was successfully established between the LUCE terminal at the NICT in Japan and the DLR ground station in Germany. The above two historic projects adopted simple and reliable DD methods and laid a technical foundation for subsequent space optical coherent detection methods.

In 2008, NICT successfully developed a communication terminal for the BPSK modulation homodyne coherent detection communication mode at a rate of 3 Gbit/s [126]. In 2009, Japan launched the Small Optical Transponder project, which is committed to developing a lightweight and compact satellite optical communication terminal. In 2011, Mitsubishi Corporation and JAXA jointly developed a homodyne coherent receiver using the BPSK modulation mode for inter-satellite optical communication [127]. By the end of 2016, Japan performed AO experiments, tracking tests, data transmission tests, and other related experiments in collaboration with international research institutions such as National Center for Space Studies (France), DLA (Germany), and ESA (Europe). NICT is the premier research institution in Japan. It is expected to realize a 10 Gbit/s two-way communication link between a GEO satellite and ground station in 2021.

In 1997, Japan set up a next-generation Leo system to realize a laser communication link between satellites and ground stations. The objective of this system is to use the 1550-nm wavelength laser, with the transmitter using the DPSK modulation mode and the receiver using the heterodyne detection mode. The communication data rate is expected to reach 2.4 Gbit/s, with the detection limit of 56 photos/bit [51, 128]. However, these objectives remain to be realized. In 2011, the University of Tokyo, Japan, released the details of the composition of the coherent optical communication system, which mainly includes a phase modulator and a 90° optical mixer. It has been reported that the receiver uses BPSK modulation at the transmitter, and its transmission data rate is 12 Gbit/s [129]. It may be noted that the digital coherent receiver is going to be the core of inter-satellite optical communication technology.

(4) China

In the 1980s, the University of Electronic Science and Technology of China took lead in the research on the space coherent optical communication system in China. It used a CO_2 laser with a communication wavelength of 1064 nm, and the feasibility of a 5-km communication link was verified using coherent detection. The research on DD has achieved remarkable progress. In August 2011, a research team led by Professor Jing Ma of Harbin University performed extensive research on the satellite-to-ground link laser communication technology in China. A communication terminal developed by the team and carried by the Ocean 2 Satellite successfully established the first satellite-to-ground laser communication link, with a single-channel data communication rate of 504 Mbit/s, in China [130–132]. In 2017, another laser communication terminal designed by the Harbin University of Technology was launched by the Practice-13 satellite, and a satellite-to-ground two-way communication link was established in January 2018 [133]. This was the first time that China had performed a high Earth orbit satellite communication experiment, supporting simultaneous interpretation of various transmission rates, with the BER being less than 10^{-6}.

Based on the research on DD from end 2016 to the beginning of 2017, a high-speed satellite-to-ground coherent optical communication terminal developed by domestic research institutions led by Shanghai Institute of Optics and Mechanics was successfully equipped with "Mozi." [134]. The satellite completed the verification of on-orbit two-way communication, in which the communication wavelength was 1550 nm, uplink used PPM, and communication rate was 20 Mbit/s. The downlink used BPSK modulation, with a communication rate of 5.12 Gbit/s. The success of this project made it is a milestone in the field of high-speed coherent optical communication in China.

In 2013, the Beijing University of Posts and Telecommunications successfully built a 20-Gbit/s QPSK spatial coherent optical experimental demonstration system in the laboratory [135]. The demonstration used a laser with a communication wavelength of 1550 nm and linewidth of 100 kHz and digital signal processing technologies such as clock recovery and frequency offset compensation to ensure the stability of the system. In 2015, the Optical Communication Research team at Nanjing University working on the BPSK modulation homodyne coherent detection system established a mathematical model of the impact of the frequency difference of the Costas optical domain PLL on the system [136]. Furthermore, the effects of the frequency offset, zenith angle, data transmission rate, and divergence angle on the BER of the system were analyzed. Research on the application of the AO system in the FSOC coherent detection system by the Chengdu Institute of Optoelectronics, Chinese Academy of Sciences has always been path breaking. Generally, the AO system is used to correct the distorted wavefront phase of the signal light and improve the accuracy of the coupling optical power. In 2018, Chen developed a set of high-sensitivity optical fiber coupling devices for a long-distance free-space coherent optical communication system to supplement the random angular error caused by receiver vibration and the adaptive optical system to correct the residual aberration of the atmospheric turbulence [137]. Their experimental results indicate that the coupling efficiency can be improved to 73.2% with a mean square error, which is reduced to 1.12%. The receiving sensitivity of the system can be improved by nearly 20 dB when compared with that of DD. In the same year, after using the aforementioned single-mode optical fiber coupling system and AO system designed, Chen and Liu also analyzed the improvement in the eye diagram and BER of the free-space coherent optical communication system of the satellite-to-ground communication link with a transmission rate of 5 Gbit/s [138].

In recent years, a research team led by Professor Xizheng Ke of Xi'an University of Technology, Shaanxi, has conducted in-depth theoretical research and experimental verification on key technologies of space optical communication, such as beam propagation characteristics in atmospheric turbulence [139, 140], coding and decoding of wireless laser communication system, channel estimation and equalization [141],

ATP [142], coherent detection [143–145], and design of optical transceiver system [146], and made significant achievements. Based on the modulation and detection mode of IM/DD, the team successfully developed a duplex FSO communicator, with a weight of less than 8 kg, a maximum transmission rate of 1.25 Gbit/s, and a BER of less than 10^{-6}, capable of high-speed transmission of audio and images. In addition, the team successfully developed a DPSK-modulated coherent detection system communication terminal, established a communication link between buildings 600 m and 1.3 km apart in March and May 2018, and realized the real-time transmission of high-definition video. Figure 1.15 depicts the video images transmitted by the 1.3-km coherent detection digital coherent receiver. In July of the same year, a 10-km-long communication link was established from the 12th floor of a building in the Jinhua campus of the Xi'an University of Technology in Xi'an, Shaanxi Province, to Bailuyuan, as depicted in Fig. 1.16. The same coherent detection method was used to accomplish the real-time transmission of video, making it another major breakthrough in the research field of coherent optical communication in China.

Fig. 1.15 Video grab transmitted using coherent detection system over a distance of 1.3 km

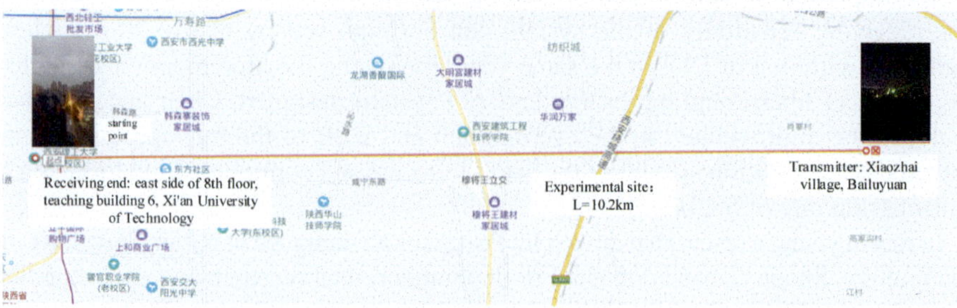

Fig. 1.16 10-km near-ground communication link

1.4 Research Status on Factors Affecting Performance of Free-Space Coherent Detection Systems

Coherent detection is used in lidar, broadcasting, and FSOC [147–153]. When compared with DD, the sensitivity of coherent detection approaches the quantum noise limit, and the receiving sensitivity can be increased by nearly 20 dB [152]. Mixing efficiency is a useful index for measuring coherence detection, and research on coherence detection has made significant progress [154–158]. To obtain the maximum mixing efficiency, it is necessary to ensure that the spatial field distribution, spot size, phase, and polarization states [157, 158] of the signal light and the local oscillator light match. Any change in the above factors may lead to a decrease in the mixing efficiency.

The signal light and local oscillator light are the light field distributions of Airy, Gauss, and Uniform in different combinations. Researchers [155–160] have studied the factors that affect the mixing efficiency of the coherent detection system. In 1975, Cohen [154] studied the impact of the spatial angle mismatch, spot size, and detector uniformity on mixing efficiency. Their results indicate that the coherent detection system corresponding to the Airy distribution signal light is insensitive to the spatial angle mismatch. In 2011, Liu [159] analyzed the effects of various amplitude distributions of signal light and local oscillator light on the mixing efficiency of the coherent detection system of inter-satellite communication links. They report that Airy + Airy and Gauss + Gauss distributions theoretically have a maximum heterodyne efficiency of 100%, whereas the Gauss + Uniform distribution has a maximum heterodyne efficiency of 82%. In the FSOC coherent detection system, the signal light at the transmitting terminal is generally a Gaussian beam; however, after long-distance transmission, the signal light at the receiving end is approximately considered to be a plane wave when it reaches the end face of the receiving antenna. After being received by the optical system, the signal light is Airy and distributed on the detector surface, and the local oscillator light with the light field distribution of Gauss is coherently mixed. Based on this mathematical model, the heterodyne efficiency of coherent detection was further studied [161–163]. In 2003, Wang [160] analyzed the effects of non-collimation, spatial angle mismatch, local oscillator, light phase bending, and other factors on the heterodyne efficiency. They concluded that the non-collimation efficiency and spatial angle mismatch were the main causes of the decline in the heterodyne efficiency. In 2013, Zhang [161] studied the effect of collimation mismatch on the heterodyne efficiency for a BPSK-modulated zero-difference coherent detection system. In 2016, Han [162] theoretically studied the impact of the waist radius of the Gaussian beam on the external aberration efficiency. Their results indicate that the optimal waist radius could be determined by adjusting the optical system to achieve the maximum mixing efficiency. In the case of light field matching between the local oscillator light and signal light, that is, both signal light and local oscillator light being Gaussian beam coherent detection mathematical models, many studies on mixing efficiency have been performed [155, 156, 163–165]. In 1978, Takenaka [156] reported that the ratio of the aperture of the detector to the minimum aperture of a signal light spot had significant impact on the SNR of the coherent detection system. In 1984, Tanaka and Saga [163] studied the effects of parameters such as the loss of supporting role, detector radius, and beam waist radius on the mixing efficiency of a background radiation coherent detection system. In 1987, Tanaka [164] studied the impact of the signal light vector tilt and offset error on the mixing efficiency. Based on the assumption that other impact factors can be ignored, their numerical results indicate that the theoretical mixing efficiency can reach more than 60% when the tilt angle is less than 10^{-4} degrees and the offset is approximately one-fifth of the incident aperture. In 2015, Li [165] theoretically studied the mixing efficiency of the coherent detection system under the Gaussian beam model. Their results indicate that, to obtain an ideal mixing efficiency, the beam waist radius of the local oscillator must be at least larger than that of the signal light, and the radius of the detector must be twice that of the local oscillator beam. In 2017, Nan [155] discussed the impact of the spot size and axis deflection on the mixing efficiency of coherent detection and determined the error range for the spot size and axis deflection when the mixing efficiency was greater than 10%.

In 2013, Yang [166] studied the impact of the aberration of the receiving optical system on the mixing efficiency of the coherent detection system. They report that the heterodyne efficiency gradually decreases with an increase in the primary aberration, and different aberration types have different influences on the external efficiency. In 2017, Kong [167] studied the impact of laser linewidth on the performance of a coherent detection system. Their results indicate that appropriately increasing the internal gain, responsivity, and response bandwidth of the detector can effectively reduce the speed of the SNR decrease with the increase in laser linewidth. In 2013, Ma [157] theoretically analyzed the effect of polarization characteristics of signal light on coherent detection. They report that circularly polarized light has a higher mixing efficiency than that of the linearly polarized light. In 2017, Yang [158] proposed a mathematical model for the mixing efficiency of coherent detection considering the polarization aberration of optical systems. This model can not only comprehensively describe coherent detection systems but also be used to evaluate and minimize the polarization aberration of optical systems.

For free-space coherent optical communication systems, coherent detection has high sensitivity and advantages of noise reduction and high spectral resolution [168–174]. It is beneficial to use the coherent detection technology to achieve the maximum mixing efficiency and SNR of the system [175, 176]. However, because of the turbulence effect, the amplitude, phase, polarization, and other characteristics of the signal light transmitted to the receiver in the atmospheric random channel do not match with those of the local oscillator light [177, 178]. This causes arrival-angle fluctuation, light intensity fluctuation, and beam drift on the end face of the receiving antenna, thereby degrading the spatial coherence of the light wave [179, 180]. The mixing efficiency of coherent detection and the SNR of the system are reduced, thereby affecting the performance of the entire coherent optical communication system.

In 1967, Fried [181] theoretically analyzed the impact of optical signals on the performance of a coherent detection system after passing through atmospheric turbulence. Their results indicate that the SNR of the system has an extreme value regardless of the size of the aperture of the detector. In 1978, Churnside [182] derived an approximate expression for the probability density function of the signal amplitude of an optical coherent detection system under turbulent conditions in sunny weather and verified it experimentally. In 1979, Churnside and Mclntyre [176] analyzed the communication performance of optical heterodyne receivers on the basis of the above research and designed an adaptive processor for optical heterodyne receivers to synchronically demodulate IF signals. In 1984, Salzman and Katzir [183] proposed a new method for calculating the heterodyne efficiency of coherent receivers exposed to atmospheric turbulence. In 1986, Winick [184] studied a simulation method that could obtain the probability distribution of signal optical attenuation based on the coherent detection system of a satellite communication link. The results indicate that the signal optical attenuation probability distribution can be used to judge the degree of performance deterioration caused by atmospheric turbulence; however, it is not useful for measuring performance deterioration by means of average signal attenuation. In 1997, Das [185] reported that the detector combination mode with a hexagonal array could effectively improve the SNR of the coherent detection system when compared with the equal-area square arrangement. In 2002, Zhu [186] reported that spatial diversity reception is helpful in suppressing the impact of atmospheric turbulence under the premise that the receiving aperture is smaller than the attenuation coherence length, and the observation time is smaller than the attenuation correlation time. In 2008, Tsiftsis [187] derived a closed-loop expression for interrupt probability and mean BER of a optical wireless heterodyne communication system after transmission through a strong atmospheric turbulent channel. The results indicated that the BER of the practical FSOC coherent detection system could not reach 10^{-9}. In 2009, Sandalidis [188] studied the effects of the attenuation and directional error of signal light transmission in atmospheric turbulent media under various atmospheric turbulence intensity fluctuations.

In 2010, Liu [189] derived the expression for average capacity after transmission through a gamma-gamma atmospheric turbulent channel for a free-space coherent detection system with DPSK modulation mode. They considered the directional error in their expression. In 2011, Liu [190] studied the satellite-to-ground link coherent optical communication system based on differential QPSK and a phase estimation algorithm using the m-power scheme. Their simulation results indicate that the phase-estimation algorithm can effectively suppress the phase noise caused by atmospheric turbulence. In 2012, Niu [191] derived closed expressions for BER and interrupt probability, enabling a high-precision performance estimation of BPSK, DPSK, and MPSK wireless coherent systems in gamma-gamma atmospheric turbulence. Their results indicate that the SNR loss of DPSK is 3 dB when compared with that of BPSK. In 2015, Bai [192] proposed a new coherent detection optical code division multiple access system model, which adopted the polarization frequency shift keying modulation mode in a free-space optical turbulence channel model. The closed expressions for error probability and interrupt probability of the system model through the gamma-gamma turbulent channel were derived. In 2015, Huang [193] derived the expression for the SNR of a free-space coherent optical communication system compensated by an AO system, in which the effects of residual aberration and amplitude disturbance on the SNR of the system were independent of each other. This result is necessary to evaluate the performance of the AO system in a coherent optical communication link. In 2015, Kong [194] discussed the impact of atmospheric turbulence on external efficiency when the signal light and local oscillator light had different light field distributions. In 2016, Park [195] derived the interrupt probability and mean error probability of a free-space optical coherent system including path loss, geometric diffusion, directional error, amplitude disturbance, and phase distortion. Furthermore, they discussed the effects of these errors on the performance of the free-space coherent optical communication system.

1.5 Research Status on Factors Affecting Partially Coherent Beam Coherent Detection System

In most applications, the light field generated by a laser source based on a physical mechanism will have random fluctuation [196], which is more accurately described by a partially coherent beam. The transmission characteristics of a partially coherent light beam through a random medium of atmospheric turbulence have been analyzed [197–199]. Furthermore, the performance of the partially coherent light coherence detection system has been studied.

In 1970, Lahti [200] calculated the average SNR of the system based on the coherent detection system of partially coherent colcolating beams, theoretically explained the aperture saturation effect, and provided the corresponding local oscillator spot size and coherent area when obtaining the optimal system SNR. In 1992, Tanaka [201] theoretically calculated the mixing efficiency of a coherent detection system, in which both the signal light and the local oscillator light were partially coherent Gaussian beams. Their results indicate that the ratio of the input aperture to the detector aperture was approximately inversely proportional to the ratio of the minimum 1/e radius of the signal beam and the minimum 1/e radius of the local oscillator beam. In 2002, Ricklin [202] used the cross spectral density function to analyze the average intensity, light wave size, radius of curvature before phase, and coherence length of partially coherent light beams transmitted through an atmospheric turbulent medium. Based on their research, in the same year, Ricklin [203] reported that the use of partially coherent beams can reduce the BER of the system caused by the intensity flicker in the focal plane of the receiver. In 2010, Salem [204] analyzed the effects of the spatial angle mismatch, detector radius, and beam overlap on mixing efficiency based on the coherent detection system of partial GSM beams. In 2012, Ren [205] derived a closed expression for the mixing efficiency of a coherent detection system in which both the signal light and the local oscillator light were partially coherent Gauss-Schel beams. When compared with fully coherent light, partially coherent light is less sensitive to turbulence intensity. In 2015, Li [206] established a coherent detection model in which both the signal light and the local oscillator light were partially coherent Gaussian Schel beams under the Tatarskii turbulence spectrum model. They report that atmospheric turbulence significantly reduces heterodyne efficiency, whereas the corresponding mixing efficiency changes stably when there is a loss of supporting role under atmospheric turbulence. In 2016, Wang [207], based on the Tatarskii turbulent power spectrum model, derived the expression for mixing efficiency and SNR when both the signal light and local oscillator light are partially coherent GSM beams with a loss of supporting role. Furthermore, they analyzed the effect of polarization change on the performance of a coherent detection system and reported that mixing efficiency depends on the initial polarization state of the beam.

1.6 Research Status of Wavefront Correction

Atmospheric turbulence has significant impact on space optical communication systems. When the laser beam reaches the receiving end through the atmospheric channel, atmospheric turbulence causes beam tilt, light intensity flicker, wavefront distortion, and other phenomena. These phenomena case the dispersion of the received light spot, decrease in the power of the received light, and increase in BER. These can even lead to communication failures in serious cases. Therefore, wavefront distortion caused by atmospheric turbulence has been extensively studied, and significant progress has been made in space communication. The AO system is the widely used correction method to mitigate the effects of these phenomena. Some other calibration devices are the deformation mirror and liquid crystal spatial light modulator (LC-SLM).

1.6.1 Research Status of Atmospheric Turbulence Compensation Technology

Several correction techniques have been proposed to suppress the effects of atmospheric turbulence. In 2002, Weyrauch applied adaptive optics to the experiment of space-light-to-single-mode fiber coupling, using a micromechanical variant mirror controlled using a large-scale integrated circuit to correct the wavefront. In this experiment, the coupling efficiency of the space-light-to-single-mode fiber was approximately 60%, and that of the multi-mode fiber was approximately 75% [208]. In 2007, Kim [209] analyzed a correction method for wavefront phase distortion caused by atmospheric turbulence. In 2009, Puryear [210] pointed out that losses caused by turbulence in atmospheric channels can be suppressed through spatial diversity reception, wavefront pre-correction, and coherent light detection. In 2009, Buske [211] studied image blotch expansion and optical resolution loss caused by wavefront distortion due to atmospheric turbulence and performed AO experiments to compensate for the phase distortion caused by the atmospheric turbulence effect. In 2010, Sinefeld [212] used an iterative feedback algorithm to perform phase compensation by estimating the 15-order Zernike polynomial coefficients. This wavefront correction method is capable of improving the coupling efficiency of the spatial optical coupling to optical fibers. In 2010, Sarmadi [213] proposed a blind optimization technology to correct the distorted wavefront by controlling the deformation mirror using a genetic algorithm to correct the wavefront and enhance the output signal. In 2010, Podanchuk [214] introduced a holographic Shack-Hartmann wavefront sensor that compensated for the phase difference through a two-channel lens array.

In 2005, Li [215] studied an adaptive compensation method for atmospheric turbulence in a space laser communication system. In 2008, Yang [216] studied the AO technology of wavefront sensorless to mitigate the effect of light intensity flicker caused by atmospheric turbulence on wavefront correction. In 2010, Chen [217] proposed an AO technology based on a stochastic parallel gradient descent algorithm, which can realize closed-loop compensation of wavefront phase distortion without a wavefront sensor. It may be noted that the convergence speed of this algorithm is slow, thereby restricting its application to systems requiring real-time performance. In 2010, Xia [218] reconstructed the distorted wavefront and corrected it using a reflection deformation mirror. In 2010, Qiao [219] analyzed the working principle and performance indexes of the deformation mirror and reported that the wavefront compensation effect of an AO system composed of a detector, controller, and deformation mirror could reach 90%. In 2010, Li [220] studied a closed-loop adaptive control optical system based on a genetic algorithm to compensate for the distortion of the laser beam. In 2011, Han [221] theoretically deduced the coupling efficiency of a single-mode fiber in slant path transmission under atmospheric turbulence and proposed the use of a stochastic parallel gradient optimization algorithm to control the deformation mirror in real time for optical phase compensation and optimize the coupling efficiency of a single-mode fiber.

The effect of atmospheric turbulence on FSOC is far from being completely mitigated and requires further study. For the near-earth atmospheric turbulent channel, atmospheric interference in optical communication has become the main obstacle for the application of the coherent light detection technology. Therefore, further research on alternative detection methods with high detection sensitivity suitable for atmospheric turbulent channels is required.

1.6.2 Research Status of Wavefront Correction Technology Abroad

Adaptive optics is a comprehensive discipline that integrates light, machinery, electricity, and control [222] to correct the wavefront distortion of a beam in real time. Adaptive optics was first proposed by American astronomer Babcock [223]. In 1953,

Babcock measured wavefront distortion information with a wavefront sensor in an on-ground astronomical telescope. He then used a deformable optical device to generate an optical phase shift conjugated with the distorted wavefront to compensate for wavefront distortion. Limited by the key technical conditions at that time, AO did not flourish. In 1972, researchers in the United States developed the first set of real-time turbulence compensation experimental systems and successfully realized wavefront distortion compensation caused by atmospheric turbulence on a 300-m horizontal transmission link [224]. Since then, many studies have been performed in the United States in the field of AO. In 1982, the US Army installed the world's first AO system on a 1.6-m telescope at the Maui Air Force Observatory in Hawaii for the detection of space targets [225] and obtained some experimental data on atmospheric disturbance compensation. Around the same time, atmospheric thermal halo compensation experiments were performed at the MIT-LL [225] and laser beam aberration correction experiments were performed at Lawrence Livermore Laboratory [226]. In 1989, the European Southern Observatory (ESO) installed a COME-ON AO system [227] on a 1.52-m astronomical telescope at the Observatory de Provence in France. The system adopted a continuous plane shape deformation mirror as a wavefront corrector, which could drive 19 units and successfully realize wavefront correction of infrared light [228]. In 1990, the system was transferred to the ESO 3.6-m Telescope in Chile, where the resolution of the system was further improved [229] and binary stars with an angular spacing of 0.22″ could be clearly distinguished. COME-ON-PLUS, which is an improved version of the COME-ON system, has increased the number of deformable mirror drive units to 52, calibration frequency to 200 Hz, and system bandwidth to 29 Hz. For the convenience of lay users, significant improvements have been made to the COME-ON system. This improved version is named the Adaptive Optics Near Infrared System.

In 2012, Roberts from The JPL of Caltech, Palomar Observatory and Cornell University, used two deformation mirrors simultaneously on the PALM-300 adaptive system of the 5-m Hale telescope to correct the low- and high-order distortion wavefronts of turbulence. This setup has 241 deformable mirror driving units correcting low-order distortion and 3388 deformable mirror driving units correcting high-order distortion, and the Strehl ratio of K-band correction reaches 0.8 correction effect [230]. In 2013, the 585 unit deformable mirror adaptive optics system developed by the University of Arizona, US, with a sampling frequency of 1 kHz and an imaging band of 0.6–1.1 μm, serving the 6.5-m Magellan II Telescope in the Atacama Desert was introduced [231].

1.6.3 Domestic Research Status of Wavefront Correction Technology

China's forays into the AO technology started in the 1970s. Notwithstanding significant efforts, there is a wide gap that must be filled to become at par with the advances in the AO field at the international level. The prominent research organizations active in this field are the following: the Institute of Optics and Electronics, Chinese Academy of Sciences; National University of Defense Technology; Beijing Institute of Technology; and Anhui Institute of Optical Machinery. The research and development activities on the deformable mirror are undertaken at the Institute of Optics and Electronics of the Chinese Academy of Sciences and Changchun Institute of Optical Machinery. Toward the end of 1980s, the Institute of Optics and Electronics established a 19-unit AO system to correct the static wavefront of the "Shenguang I" device and used the mountain climbing algorithm to perform closed-loop control. It was the first closed-loop control AO system capable of being successfully used among similar devices worldwide [232]. The 21-unit AO system developed based on Chinese Academy of Sciences photoelectric technology research was installed on the 2.16-m telescope of the Beijing Astronomical Observatory [233]. This AO system was the first set of infrared adaptive optical systems developed by China. In addition, the 127-unit deformable mirror developed by the Institute of Optics and Electronics was applied to the 1.8-m telescope at the Yunnan Observatory with a correction of ± 3 μm, corresponding to weak turbulence with r_0 of 15 cm and the first 35 Zernike wavefront [234, 235]. The 37-unit AO system was applied to the 1-m solar telescope at Kunming [236].

The Changchun Institute of Optics, Fine Mechanics and Physics developed a 97-unit deformable mirror in 2010, and a 137-unit deformable mirror in 2011. These mirrors were applied to a 1.23-m telescope with a working wavelength of 700–900 nm. In 2012, a 961-unit deformable mirror was successfully developed [237]. In addition, the Changchun Institute of Optics, Fine Mechanics and Physics has conducted extensive research on the AO of LC-SLMs. The liquid crystal AO system developed by the Changchun Institute of Optics, Fine Mechanics and Physics has been continuously improved in terms of its performance. The response frequency increased from tens to hundreds of hertz at the beginning, and the optical power loss decreased to less than 20%; thus, achieving encouraging experimental results. In the experiment at the Changchun Institute of Optics, Fine Mechanics and Physics, the S–H sensor and LC-SLM were used as wavefront correctors to realize atmospheric turbulence compensation in the horizontal direction up to 500 m, and the wavefront mean square error after compensation was reduced to 0.06λ [238].

The Anhui Institute of Optics and Electronics has been engaged in research on the measurement of atmospheric turbulence for many years. The blind optimization algorithm of stochastic parallel gradient descent (SPGD) was used by Ma et al. to directly optimize the system performance index to compensate for wavefront distortion. This approach achieved a desirable correction effect after experimental verification [239, 240].

The National University of Defense Technology has performed comprehensive research on the theory and experiment of the AO system. In 2002, Hou built an AO system based on a common optic/common module (CP/CM). In this system, a 37-unit deformable mirror was used as a wavefront corrector, and the double Shack-Hartmann information fusion technology was applied. A closed-loop control experiment of an AO system based on the CP/CM was realized [241]. In 2004, Hu used the biconjugate gradient stabilized algorithm to reconstruct the wavefront. The reconstructed wavefront was applied in the AO system, achieving a 0.01% accuracy in 70 iterations [242]. In 2011, Liu applied holography technology to the wavefront sensing technology and achieved encouraging results [243].

Xia of the University of Electronic Science and Technology of China used the Zernike polynomial mode method to perform an AO wavefront correction experiment. Their experimental results indicate that when the AO technology is used in an FSOC system, the optical energy concentration of the receiving end is significantly improved and the system BER is reduced. Under the condition that the BER remains unchanged, the transmitting power of the FSOC with the introduction of AO technology is reduced by 50% when compared with that without the AO system [244].

Ke of the Xi'an University of Technology performed correction experiments on the wavefront distortion of a ground laser beam and vortex light caused by atmospheric turbulence. Their experimental results indicate that the stronger the atmospheric turbulence, the more serious the ground beam speckle and the beam center drift caused by the turbulence. For the distorted wavefront, the AO system without the wavefront sensor based on the SPGD algorithm was used to correct the aberrant wavefront, and the Strehl ratio of the far-field light intensity increased from 0.15 before correction to 0.81 after correction, and the spot center convergence and center light intensity increased significantly [245]. Similarly, the wavefront distortion of the vortex beam is corrected using the AO system without a wavefront sensor based on the SPGD algorithm, which can improve the mode purity of the single-mode vortex and multi-mode multiplexed vortex. Experimental results indicate that after correction, the Strehl ratios of the single-mode vortex and multi-mode multiplexed vortex can be improved to approximately 0.85 and 0.72, respectively. Furthermore, the wavefront correction effect of the single-mode vortex is due to the multi-mode multiplexed vortex [246].

Although many achievements have been made in the wavefront correction technology in China, the technical mastery of the core field is still relatively weak. When compared with the advances in this field at the international level, the domestic research is significantly backward; however, the progress is still relatively fast. The advances in wavefront correction with a deformation mirror have been successfully applied in astronomical observation and laser communication. In addition, the wavefront correction technology based on a blind optimization algorithm without a wavefront sensor is also developing vigorously, and some research achievements have been made.

To mitigate the impact of the atmospheric turbulence on the performance of the optical communication system, wavefront correction technology has been studied at home and abroad. Furthermore, the wavefront correction technology based on the LC-SLM has become a focus of current research. However, all the aforementioned research achievements are aimed at free-space optical communication systems. In the field of spatial coherent optical communication, there are few reports on the effects of wavefront distortion on the quality of the communication system, namely, heterodyne efficiency and BER. Therefore, the wavefront correction of a space coherent optical communication system is investigated in this study. Wavefront correctors widely used at present include deformable mirrors and LC-LSMs. According to the theory of physical optics, the so called wavefront distortion is actually the optical path, and the optical path difference is the product of the optical transmission path and the refractive index of the transmission medium. Correction of the distorted wavefront can be realized by changing the geometric path of optical transmission using the wavefront correction with a deformable mirror. The correction of the distorted wavefront can be realized by changing the refractive index of the transmission medium with LC-LSM. LC-LSM is based on the electrically controlled birefringence principle of liquid crystals. It realizes continuous modulation of the light wave phase by changing the electric drive signal of the liquid crystal. LC-LSM has been widely used in fields such as AO, wavefront sensing, and beam shaping because it offers the advantages of low price, fast response speed, low energy consumption, and programmable control.

References

1. Lawrence RS, Strohbehn JW (1970) A survey of clear air propagation effects relevant to optical communications. Proc IEEE 58(10):1523–1545
2. Zhu X (2002) Free-space optical communication through atmospheric turbulence channels. IEEE Trans Commun 50(8):1293–1299
3. Xiao SL, Zeng QJ (2003) Applications of optical coherent detection. Optical Coherent Technol 27(11):28–31
4. Su YK, Li Z (1995) Sensitivity analysis of FSK subcarrier fiber-optic transmission system. Beijing Univ Aeronautics Astronautics 21(1):35–40
5. Huang W, Takayanagi J (1993) Atmospheric optical communication system using subcarrier PSK modulation. IEICE Trans Commun E 76-B(9):1169–1177
6. Wang L, Feng Y (2007) Progress and trend of satellite coherent optical communication. Laser Optoelectronics Progr 44(6):49–53
7. Yao ZS, Hu Y (2005) Progress in Inter-satellite coherent optical communications. Optical Commun Technol 29(8):44–46
8. Zoran S, Bernhard F, Hanspeter L (2010) Optical intersatellite communication. IEEE J Sel Top Quantum Electron 16(5):1051–1057
9. Michael GT (2009) Phase estimation methods for optical coherent detection using digital signal processing. J Lightwave Technol 27(7):901–914
10. John RB, Edward AL (1990) Performance of coherent optical receivers. Proc IEEE 78(8):1369–1394
11. Pfau T, Wordehoff C, Peveling R et al (2008) Ultra-fast adaptive digital polarization control in a realtime coherent polarization-multiplexed QPSK receiver. In: Optical fiber communication/national fiber optic engineers conference, pp 1–3
12. Tyson RK (1996) Adaptive optics and ground-to-space laser communications. Appl Opt 35(19):3640–3646
13. Pierre RB, David WR, Mark LP et al (1998) Performance improvement of a laser communication link incorporating adaptive optics. Proc SPIE 3432:93
14. Chen CY, Yang HM, Jiang HL (2009) Research progress of mitigation technologies of turbulence effects in atmospheric optical communication. Acta Armamentarii 30(6):779–791
15. Xu N, Liu LR, Liu DA et al (2007) Technologies and recent progress of coherent optical communications in free space. Laser Optoelectronics Progr 44(8):44–51
16. Lesh JR, DePaula RP (1995) Overview of NASA R&D in optical communications. Proc SPIE 2381:4–11

17. Isaccl K (1999) Horizontal link performance of the STRV2 laser experiment ground terminals. Proc SPIE 3615:11–22
18. Thomas B (2003) MCAO at the vacuum tower telescope tenerife. Proc SPIE 4839:544–553
19. Boroson DM, Robinson BS (2014) The lunar laser communication demonstration: NASA's first step toward very high data rate support of science and exploration missions. Space Sci Rev 185(1–4):115–128
20. Fletcher GD, Hicks TR, Laurent B (1991) The SILEX optical interorbit link experiment. Electronic Commun Eng J 3(6):273–279
21. Baister G (1994) Small optical user terminal for intersatellite communications. In: Proceeding of laser 93. Laser in Engineering, pp 780–783
22. Gatenby P (1995) Small laser terminal for operational intersatellite links. Space Commun 13:257–267
23. Nielsen TT, Oppenhaeuser G (2002) In orbit test result of an operational optical intersatellite link between ARTEMIS and SPOT4, SILEX. Proc SPIE 4635:1–15
24. Lange R, Smutny B (2004) BPSK Laser communication terminals to be verified in space. In: Military communications conference, Monterey, CA, USA, pp 441–444
25. Trisno S, Davis CC (2006) Performance of free space optical communication systems using polarization shift keying modulation. Proc SPIE 6304:63040–63049
26. Popoola WO, Zabin G (2009) BPSK subcarrier intensity modulated free-space optical communications in atmospheric turbulence. J Lightwave Technol 27(8):967–973
27. Seels, Troendled, Heinef et al (2014) Alphasat laser terminal commissioning status aiming to demonstrate GEO -relay for sentinel SAR and optical sensor data. In: Proceedings of 2014 IEEE international geoscience and remote sensing symposium, Quebec City, QC, IEEE, pp 100–101
28. Toyoshima M, Takizawa K, Kuri T et al (2006) Ground-to-OICETS laser communication experiments. Proc SPIE 6304:1–8
29. Jono T, Takayama Y, Shiratama K et al (2007) Overview of the inter-orbit and orbit-to-ground laser communication demonstration by OICETS. Proc SPIE 6457(02):645702–1~645702–10
30. Yamamoto A, Hon T, Shimizu T et al (1994) Japanese first optical inter-orbit communications engineering test satellite (OICETS). Proc SPIE 2210:30–38
31. Zhang YH, Huo ZR, Wang HF (2004) Status quo and development trend of free space optical communication. China Data Commun 12:78–82
32. Li XF, Zhu B (2006) Design of signal beam width in space-to-ground laser communication links. Optical Commun Technol 29(10):34–35
33. Zhi XJ, Ai Y (2004) Design and analysis of a kind of optical system for free space optical communication. Semiconductor Optoelectronics 24(3):197–199
34. Cao Y, Ai Y, Ye DM (2008) Ground simulation experiment of acquisition and tracking with beacon between the spatial movement platform. J Optoelectronics Laser 19(3):365–368
35. Xiong Z, Ai Y, Shan X (2011) Experiment of 5 Gb/s DWDM free space optical communication over 4.6 km. Infrared Laser Eng 40(10):1959–1963
36. Zeng ZL, Wang ZY, Lei LJ (2009) The development of a new type of 155Mb/s atmosphere laser communicator. Optical Commun Technol 33(6):28–30
37. Jiang SQ (2014) Research on method for precision tracking departure angle in satellite-ground laser links. Harbin Institute of Technology
38. Jiang HL, An Y, Zhang YL et al (2015) Analysis of the status quo, development trend and key technologies of space laser communication. J Spacecraft TT C Technol 34(3):207–217
39. Chen D, Ke XZ (2012) Research on diversity receive technology on wireless optical communication using subcarrier modulation. J Commun 33(8):128–131
40. Chatzidiamantis ND, Lioumpas AS, Karagiannidis GK et al (2010) Optical wireless communications with adaptive subcarrier PSK intensity modulation. In: IEEE global telecommunications conference (GLOBECOM 2010), pp 1–6
41. Fernández MM, Cheung KM, Mukai R (2006) Coherent optical focal plane array receiver for PPM signals: investigation and applications. Proc SPIE 630401(8):1–11

42. Shalaby HMH (1995) Maximum achievable throughputs for uncoded OPPM and MPPM in optical direct-detection channels. J Lightwave Technol 13(11):2121–2128
43. Shiu D, Kahn JM (1999) Differential pulse-position modulation for power-efficient optical communication. IEEE Trans Commun 47(8):1201–1210
44. Koepf GA, Marshialek RG, Begley DL (2002) Space laser communications: a review of major programs in the United States. Int J Electron Commun 56(4):232–242
45. Fernández MM, Vilnrotter VA (2004) Coherent optical receiver for PPM signals received through atmospheric turbulence: performance analysis and preliminary experimental results, SPIE. In: Proceeding of the international society for optical engineering (SPIE), Bellingham Washiington, USA, pp 151–163
46. Lange R, Smutny B, Wandernoth B et al (2006) 142km 5.625Gbps free-space optical link based on homodyne BPSK modulation. SPIE. In: Proceeding of the international society for optical engineering (SPIE), Bellingham Washiington, USA, pp 61050A1–61050A9
47. Smutny B, Kaempfner H, Muehlnikel G et al (2009) 5.6 Gbps optical intersatellite communication link, SPIE. In: Proceeding of the international society for optical engineering (SPIE), Bellingham Washiington, USA, pp 719906-1–719906-8
48. Seel S, Kaempfner H, Heine F et al (2007) Space to ground bidirectional optical communication link at 5.6 Gbps and EDRS connectivity outlook. In: IEEE aerospace conference. IEEE Computer Society, Washiington, DC, USA, pp 1–7
49. Fields R, Kozlowski D, Yura H et al (2011) 5.625 Gbps Bidirectional laser communications measurements between the NFIRE satellite and an optical ground station. In: Space optical systems and applications (ICSOS). IEEE Computer Society, Santa Monica, CA, pp 44–53
50. Atia WA, Bondurant RS (1999) Demonstration of return-to-zero signaling in both OOK and DPSK formats to improve receiver sensitivity in an optically preamplified receiver. In: Proceedings of LEOS 12th annual meeting, issue 1, pp 226–227
51. Koyama Y, Morikawa E, Shiratama K et al (2004) Optical terminal for neLS in-orbit demonstration. Proc SPIE 5338:29–36
52. Sasaki T, Toyoshima M, Takenaka H (2011) Digital coherent optical receiver for satellite laser. In: Space optical systems and applications (ICSOS). IEEE Computer Society, Santa Monica, CA, pp 245–247
53. Klaus P (1998) Laser communication terminals: a key building block for the new broadband satellite networks. Proc SPIE 3408:172–177
54. Zu JF, Liu LR, Luan Z et al (2003) Progress and trends in inter-satellite optical links. Laser Optoelectronics Progr 40(3):7–10
55. Wang W (2007) Research on atmospheric optical communication system. Huazhong University of Science and Technology
56. Zhao CZ (2007) The primary scheme of coherent optical space communication and the research of heterodyne efficiency. Yanshan University
57. Zhang WR, Zeng XD, Man XK (2008) Study on optical heterodyne detection. Infrared Laser Eng 37(9):146–148
58. Wang H (2009) Design of coherent optical communication zero difference BPSK system. University of Electronic Science and Technology of China
59. Zhao CY (2010) The study of coherent optical communication of heterodyne asynchronous demodulation technology. Changchun University of Science and Technology
60. Luo BB (2014) Coherent space optical digital communication. Beijing University of Posts and Telecommunications
61. Liu JH (2012) The effects of laser phase noise on the performance of coherent optical fiber communication systems. J Xi'an Inst Posts Telecommun 17(5):25–28
62. Taylor MG (2004) Coherent detection method using DSP for demodulation of signal and subsequent equalization of propagation impairments. IEEE Photonics Technol Lett 16(2):674–676
63. Zhu R (2009) Research on modulation schemes and automatic polarization control technology in coherent optical communication system. Beijing University of Posts and Telecommunications, Beijing

64. Sun H, Wu KT, Robers K (2008) Real-time measurements of a 40 Gb/s coherent system. Optical Express 16(2):873–879
65. Zhou X, Yu J, Qian D et al (2008) 8×114 Gb/s, 25-GHz-spaced, PolMux-RZ-8PSK transmission over 640 km of SSMF employing digital coherent detection and EDFA-only amplification. In: Optical fiber communication conference/national fiber optic engineers conference, San Diego, USA, PDP1
66. Okamoto S, Omiya T, Kasai K et al (2010) 140 Gbit/s coherent optical transmission over 150 km with a 10 Gsymbol/s polarization-multiplexed 128 QAM signal. In: Proceeding of the optical fiber communication /national fiber optic engineers conference, San Diego, USA, pp 1–4
67. Jensen JB, Rodes R, Zibar D et al (2011) Coherent detection for 1550 nm, 5 Gbit/s VCSEL based 40 km bidirectional PON transmission. In: OFC/NFOEC. Proceeding of the optical fiber communication and conference, Los Angels California, USA, pp 1–3
68. Raybon G, Winzer PJ, Adamiecki AA et al (2011) Transmission over 2400 km using an all-ETDM 80-Gbaud (160-Gb/s) QPSK transmitter and coherent receiver. In: European conference and exposition on optical communications 2011 (ECOC 2011). European Conference and Exposition on Optical Communications, Geneva, Switzerland, pp 1–3
69. Feng Y (2010) Digital optical coherent detection and its application. Tsinghua University, Beijing
70. Zeng Y (2008) Research on the optical receiver in coherent optical communication system. Beijing University of Posts and Telecommunications, Beijing
71. Luo Y, Wang Y, Zhang F (2011) Linear compensation based on manakov equation for coherent 40 Gbit/s polarization-multiplexed QPSK transmission. Acta Photonica Sinica 883–887
72. Yagi M, Satomi S, Ryu S (2018) Field trial of 160-Gbit/s, polarization-division multiplexed RZ-DQPSK transmission system using automatic polarization control. In: Optical fiber communication conference (OFC). In: Proceeding of the optical fiber communication and conference, San Diego, CA, pp 1–3
73. Mamdoohi G, Esmailian A, Abas AF (2009) Realization of microcontroller-based polarization control system with genetic algorithm. In: IEEE Malaysia International Conference on Communications (MICC 2009) 9th. Kuala Lumpur, Malaysia. In: Proceedings of the 9th IEEE Malaysia International Conference on Communications, pp 774–779
74. Katz G, Sadot D, Tabrikian J (2006) Electrical dispersion compensation equalizers in optical direct- and coherent-detection systems. IEEE Trans Commun 54(11):2045–2050
75. Bulow H (2007) Electronic dispersion compensation. In: Proceeding of the optical fiber communication and conference Anaheim, CA, USA, Paper OMG5, pp 1–3
76. Savory SJ (2010) Digital coherent optical receivers: algorithms and subsystems. Quantum Electronic 16(5):1164–1179
77. Zhong YZ (2011) Research on equalization technology of electric channel in coherent polarization multiplexing fiber communication system of 100Gbps and above. University of Electronic Science and Technology of China
78. Koepf GA, Marshalek RG, Begley DL (2002) Space laser communications: a review of major programs in the United States. Int J Electron Commun 56(4):232–242
79. Hemmati H (2003) Optical space communications at JPL. In: The 16th annual meeting of the IEEE lasers and electro-optics society, Tucson, AZ, USA, pp 81–82
80. Vilnrotter V, Lau CW (2002) Quantum detection and channel capacity for communications applications. Proc SPIE 4635:103–115
81. Chan VWS, Kaufmann JE (1988) Coherent optical intersatellite crosslink systems. Components for fiber optic applications III and coherent lightwave communications. Proc SPIE 988:325–335
82. Arono S (1999) Power versus stabilization for laser satellite communication. Appl Opt 38(15):3229–3233
83. Boroson DM, Robinson BS, Murphy DV et al (2014) Overview and results of the Lunar laser communication demonstration. International society for optics and photonics. Proc SPIE 8971:89710S (11)

84. Robinson BS, Boroson DM, Burianek DA et al (2011) The lunar laser communications demonstration. International conference on space optical systems and applications. Proc IEEE 7923:54–57
85. Robinson BS, Boroson DM, Burianek D et al (2014) The NASA lunar laser communication demonstration successful high rate laser communications to and from the moon. International conference on space operations. Proc IEEE 1685:1–7
86. Li XL (2018) Research on several key technologies of free space optical communication based on digital coherent detection. University of Chinese Academy of Sciences, Beijing
87. Keizo N, Akio Y (2000) Performance test result of LUCE (laser utilizing communications equipment) engineering model. Proc SPIE 3932:68–76
88. Allen D, Pillsbury (1993) LITE engineering model-II: the flight- qualified subsystems. Proc SPIE 1866:58–72
89. Don M, Boroson (1993) LITE engineering model-I: operation and performance of the communications and beam control subsystems. Proc SPIE 1866:73–82
90. Fernández MM, Vilnrotter VA (2004) Coherent optical receiver for PPM signals received through atmospheric turbulence: performance analysis and preliminary experimental results. Proc SPIE 151–163
91. Dreischer T, Maerki A, Weigel T et al (2002) Operating in sub-are seconds: high precision laser terminals for intersatellite communications. Optomechatronic Syst III Proc SPIE 4902:87–98
92. Pribil K, Serbe C, Wandernoth B et al (1995) SOLACOS YKS: an optical high-data-rate communication system for intersatellite link applications. Free-space laser communication technologies VII. Proc SPIE 2381:83–88
93. Pribil K, Fleming AJ (1995) SOLACOS: system implementation. Free-space laser communication technologies VII. Proc SPIE 2381:143–150
94. Zhou Y, Shi L (1998) SOLACOS: German space laser communication system. J Univ Electronic Sci Technol China 27(5):557–560
95. Pribil K, Flemming J, GmbH D (1994) Solid state laser communications in space (SOLACOS) high data rate satellite communication system verification program. Space optics 1994: space instrumentation and spacecraft optics. Proc SPIE 2210:39–48
96. Flemming J, Pribil K, GmbH D (1994) Solid state laser communications in space (SOLACOS) position, acquisition, and tracking (PAT) subsystem implementation. Space optics 1994: space instrumentation and spacecraft optics. Proc SPIE 2210:164–172
97. Eineder M, Runge H, Boerner E et al (2004) SAR interferometry with TerraSAR-X. In: Proceedings of the Fringe 2003 workshop (ESA SP-550), Frascati, Italy, vol 45, pp 1–6
98. Werninghaus R (2004) TerraSAR-X-mission. SAR image analysis, modeling, and techniques VI. Proc SPIE 5236:9–16
99. Lange R, Smutny B (2004) BPSK laser communication terminals to be verified in space. IEEE Military communications conference. Proc IEEE 1:441–444
100. Guo YF, Wang HM (2013) Survey on SILEX and subsequent development of space laser communication in Europe. Spacecraft Eng 22(2):88–93
101. Tolker-Nielsen T, Oppenhaeuser G (2002) In-orbit test result of an operational optical intersatellite link between ARTEMIS and SPOT4, SILEX. Proc SPIE 4635:1–15
102. Pribil K (1998) Laser communication terminals: a key building block for the new broadband satellite networks. Broadband European Networks and Multimedia Services. Proc SPIE 3408:172–177
103. Baister G, Dreischer T, Rugi Grond E et al (2003) The OPTEL terminal development programme enabling technologies for future optical crosslink applications. American Institute of Aeronautics and Astronautics, Long Beach, California, vol 6229, pp 1–8
104. Zhang RL (2014) Research on the key technology of 2–6GHz broadband microwave signal coherent optical transmission. University of Electronic Science and Technology of China
105. Lunde CT, Fields R, Jordan J et al (2008) The near-field infrared experiment (NFIRE) satellite program laser communication terminal (LCT): international cooperation for joint LCT experiments. Laser Sci 34(11):4073–4075

106. Pitz W, Miller D (2010) The TerraSAR-X satellite. IEEE Trans Geosci Remote Sens 48(2):615–622
107. Werninghaus R, Buckreuss S (2009) The TerraSAR-X mission and system design. IEEE Trans Geosci Remote Sens 48(2):606–614
108. Buckreuss S, Werninghaus R, Pitz W (2009) The German satellite mission TerraSAR-X. IEEE Aerosp Electron Syst Mag 24(11):4–9
109. Buckreuss S, Schättler B (2010) The TerraSAR-X ground segment. IEEE Trans Geosci Remote Sens 48(2):623–632
110. Fields R, Kozlowski D, Yura H et al (2011) 5.625Gbps bidirectional laser communications measurements between the NFIRE satellite and an optical ground station. In: 2011 international conference on space optical systems and applications. Proc SPIE 8184(4):44–53
111. Gregory M, Heine F, Kämpfner H et al (2010) Inter-satellite and satellite-ground laser communication links based on homodyne BPSK. Free-space laser communication technologies XXII. Proc SPIE 7587:75870E(5)
112. Miglior R, Duncan J, Pulcino V et al (2016) Outlook on EDRS-C. In: International conference on space optics—ICSO. Proc SPIE 10562:105622S(9)
113. Arimoto Y, Toyoshima M, Toyoda M et al (1995) Preliminary result on laser communication experiment using Engineering Test Satellite-VI (ETS-VI). Free-space laser communication technologies VII. International society for optics and photonics. Proc SPIE 2381:151–158
114. Fletcher GD, Hicks TR, Laurent B (1991) The SILEX optical interorbit link experiment. Electron Commun Eng 3(6):273–279
115. Akioka M, Kubo-oka T, Suzuki K et al (2017) Development of a breadboard model of space laser communication terminal for optical feeder links from GEO. In: Society of photo-optical instrumentation engineers conference series, Biarritz, France, 10562
116. Inaoka K, Tashima Y, Ichikawa S, Yajima M (2014) The Approach to realize a higher speed, more compact and longer-lasting operational optical space communications system. In: Proceedings of International Conference on Space Optical Systems and Applications (ICSOS)
117. Lange R, Smutny B, Wandernoth B et al (2006) 142km 5.625Gbps free-space optical link based on homodyne BPSK modulation. SPIE 61050A(9)
118. Mark G, Frank H, Kämpfner H et al (2011) Coherent inter-satellite and satellite-ground laser links. SPIE (7923):792303(8)
119. Smutny B, Kaempfner H, Muehlnikel G et al (2009) 5.6 Gbps optical intersatellite communication link. SPIE 719906(8)
120. Böhmer K, Mark G, Frank H et al (2014) Laser communication terminals for the European data relay system. SPIE 8246:82460D(6)
121. Seel S, Kaempfner H, Heine F et al (2011) Space to ground bidirectional optical communication link at 5.6 Gbps and EDRS connectivity outlook. IEEE 1–7
122. Fields R, Kozlowski D, Yura H et al (2011) 5.625 Gbps Bidirectional laser communications measurements between the NFIRE satellite and an optical ground station. IEEE 44–53
123. Takayama Y, Jono T, Toyoshima M et al (2007) Tracking and pointing characteristics of OICETS optical terminal in communication demonstrations with ground stations. In: Free-space laser communication technologies XIX and atmospheric propagation of electromagnetic waves. Proc SPIE 6457:645707(9)
124. Toyoshima M, Takahashi T, Suzuki K et al (2007) Laser beam propagation in ground-to-OICETS laser communication experiments. Atmospheric propagation IV. Proc SPIE 6551:65510A (12)
125. Toyoshima M, Sasaki T, Takenaka H et al (2011) Research and development of free space laser communications and quantum key distribution technologies at NICT. In: 2011 international conference on space optical systems and applications, Koganei, Japan, pp 1–7
126. Toyoshima M, Shoji Y, Kuri T et al (2008) Development of real time 3Gbps BPSK optical coherent receiver using field programmable gate array for free space laser communications. In: The 26th international communications satellite systems conference, San Diego, CA, pp 1–8

127. Yuksel H, Davis CC (2006) Aperture averaging analysis and aperture shape invariance of received scintillation in free space optical communication links. Free space laser communications VI. Proc SPIE 6304:63041E (11)
128. Atia WA, Bondurant RS (1999) Demonstration of return-to-zero signaling in both OOK and DPSK formats to improve receiver sensitivity in an optically preamplified receiver. Proc LEOS (1):226–227
129. Sasaki T, Toyoshima M, Takenaka H (2011) Digital coherent optical receiver for satellite laser. IEEE 245–247
130. Li X, Geng T, Ma S et al (2017) Performance improvement of coherent free-space optical communication with quadrature phase shift keying modulation using digital phase estimation. Appl Optics 56(16):4695–470 1
131. Yu SY, Yan S, Tan LY et al (2015) Time estimation of the link stability in inter-satellite optical communications. Chin J Lasers 42(11):137–143
132. Wu F, Yu SY, Ma ZT et al (2014) Correction of pointing angle deviation and in-orbit validation in satellite-ground laser communication links. Chin J Lasers 41(6):154–215
133. Song C (2017) China's first high-throughput communication satellite Shijian-13 was successfully launched. Chin Equipment Eng (8):6
134. Wang JL (2017) "Micius" quantum satellite successfully achieved all established scientific objectives. Science 69(5):16
135. Luo BB, Li Y, Xu T et al (2013) 20Gb/s coherent free-space optical communication system with QPSK modulation. In: Asia communications and photonics conference. Optical society of America, AF2F, 56
136. Li M, Li B, Song Y et al (2015) Investigation of costas loop synchronization effect on BER performance of space uplink optical communication system with BPSK scheme. IEEE Photonics J 7(4):1–9
137. Chen M, Liu C, Rui D et al (2019) Highly sensitive fiber coupling for free space optical communications based on an adaptive coherent fiber coupler. Optics Commun 430:223–226
138. Chen M, Liu C, Rui D et al (2018) Experimental results of 5-Gbps free-space coherent optical communications with adaptive optics. Optics Commun 418:115–119
139. Ke XZ, Wang J (2015) Intensity of reflected wave from corner reflector illuminated by partially coherent beam in the atmospheric turbulence. Acta Optica Sinica 42(10):9–17
140. Ke XZ, Wang J (2015) Comparison of polarization property of partially coherent beam between propagating along an uplink path and a downlink path in atmospheric turbulence. J Phys 64(22):224204
141. Kang Y, Ke XZ (2015) Multi-dimensions encoding scheme for visible light communication system. Chin J Lasers 42(2):130–136
142. Ke XZ, Lei SC, Yang PS (2016) Beam coaxial alignment detection in atmospheric laser communication. Chin J Laser 43(6):181–190
143. Kong YX, Ke XZ, Yuan Y (2016) Impact of local oscillator power on SNR in space coherent optical communications. Infrared Laser Eng 45(2):0222002
144. Tan ZK, Ke XZ (2017) Mixing efficiency in coherent detection system. Laser Optoelectronics Progr 54(10):126–134
145. Ma BB, Ke XZ, Zhang Y (2019) Polarization control and control algorithm of beams in coherent optical communication system. Chin J Lasers 46(1):0106002
146. Ke XZ, Lei SC (2016) Spatial light coupled into a single-mode fiber by a Maksutov-Cassegrain antenna through atmospheric turbulence. Appl Opt 55(15):3897–3902
147. Jia XD, Sun DS, Shu ZF et al (2015) Optimal design of the telescope in coherent lidar and detection performance analysis. Acta Optica Sinica 35(3):0301001
148. Zhao Y, Tong SF, Song YC et al (2015) Research progress of optical phase locked loop in space laser communication. Laser and Optoelectronics Progress 52(9):080002
149. Xu YX, Xu MM, Sun JF et al (2016) Integrated technology of communication and velocity measurement in satellite coherent optical communication. Laser Optoelectronics Progr 53(12):120603

150. Tan LY, Yang YQ, Ma J (2008) Pointing and tracking errors due to localized deformation in inter-satellite laser communication links. Opt Express 16(17):13372–13380
151. Das KK, Iftekharuddin KM, Karim MA (1997) Improved heterodyne mixing efficiency and signal-to noise ratio with an array of hexagonal detectors. Appl Optics 36(27):7023–7026
152. Belmonte A, Kahn JM (2008) Performance of synchronous optical receivers using atmospheric compensation techniques. Opt Express 16(18):14151–14162
153. Wei G, Zhou J, Long XW (2012) Analysis of signal-to-noise ratio and heterodyne efficiency for reference-beam laser Doppler velocimeter. Optics Laser Technol 44(1):108–113
154. Cohen SC (1975) Heterodyne detection: phase front alignment, beam spot size, and detector uniformity. Appl Opt 14(8):1953–1959
155. Nan H, Zhang P, Tong SF et al (2017) Analysis of the effect of light spot size and ray axis deflection on heterodyne efficiency of space optical hybrid. Infrared Laser Eng 46(4):0422001
156. Takenaka T, Tanaka K, Fukumitsu O (1978) Signal-to-noise ratio in optical heterodyne detection for Gaussian fields. Appl Opt 17(21):3466–3471
157. Ma TT, Shou F, Tong H, Nan et al (2017) Effect of polarization property of signal light on mixing efficiency of space coherent detection. Laser Optoelectronics Progr 54(7):020604
158. Yang YF, Yan CX, HCH et al (2017) Modified heterodyne efficiency for coherent laser communication in the presence of polarization aberrations. Optics Express 25(7):7567–7591
159. Liu HZ, Ji YF, Xu N et al (2011) Effect of amplitude profile difference of signal and local oscillator wave on heterodyne efficiency in the inter-satellite coherent optical communication system. Acta Optica Sinica 31(10):38–63
160. Wang Q, Wang CH, Shang TL (2003) Heterodyne efficiency for coherent detection of gaussian local oscillator and erie spot signa. Chin J Lasers 30:183–186
161. Zhang T, Tong SF (2013) Effect of collimating mismatch on mixing efficiency for space coherent laser communication. J Changchun Univ Sci Technol (Nat Sci Edn) 36(1–2):13–15
162. Han D, Liu YQ, Zhao X (2016) Effect of beam waist radius on heterodyne efficiency of spatial coherent optical communication. J Changchun Univ Technol (Nat Sci Edn) 39(3):36–40
163. Tanaka K, Saga N (1984) Maximum heterodyne efficiency of optical heterodyne detection in the presence of background radiation. Appl Opt 23(21):3901–3904
164. Tanaka K, Ohta N (1987) Effects of tilt and offset of signal field on heterodyne efficiency. Appl Opt 26(4):627–632
165. Li XY, Ma ZF, Shi DL (2015) Effect of Gaussian fields distribution on mixing efficiency for coherent detection. Infrared Laser Eng 44(2):539–543
166. Yang QB, Tan LY, Ma J (2011) Influence of aberrations of receiver antenna on heterodyne efficiency in free space optical communications based on heterodyne detection. In: International ICST conference on communications and networking in China, Beijing, China, pp 776–779
167. Kong YX, XZK, Yang Y (2017) Influence of laser linewidth on the performance of space coherent optical communication system. Chin J Sci Instrument 38(7):1668–1674
168. Wang EH, Hu YH, Li L (2011) Influence analysis of atmosphere on heterodyne detection lidar. Infrared Laser Eng 40(10):1896–1899
169. Belmonte A, Rye BJ (2000) Heterodyne lidar returns in the turbulent atmosphere: performance evaluation of simulated systems. Appl Opt 39(15):2401–2411
170. Frehlich RG, Kavaya MJ (1991) Coherent laser radar performance for general atmospheric refractive turbulence 30(36):5325–5352
171. Belen'kii MS (1993) Effect of atmospheric turbulence on heterodyne lidar performance. Appl Optics 32(27):5368–5372
172. Chambers DM (1997) Modeling heterodyne efficiency for coherent laser radar in the presence of aberrations. Opt Express 1(3):60–67
173. Leeb WR, Winzer PJ, Kudielka KH (1998) Aperture dependence of the mixing efficiency, the signal-to-noise ratio, ant the speckle number in coherent lidar receivers. Appl Opt 37(15):3143–3148
174. Chen CC, Gardner CS (1989) Impact of random pointing and tracking errors on the design of coherent and incoherent optical intersatellite communication links. IEEE Trans Commun 37(3):252–326

175. Li C, Mao HM, Zhen SL et al (2007) Influence of atmospheric turbulence on laser heterodyne detection. J Atmos Environ Optics 2(4):247–250
176. Churnside JH, McIntyre CM (1980) Heterodyne receivers for atmospheric optical communications. Appl Optics 19(4):582–590
177. Strohbehn JW, Clifford SF (1967) Polarization and angle-of-arrival fluctuation for a plane wave propagated through a turbulent medium. IEEE Trans Antennas Propag 15(3):416–421
178. Belmonte A (2003) Analyzing the efficiency of a practical heterodyne lidar in the turbulent atmosphere: telescope parameters. Opt Express 11(17):2041–2046
179. Xu J, Mao HM, Zhen SL et al (2007) The research for the deterioration of spatial coherence of laser heterodyne detection caused by atmospheric turbulence. Laser Infrared 37(12):1245–1249
180. Sun YY (1985) Laser heterodyne detection in turbulent atmosphere. Chin J Lasers 13(3):174–177
181. Fried DL (1967) Optical heterodyne detection of an atmospherically distorted signal wavefront. Proc IEEE 55(1):57–67
182. Churnside JH, McIntyre CM (1987) Signal current probability distribution for optical heterodyne receivers in the turbulent atmosphere 1: theory. Appl Opt 17(14):2141–2147
183. Salzman J, Katzir A (1984) Heterodyne detection SNR: calculations with matrix formalism. Appl Opt 23(7):1066–1074
184. Winick KA (1986) Atmospheric turbulence-induced signal fades on optical heterodyne communication links. Appl Opt 25(11):1817–1825
185. Das KK, Iftekharuddin KM, Karim MA (1997) Improved heterodyne mixing efficiency and signal-to-noise ratio with an array of hexagonal detectors. Appl Opt 36(27):7023–7026
186. Zhu XM, Kahn JM (2002) Free-space optical communication through atmospheric turbulence channels 50(8):1293–1300
187. Tsiftsis TA (2008) Performance of heterodyne wireless optical communication systems over gamma-gamma atmospheric turbulence channels. Electron Lett 44(5):373–374
188. Sandalidis HG, Tsiftsis TA, Karagiannidis GK (2009) Optical wireless communications with heterodyne detection over turbulence channels with pointing errors. J Lightwave Technol 27(20):4440–4445
189. Liu C, Yao Y, Sun Y et al (2010) Average capacity for heterodyne FSO communication systems over gamma-gamma turbulence channels with pointing errors. Electron Lett 46(12):851–852
190. Liu P, Yan JJ, Zhang Z (2011) Analysis of improved performance for a satellite-to-ground coherent optical communication system with DQPSK modulation due to phase estimation. In: International ICST conference on communications and networking in China, Beijing, China, pp 1–5
191. Niu MB, Song XG, Cheng JL et al (2012) Performance analysis of coherent wireless optical communications with atmospheric turbulence. Opt Express 20(6):6515–6520
192. Bai F, Su YW, Sato T (2015) Performance analysis of heterodyne detected OCDMA systems using PolSK modulation over a free space optical turbulence channel. Electronics 4(4):785–798
193. Huang J, Mei HP, Deng K et al (2015) Signal to noise ratio of free space homodyne coherent optical communication after adaptive optics compensation. Optics Commun 356:574–577
194. Kong YX, Ke XZ, Yang Y (2015) Influence research of atmospheric turbulence on space coherent optical communications. Laser Optoelectronics Progr 52(8):080601
195. Park J, Lee E, Chas BC et al (2016) Impact of pointing errors on the performance of coherent free space optical systems. IEEE Photonics Technol Lett 28(2):181–184
196. Born M, Wolf E (2000) Principles of optics. Cambridge University Press
197. Baykal Y, Eyyuboglu HT, Cai YJ (2009) Scintillations of partially coherent multiple Gaussian beams in turbulence. Appl Opt 48(10):1943–1954
198. Lu W, Liu L, Sun JF et al (2007) Change in degree of coherence of partially coherent electromagnetic beams propagating through atmospheric turbulence. Optics Commun 271:1–8
199. Ricklin JC, Davidson FM (2002) Point to point wireless communication using partially coherent optical fields. Free space laser communication and laser imaging. Proc SPIE 4489:156–165

200. Lahti JN, Nagel CM (1970) Mixing partially coherent fields with Gaussian irradiance profiles optimization criteria. Appl Opt 9(1):115–124
201. Tanaka T, Taguchi M, Tanak K (1992) Heterodyne efficiency for a partially coherent optical signal. Appl Opt 31(25):5391–5394
202. Ricklin JC, Davidson FM (2002) Atmospheric turbulence effects on a partially coherent Gaussian beam: implications for free-space laser communication. J Opt Soc Am A 19(9):1794–1802
203. Ricklin JC, Davidson FM, Weyrauch T (2002) Free space laser communication using a partially coherent laser source. Optics in atmospheric propagation and adaptive systems IV. Proc SPIE 4538:13–23
204. Salem M, Rolland JP (2010) Heterodyne efficiency of a detection system for partially coherent beams. J Opt Soc Am A 27(5):1111–1119
205. Ren YX, Dang AH, Liu L (2012) Heterodyne efficiency of a coherent free-space optical communication model through atmospheric turbulence. Appl Opt 51(30):7246–7254
206. Li CQ, Wang TF, Zhang HY et al (2015) The performance of heterodyne detection system for partially coherent beams in turbulent atmosphere. Optics Commun 356:620–627
207. Wang Y, Li CQ, Wang TF et al (2016) The effects of polarization changes of stochastic electromagnetic beams on heterodyne detection in turbulence. Laser Phys Lett 13(11):116006
208. Liu CH (2011) Research on holographic mode wavefront sensing Technology. National University of Defense Technology, Changsha, pp 83–95
209. Kim I, Goldfarb G, Li G (2007) Electronic wavefront correction for PSK free-space optical communications. Electron Lett 43:1108–1109
210. Puryear A, Chan VWS (2009) Coherent optical communication over the turbulent atmosphere with spatial diversity and wavefront predistortion. In: Global telecommunications conference, pp 1–8
211. Buske I, Riede W (2009) A mobile adaptive optics system for compensation of atmospheric turbulence-induced phase distortions. In: CLEO/Europe—EQEC 2009—European conference on lasers and electro-optics and the European quantum electronics conference, European 2009
212. Sinefeld D, Ella R, Zaharan O et al, (2010) Wavefront aberration correction in a free-space optical communication link using only the fiber-coupled optical power as a feedback mechanism. Electrical and Electronics Engineers in Israel (IEEEI), pp 890–891
213. Avanaki MRN, Hojjatoleslami SA, Ebrahimpour R et al (2010) Genetic algorithm for optimization of optical systems. Electrical Engineering (ICEE), pp 172–176
214. Podanchuk DV, Dan'ko VP, Kotov MM (2010) Holographic shack-hartmann wavefront sensor. In: Advanced Optoelectronics and Lasers (CAOL), pp 208–210
215. Li H, Zhang HT, Yin FC (2008) The adaptive optics technique for atmospheric turbulence in laser communication system. J Changchun Univ Sci Technol (Nat Sci Edn) 2:1–3
216. Yang HZ, Cai DM, Chen B et al (2008) Analysis of adaptive optics techniques without a wave-front sensor and its application in atmospheric laser communications. Chin J Lasers 35(5):681–684
217. Chen B, Li XY, Jiang WH (2010) Optimization of stochastic parallel gradient descent algorithm for adaptive optics in atmospheric turbulence. Chin J Lasers 37(4):959–964
218. Xia LJ, Li XF (2010) Transmission wave-front correction of atmospheric optical communication based on adaptive optics technology. J Terahertz Sci Electronic Inf Technol 08(003):331–335
219. Qiao RJ, Chen MS (2010) Research of DM in adaptive optics. Electronic Des Eng 18(3):74–78
220. Li J, Xiao J (2010) A closed-loop adaptive optical system based on genetic algorithm. J Wuhan Inst Technol 29(01):47–49
221. Han LQ, Wang Q, Tai KG et al (2010) Study on the performance of free space optical communication System based on adaptive optical compensation. J Appl Optics 31(2):301–303
222. Hardy JW (1998) Adaptive optics for astronomical telescopes. Oxford University Press, New York

223. Babcock HW (1953) The possibility of compensating astronomical seeing. Publ Astronmical Soc Pacific 65(386):229–236
224. Popoola WO, Ghassemlooy Z (2008) Performance of sub-carrier modulated freespace optical communication link in negative exponential atmospheric turbulence environment. Int J Autonomous Adaptive Commun Syst 1(3):342–355
225. Greenwood DP, Primmerman CA (1992) Adaptive optics research at Lincoln laboratory. Lincoln Lab J 5(1):3–24
226. Salmon JT (1991) Real time wavefront correction system using a zonal deformable mirror and a Hartmann sensor. Proc SPIE 1542:2–17
227. Rousset G, Fontanella JC, Kern P (1990) First diffraction-limited astronomical images with adaptive optics. Astron Astrophys 230(2):L29–L32
228. Kern P, Lena P, Gigan P (1989) Come-on: an adaptive optics prototype dedicated to infrared astronomy. Proc SPIE 1130:17
229. Rigaut F, Rousset G, Kern P (1991) Adaptive optics on a 3.6-M telescope: results and performance. Astron Astrophys 250(1):280–290
230. Roberts JE, Richard G et al (2012) Results from the PALM-3000 high-order adaptive optics system. Proc SPIE 8447:84470Y(8)
231. Genetron E, Cuby J, Rigaut F et al (1991) Come-on-plus project: an upgrade of the come-on adaptive optics prototype system. Proc SPIE 1542:297–307
232. Jiang WH (2006) Adaptive optical technology. Chin J Nat 28(1):7–13
233. Rao CH, Jiang WH (1996) Error and performance analysis of an infrared adaptive optics system at the 2.16 m telescope. Res Astron Astrophys 16(4):428–437
234. Rao CH, Jiang WH, Zhang YD et al (2008) Progress on the 127-element adaptive optical system for 1.8m telescope. Proc SPIE 7015:70155Y
235. Rao CH, Wei K, Zhang XJ et al (2010) First observations on the 127-element adaptive optical system for 1.8m telescope. Proc SPIE 7654:76541H
236. Zhu L, Gu N, Chen S et al (2012) Real time controller for 37-element low-order solar adaptive optics system at 1m new vacuum solar telescope. Proc SPIE 8415:84150V
237. Liu XD, Liu XY, Wang JL et al (2013) Development and performance test of the 961-element deformable mirror. Acta Optica Sinica 3(6):0601001(6)
238. Zhang XY, Hu LF, Cao ZL et al (2014) Improve the accuracy of interaction matrix measurement for liquid-crystal adaptive optics systems. Optics Express 22(12):14221–14228
239. Ma HM, Zhang PF, Zhan JH et al (2010) Stochastic parallel gradient descent algorithm for adaptive optics system. High Power Laser Particle Beams 22(6):1206–1210
240. Li YK, Chen DQ, Du XW (2000) Simulation of full field correction with two-deformable-mirror adaptive optics. High Power Laser Particle Beams 12(6):665–669
241. Hou J (2002) Research on new concept of wavefront detection based on adaptive optics. National University of Defense Technology, Changsha
242. Hu MF, Chen ZP (2004) Application of BICGSTAB algorithm in wave-front reconstruction and control. Opto-Electronic Eng 31(11):1–4
243. Liu CH (2011) Research on holographic mode wavefront sensing technology. National University of Defense Technology, Changsha
244. Xia LJ, Li XF (2010) Transmission wave-front correction of atmospheric optical communication based on adaptive optics technology. J Terahertz Sci Electronic Inf Technol 8(3):331–335
245. Wu JL, Ke XZ (2018) Adaptive optics correction of wavefront sensorless. Laser Optoelectronics Progr 55(3):133–139
246. Ke XZ, Wang XY (2018) Experimental study on the correction of wavefront distortion for vortex beam. Acta Optica Sinica 38(3):204–210

Chapter 2
Coherent Optical Communication

Contents

X. Ke and J. Wu, *Coherent Optical Wireless Communication Principle and Application*, Optical Wireless Communication Theory and Technology,
https://doi.org/10.1007/978-981-19-4823-7_2

Compared to intensity modulation/direct detection (IM/DD), coherent optical communication systems can achieve a detection sensitivity gain of approximately 20 dB (homodyne detection can reach 23 dB), allowing for longer distance transmission under the same power. Furthermore, coherent optical communication systems are more conductive to the suppression of atmospheric turbulence and channel fading. This chapter introduces the concepts behind coherent optical communication.

2.1 Basic Principles of Coherent Optical Communication

2.1.1 Fundamentals

The coherent optical communication system is summarized in Fig. 2.1. Light emitted from the laser is the carrier, and the signal is modulated using direct or external modulation (amplitude, frequency, phase, or polarization). At the receiving antenna, the spatial optical signal is coupled into the optic fiber through the fiber coupler; then, the signal is generated by the local oscillator laser and the light wave is superimposed with the received signal in the mixer. An intermediate frequency (IF) signal is generated from the balanced detector output, and the baseband signal can be obtained directly during homodyne detection.

In Fig. 2.2, the modulated signal is transmitted to the receiving end and mixed with the optical signal generated by the local oscillator laser; then, the IF signal is obtained and sent to the balance detector for detection. After the electrical signal output from the balance detector is amplified, filtered, and demodulated using the IF, it is restored to the digital signal at the transmitting end. The electric fields of the

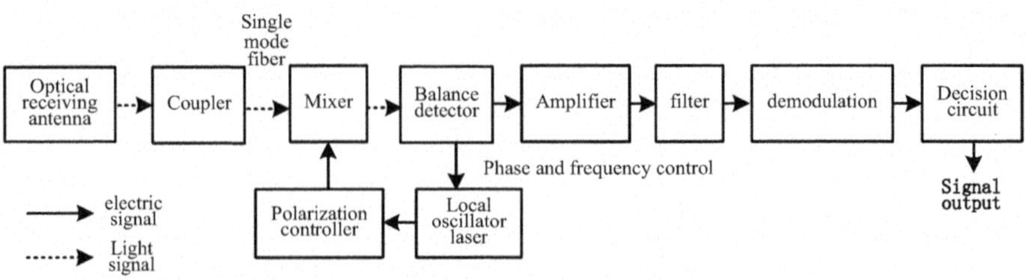

Fig. 2.1 Flowchart summarizing the coherent optical communication system

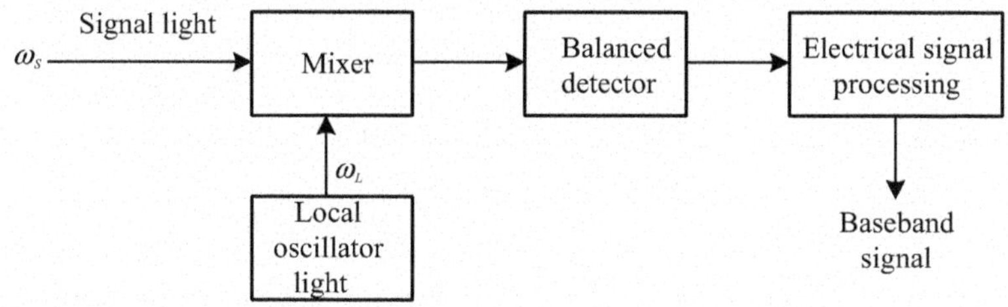

Fig. 2.2 Principal diagram of coherent detection [1]

signal light (E_S) and the local oscillator light (E_L) can be expressed as follows:

$$E_S(t) = A_S(t)\cos(\omega_s t + \varphi_s) \quad (2.1)$$

$$E_L(t) = A_L\cos(\omega_L t + \varphi_L) \quad (2.2)$$

where, A_L, ω_L, and φ_L are the amplitude, central angular frequency, and phase of the local oscillator light, respectively, and A_S, ω_S, and φ_s are the amplitude, central angular frequency, and phase of the signal light, respectively. Assuming that the signal and local oscillator lights have the same polarization direction and are incident on the optical mixer in parallel, then the incident light power is $\left[E_{S(t)} + E_{L(t)}\right]^2$, and the IF current signal output (i_p) from the mixer can be expressed as [1]

$$\begin{aligned} i_p = \alpha P_{(t)} &= \alpha\overline{\left[E_{S(t)} + E_{L(t)}\right]^2} \\ &= \alpha\Big\{A_S^2\overline{\cos^2(\omega_{St} + \varphi_S)} + A_L^2\overline{\cos^2(\omega_{Lt} + \varphi_L)} \\ &+ A_S A_L\overline{\cos[(\omega_S + \omega_L)t + (\varphi_S + \varphi_L)]} \\ &+ A_S A_L\overline{\cos[(\omega_S - \omega_L)t + (\varphi_S - \varphi_L)]}\Big\} \end{aligned} \quad (2.3)$$

In Eq. (2.3), $\alpha = en/hv$; the first and second terms are cosine squared terms; the time average value in the integer period is 1/2; and the sum is $\left(A_S^2 + A_L^2\right)/2$, equivalent to the direct current (DC) component of the detector output. The third term is the sum frequency term, which can be divided into two situations: (1) the time average value corresponding to the sum frequency term in the integer period is zero and (2) in a non-integer period, the frequency corresponding to the sum frequency term is very high and the photodetector cannot respond under normal circumstances. The fourth term is the difference frequency term, and because its rate of change is slower than that of the light field, it can be regarded as a constant. After filtering the DC and sum-frequency terms through a band-pass filter, the IF photocurrent output (i_{IF}) from the photodetector can be expressed as [2]

$$\mathrm{i}_{IF} = \alpha A_S A_L\cos[(\omega_S - \omega_L)\mathrm{t} + (\varphi_S - \varphi_L)] \quad (2.4)$$

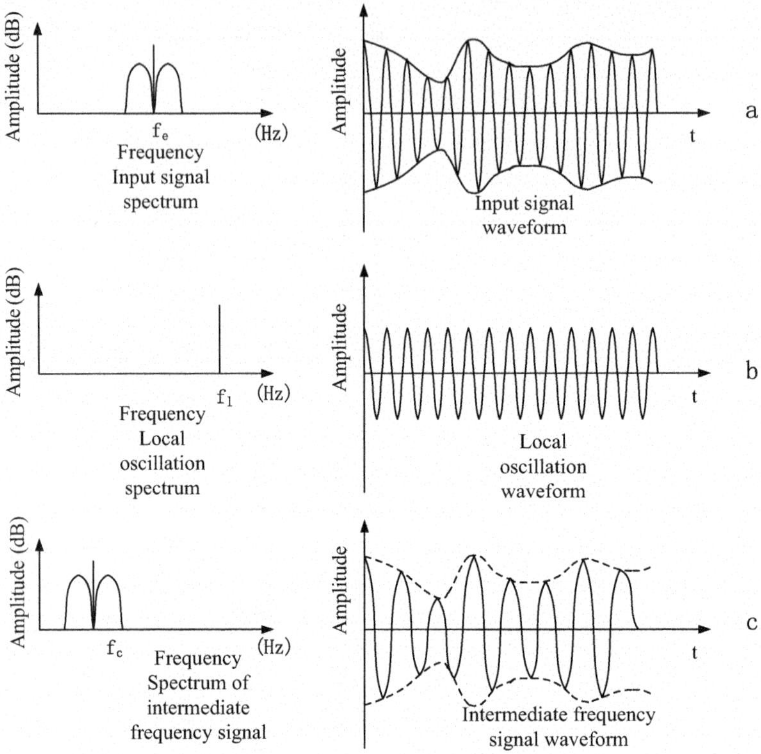

Fig. 2.3 Heterodyne detection frequency spectrum and waveform [3]

The frequency spectrum and waveform of heterodyne detection are shown in Fig. 2.3.

Using the average optical power, Eq. (2.4) can be expressed as [2]

$$\mathrm{i}_{IF} = 2\alpha\sqrt{P_S P_L}\cos[(\omega_S - \omega_L)\mathrm{t} + (\varphi_S - \varphi_L)] \tag{2.5}$$

Here, $P_S = A_S^2/2$ is the average optical power of the signal light and $P_L = A_L^2/2$ is the average optical power of the local oscillator light. When restoring the baseband signal, the received optical signal carrier frequency is transformed into a signal f_{IF}, and the IF signal is transformed into a baseband signal. This detection method is called heterodyne detection.

2.1.2 Homodyne Detection

When the local oscillator light frequency ω_L is the same as the signal light frequency ω_S (at this time $\omega_{IF} = 0$), it is referred to as homodyne detection. From Eq. (2.5), the photocurrent output by the balanced detector can be expressed as [2]

$$i_{IF} = 2\alpha\sqrt{P_S P_L}\cos(\varphi_S - \varphi_L) \tag{2.6}$$

Considering that the optical phase of the local oscillator is locked on the signal optical phase ($\varphi_S = \varphi_L$), the output photocurrent during homodyne detection can be expressed as

$$i_{IF} = 2\alpha\sqrt{P_S P_L}. \tag{2.7}$$

Other coherent optical communication systems exist. For example, Fig. 2.4 shows the principle diagram of auto-coherent detection using differential phase shift keying (DPSK) modulation. The received optical signal is coupled to the single-mode fiber through the reflective receiving antenna and enters the DPSK demodulator after being amplified by the erbium doped optical fiber amplifier. The signal is divided into two channels in the demodulator; and, with the exception of a 1-bit delay, the channels remain consistent. This information is then converted into intensity information, and the optical signal is converted into an electrical signal by the photodetector.

Figure 2.5. shows the heterodyne detection system of subcarrier modulation. The square term is constructed in the signal processing unit. Owing to the local laser, the heterodyne gain of this scheme is in the range 18–23 dB.

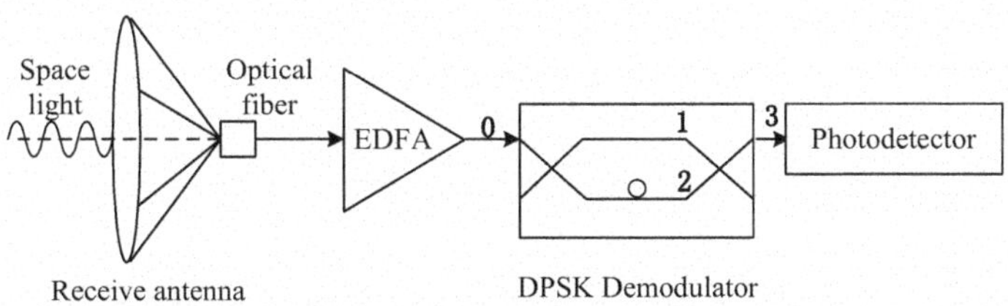

Fig. 2.4 Auto-coherent (self-homodyne) detection using differential phase-shift keying (DPSK) modulation. EDFA stands for erbium doped optical fiber amplifier [4]

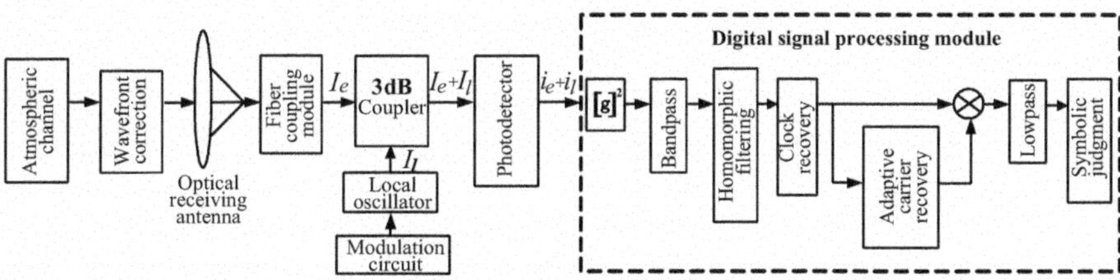

Fig. 2.5 Subcarrier intensity modulation heterodyne detection system [5]

2.1.3 Heterodyne Detection

In the case of heterodyne detection, ω_L differs from ω_S and the heterodyne current generated by the balanced detector can be expressed as [2]:

$$i_{IF} = 2\alpha\sqrt{P_S P_L}\cos[\omega_{IF}t + (\varphi_S - \varphi_L)]. \tag{2.8}$$

Equation (2.8) shows that.

(1) The output IF current is proportional to P_L. Even if the received optical signal power P_s is small, a sufficiently large output current can be obtained by increasing P_L. The local oscillator light is used for optical amplification in coherent detection, and sensitivity is improved because the system obtains a mixing gain.
(2) Because $\omega_s - \omega_L$ remains constant (or $\omega_{IF} = 0$) in coherent detection, it is necessary for the signal and local oscillator light sources have high frequency stability, a narrow spectral width, and a specific frequency tuning range.
(3) In both homodyne and heterodyne detection, the system must maintain phase lock and polarization direction matching between the received signal and local oscillator lights.

Optical heterodyne detection is holographic detection technology. The information carried by the amplitude E_S, frequency $\omega_S = \omega_{IF} + \omega_L$ (ω_L is known, ω_{IF} can be measured), and phase φ_S of the light field can be detected.

2.1.4 Detection of an Amplitude Modulated Signal

If a signal with a modulation frequency of Ω is loaded on the amplitude of a light wave with a frequency of ω_S, the amplitude-modulated light wave can be expressed as [6]

$$\begin{aligned} E_S(s) =& A_0\left[1 + \sum_{n=1}^{\infty}\cos(\Omega_n + \varphi_n)\right]\cos(\omega_s t + \varphi_s) \\ &= A_0\cos(\omega_s t + \varphi_s) + \sum_{n=1}^{\infty}\frac{m_n A_0}{2}\cos[(\omega_s + \Omega_n)t + (\varphi_s - \varphi_n)] \\ &+ \sum_{n=1}^{\infty}\frac{m_n A_0}{2}\cos[(\omega_s - \Omega_n)t + (\varphi_s - \varphi_n)] \end{aligned} \tag{2.9}$$

Here, A_0 is the average amplitude of the amplitude modulation (AM) wave, Ω_n and φ_n are the angular frequency and initial phase of the nth harmonic component, respectively, and m_n is the AM coefficient. If the AM signal $E_S(t)$ is coherent with

Fig. 2.6 Amplitude modulated (AM) light wave signal spectrum [6]

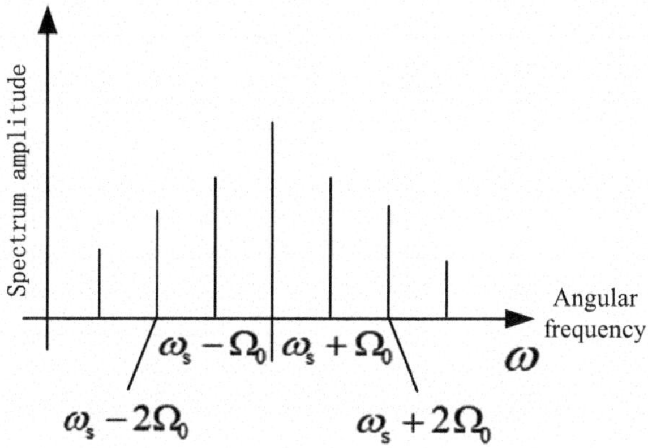

the local oscillator light $E_L(t)$, the instantaneous intermediate-frequency current can be expressed as [6]

$$\begin{aligned} i_{IF} =& \alpha A_0 A_L \cos[(\omega_L-\omega_S)t+(\varphi_L-\varphi_S)] \\ &+\alpha A_L\sum_{n=1}^{\infty}\frac{m_n A_0}{2}\cos[(\omega_L-\omega_s-\Omega_n)t+(\varphi_L-\varphi_s-\varphi_n)] \\ &+\alpha A_L\sum_{n=1}^{\infty}\frac{m_n A_0}{2}\cos[(\omega_L-\omega_s-\Omega_n)t+(\varphi_L-\varphi_s-\varphi_n)] \end{aligned} \quad (2.10)$$

The signal converted by the photodetector is proportional to the instantaneous IF current. The modulated signal carried on the amplitude of the light wave is transferred to the frequency current without $\omega_{IF}=\omega_L-\omega_S$ distortion (frequency and phase modulated coherent detection use a similar process). Figures 2.6 and 2.7 show the instantaneous IF electrical signal spectrum following AM signal light and coherent detection.

For AM signals, the photocurrent signal obtained using homodyne detection is the original shape of the light wave modulation signal; however, during homodyne detection, small changes in the amplitude of the local oscillator light will directly impact the signal spectrum and cause signal distortion. Consequently, homodyne detection has a greater requirement for local oscillator light amplitude stability.

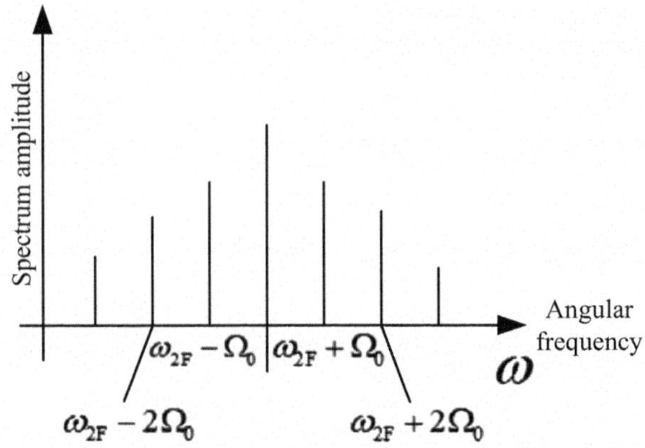

Fig. 2.7 Coherent detection instantaneous intermediate frequency electrical signal spectrum [6]

2.2 Coherent Modulation and Demodulation

2.2.1 Optical Modulation

In a coherent optical communication system, the transmitter adopts a direct (or external) approach to modulate the optical carrier amplitude, frequency, or phase. In terms of digital modulation, three basic forms are available: amplitude shift keying (ASK), frequency shift keying (FSK), and phase shift keying (PSK) [7, 8].

(1) ASK modulation

In ASK modulation, frequency and phase are constants, amplitude is a variable, and information bits are transmitted via the carrier wave amplitude. Because the binary amplitude keying (2ASK) modulation signal has only two levels (0 or 1), the multiplication result is equivalent to turning the carrier frequency on or off. This means that when the modulated digital signal is "1," the carrier is transmitted, and when the modulated digital signal is "0," the carrier is not transmitted. This principle is shown in Fig. 2.8a, where $s(t)$ is the baseband rectangular pulse. Generally, the carrier signal is represented by a cosine signal, and the modulation signal converts the digital sequence into a unipolar baseband rectangular pulse sequence. Then, the on/off keying function is used to multiply the output with the carrier and move the spectrum to the vicinity of the carrier frequency. The 2ASK modulation waveform is shown in Fig. 2.8b.

Figure 2.9 shows that, during PSK modulation, the phase of the carrier changes with the state of the modulated signal. If two carriers with the same frequency start to oscillate at the same time, they are considered to be in-phase; that is, they reach their peak, zero, and nadir values at the same time. A phase inversion occurs when one carrier reaches its peak positive value at the same time as the second carrier reaches its peak negative value. Generally, a signal oscillation is equal to 360°. If one wave is half an oscillation (180°) away from the other wave, the waves are in antiphase.

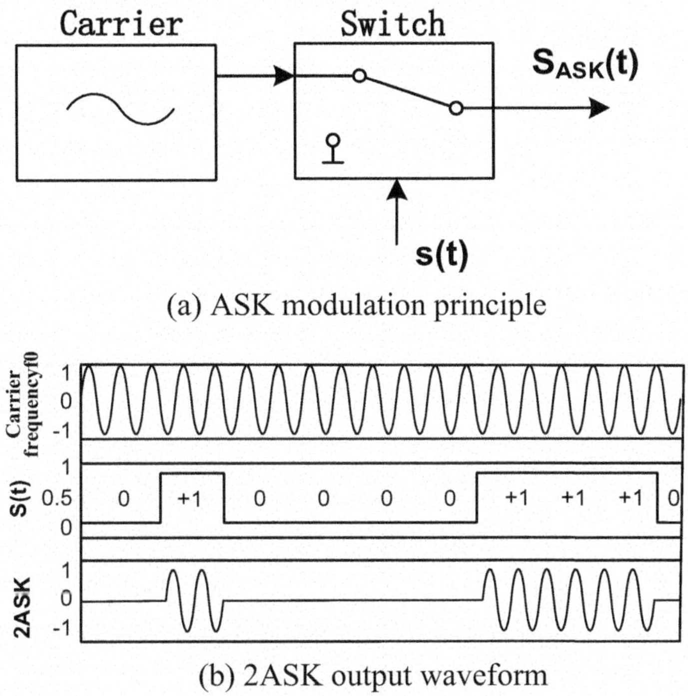

Fig. 2.8 Binary amplitude shift keying (2ASK) modulation principal diagram [8]

When transmitting digital signals, the "1" code controls the 0° phase and the "0" code controls the 180° phase.

(2) FSK modulation

The earliest modulation method used in digital signal transmission, FSK uses digital signals to modulate the carrier frequency. In FSK modulation, amplitude and phase are constants, frequency is a variable, and information bits are transmitted via the carrier wave frequency, as shown in Fig. 2.10a. The signals transmitted by FSK are binary (0 and 1), whereas in MFSK, M symbols are represented by M frequencies. The 2FSK waveform is shown in Fig. 2.10b. This method is relatively easy to implement, has good anti-noise and anti-attenuation performance, is stable and reliable, and is the best choice for medium- and low-speed data transmission.

2.2.2 *Coherent Demodulation*

There are two demodulation methods for coherent detection: synchronous and asynchronous demodulation [9]. In homodyne detection, the optical signal is directly demodulated into a baseband signal and the frequency and phase of the local

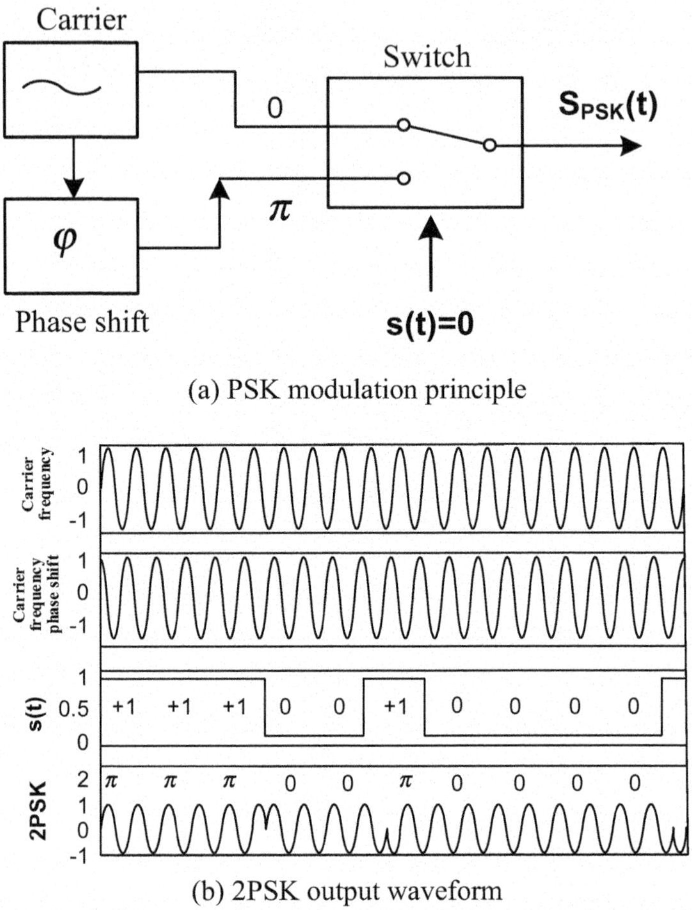

Fig. 2.9 Binary phase shift keying (2PSK) modulation principle diagram [8]

oscillator light must be the same as those of the signal light. Because of this, synchronous demodulation is predominantly used in homodyne detection, as shown in Fig. 2.11. Synchronous demodulation appears simple; however, the actual technology is highly complex. In heterodyne detection, phase and frequency matching between the signal light and local oscillator light is not required; hence, synchronous or asynchronous demodulation can be used. Synchronous demodulation requires the recovery of the intermediate frequency (IF, ω_{IF}), and therefore requires an electric phase-locked loop. Conversely, asynchronous demodulation simplifies the receiver design and is easier to implement. Homodyne and heterodyne synchronous and asynchronous signal demodulation methods share the principles and implementation methods of synchronous and asynchronous demodulation in radio technology, as shown in Fig. 2.12.

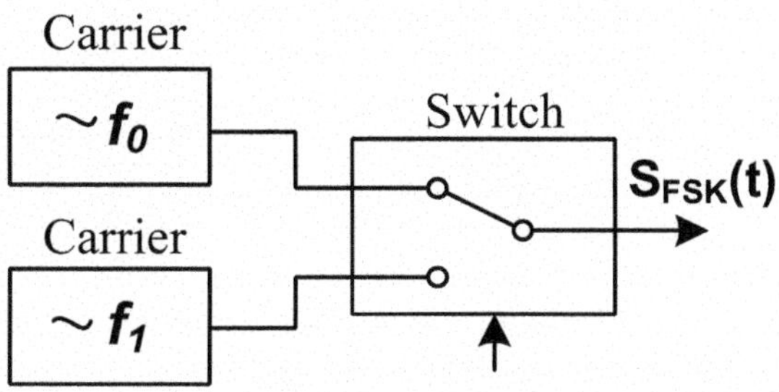

(a) FSK modulation principle

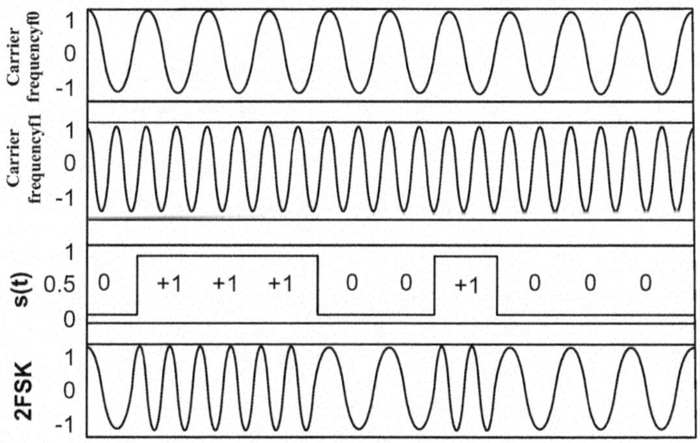

(b) 2FSK output waveform

Fig. 2.10 Binary frequency shift keying (2FSK) modulation principal diagram [8]

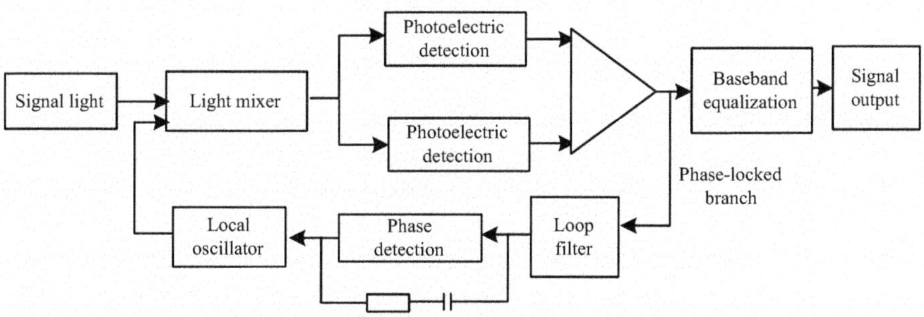

Fig. 2.11 Homodyne synchronous demodulation [9]

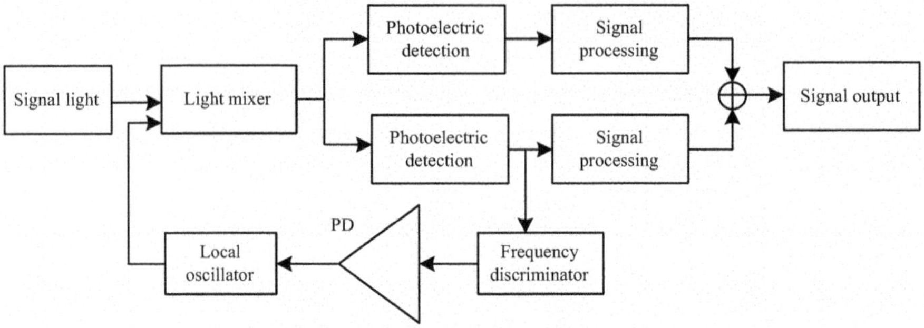

Fig. 2.12 Homodyne asynchronous demodulation [9]

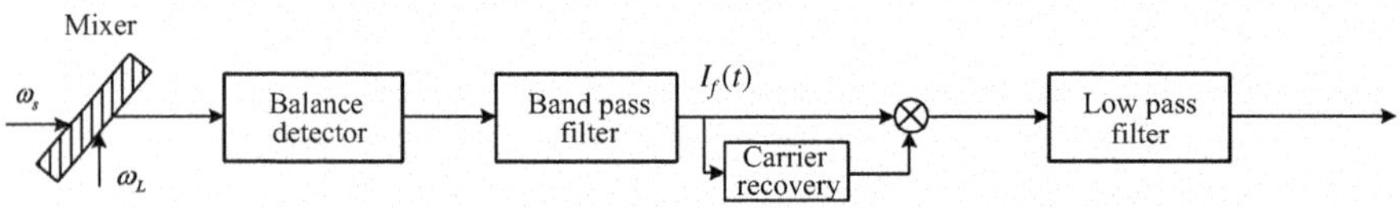

Fig. 2.13 Process diagram of the heterodyne synchronous receiver [10]

(1) Heterodyne synchronous demodulation.

Figure 2.13 shows the processes performed by the heterodyne synchronous demodulation receiver. The photocurrent generated by the balanced detector passes through the band-pass filter at the IF between the signal and local oscillator light frequencies. When noise is not considered, the signal passes through the frequency band. The filter can be written as [10]

$$I_f(t) = 2\alpha\sqrt{P_S P_L}\cos(\omega_{IF}t - \varphi). \tag{2.11}$$

Here, φ is the phase difference between the local oscillator and signal lights. For synchronous demodulation, we multiply $\mathrm{I}_f(t)$ and $\cos(\omega_{IF}t)$, and use a low-pass filter to obtain the baseband signal as follows [10]:

$$I_d = \frac{1}{2}\Big[2\alpha\sqrt{P_S P_L}\cos(\varphi) + i_C\Big]. \tag{2.12}$$

Here, i_C is the in-phase Gaussian random noise with a zero mean. Equation (2.12) shows that only the in-phase noise component affects the performance of the synchronous heterodyne receiver. Synchronous demodulation requires the recovery of an IF microwave carrier ω_{IF}, and there are several ways to achieve this; however, all methods require the use of an electric phase-locked loop. Commonly used phase-locked loops include square and Costas loops.

(2) Carrier recovery

To achieve coherent demodulation, the receiver must recover a coherent carrier with the same frequency and phase as the transmitted modulated carrier, referred to as carrier recovery. Carrier recovery methods can use a pilot frequency or a direct approach. The pilot method inserts a pilot frequency in a specified position while sending a useful signal, enabling the receiver to extract the carrier from the pilot frequency. The direct method does not send a pilot frequency, and the receiver extracts the carrier directly from the transmitted signal.

The pilot frequency can be inserted into the frequency or time domain. Frequency domain insertion means that the pilot signal is inserted while sending useful information, whereas carrier information is transmitted at a set time in time domain insertion. Using the time domain insertion pilot method, the inserted pilot signal and transmitted information are distinguished in time and the digital signals are grouped into frames. The pilot signal is then sent in a specified time interval in a set time sequence. Time domain insertion is discontinuous and has a short duration; hence, it is not possible to use a narrow band filter to extract the coherent carrier using this method.

Direct carrier recovery methods can be classified as square transformation, square ring, and special phase-locked loop methods. They all perform nonlinear processing of the received signal and then extract the carrier component. The special phase-locked loop method eliminates modulation and denoising, and can identify the phase error between the suppressed carrier component in the received signal and the output signal of the local voltage-controlled oscillator, thereby extracting the coherent carrier. Special phase-locked loops can be in-phase quadrature, inverse modulation, or decision phase-locked loops.

When the direct method is used to extract the carrier, the power signal-to-noise ratio is larger because the pilot power is not occupied at the transmitting end. This can prevent mutual interference between the pilot and the signal, which is common when using the pilot method, and pilot phase errors caused by an imperfect channel. However, some modulation systems, such as single sideband modulation, cannot use direct methods. The insert pilot method uses a separate pilot signal, which can extract the carrier signal, and it can also be used for automatic gain control. In the modulation system, where the direct method cannot be used, only the insert pilot method can be used. It needs to consume power during insertion. Therefore, under the condition of the same total power, compared to the direct method, the power signal-to-noise ratio of the inserted pilot is smaller.

Carrier synchronization has the following basic requirements: a small synchronization error, long hold time, short setup time, and the power band occupied by the synchronization should be as small as possible. Carrier synchronization error distorts the received signal waveform, decreases the signal-to-noise ratio, and increases the bit error rate (BER).

(3) Heterodyne asynchronous demodulation

Figure 2.14 shows the processes performed by heterodyne asynchronous demodulation receiver. Because it does not require IF microwave carrier recovery, the receiver design can be simplified. The receiver uses envelope detection and a low-pass filter to transform the band-pass filtered signal $I_f(t)$ into a baseband signal, and the signal current is determined as follows [10]:

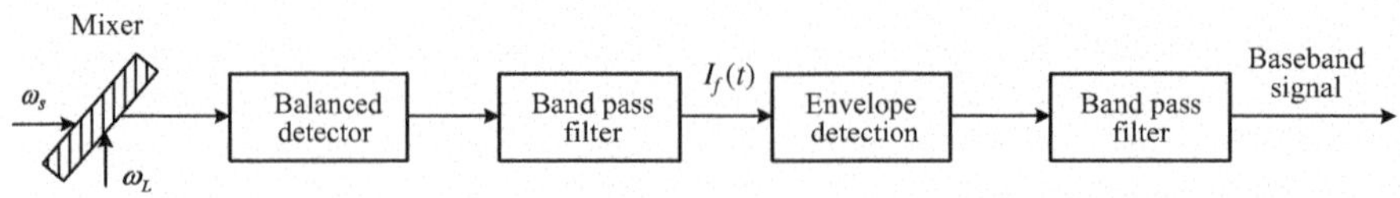

Fig. 2.14 Process diagram of the external difference step receiver [10]

$$\mathrm{I}_d = \left[\left(i_C + 2\alpha\sqrt{P_S P_L}\cos\varphi\right)^2 + \left(i_S + 2\alpha\sqrt{P_S P_L}\sin\varphi\right)^2\right]^{1/2} \tag{2.13}$$

Here, i_C and i_S are Gaussian random noise components, and i_S is the current fluctuation due to shot noise. Shot noise is caused by the dispersion of current-forming carriers in semiconductor devices. At low and intermediate frequencies, shot noise has nothing to do with frequency and shows white noise characteristics. The step demodulation receiver differs from the heterodyne synchronous receiver in that the in-phase and quadrature components of the receiver noise affect the signal output, and the sensitivity of the external difference step receiver is slightly reduced (by approximately 0.5 dB). Furthermore, no synchronous receiver is required for frequency carrier recovery, and the line width requirements of the signal light source and the local oscillator light source are lower. Therefore, the outer differential step demodulation scheme is often used in coherent optical communication systems.

2.2.3 *System Performance*

(1) Signal-to-Noise Ratio of Optical Heterodyne Detection

A core problem with optical heterodyne detection systems is improving the system signal-to-noise ratio. The signal-to-noise ratio at the output of the IF filter can be expressed as [10]

$$\left(\frac{S}{N}\right)_{IF} = \frac{P_{IF}}{N_P} = \frac{\alpha^2 M^2 P_S P_L R_L}{M^2 e\Delta f_{IF}[\alpha(P_S + P_L + P_B) + I_D]R_L + 2kT\Delta f_{IF}}. \tag{2.14}$$

Here, M is the internal gain of the mixer, P_B is the background radiation power, I_D is the dark current of the optical mixer, and α is the responsivity. The first term in the denominator is shot noise, and the second term is the thermal noise caused by the internal resistance of the mixer and the load resistance of the preamplifier. When the vibration power P_L is sufficiently large, the shot noise caused by the local vibration in the denominator is much larger than all other noises, and Eq. (2.14) becomes [10]

$$\left(\frac{S}{N}\right)_{IF} = \frac{\eta P_S}{h\nu\Delta f_{IF}}. \tag{2.15}$$

This is the maximum signal-to-noise ratio that the optical heterodyne detection system can achieve, and is called the quantum detection limit or quantum noise limit of optical heterodyne detection. From Eq. (2.15), the conditions for realizing quantum noise detection can be derived. For systems where the main noise source is thermal noise, the following conditions must be met to achieve quantum noise limit detection [10]:

$$\frac{e\eta R_L}{h\nu} P_L \Delta i_{IF} > 2kT\Delta f_{IF}.$$

From this we derive [10]

$$P_L > \frac{2kTh\nu}{e^2\eta R_L} \tag{2.16}$$

Equation (2.16) shows that increasing the power of the local oscillator light is beneficial to suppress all other noises, except that caused by the signal light, thereby obtaining a high conversion gain. Because the local oscillator light itself also produces noise, a larger P_L is not always better. When the power of the local oscillator light is sufficiently large, the shot noise generated by the local oscillator light will be greater than all other noises. If the power of the local oscillator light is further increased, then the signal-to-noise ratio of the optical heterodyne detection system will be reduced. However, if the local oscillator light power is too small, the IF conversion gain will decrease. In an actual optical heterodyne detection system, the power of the local oscillator light should be selected to obtain the best IF signal-to-noise ratio and largest IF conversion gain.

(2) Noise Equivalent Power of Optical Heterodyne Detection

In the quantum noise limit formula (Eq. 2.15), if $(S/N)_{IF} = 1$, then the noise equivalent power *NEP* of optical heterodyne detection can be obtained as follows [10]:

$$NEP = \frac{h\nu\Delta f_{IF}}{\eta} \tag{2.17}$$

Equation (2.17) is sometimes referred to as the sensitivity of optical heterodyne detection, and it is the theoretical limit of optical heterodyne detection. If the detector has a quantum efficiency of $\eta = 1$ and a measurement bandwidth of $\Delta f_{IF} = 1\text{Hz}$, then the theoretical sensitivity limit of optical heterodyne detection is one photon.

In practice, the sensitivity of optical heterodyne detection is restricted by many factors. For example, thermal and amplifier noise introduced by the optical mixer itself and load resistance will reduce the sensitivity of optical heterodyne detection, even if (NEP) increases. Therefore, the detection sensitivity of an actual optical heterodyne system can only approach the theoretical limit.

2.3 Factors Affecting Detection Sensitivity

2.3.1 *Phase Noise*

In a coherent optical detection system, the phase noise of the emitting laser and the local oscillator laser are the main factors leading to decreased sensitivity. Therefore, the homodyne and heterodyne signal phase φ_S, and the local oscillator light phase φ_L must be relatively stable to reduce sensitivity loss. Phase instability in the photoelectric detection process leads to current instability, thereby reducing the signal-to-noise ratio.

2.3.2 *Intensity Noise*

For Eq. (2.14), if the intensity noise $\delta_1 = RP_L r_l$ is increased, r_l is the parameter related to the local oscillator laser intensity noise (RIN). If the RIN spectrum is consistent with the receiver bandwidth Δf_{IF}, then Δf_{IF} can be approximated by 2 (RIN) Δf_{IF}, and the signal-to-noise ratio can be expressed as [10]

$$\left(\frac{S}{N}\right)_{IF} = \frac{P_{IF}}{N_P} = \frac{\alpha^2 M^2 P_S P_L R_L}{M^2 e\Delta f_{IF}[\alpha(P_S + P_L + P_B) + I_D]R_L + \sigma_I + 2kT\Delta f_{IF}} \tag{2.18}$$

If the receiver works under the limitation of shot noise ($\sigma_S^2 = 2q(I + I_d)\Delta f_{IF}$, q is electronic charge), then the local oscillator light power P_L should be sufficiently large. Equation (2.18) shows that when P_L increases, intensity noise also increases. It will become as large as shot noise, and the signal-to-noise ratio will decrease unless the signal light power P_S is also increased to offset the increase in receiver noise. The increase in P_S is the power penalty caused by the intensity noise of the local oscillator light. If the system is under the limit of shot noise, then both the dark current I_D and thermal noise in Eq. (2.18) can be ignored. The power penalty δ_1(expressed in dB) caused by intensity noise can be expressed as [10]

$$\delta_1 = 10\lg[1 + (\eta P_L(RIN))/h\nu], \tag{2.19}$$

and the BER can be expressed as [11]

$$SNR = \frac{1}{4}\left[erfc\left(\frac{I_I - I_D}{\sqrt{2}\sigma_1}\right) + erfc\left(\frac{I_D - I_0}{\sqrt{2}\sigma_0}\right)\right] \tag{2.20}$$

Here,I_D is the judging current; I_1 and I_0 represent the current of codes “1” and “0,” respectively; σ_1 and σ_0 represent the variance of codes “1” and “0,” respectively; and $erfc(\cdot)$ is the error function. When $I_D = \frac{\sigma_0 I_1 - \sigma_1 I_0}{\sigma_1 + \sigma_0}$, the BER is the lowest.

2.3.3 Polarization Noise

In heterodyne detection, the signal light is first mixed with the local oscillator light to obtain the IF signal, whereas the baseband signal is obtained directly in homodyne detection. During this process, it is necessary for the signal and local oscillator lights to maintain the same polarization state. The depolarization effect of the atmospheric channel causes the polarization state of the signal light to change over time; hence, random polarization noise is generated when the signal and local oscillator lights are mixed. When the signal light is orthogonal to the local oscillator light, the signal disappears. Consequently, polarization noise must be suppressed in a coherent optical communication system.

2.3.4 Key Technologies of Coherent Optical Communication Systems

Compared to the IM/DD system, successful coherent optical communication requires that signal and local oscillator light sources use lasers with a high-frequency stability and spectral purity [12]. In coherent optical communication systems, the IF is generally 2×10^8 to 2×10^9 Hz, whereas the coherent optical carrier frequency at 1550 nm is approximately 2×10^{14} Hz. The IF is multiplied 10^{-6} to 10^{-5} by the optical carrier frequency; hence, the light source frequency stability must be better than 10^{-8}. Furthermore, the signal and local oscillator light sources must meet strict matching conditions during mixing to obtain a high mixing efficiency. Matching approaches include spatial phase, wavefront, and polarization direction matching [13].

2.4 Spatial Phase Conditions for Optical Heterodyne Detection

2.4.1 Spatial Phase Difference Conditions

Figure 2.15 shows the spatial relationships involved in optical heterodyne detection. To study the influence of the non-coincidence of the two light wavefronts on heterodyne detection, we assume that the signal and the local oscillator lights are both plane waves, and there is an angle θ between their wavefronts. To simplify the analysis, we assume that the photosensitive surface of the optical mixer is a square with a side length d. The incident of the local oscillator light is perpendicular to the surface of the mixer, and the signal and the local oscillator lights have a mismatch angle at the wavefront; hence, the incident of the signal light is oblique to the mixer surface, and the same wavefront reaches the surface of the optical mixer at different times. This

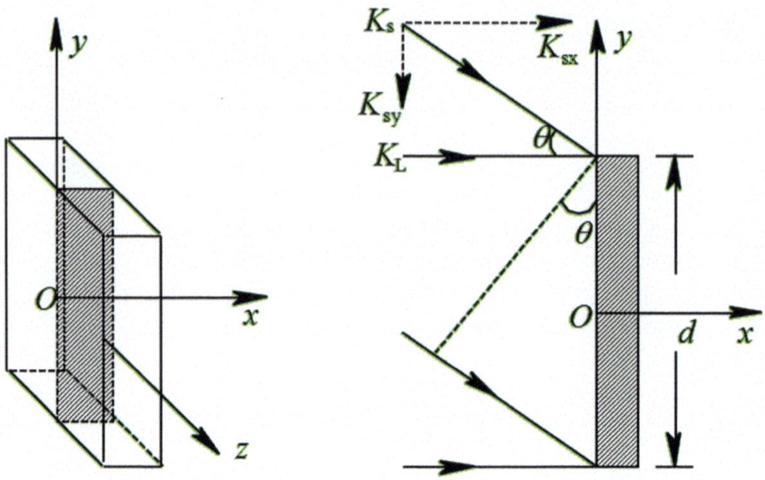

Fig. 2.15 Spatial relationship of optical heterodyne detection [14]

is equivalent to proceeding at speed v_x in the direction of x, so that the wavefront phase difference occurs at different points on the surface of the optical mixer. The expression of the light field of the local oscillator and signal lights, respectively, can be written as [14]

$$E_L(t) = A_L \cos(\omega_L t + \varphi_L) \tag{2.21}$$

$$E_s(t) = A_s \cos\left(\omega_s t + \varphi_s - \frac{\omega_s}{v_x} \cdot x\right) \tag{2.22}$$

Here, $\omega_s / v_x = k_x$ is the component of the signal light wave vector $\boldsymbol{k}$ in the direction of x. From Fig. 2.15, we know that $k_x = k \sin\theta = (\omega_S / c) \sin\theta$; hence, $v_x = c / \sin\theta$, where c is the speed of light. The signal light field can be expressed as [14]

$$E_s(t) = A_s \cos(\omega_s t + \varphi_s - \frac{2\pi \sin\theta}{\lambda_s} \cdot x) \tag{2.23}$$

The total electric field incident on the optical mixer is

$$E_t(t) = E_s(t) + E_L(t) \tag{2.24}$$

The instantaneous photocurrent output following optical mixing can be expressed as [10]

$$i_P(t) = \frac{G\beta}{d^2}\int_{-d/2}^{d/2}\int_{-d/2}^{d/2}\left[A_S\cos\left(\omega_S t + \varphi_S - \frac{2\pi\sin\theta}{\lambda_S}\right) + A_L\cos(\omega_L t + \varphi_L)\right]^2 dxdy$$

$$\begin{aligned} &= \frac{G\beta}{d^2}\int_{-d/2}^{d/2}\int_{-d/2}^{d/2}\{A_S^2\cos^2\left(\omega_S t + \varphi_S - \frac{2\pi\sin\theta}{\lambda_S}\right) + A_L^2\cos^2(\omega_L t + \varphi_L) \\ &+ A_S A_L\cos\left[(\omega_L - \omega_S)t + (\varphi_L - \varphi_S) + \frac{2\pi\sin\theta}{\lambda_S}\right] \\ &+ A_S A_L\cos\left[(\omega_L + \omega_S)t + (\varphi_L + \varphi_S) + \frac{2\pi\sin\theta}{\lambda_S}\right]\Bigg\}dxdy \end{aligned} \tag{2.25}$$

The instantaneous IF current output of the IF filter is [10]

$$i_{IF} = \frac{G\beta}{d^2}\int_{-d/2}^{d/2}\int_{-d/2}^{d/2}\left\{A_S A_L\cos\left[(\omega_L - \omega_S)t + (\varphi_L - \varphi_S) + \frac{2\pi\sin\theta}{\lambda_S}\right]\right\}dxdy \tag{2.26}$$

Following integration, the instantaneous IF current output from the optical mixer is [10]

$$i_{IF} = G\beta A_S A_L\cos[(\omega_L - \omega_S)t + (\varphi_L - \varphi_S)]\frac{\sin(\omega_S d/2v_x)}{\omega_S d/2v_x} \tag{2.27}$$

Because $v_x = c/\sin\theta$, the instantaneous intermediate frequency current is related to the mismatch angle θ. When this formula is factored into Eq. (2.27), the IF current reaches its maximum value. This requires $\omega_S d/2v_x = 0$, which means $\sin\theta = 0$, and therefore $\theta = 0$. In practice, it is difficult to adjust the θ angle to zero because the IF output is generally less than the maximum value. To maximize the IF output, the factor $\sin(\omega_S d/2v_x)/(\omega_S d/2v_x)$ should be as close to 1 as possible. If the permitted IF output is 10% lower than the IF output a t $\theta = 0$, then $\omega_S d/2v_x$ must be equal to or less than 0.8 radians. According to this requirement, the mismatch angle θ can be expressed as [10]

$$\frac{\sin(\omega_S d/2v_x)}{\omega_S d/2v_x} = 1$$

$$\theta \leq \frac{\lambda_S}{4d} \tag{2.28}$$

The mismatch angle θ is proportional to λ_S and inversely proportional to the size of the optical mixer d; that is, the longer the wavelength and the smaller the size of the optical mixer, the larger the allowable mismatch angle. It is apparent that

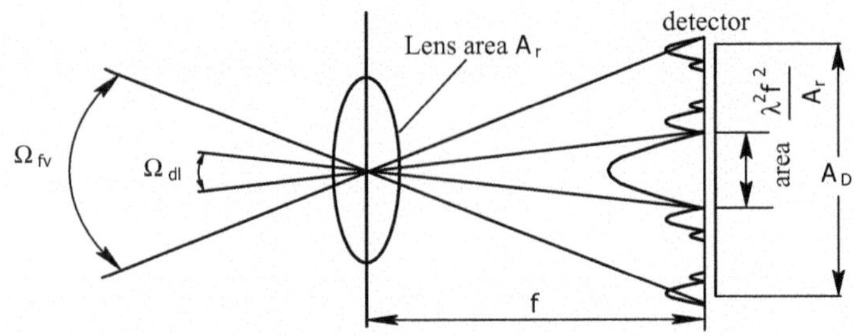

Fig. 2.16 The Airy disk principle in an optical lens antenna [15]

the spatial collimation requirements of coherent detection are strict, and the shorter the wavelength, the stricter the spatial collimation. Consequently, optical heterodyne detection in the infrared band is more advantageous compared to heterodyne detection in the visible light band. Because of the strict spatial collimation requirements, optical heterodyne detection has good filtering performance, another important feature of the system.

To ensure that the signal and local oscillation light waves are spatially coherent on the light mixing surface, the two beams must be as close to parallel as possible. This relatively strict requirement presents certain difficulties in the realization of optical heterodyne detection. Currently, the most promising means of addressing this issue is by using the Airy disk principle method, as shown in Fig. 2.16.

According to physical optics, a normally incident light wave will form a diffraction spot on the detector at the focal plane after passing through a lens with an area of A_r. The area $\lambda^2 f^2/A_r$ of the largest peak in the diffraction spot is called the Airy spot area, and this area determines the diffraction-limited field of view of the receiving system. If it is represented by a solid angle of Ω_{dl}, then

$$\Omega_{dl} \approx \frac{\lambda^2 f^2}{A_r} \cdot \frac{1}{f} = \frac{\lambda^2}{A_r} \tag{2.29}$$

If expressed by a plane angle, then

$$\theta_{dl} \approx = \frac{\lambda}{D_\tau} \tag{2.30}$$

Here, D_r is the lens diameter. A schematic diagram of the Airy disk principle is shown in Fig. 2.17. A lens is used to focus the signal light onto the surface of the optical mixer, and the effective area of the optical mixer is the area of the Airy disk. At the same time, the local oscillator light is irradiated on the Airy spot, and light mixing can occur. As the temperature D_r/f increases, the requirements for mismatch angles become wider.

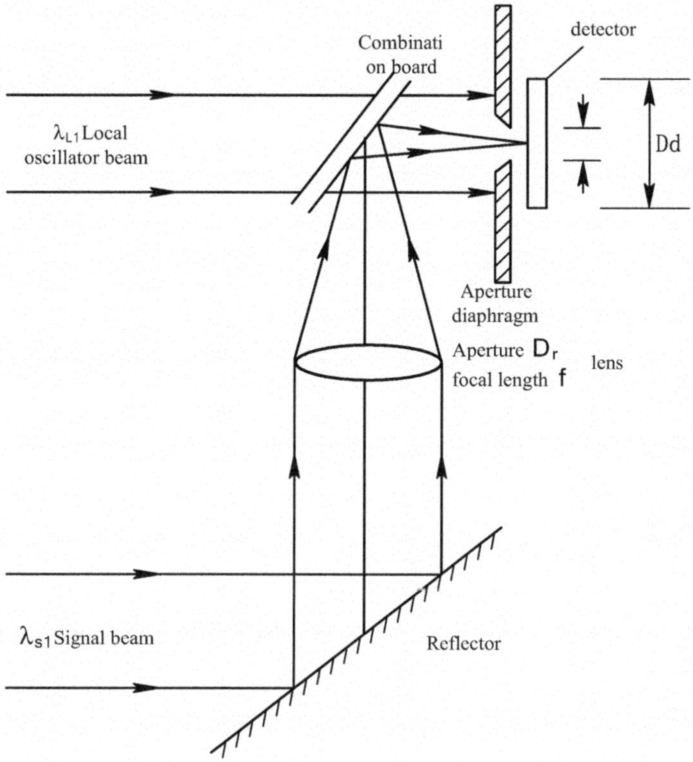

Fig. 2.17 Schematic diagram of the Airy disk principle [16]

2.4.2 *Frequency Conditions*

(1) Good Monochromaticity

To obtain high-sensitivity coherent detection, it is necessary for the signal and local oscillator lights to have high monochromaticity and frequency stability. In terms of physical optics, optical heterodyne detection (or coherent detection) is the result of interference between two light waves after superposition, and this is heavily dependent on the monochromaticity of both the signal and local oscillator beams. Light with only a single wavelength is called monochromatic light, and monochromaticity means that the beam is comprised of light within a very narrow frequency or spectral line. The color of light is determined by its frequency, and this is generally determined by the light source. Fully monochromatic light is difficult to produce because emissions from any light source have a certain frequency range, and the intensity of light for each frequency is different. A high degree of monochromaticity is an important characteristic of lasers; however, their excited state is always comprised of a certain energy level width, and consequently, the laser beam width $\Delta\nu$ cannot be zero. Generally, narrower values of $\Delta\nu$ mean greater light monochromaticity. To obtain laser output with good monochromaticity, the laser must operate in a single longitudinal mode (single frequency) by using a short cavity structure or other mode selection technologies.

(2) Small Frequency Drift [16]

The frequency drift of the signal and local oscillator lights must be limited within the allowable range to prevent performance deterioration of the optical heterodyne detection system. If the relative drift of the signal and local oscillator lights is large, then the difference between the two frequencies may exceed the bandwidth of the IF filter. If this occurs, the pre-amplification and IF amplifying circuit following the optical mixer cannot apply the IF signal. In optical heterodyne detection, special measures are required to stabilize the frequencies of the signal and the local oscillator lights. This is an important reason why optical heterodyne detection is more complicated than the direct detection method.

2.4.3 *Polarization Conditions*

The essence of optical heterodyne detection is the interference between the local oscillator and signal lights on the photosensitive surface of the photodetector. According to the difference in the polarization state of the light wave during the propagation process, the electric field distribution of the local oscillator and signal lights, respectively, can be expressed as [17]

$$E_L t = \hat{e}_L A_L \cos(\omega_L \mathrm{t} + \varphi_L) \tag{2.31}$$

$$E_S(t) = \hat{e}_S A_S \cos(\omega_S t + \varphi_S) \tag{2.32}$$

Here, $\hat{e}_L$ and $\hat{e}_S$ are the unit vectors of the polarization directions of the local oscillator and signal lights, respectively. By incorporating Eqs. (2.31) and (2.32) into Eq. (2.3), the IF current can be simplified as follows [13]

$$\mathrm{i}_{IF} = \alpha \hat{e}_S \hat{e}_L A_S A_L \cos[(\omega_S - \omega_L)\mathrm{t} + (\varphi_S - \varphi_L)] \tag{2.33}$$

When the polarization directions of the two lights are the same, $\hat{e}_L \hat{e}_S = 1$, and when the polarization directions of the two beams have an angle of θ, Eq. (2.33) becomes [17]

$$\mathrm{i}_{IF} = \alpha \cos\theta \left|\hat{e}_S\right| \left|\hat{e}_L\right| A_S A_L \cos[(\omega_S - \omega_L)\mathrm{t} + (\varphi_S - \varphi_L)] \tag{2.34}$$

Here, $\cos\theta = \hat{e}_L \hat{e}_S / \left|\hat{e}_L\right| \left|\hat{e}_S\right|$ represents the matching influence of the polarization directions of the two beams on the heterodyne signal, and its value is between 0 and 1. When the polarization directions of the two beams are the same; that is $\theta = 0°$, the heterodyne output IF current is the largest. Conversely, when the polarization directions of the two beams are perpendicular; that is $\theta = 90°$, the heterodyne output IF current is the smallest.

In summary, the spatial, frequency, and polarization matching conditions are all important to realize successful optical heterodyne detection. To obtain the maximum IF output, the wavefronts of the signal and local oscillator light waves must match. To perform effective mixing, the signal and local oscillator lights must be monochromatic, have the same polarization, and their frequencies must be extremely stable.

2.5 Homodyne Detection and Heterodyne Detection

Coherent detection is the coherent effect of two beams on the photosensitive surface of a detector, a schematic diagram of coherent detection is shown in Fig. 2.18. The signal light $E_s(\mathbf{r})$ interferes with the local oscillator light $E_l(\mathbf{r})$ on the effective area of the optical mixer, and the synthesized light enters the surface of the photodetector perpendicularly. The detector can detect interference fringes fluctuate at a certain angular frequency to form a beat frequency signal.

Assuming that the polarization direction and wavefront of the signal light and local oscillator light are identical, the two beams coincide with each other, and the transmission direction is parallel to the optical axis and perpendicular to the detector surface, the corresponding photoelectric fields can be expressed as [18]

$$E_s = A_s \exp[-j(\omega_s t + \varphi_s)] \tag{2.35}$$

$$E_l = A_l \exp[-j(\omega_l t + \varphi_l)] \tag{2.36}$$

where, A_s and A_l, ω_s and ω_l, φ_s and φ_l represent the amplitude and the angular frequency and initial phases of signal light and local oscillator light respectively. The photosensitive surface area of the optical mixer is S, and the total light intensity of the coherent synthesis of the two beams on the photosensitive surface is $|E_s + E_l|$. Because the detector current i_p is proportional to the square of the total light intensity, the current on the photosensitive surface of the detector can be expressed as [18]

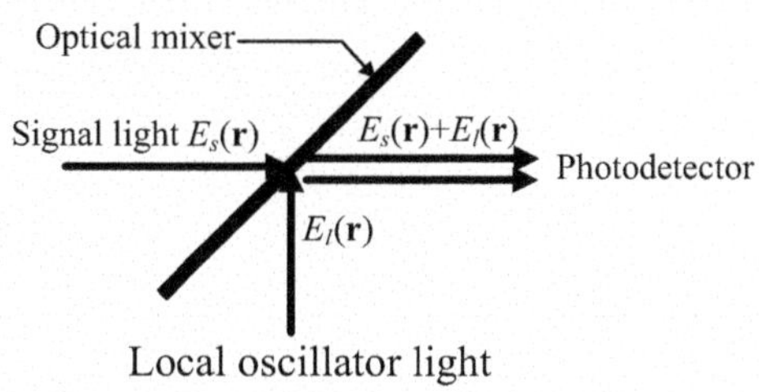

Fig. 2.18 Schematic diagram of coherent detection [2]

$$
\begin{aligned}
i_{\mathrm{p}} &= R\int_S |E_s(t) + E_l(t)|^2 ds \\
&= R\int_S \left\{\overline{A_s^2\cos^2(\omega_s t + \varphi_s)} + \overline{A_l^2\cos^2(\omega_l t + \varphi_l)}\right. \\
&+ A_s A_l \cos[(\omega_s + \omega_l)t + (\varphi_s + \varphi_l)] \\
&+ A_s A_l \cos[(\omega_s - \omega_l)t + (\varphi_s - \varphi_l)]\}ds
\end{aligned} \tag{2.37}
$$

where, $R = e\eta/h\nu$ is the responsivity of the detector, η is the quantum efficiency and is set to zero on the entire photosensitive surface of the detector, e is the electron charge, h is the Planck constant, and ν is the carrier frequency. The first and second terms in Eq. (2.37) are DC terms, the mean square of the cosine function is 1/2, the third term is the sum frequency term, the detector cannot respond to the high-frequency light wave, and the fourth term is the frequency difference term, which makes the angular frequency difference $\Delta\omega = \omega_s - \omega_l$. The initial phase difference $\Delta\varphi = \varphi_s - \varphi_l$When $\Delta\omega$ is below the cut-off frequency of the detector, the photodetector's photocurrent output has a frequency of $\Delta\omega/2\pi$, that is, the output photocurrent is composed of a DC term and a differential frequency term:

$$
i_p = R\int_S \left[\frac{A_s^2}{2} + \frac{A_l^2}{2} + A_s A_l \cos(\Delta\omega t + \Delta\varphi)\right]ds \tag{2.38}
$$

If the frequency of the intermediate frequency (IF) signal is limited to the differential passband range, the DC term in Eq. (2.38) is filtered out, and the IF current output is passed through the band-pass filter with central frequency $\Delta\omega$, as [2]

$$
\begin{aligned}
i_{IF} &= 2R\int_S A_s A_l \cos(\Delta\omega t + \Delta\varphi)ds \\
&= 2R\int_S A_s A_l \cos[(\omega_s - \omega_l)t + (\varphi_s - \varphi_l)]ds
\end{aligned} \tag{2.39}
$$

It can be seen from Eq. (2.39) that the amplitude $2RA_sA_l$, frequency $\Delta\omega$, and phase $\Delta\varphi$ of the IF current for A_s, ω_s, and φ_l vary, which shows that the amplitude, frequency, and phase of the signal light can be detected by coherent detection. In other words, coherent detection technology can demodulate optical signals exhibiting amplitude, frequency, and phase modulation. Equation (2.39) can also be expressed in terms of the power [2]

$$
i_{IF} = 2R\int_S \sqrt{P_s P_l}\cos(\Delta\omega t + \Delta\varphi)ds
$$

$$= 2R \int_S \sqrt{P_s P_l} \cos[(\omega_s - \omega_l)t + (\varphi_s - \varphi_l)]ds \tag{2.40}$$

It can be seen from Eqs. (2.39) and (2.40) that the amplitude of the IF signal is related to the amplitude of the local oscillator light and signal light. When the signal light power is very small (generally on the order of nW), and the local oscillator light power (generally on the order of mW) is sufficiently large, the demodulation information of the IF signal can still be detected using the method of coherent detection, which illustrates that the local oscillator light can amplify the IF signal in coherent reception, and that coherent detection has a higher detection sensitivity than direct detection; in addition, the coherent detection can filter the DC component directly, which indicates that the technology has good filtering characteristics; at the same time, in a free-space optical coherent detection system, the transmitting terminal can output the IF signal from the coherent receiver at the receiving end and recover the source information regardless of the modulation mode. Depending on the frequency of the frequency difference term $\Delta\omega$, coherent detection can be classified into two types: homodyne coherent detection and coherent detection.

2.5.1 Homodyne Coherent Detection

In Eq. (2.39) or Eq. (2.40), when the difference frequency term $\Delta\omega = 0$, the coherent detection method is referred to as homodyne coherent detection. The IF current of the homodyne coherent detection output is differs from that in Eqs. (2.39) and (2.40), it can be expressed as:

$$\begin{aligned} i_{IF} &= 2R \int_S A_s A_l \cos(\Delta\varphi)ds \\ &= 2R \int_S \sqrt{P_s P_l} \cos(\Delta\varphi)ds \end{aligned} \tag{2.41}$$

for the homodyne coherent detection method, then in addition to matching the polarization state consistency, spatial collimation and wavefront curvature between the signal light and local oscillator light, the frequencies of the two light waves must be completely consistent. Generally speaking, the homodyne coherent detection method is very sensitive to the phase of the signal light and local oscillator light, which ensures high detection sensitivity; however, it is unsuitable for frequency modulation, and only intensity and phase modulation can be realized. Generally speaking, for a homodyne coherent detection system, the local oscillator laser should have a tuning function with high precision and high rate. An optical phase-locked loop is used to stabilize the frequency and phase of the signal light and local oscillator light [19, 20]; this increases the coherent receiver design difficulty, equipment complexity, and cost.

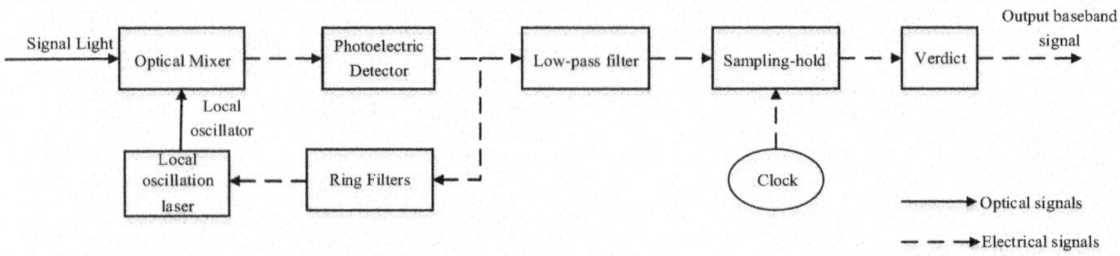

Fig. 2.19 Demodulation principle block diagram of homodyne coherent detection system

In contrast to the coherent detection system, the signal and local oscillator lights in the homodyne coherent detection system pass through the mixer and photodetector, and the output electrical signal is directly converted to the baseband signal without secondary demodulation in the electrical domain. A high-precision optical phase-locked loop is introduced to control the phase difference between the signal light and local oscillator light, this is also referred to as synchronous demodulation. The principle block diagram is shown in Fig. 2.19.

2.5.2 *Heterodyne Detection*

In Eq. (2.39), when the difference frequency term $\Delta\omega \neq 0$, the coherent detection method is also referred to as coherent detection. The expression of the IF current in the coherent detection output is consistent with Eqs. (2.39) and (2.40). For coherent detection technology, the polarization direction and wavefront curvature of the two beams must be strictly consistent for the signal light, and the local oscillator light must be perpendicular to the optical mixer. In addition, it is necessary to ensure the monochromaticity of the two beams. Generally, the narrower the spectral width, the better the monochromaticity.

The heterodyne detection system must demodulate the IF signal from the detector twice, to recover the baseband signal. According to the different modulation modes of the transmitter, the demodulation methods for designing a coherent receiver can be classified into synchronous demodulation and asynchronous demodulation. The principle block diagram is shown in Fig. 2.20. Figure 2.20a depicts a the block diagram of the synchronous demodulation principle of coherent detection. The carrier recovery of the IF signal is performed by the electric domain phase-locked loop (PLL); then, the IF signal and recovered carrier signal are inputted to the multiplier; finally, the baseband signal is output through a low-pass filter. The synchronous demodulation method requires a constant phase difference between the signal light and local oscillator light, and it does not require the frequency of the two beams to be strictly consistent.

In addition, the coherent receiver can also apply asynchronous demodulation, also referred to as envelope demodulation; its principle block diagram is shown in Fig. 2.20b. Envelope demodulation involves directly obtaining the baseband signal

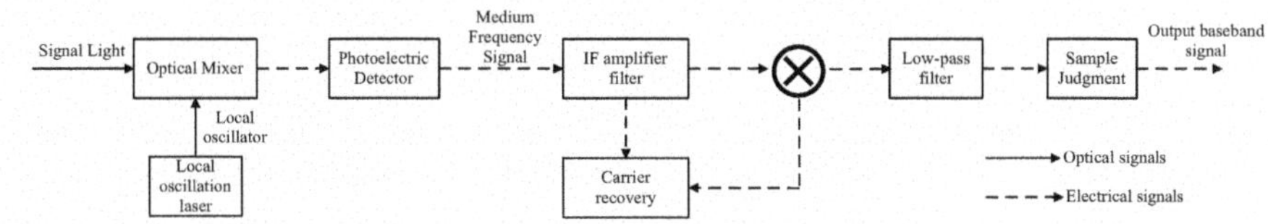

(a) Coherent-detection synchronous demodulation

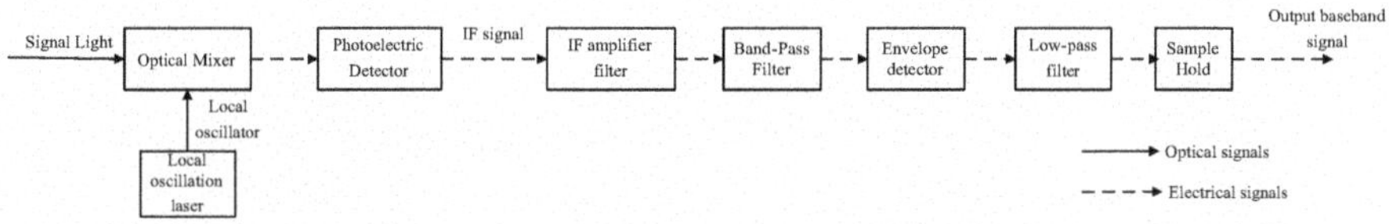

(b) Coherent-detection asynchronous demodulation

Fig. 2.20 Demodulation principle block diagram of coherent detection system

after passing the IF signal through the envelope detector and low-pass filter; however, this method must ensure that the frequency of the signal light and the local oscillator light are strictly matched, and the phases of the two beams should also be locked.

2.6 Composition of Heterodyne Detection System

Based on the principles and characteristics of coherent detection, we developed a free-space optical heterodyne detection communication system with differential phase-shift keying (DPSK) modulation. Figure 2.21 presents a schematic diagram of the coherent detection system. The entire system selects a narrow-linewidth laser as the signal light and local oscillator light, respectively, and it uses an electro-optic phase modulator to load the source information into the signal light and connect it to the transmitting optical system for transmission. The signal is then transmitted to the receiving end through an atmospheric turbulence random channel, which applies adaptive optics (AO) technology, polarization control, laser frequency stabilization technology, coherent detection, and high-speed digital signal processing technology to complete the electrical signal recovery.

At the transmitting terminal, the signal light and source information are simultaneously loaded on the phase modulator to realize phase modulation, and the signal light (containing modulation information) is sent out by the optical antenna. After the signal light is transmitted through the free-space atmospheric random channel, the wavefront and polarization directions are naturally affected by atmospheric turbulence, and the coherence with the local oscillator light is reduced.

At the terminal of the heterodyne receiver, to obtain the maximum heterodyne efficiency, AO technology is used to correct the distorted wavefront of the signal

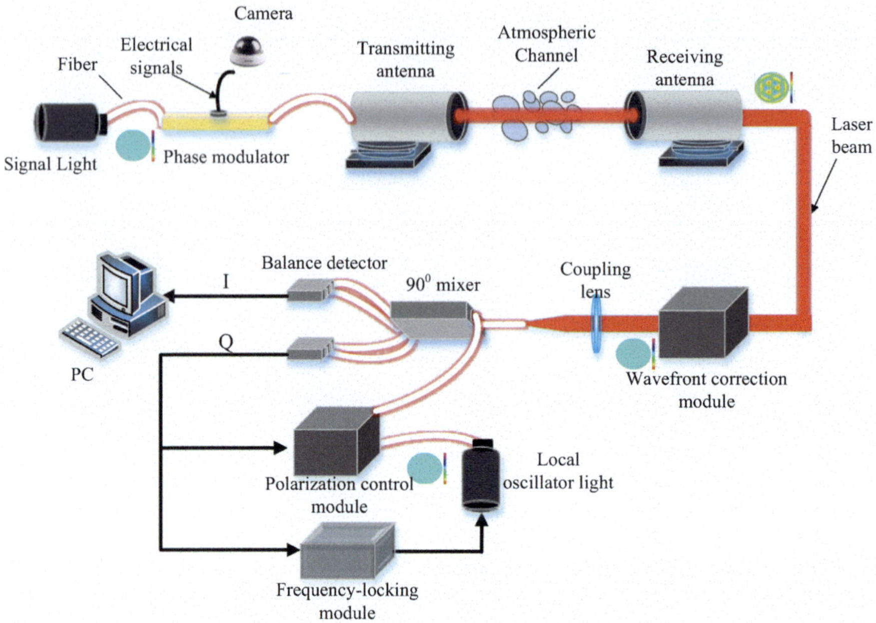

Fig. 2.21 Schematic diagram of free space coherent detection communication system

light, and automatic polarization control technology is used to achieve real-time matching of the polarization state between the signal light and local oscillator light.

Simultaneously, to meet the requirements of coherent demodulation for carrier frequency stability, laser frequency stabilization technology is applied to control the frequency of the IF signal within $\pm$ 2 M. Using spatial optical fiber coupling technology, the corrected signal light is coupled to the single-mode fiber, connected with the local oscillator light, and sent to the 90° mixer for coherent mixing. Four optical signals with an output phase difference of 90° are connected to the balance detector.

After photoelectric conversion, these are connected to a high-speed digital signal processing system in the form of IF current; finally, the baseband signal can be recovered in real time.

2.6.1 Wavefront Correction Module

AO technology is the main method for mitigating the impact of atmospheric turbulence upon beam propagation [21]. The basic principle of wavefront correction is phase conjugation, in which the wavefront corrector generates a compensation phase conjugated with the distorted wavefront aberration. The objects of correction generally include low-order and high-order aberrations. The so-called low-order aberration

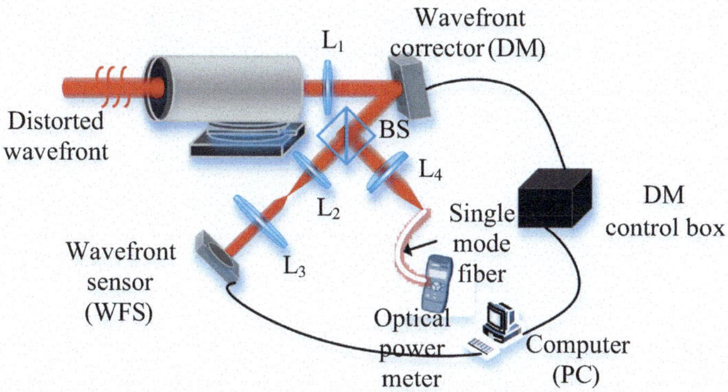

Fig. 2.22 Schematic diagram of AO system based on wavefront sensor

refers to the phase tilt produced by atmospheric turbulence, which is mostly corrected by the acquisition, pointing and tracking (APT) system [22]. Aberrations other than tilt aberrations are collectively referred to as higher-order aberrations (HOAs). The principle block diagram of a traditional AO system with a wavefront sensor is shown in Fig. 2.22.

In Fig. 2.22, the signal light with wavefront distortion is received by the optical antenna, compressed to a 10 mm spot by Lens L1, and enters the deformation mirror (DM) aperture. After DM correction, the reflected beam (with residual aberration) passes through the center of the beam splitter (BS), Lens L2, Lens L3, and a Hartmann wavefront sensor (WFS) oriented parallel to the optical axis; finally, the light spot completely enters the pupil surface of the WFS. The control voltage to be loaded into the DM is calculated by the control system, to completely compensate for the detected wavefront error; this forms a wavefront correction closed-loop system. The other sub-beams was used for spatial optical fiber coupling. An optical power meter was used to monitor the optical power coupled to the single-mode fiber, to ensure that the coupled optical power met the requirements of coherent detection and demodulation.

The distorted wavefront of the signal light disturbed by atmospheric turbulence is assumed to be $\varphi\ (\rho,\theta)$, which can be expressed as a linear combination of Zernike polynomials, as [23]

$$\varphi(\rho,\theta)=\sum_{i=1}^{N}a_i Z_i(\rho,\theta) \tag{2.42}$$

where α_i is the i-th order aberration coefficient and $Z_i(\rho,\theta)$ represents the i-th term of the Zernike polynomial, for which $i=1,2$ in $Z_i(\rho,\theta)$ describes the x-tilt and y-tilt aberrations, respectively, and $i\geq 3$ in $Z_i(\rho,\theta)$ describes higher-order aberrations.

The performance of the AO correction system is generally measured by the root mean square (RMS), peak-to-valley (PV) ratio of wavefront aberration, and Strehl ratio (SR) of far-field images before and after correction. The SR of the far-field image represents the far-field-received energy concentration, which is also an important index for evaluating the system performance. This index plays an important role in the performance analysis of an AO-based coherent detection system.

The RMS value of wavefront aberration can be defined as [24]

$$RMS = \sqrt{\langle[\varphi(\rho,\theta) - \langle\varphi(\rho,\theta)\rangle]^2\rangle} = \sqrt{\langle\varphi^2(\rho,\theta)\rangle - \langle\varphi(\rho,\theta)\rangle^2} \tag{2.43}$$

where, $\varphi(\rho,\theta)$ is the wavefront aberration within the effective aperture. The wavefront PV value can be defined as [24]

$$PV = \max[\varphi(\rho,\theta)] - \min[\varphi(\rho,\theta)] \tag{2.44}$$

In addition, SR can be described as [23]

$$SR = \frac{I(x,y)_{\max}}{I(x,y)_{ideal}} \tag{2.45}$$

where $I\ (x,y)_{\max}$ and $I\ (x,y)_{ideal}$ are the maximum light intensity value and the ideal light intensity value, respectively, in the collected image (x, y) region. The greater the ratio of the RMS and PV values, the more serious the wavefront distortion. The SR is expressed as a number between zero and one. The greater the value, the higher the energy concentration of the spot center. To a certain extent, this indicates that the effect of wavefront correction is superior. In the practical system, the laser used is a Gaussian beam; thus, SR can be approximately expressed as [24]

$$SR \approx \exp(-RMS^2) = \exp\left(-\sum_{i=1}^{N} a_i^2\right) \tag{2.46}$$

Figure 2.23 shows the numerical relationship between SR and RMS. It can be seen from the figure that under an increase in RMS, the SR value decreases. When RMS reaches 1 rad, it continues to increase, and the corresponding SR value changes little. Therefore, for a coherent detection system with AO correction, when the wavefront aberration is close to 0, the SR value can be approximately expressed as [25]

$$SR \approx 1 - \sum_{i=1}^{N} a_i^2 \tag{2.47}$$

Figure 2.24a and b show the changes in the RMS values of the open-loop and closed-loop AO systems, respectively; (c) and (d) show the changes in the PV values in the same systems, respectively; (e) shows the change in system SR under a closed-loop configuration. As can be seen from (a), when the AO system is open-loop, the

Fig. 2.23 The relationship between SR and RMS

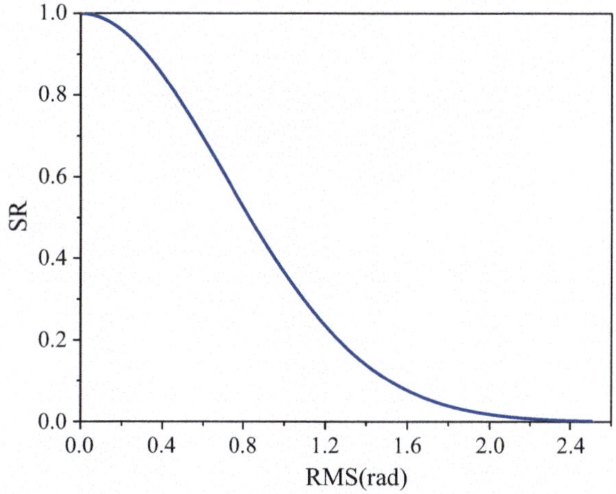

RMS value registers a 6 μm fluctuation up and down. When the AO system is closed-loop, the RMS value decreases to 0.04–0.13 μm; compared with the open-loop data, the RMS under an AO closed-loop is significantly reduced [25].

The corresponding PV value is often stable at 0.3 after the system realizes a closed loop at ~0.8 μm; compared with the PV value before the open loop, the PV value after the closed-loop is established also decreases significantly. In (e), the SR value is maintained above 0.8. Combining the RMS, PV, and SR values after the closed loop, it can be concluded that the distorted wavefront simulated in the laboratory is well corrected, because the aberration in many AO systems is considered well compensated when the SR value reaches 0.8.

2.6.2 Polarization Control Module

When the signal light and local light at the receiving end are coherently mixed, the mixer is sensitive to the polarization state of both. The local oscillator light provided by the local laser is stable and polarized, and the polarization state of the signal light changes after passing through the atmosphere and optical system. Therefore, the polarization state of the signal light is uncertain upon reaching the mixer, and the optical fiber birefringence effect produced by temperature slowly shifts its polarization state [26–28]. To obtain a high mixing efficiency, the polarization state of the signal light must be controlled in a coherent optical communication system [29, 30].

The polarization control module generally includes three components: a polarization controller, a detection module, and a data-processing module. The polarization controller controls the polarization state of the beam by varying the input signal. The detection module is used to realize the real-time detection of the beam's polarization state, to provide a feedback signal. The data processing module mainly completes

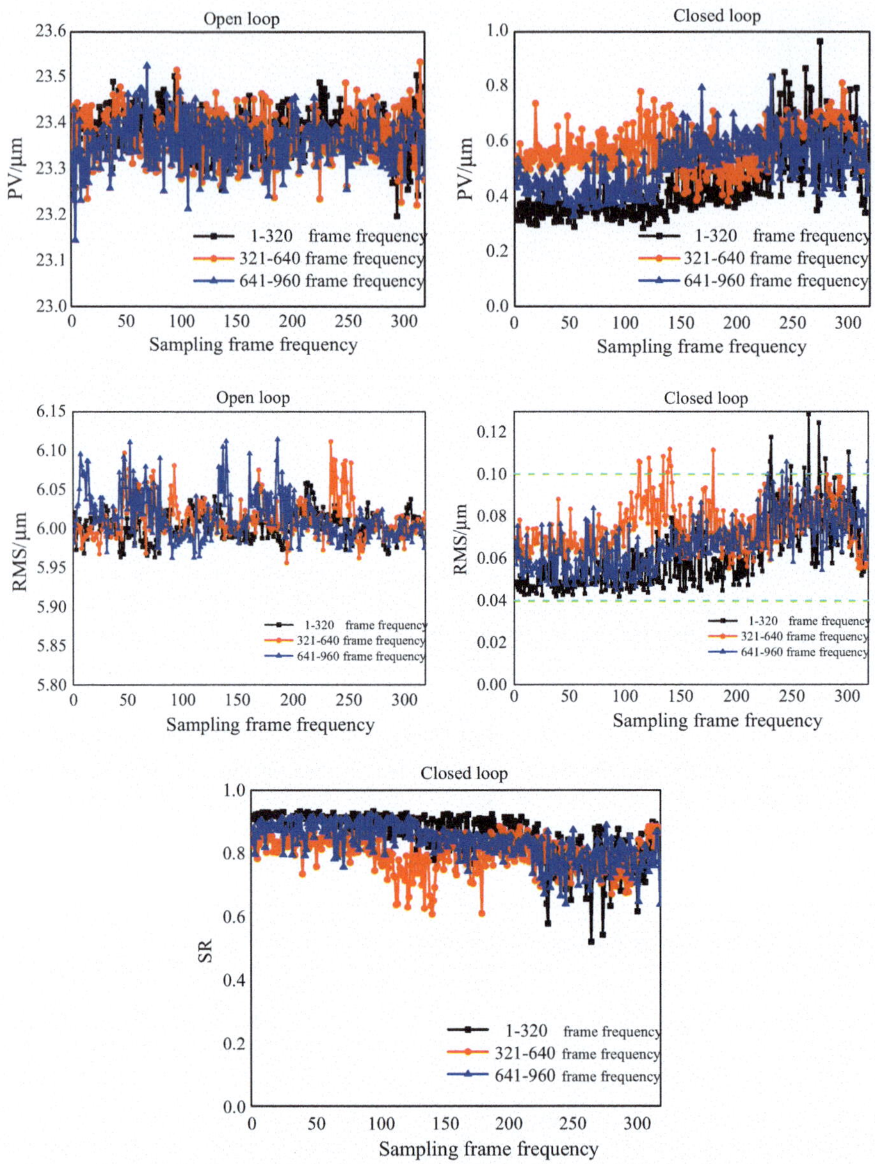

Fig. 2.24 RMS, PV and SR changes of AO system in open-loop and closed-loop configurations

the feedback signal processing and control signal generation. A schematic diagram of the polarization control module for coherent detection is shown in Fig. 2.25. The signal light received through the optical system is mixed with local oscillator light. After coherent mixing in the mixer, four beams with a phase difference of 90° are outputted and enter the balanced detector. In the balanced detector, the optical signal

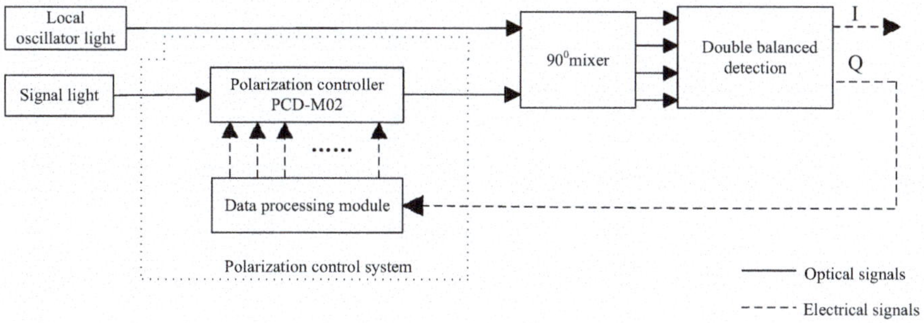

Fig. 2.25 Schematic diagram of polarization control

is converted into a current through the photodiode, and one of the two orthogonal piezoelectric signals is used as the feedback signal for polarization control.

The polarization control uses an extruded optical fiber electric polarization controller. The internal structure diagram is shown in Fig. 2.26. It is composed of three optical fiber extruders that are aligned at 45° to each other. By controlling the voltages V1, V2, and V3 of the three channels of the polarization controller, forces F1, F2, and F3 are generated. The optical fiber birefringence effect caused by the pressure introduces a variable phase delay in the orthogonal direction, thereby changing the polarization state of the signal light.

Figure 2.27a shows the change of the signal light polarization state on the Banga ball during the polarization control of the laboratory environment, in which the red dots represent the polarization state occupied by the signal light during the control process; (b) shows the change of the corresponding Stokes parameters. The initial polarization state S_s of the signal light was randomly generated, and $S_s = (0.03, 0.53, -0.85)$. As shown in Fig. 2.27a and b, the polarization state of the controlled signal light fluctuates around $S_{s'} = (1, 0, 0)$. At this time, the signal light is linearly polarized along the X-axis, exhibiting the highest mixing efficiency and largest amplitude of the IF signals.

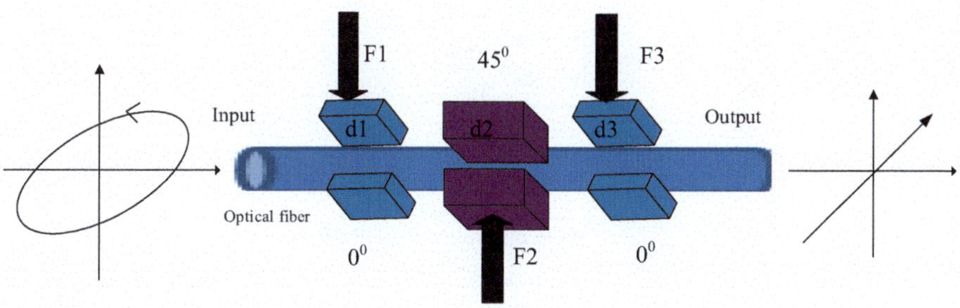

Fig. 2.26 Structure of the extruded optical fiber electric polarization controller

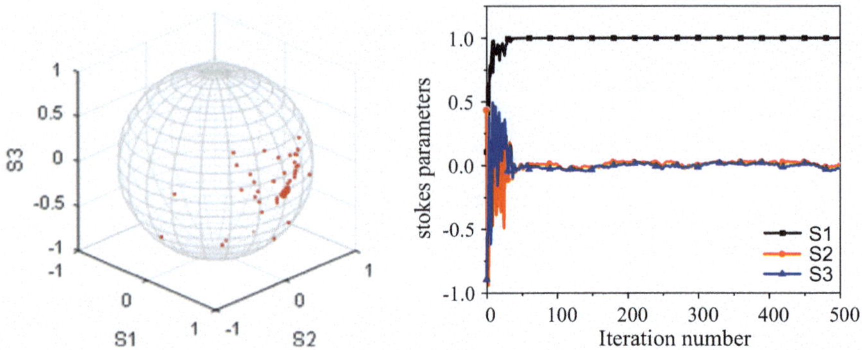

(a) Change in Banga sphere polarization state. (b) Change in the Stokes parameters.

Fig. 2.27 Change of signal light polarization state

2.6.3 Laser Frequency Stabilization Module

When designing a coherent detection system receiver, the frequency difference between the signal light and local oscillator light must be constant. The working temperature, external vibration, heat dissipation, and other factors of the laser can cause the laser frequency emitted by the laser to drift randomly within a small range under the increase in working time Δf. The frequency of the IF signal outputted by the balanced detector fluctuates randomly, which is not conducive to subsequent synchronous demodulation [31]. Based on the above problems, a frequency control loop was designed according to a practical coherent detection communication system. The principle block diagram is shown in Fig. 2.28.

An external phase detector (ADF4002) was used to convert the frequency value of the IF carrier signal into a voltage value; this was passed through the phase detector for processing, and a single-chip microcomputer (STM32) was used to judge the

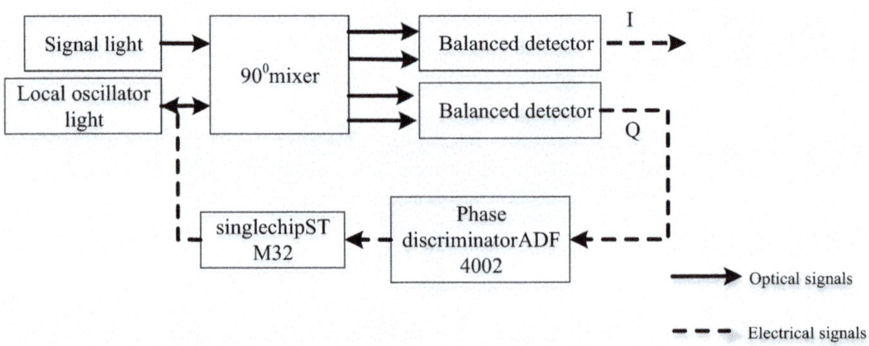

Fig. 2.28 Frequency control schematic diagram of coherent detection system

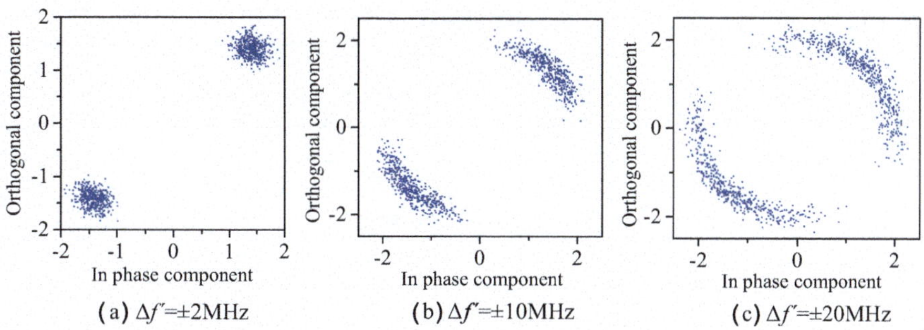

Fig. 2.29 IF signal constellation corresponding to different frequency drift errors

input voltage; the output control signal was applied to the internal voltage-controlled oscillator of the local oscillator laser, and the tracking of the signal optical frequency was realized by changing the frequency of the local oscillator laser. The entire control process is referred to as the automatic frequency control (AFC) of the IF signal carrier frequency.

Let the noise of the transmission channel be Gaussian white noise; the remaining noises are ideal, the laser frequency drift is evenly distributed within a certain range, and the target value of the IF carrier frequency $\Delta f = f_s - f_l = 100$ MHz; a binary phase-shift keying (BPSK) modulation mode is adopted, and SNR = 20 dB; when the carrier frequency drift error $\Delta f'$ differs, the signal constellation diagram of the receiving end is shown in Fig. 2.29. As can be seen from Fig. 2.29a–c, when the carrier frequency drift error of the IF signal increases, the axial rotation at the modulation point position on the constellation becomes more notable.

Taking different carrier frequency drift errors $\Delta f'$, the variation curve of the bit-error rate (BER) with respect to SNR is shown in Fig. 2.30. As can be seen from the figure, under an increase in the frequency drift $\Delta f'$, the BER increases gradually; when the frequency drift error $\Delta f'$ is ±40 MHz, the BER is maintained at ~14% under the increase in SNR, which fails to meet actual communication requirements. The designed frequency control loop is connected to the coherent detection system in the laboratory; the carrier frequency adjustment process is shown in Fig. 2.31.

In Fig. 2.31, the left-hand side of the dotted line represents the carrier frequency adjustment process. After approximately 220 s, the carrier frequency of the IF signal stabilizes. On the right-hand side of the dotted line is the change process with respect to time after the carrier frequency has been stabilized. It can be seen that the carrier frequency of the IF signal can be stabilized at 100 MHz for a long time, the upper and lower frequency drift errors do not exceed 2 MHz, and the corresponding mean carrier frequency is $\mu = 99.9631$ MHz with a variance of $\sigma^2 = 0.3390$ MHz^2. According to the results obtained in Figs. 2.29 and 2.30, when the frequency drift error $\Delta f'$ is no more than 2 MHz, the constellation exhibits good aggregation and a small bit error.

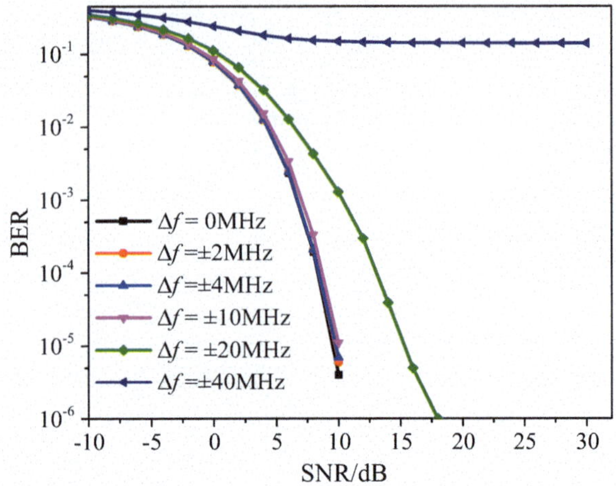

Fig. 2.30 BER variation curve corresponding to different carrier frequency drift errors

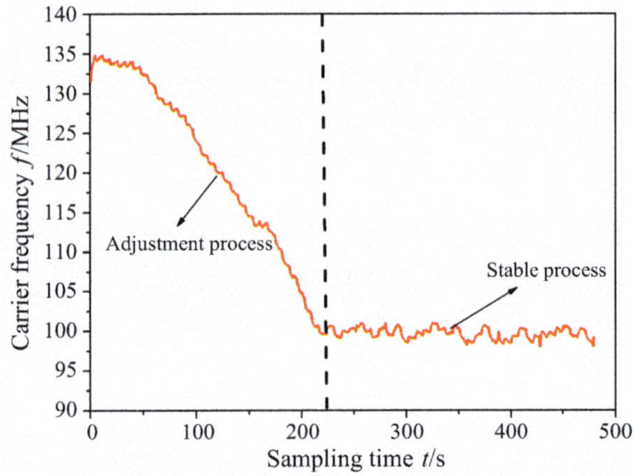

Fig. 2.31 Change of IF signal carrier frequency when AFC is connected to heterodyne system

2.6.4 Balanced Detection Module

In the FSO-HD system, a relatively mature polarization beam splitter-type 90° optical mixer and a series-balanced detector are used to form a coherent detection system. The principle block diagram is shown in Fig. 2.32. Upon entering the optical mixer, both the signal light and local oscillator light pass through Polarizers P1 and P2 with an x-axis transmission direction; then, the local oscillator light passes through the 1/4 wave plate (QWP) with the fast axis at 45° to the x-axis, and the signal light passes through the 1/2 wave plate (HWP) aligned at 22.5° to the x-axis; next, the signal light and the local oscillator light ensure the output of the four optical signals with a phase difference of 90° after passing through BS and PBS. Finally, four optical signals output I and Q electrical signals through the balanced detector.

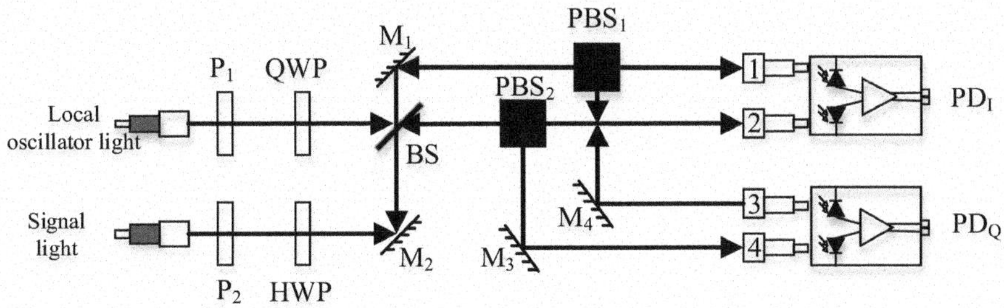

BS:50/50Spectroscope QWP:1/4wave plate
P:Polarizer HWP:1/2wave plate
M:lens PBS: polarized beam splitter

Fig. 2.32 Schematic block diagram of optical fiber mixer and balance detection

Here, the Jones matrix is used to represent the Jones vector of the local oscillator light and signal light after passing through P1 and P2; this can be expressed as

$$E_l = A_l \exp(i\omega_l t)\begin{pmatrix} 1 \\ 0 \end{pmatrix}, \tag{2.48}$$

$$E_s = A_s \exp(i\omega_s t)\begin{pmatrix} 1 \\ 0 \end{pmatrix}. \tag{2.49}$$

The Jones matrix can be used to represent the change in the polarization state of the wave plate. The Jones matrix corresponding to the 1/4 and 1/2 wave plates in optical mixing used in Fig. 2.32 can be expressed as

$$J_{1/4}\left(\frac{\pi}{4}, \frac{\pi}{2}\right) = \frac{\sqrt{2}}{2}\begin{pmatrix} 1 & -j \\ -j & 1 \end{pmatrix}, \tag{2.50}$$

$$J_{1/2}\left(\frac{\pi}{8}, \pi\right) = \frac{\sqrt{2}}{2}\begin{pmatrix} 1 & 1 \\ 1 & -1 \end{pmatrix}. \tag{2.51}$$

The Jones vectors of the local oscillator light and signal light passing through the 1/4 and 1/2 wave plates are

$$E_l' = J_{1/4}\left(\frac{\pi}{4}, \frac{\pi}{2}\right) E_l = \frac{A_l \exp(i\omega_l t)}{\sqrt{2}}\begin{pmatrix} 1 \\ -\exp(-i\pi/2) \end{pmatrix}, \tag{2.52}$$

$$E_s' = J_{1/2}\left(\frac{\pi}{8}, \pi\right) E_s = \frac{A_s \exp(i\omega_s t)}{\sqrt{2}}\begin{pmatrix} 1 \\ 1 \end{pmatrix}, \tag{2.53}$$

respectively. The Jones matrix of the polarization beam splitter used in Fig. 2.32 can be expressed as

$$\begin{cases} R = \begin{bmatrix} 0 & 0 \\ 0 & r_{\perp} \exp(i\rho_{\perp}) \end{bmatrix} \\ T = \begin{bmatrix} t_{\parallel} \exp(i\tau_{\parallel}) & 0 \\ 0 & 0 \end{bmatrix} \end{cases} \tag{2.54}$$

where R and T are the reflection and transmission matrices of the polarization beam splitter, respectively; $r_{\perp}$ and $r_{\perp}$, and $t_{\parallel}$ and $\tau_{\parallel}$ represent the amplitude and phase in the reflection and transmission directions, respectively. Ideally, $r_{\perp} = t_{\parallel} = 1$.

Four optical signals with a phase difference of 90° are reflected by the polarization beam splitter, and their Jones vectors can be expressed as

$$E_{out1} = \frac{1}{2}\begin{pmatrix} \overline{S} + \overline{LO}\exp(-i\pi/2) \\ 0 \end{pmatrix} \tag{2.55}$$

$$E_{out2} = \frac{1}{2}\begin{pmatrix} \overline{S}\exp(-i\pi/2) + \overline{LO} \\ 0 \end{pmatrix} \tag{2.56}$$

$$E_{out3} == \frac{1}{2}\begin{pmatrix} 0 \\ \overline{S} + \overline{LO}\exp(-i\pi) \end{pmatrix} \tag{2.57}$$

$$E_{out4} = \frac{1}{2}\begin{pmatrix} 0 \\ (\overline{S} + \overline{LO})\exp(-i\pi/2) \end{pmatrix} \tag{2.58}$$

Finally, the IF signals of the I and Q channels outputted by the balanced detector are

$$\mathrm{i}_I = A_s A_l \sin[(\omega_s - \omega_l)t] = A_s A_l \sin(\Delta\omega t), \tag{2.59}$$

$$\mathrm{i}_Q = A_s A_l \cos[(\omega_s - \omega_l)t] = A_s A_l \cos(\Delta\omega t). \tag{2.60}$$

2.6.5 Coherent Demodulation Module

In the coherent detection system, the carrier frequency dynamic offset of the IF signal outputted by the balanced detector is large, and the traditional Costas PLL cannot complete the phase recovery when the carrier frequency changes significantly; hence, other auxiliary loops are required to achieve stable tracking [32, 33].

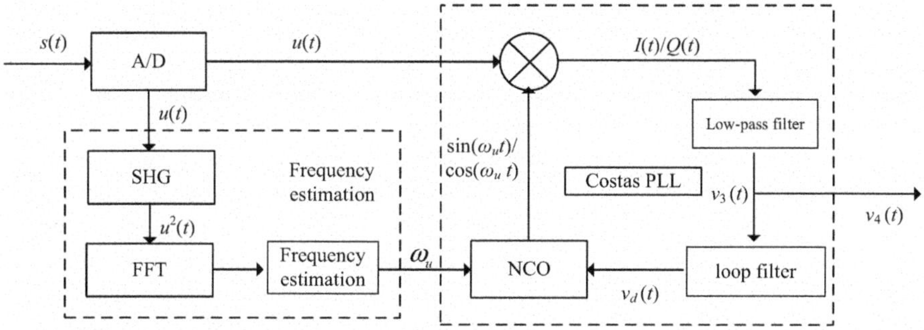

Fig. 2.33 Demodulation principle block diagram based on FPGA

The demodulation principle block diagram is shown in Fig. 2.33. The Costas PLL frequency estimation was used to demodulate the IF signal, and the demodulation scheme was verified on the field-programmable gate array (FPGA) demodulation platform. After A/D undersampling the received IF signal $s(t)$, the IF sampling signal $u(t)$ was obtained, which was divided into two channels and sent to the FPGA for processing. One channel performed frequency doubling and a fast Fourier transform (FFT) on the IF sampling signal through a multiplier, and it estimated the frequency of the IF sampling signal through the spectral line position ω_u. The digitally controlled oscillator (NCO) in the Costas ring was controlled by this value; the other was sent to the Costas ring for carrier recovery and signal demodulation. When the system functioned stably, the baseband signal $v_3(t)$ could be obtained by decomposing $v_4(t)$-carrying modulation information.

2.7 Performance Analysis of Heterodyne Detection System

2.7.1 *Signal to Noise Ratio and Detection Sensitivity of Heterodyne Detection System*

The power of the I-channel (or Q-channel) of the current outputted by the balanced detector can be expressed as [18]

$$\begin{aligned}\left\langle i_{IF}^2\right\rangle &= \frac{1}{2Z_0}(2R)^2\left|\int_S A_s A_l \cos[(\omega_s-\omega_l)t+(\varphi_s-\varphi_l)]ds\right|^2 \\ &= \frac{2}{Z_0}R^2\left|\int_S E_l\cdot E_s^* ds\right|^2\end{aligned} \tag{2.61}$$

where Z_0 is the free-space impedance. When the local oscillator light power is relatively large, the shot noise produced by the local oscillator light is significantly larger than that of the other noises. Generally, the noise at the receiver primarily originates from local oscillator light, as [18]

$$\langle i_N^2 \rangle = 2e\Delta f_{IF} I_l = 2e\Delta f_{IF} R P_l = \frac{2e\Delta f_{IF} R \int_S A_l^2 ds}{Z_0} \tag{2.62}$$

where Δf_{IF} is the effective noise bandwidth (generally, $\Delta f_{IF} = \mathrm{B}/2$) and P_l is the local oscillator light power. By definition, the SNR of the photodetector output can be expressed as

$$SNR_{IF} = \frac{\langle i_{IF}^2 \rangle}{\langle i_N^2 \rangle} = \frac{\eta \left| \int_S E_l \cdot E_s^* ds \right|^2}{\mathrm{h}\upsilon \Delta f_{IF} \cdot \int_S A_l^2 ds} = \frac{\eta \left| \int_S E_l \cdot E_s^* ds \right|^2}{\mathrm{h}\upsilon \Delta f_{IF} \cdot \int_S E_l \cdot E_l^* ds} \tag{2.63}$$

wherein the signal light power can be written as $P_s = \left(\int_S E_s \cdot E_s^* ds \right) / 2Z_0$; thus, Eq. (2.63) can be rewritten as

$$SNR_{IF} = \frac{P_s \eta}{\mathrm{h}\upsilon \Delta f} \frac{\left| \int\limits_S E_l \cdot E_s^* ds \right|^2}{\int_S E_l \cdot E_l^* ds \cdot \int_S E_s \cdot E_s^* ds} \tag{2.64}$$

According to Eq. (2.64), the mixing efficiency of the coherent detection can be defined as [18]

$$\eta_{het} = \frac{\left| \int_S E_l \cdot E_s^* ds \right|^2}{\int_S E_l \cdot E_l^* ds \cdot \int\limits_S E_s \cdot E_s^* ds} \tag{2.65}$$

When the mixing efficiency reaches an ideal extreme value, Eq. (2.64) can be reduced to

$$SNR_{IF} = SNR_0 = \frac{\eta P_s}{h\nu \Delta f} = \frac{2\eta P_s}{h\nu B} \tag{2.66}$$

Equation (2.66) is the maximum SNR achievable by an optical coherent detection system. Generally, this value is referred to as the quantum detection limit or quantum noise limit of optical coherent detection. Assuming $SNR_0 = 1$, the noise equivalent power (NEP) $= P_s = h\nu \Delta f / \eta$. This quantity is sometimes called the sensitivity of optical coherent detection, which is the theoretical limit of optical coherent detection; that is, when the quantum efficiency of the optical mixer is 1, $\Delta \mathrm{f} = 1$ Hz. Then, the limit of optical coherent detection sensitivity is a single photon. Although we are far from reaching such high detection sensitivities in practice, the optical coherent detection method can detect very weak light signals, generally of the order of nanowatts [18].

If other noises of the system are also considered, the corresponding noise can be expressed as [18]

$$N_{p1} = \frac{2e^2\eta}{h\upsilon} P_s \Delta f_{IF} R_L + \frac{2e^2\eta}{h\upsilon} P_l \Delta f_{IF} R_L + \frac{2e^2\eta}{h\upsilon} P_b \Delta f_{IF} R_L \tag{2.67}$$

$$N_{P2} = 4KT_b \Delta f_{IF} \tag{2.68}$$

The three terms in Eq. (2.67) represent the shot noise produced by signal light, intrinsic light, and background radiation, respectively; Eq. (2.68) represents the thermal noise produced by the system, where P_b is the background radiation power, K is the Boltzmann constant, T_b is the working temperature, and R_L is the effective resistance of the photodetector. According to the definition given in Eq. (2.63), the SNR ratio of the coherent detection system, considering the above noise factors, can be expressed as

$$SNR_{IF} = \frac{\left(\frac{e\eta}{h\upsilon}\right)^2 P_s P_l R_L}{e\left[\frac{e\eta}{h\upsilon}(P_s + P_l + P_b)\right]\Delta f_{IF} R_L + 2KT_b \Delta f_{IF}} \tag{2.69}$$

Here, the electronic charge e = 1.6×10^{-19}c, quantum efficiency $\eta = 0.8$, Planck's constant $h = 6.626 \times 10^{-34}$ J·s, carrier frequency $\upsilon = 1.9 \times 10^{14}$ m/s, $R_L = 50\ \Omega$, background radiation power $P_b = 10^{-9}$ W, effective noise bandwidth $\Delta f_{IF} = 100$ MHz, K = 1.38×10^{-23} J/K, and $T_b = 300$ K. When the local oscillator optical power is given different values, the change in the SNR of the coherent detection system with signal optical power is as shown in Fig. 2.34.

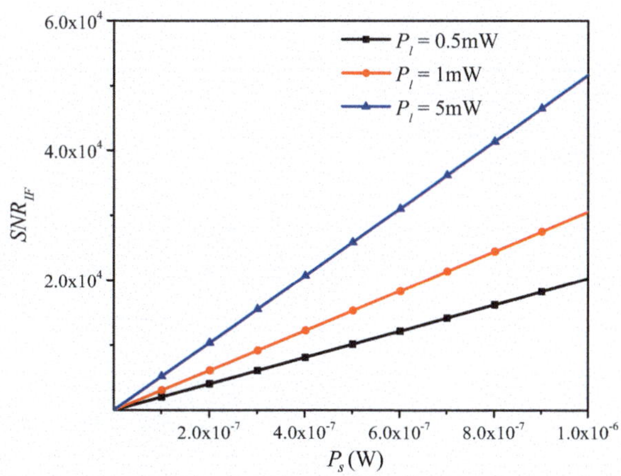

Fig. 2.34 Variation of SNR for coherent detection system with signal optical power

As shown in Fig. 2.34, the received signal optical power has a positive linear relationship with the SNR of the coherent detection system. To suppress the shot noise and system thermal noise generated by signal light and background radiation, we can increase the local oscillator light power to effectively improve the system's SNR. However, the larger the P_L, the better. If the local oscillator light power is sufficiently large, the shot noise caused by the local oscillator light will markedly exceed that of other noises, and the SNR of the system will decrease.

If the main noise source of the coherent detection system is thermal noise, then to detect at the quantum noise limit, the local oscillator light power must meet the conditions of Eq. (2.35). If the same parameters as those in Fig. 2.34 are selected and the quantum efficiency of the detector differs, the variation of the minimum local oscillator light power satisfying the conditions for the background temperature is as shown in Fig. 2.35.

$$P_l \geq \frac{2KT_b h\upsilon}{e^2\eta R_L}. \tag{2.70}$$

It can be seen from Fig. 2.35 that the minimum local oscillator light power required by the heterodyne system to reach the quantum noise detection limit increases under an increase in temperature. If the quantum efficiency of the detector is increased, the requirement for low power will be reduced at the same temperature. Therefore, according to the response of the actual optical mixer and detector, the local oscillator light power should be reasonably selected so that the coherent detection system can optimize the SNR.

In an actual coherent detection system, the thermal noise introduced by the optical mixer itself and the detector payload resistance also reduces the sensitivity of the coherent detection system; that is, the NEP value will increase. The corresponding detection sensitivity of the coherent detection system described in Eq. (2.69) can be expressed as

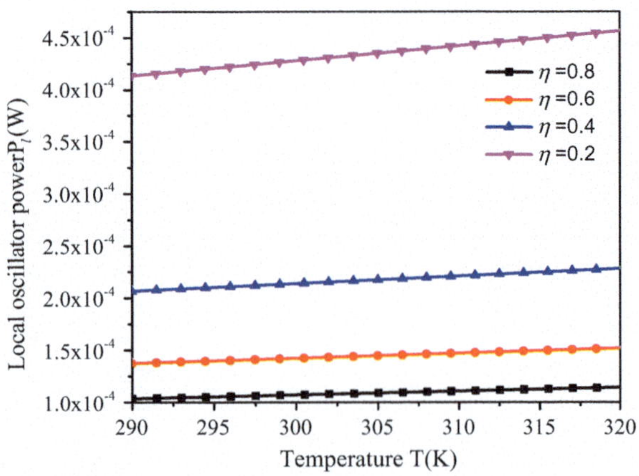

Fig. 2.35 Variation of local oscillator light power with respect to background temperature

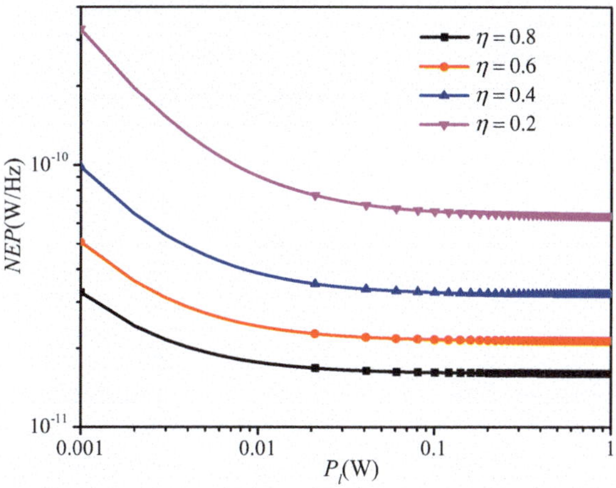

Fig. 2.36 Relationship between NEP and local oscillator light power P_l of coherent detection system

$$NEP = \frac{\left[\frac{e^2\eta}{h\upsilon}(P_l + P_b)\right]\Delta f R_L + 2KT_b\Delta f}{\left[\left(\frac{e\eta}{h\upsilon}\right)^2 P_l R_L - \frac{e^2\eta}{h\upsilon}\Delta f R_L\right]} \tag{2.71}$$

When the same parameters as those in Fig. 2.34 are selected and the detector quantum efficiency differs, the change in sensitivity NEP of the coherent detection system under local oscillator light power is as shown in Fig. 2.36.

It can be seen from the figure that the quantum noise limit is not reached; thus, increasing the local oscillator light power will reduce the NEP value. When the local oscillator light power remains unchanged, the NEP value of coherent detection systems composed of different types of optical mixers and photodetectors are different; that is, increasing the quantum efficiency of the detector can reduce the NEP value of the coherent detection system. The local oscillator light power P_l= 1 mW was selected, and the values of the other parameters were the same as those in Fig. 2.34. The variation of the heterodyne system NEP with respect to temperature is as shown in Fig. 2.37.

As can be seen from Fig. 2.37, the NEP value of the optical coherent detection system increases under an increase in the working background temperature. This is because under the increase in operating temperature of the optical mixer, the corresponding dark current increases, and the shot noise produced by the dark current likewise increases.

Therefore, even for the same coherent detection system, if the working conditions are different, the corresponding system NEP values also differ. When combining the above theoretical results with the actual optical coherent detection system, and selecting the reasonable maximum local oscillator light power to ensure that the detection system is in the quantum noise limit working state, the appropriate working temperature should also be selected to minimize the NEP value of the system.

Fig. 2.37 Variation of NEP of coherent detection system with respect to temperature

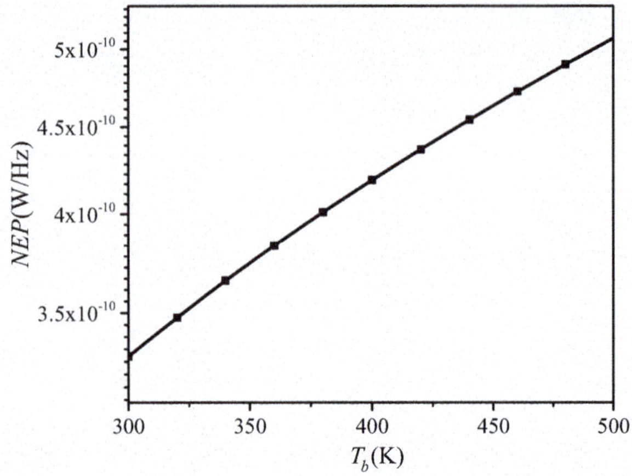

2.7.2 *Performance Analysis of Heterodyne Detection System Under Ideal Conditions*

It is assumed that the signal light and local oscillator light have the same polarization state, they all transmit parallel to the optical axis and incident perpendicular to the surface of the detector. The photosensitive surface of the detector coincides with the focal plane of the receiving antenna. When the aberration of the receiving optical system is not considered, after Fresnel diffraction, the signal light is airy distribution on the detector surface and local oscillator light is Gaussian distribution, as shown in Fig. 2.38.

According to the Fresnel diffraction principle, the light field distribution of the incident plane wave on the focal plane of the receiving antenna is [34]

$$U_{\mathrm{Sa}}(\boldsymbol{r},\varphi)=\frac{1}{\mathrm{i}\lambda f}\exp(\mathrm{i}\boldsymbol{k}f)\exp\left(\frac{\mathrm{i}\boldsymbol{k}r^2}{2f}\right)\pi\left(\frac{D}{2}\right)^2\left[\frac{2\mathrm{J}_1\left(\frac{\pi Dr}{\lambda f}\right)}{\frac{\pi Dr}{\lambda f}}\right] \tag{2.72}$$

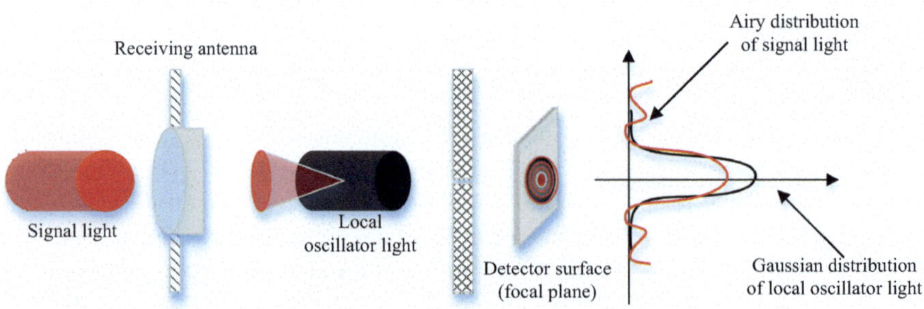

Fig. 2.38 Light-field distribution of signal light and local oscillator light

The field distribution of the local light on the focal plane can be approximately expressed via a Gaussian distribution, as [34]

$$U_{\mathrm{La}}(\boldsymbol{r}, \varphi) = \sqrt{\frac{2}{\pi \omega_0^2}} \exp\left(-\frac{r^2}{\omega_0^2}\right) \tag{2.73}$$

In Eqs. (2.72) and (2.73), λ is the wavelength of the signal light; k = 2 π/ λ is the light wave number; D and f are the diameter and focal length of the receiving optical system, respectively; *D/f* is the relative aperture; $\boldsymbol{r}$ is the radial vector from any point on the photosensitive surface of the detector to the center; ω_0 is the beam waist radius of the local oscillator light; and J_1 is a Bessel function of the first kind.

It is assumed that the area of the detector's photosensitive surface is much larger than the area of the signal light diffraction spot, and the absorption loss of the antenna and detector is neglected, with the ideal condition $\iint U_{\mathrm{La}}(r)U_{\mathrm{La}}^*(r)rdrd\theta = 1$. By substituting Eqs. (2.72) and (2.73) into Eq. (2.65), we can simplify this as

$$\eta_{het} = \frac{4}{\omega_0^2} \frac{\left| \int\limits_0^\infty \exp(-ikf) \exp\left(-\frac{ikr^2}{2f}\right) \exp\left(-\frac{r^2}{\omega_0^2}\right) \mathrm{J}_1\left(\frac{\pi Dr}{\lambda f}\right) dr \right|^2}{\int\limits_0^\infty \left[\mathrm{J}_1\left(\frac{\pi Dr}{\lambda f}\right)\right]^2 \Big/ r\mathrm{d}r} \tag{2.74}$$

If D in Eq. (2.74) is 10 mm, λ = 1550 nm (ω_0 is 2.6λ, 3.4λ, 3.6λ, respectively) or 1310 nm (ω_0 is 2.9λ, 3.7λ, 4.2λ, respectively), provided D/F is in the range of 0–0.5η; then, the relation curve between mixing efficiency and D/F is as shown in Fig. 2.39.

It can be seen from the figure that for different wavelengths, there always arises an extreme value of mixing efficiency with respect to the change in relative aperture, and

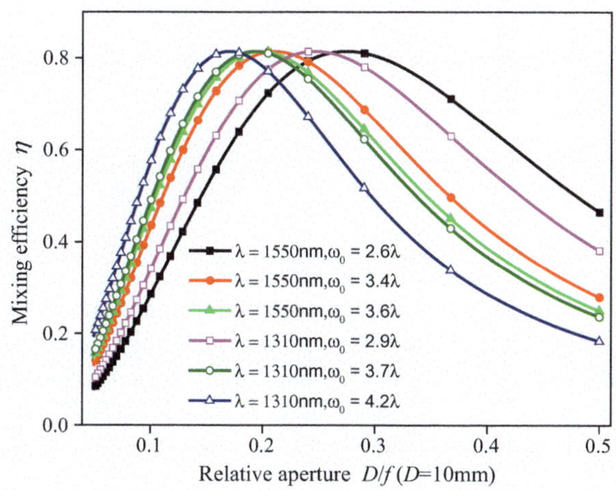

Fig. 2.39 Variation of η_{het} with respect to *D/f* when D = 10 mm and λ and ω_0 take different values

Table 2.1 Extreme values of mixing efficiency, corresponding to different wavelengths and waist radii

Wavelength/nm	Waist radius/nm	Relative aperture	Mixing efficiency (%)
1550	2.6λ	0.27	81.09
	3.4λ	0.21	81.45
	3.6λ	0.19	81.21
1310	2.9λ	0.26	81.06
	3.7λ	0.20	81.45
	4.2λ	0.17	81.31

the corresponding relative aperture differs when the extreme value is obtained. This is because when the relative aperture is small and the focal length of the receiving antenna is long, the area of the incident spot and photosensitive surface of the detector are mismatched, and the mixing efficiency is low. Under an increase in the relative aperture and a shortening of the antenna focal length, the spot energy received by the detector surface gradually increases, and the mixing efficiency gradually increases. When the relative aperture reaches a certain value, the mixing efficiency reaches its extreme value. If the relative aperture continues to increase, the focal length continues to shorten, the second ring of the Airy spot affected by the diffraction effect darkens, and the signal (after mixing with the signal light) become weaker, which lowers the mixing efficiency again. When the wavelength is constant, the change in the beam waist radius also affects the extreme value of the mixing efficiency and the corresponding relative aperture when the extreme value is obtained.

Table 2.1 shows the extreme value of each curve and the corresponding relative aperture for different values of the wavelength and beam waist radius in Fig. 2.39. It can be seen from the table that when $\lambda = 1550$ nm, ω_0 is 3.4 λ, and selecting the lens with $D/f = 0.21$ means that η_{het} is maximized at 81.45%. When $\lambda = 1310$ nm and $\omega_0 = 3.7\lambda$, selecting the lens with $D/f = 0.21$ can also make the mixing efficiency reach an extreme value of 81.45%.

Combined with the actual coherent detection system, the bit error rate of the coherent detection system with BPSK modulation is [35].

$$BER = \frac{1}{2} erfc\left(\sqrt{\frac{SNR_{IF}}{2}}\right) \tag{2.75}$$

A wavelength of 1550 nm was selected as the communication wavelength, and the waist radius of the local oscillator light was calculated as $\omega_0 = 3.4\lambda$; the noise bandwidth was 200 MHz, and the quantum efficiency on the photosensitive surface of the detector was 95%. According to Eqs. (2.64) and (2.75), by considering only the presence of shot noise generated by local light, the SNR and BER of the system was found to be changed with respect to the relative aperture, as shown in Fig. 2.40.

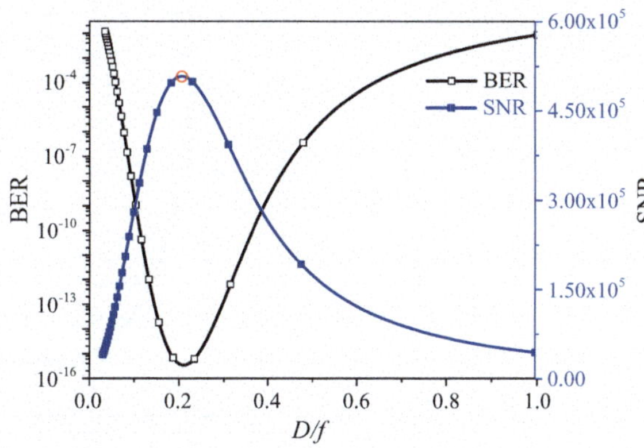

Fig. 2.40 Variation of SNR and BER with respect to *D*/*f*

As can be seen from Fig. 2.40, the maximum and minimum values of the SNR and BER, respectively, are observed at the relative aperture $D/f = 0.21$, which is consistent with both the relative aperture (corresponding to the extreme value of mixing efficiency calculated above) and the change of mixing efficiency. In addition, it can be seen from the figure that under ideal conditions, the lowest BER of the coherent detection system can reach the order of 10^{-16}, and the corresponding maximal SNR is of the order of 10^5.

The change in sensitivity of the coherent detection system with respect to the relative radius *D*/*f* is shown in Fig. 2.41. It can be seen from the figure that the detection sensitivity NEP of the coherent detection system decreases rapidly in the range of 0–0.21 *D*/*f*, and the lowest detectable signal optical power is ~ 3.4 nW. Under an increase in *D*/*f*, the value of NEP remains unchanged. Combining this with the theoretical analysis parameters given for the coherent detection system in the laboratory, we find that in a coherent detection system without any error, the detection sensitivity can reach approximately 3.4 nW in theory, which lays a theoretical foundation for experiments upon an actual coherent detection system.

2.7.3 Performance of Heterodyne Detection System with Optical Alignment Error

When the signal light and local oscillator light are vertically incident to the detector, the optical axis radial error is expressed as the deviation between the signal light optical axis and local oscillator light optical axis when the focal plane of the receiving antenna coincides with the surface of the detector. It is assumed that the relative amplitude offset of the local oscillator light and signal light is Δ; taking Δ into Eq. (2.73), the field distribution of the local oscillator light can be expressed as

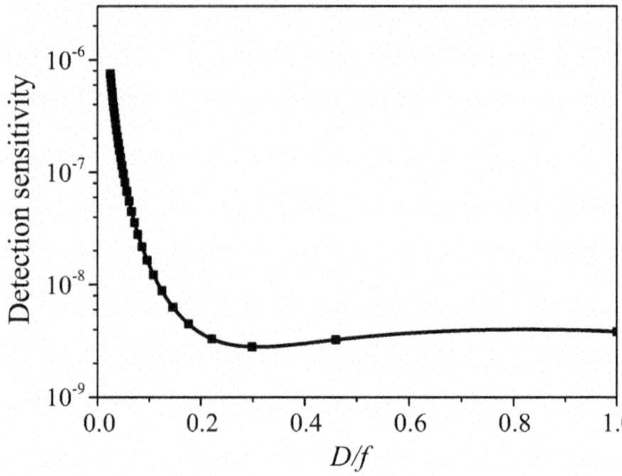

Fig. 2.41 Variation of detection sensitivity with respect to *D*/*f*

$$U_{La}(r,\varphi)=\sqrt{\frac{2}{\pi\omega_0^2}}\exp\left[-\left(\frac{r+\Delta}{\omega_0}\right)^2\right] \tag{2.76}$$

The field distribution of the signal light in the focal plane remains unchanged; taking Eqs. (2.72) and (2.76) into Eq. (2.65), the expression of the mixing efficiency produced by the optical axis radial error is

$$\eta_{het\Delta}=\frac{\left|\int_0^\infty \frac{1}{-i}\exp(-ikf)\exp\left(\frac{-ikr^2}{2f}\right)\exp\left[-\left(\frac{r+\Delta}{\omega_0}\right)^2\right]J_1\left(\frac{\pi Dr}{\lambda f}\right)dr\right|^2}{\int_0^\infty\left[J_1\left(\frac{\pi Dr}{\lambda f}\right)\right]^2\Big/ r\mathrm{d}r\int_0^\infty\exp\left[-2\left(\frac{r+\Delta}{\omega_0}\right)^2\right]r\mathrm{d}r} \tag{2.77}$$

The parameters in Eq. (2.77) were set as follows: $\lambda=1550$ nm, $\omega_0=3.4\lambda$, and $D=$ 10 mm. Figure 2.43 shows the variation of the mixing efficiency $\eta_{het\Delta}$ with respect to *D*/*f* under different radial offset errors Δ. As shown, for different radial offset errors, the mixing efficiency $\eta_{het\Delta}$ exhibited an extreme value under the increase in *D*/*f*; with the increase in radial offset error Δ, the extreme value of the mixing efficiency $\eta_{het\Delta}$ decreased gradually, and the *D*/*f* corresponding to the extreme value increased. This was because the optical axis radial offset error means that the diffraction spot on the photosensitive surface of the detector requires strong light energy to achieve coherent mixing, the focal length of the lens becomes longer, and the corresponding relative aperture and equivalent radius also gradually increase.

To analyze the influence of radial error on the coherent detection system more clearly, Fig. 2.44 shows the variation of the system's BER and SNR with respect to the radial error Δ. It can be seen from the figure that when the radial error Δ increases, the BER of the system increases from to 10^{-19} to 10^{-9}, and the corresponding SNR decreases from 6×10^5 to 2.5×10^5.

(2) **Angle offset error**

As shown in Fig. 2.42b, when the focal plane of the receiving antenna coincides with the surface of the detector and the local light is incident perpendicular to the detector surface, the angle θ between the signal light and the optical axis of the local light is referred to as the mismatch angle. The direction of the signal light wave vector forms an angle θ with the optical Z axis, where $\boldsymbol{K}_L$ is the optical loss of the vertically incident local oscillator light, and $\boldsymbol{K}_S$ is the optical loss of the obliquely incident signal light. Assuming that the phase of the obliquely incident signal light at any point on the detector's photosensitive surface can be expressed as $kr\cos\varphi\sin\theta$, we obtain the integral formula $\int_0^{2\pi} \exp(ikr\cos\varphi\sin\theta)\mathrm{d}\varphi = 2\pi J_0(kr\sin\theta)$; because the mismatch angle is small, sinθ can be approximated as θ. Finally, the mixing efficiency expression containing the mismatch angle can be obtained by simplification, as

$$\eta_{het\theta} = \frac{4}{\omega_0^2}\frac{\left|\int_0^\infty \exp(-ikf)\exp\left(-\frac{ikr^2}{2f}\right)\exp\left(-\frac{r^2}{\omega_0^2}\right)\mathrm{J}_1\left(\frac{\pi Dr}{\lambda f}\right)\mathrm{J}_0(kr\theta)\mathrm{d}r\right|^2}{\int_0^\infty \left[\mathrm{J}_1\left(\frac{\pi Dr}{\lambda f}\right)\right]^2/r\mathrm{d}r} \tag{2.78}$$

In Eq. (2.78), $\lambda = 1550$ nm, $\omega_0 = 3.4\lambda$, and D = 10 mm are also selected, and when θ differs, the relationship curve of $\eta_{het\theta}$ with *D/f* is as shown in Fig. 2.45. It can be seen from the figure that when θ increases, the extreme value of $\eta_{het\theta}$ gradually

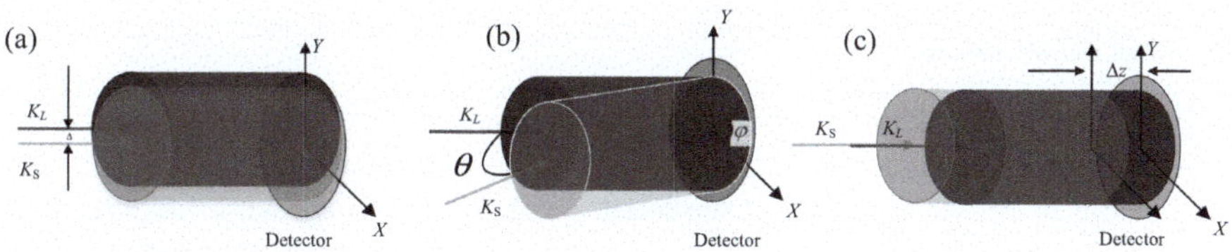

Fig. 2.42 **a** Radial offset error, **b** angle offset error, and **c** axis offset error

Fig. 2.43 Variation of ηhetΔ with respect to D/f when Δ takes different values

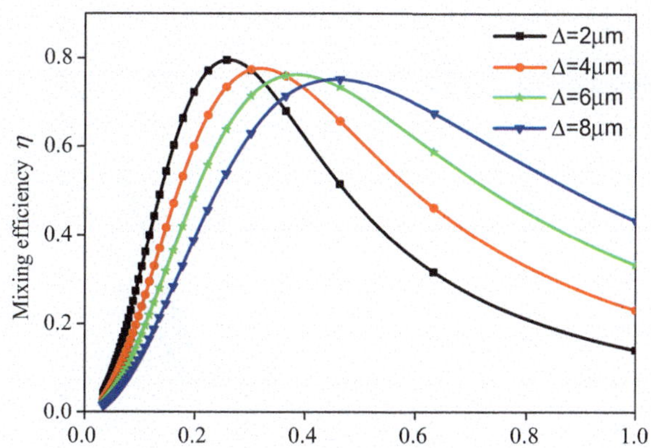

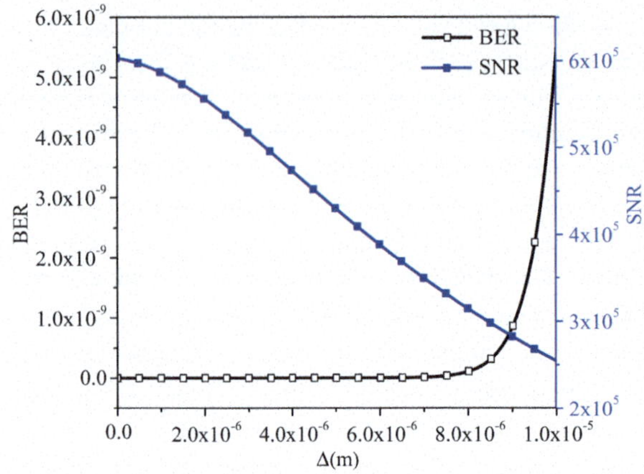

Fig. 2.44 Variation of BER and SNR with respect to Δ

decreases, and the corresponding *D/f* gradually increases when the extreme value is reached. The alignment accuracy of the APT system can be improved to reduce the influence of mismatch angle on the mixing efficiency.

For the same parameters as in Fig. 2.44. Figure 2.46 shows the variation in the system BER and SNR with respect to the mismatch angle θ. It can be seen from the figure that when the mismatch angle θ increases, the system's BER increases from 10^{-19} to 10^{-2}, and the corresponding SNR decreases from 6×10^5 to 1×10^4.

(3) **Axis offset error**

As shown in Fig. 2.42c, the axial defocusing means that the local oscillator light and signal light are incident perpendicular to the surface of the detector, and the plane of the beam waist radius of the local oscillator coincides with the detector's photosensitive surface. The axial error Δz is present when the focal plane and beam waist radius plane do not coincide. Considering the aperture limitation, the light field

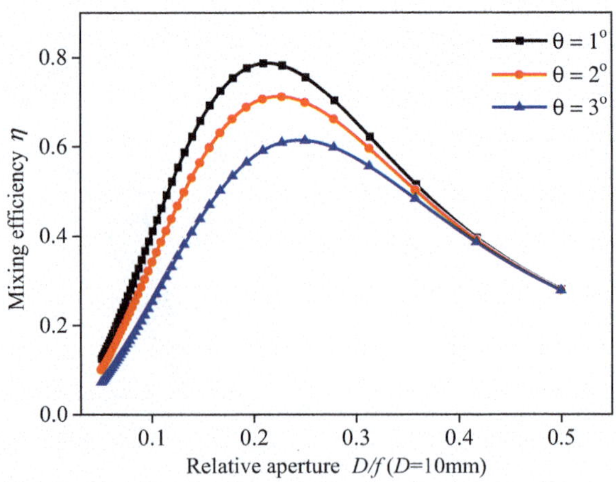

Fig. 2.45 Change of $\eta_{het\theta}$ with respect to D/f when θ takes different values

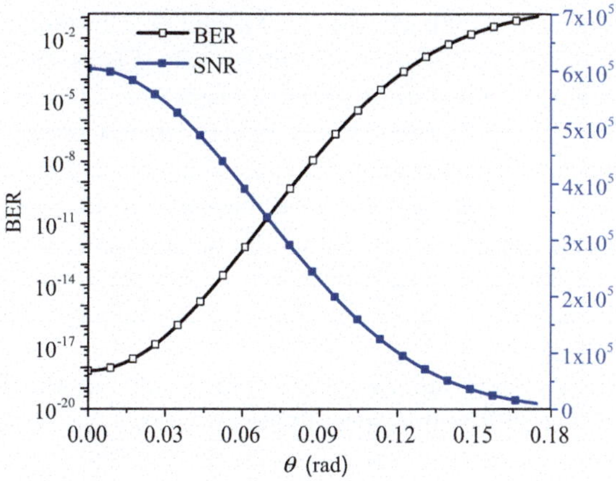

Fig. 2.46 Variation of BER and SNR with respect to θ

distribution of the signal light at Δz from the photosensitive surface of the detector can be expressed as

$$U_{\mathrm{Sa}}(r,\Delta z)=\frac{1}{\mathrm{i}\lambda(f+\Delta z)}\exp\left[ik(f+\Delta z)+\frac{ikr^2}{2(f+\Delta z)}\right]\times B\left\{\mathrm{circ}\left(\frac{r_0}{D/2}\right)\exp\left[\frac{ikr_0^2}{2(f+\Delta z)}\right]\exp\left(-\frac{ikr_0^2}{2f}\right)\right\} \tag{2.79}$$

Thus,

$$G(\rho)=B\left\{\mathrm{circ}\left(\frac{r_0}{D/2}\right)\exp\left[\frac{ikr_0^2}{2(f+\Delta z)}\right]\exp\left(-\frac{ikr_0^2}{2f}\right)\right\}. \tag{2.80}$$

For Fourier–Bessel function transformation [35], we can use the properties of the Bessel function and calculation simplifications to simplify G(ρ) as

$$G(\rho)\approx\pi D^2\frac{\mathrm{J}_1(\pi\rho D)}{\pi\rho D}-\left(\frac{D}{2}\right)^4\frac{ik\Delta z}{f(f+\Delta z)}\frac{\mathrm{J}_1(\pi\rho D)}{\pi\rho D}+\left(\frac{D}{2}\right)^3\frac{ik\Delta z}{\pi\rho f(f+\Delta z)}\frac{\mathrm{J}_2(\pi\rho D)}{\pi\rho D} \tag{2.81}$$

where $\rho = r/[\lambda(f + \Delta z)]$. Substituting Eq. (2.81) into Eqs. (2.79) and (2.73) into Eq. (2.65), we can simplify and obtain

$$\eta_{het\Delta z}=\frac{4\left|\int_0^{\infty}\frac{r}{\mathrm{i}\lambda(f+\Delta z)}\exp\left(-\frac{r^2}{\omega_0^2}\right)\exp[ik(f+\Delta z)]\exp\left[\frac{ikr^2}{2(f+\Delta z)}\right]G(\rho)\mathrm{d}r\right|^2}{\omega_0^2\int_0^{\infty}\frac{r}{\lambda^2(f+\Delta z)^2}G(\rho)^2\mathrm{d}r} \tag{2.82}$$

When λ = 1550 nm, $\omega_0 = 3.4\lambda$, and D = 10 mm, the relationship curve of $\eta_{het\Delta z}$ with respect to *D/f* for different Δz values is as shown in Fig. 2.47. It can be seen from Fig. 2.47 that the distance Δz between the focal plane of the receiving antenna and the photosensitive surface of the detector increases from 0 to 80 μm, and the corresponding extreme values of $\eta_{het\Delta z}$ and *D/f* do not significantly change. The position of the signal light focal plane can be adjusted by controlling the parameters of the optical system such that the focal plane and detector surface are coincident, thereby reducing the effect of axial defocusing upon the mixing efficiency and further realizing ideal coherent mixing between the signal light and local oscillator light frequency.

For the same parameters as in Figs. 2.44, 2.48 shows the variation in the system's BER and SNR with respect to the axial error Δz. It can be seen from the figure that when the axial error Δz increases, the BER increases from 10^{-19} to 10^{-18}, and the corresponding SNR decreases from 6×10^5 to 5.6×10^5.

Fig. 2.47 Change of ηhetΔz with respect to *D/f* when Δz takes different values

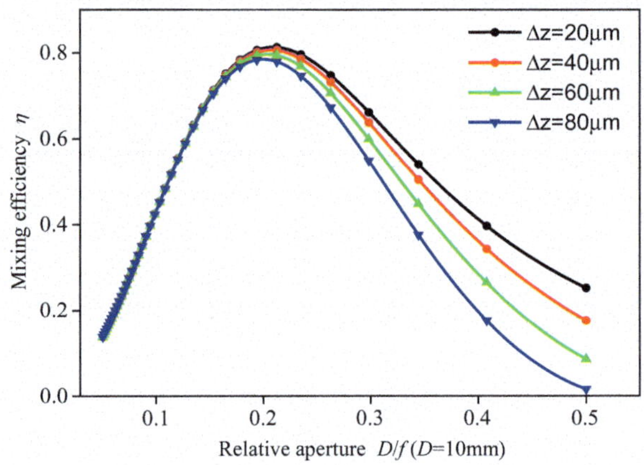

Fig. 2.48 Change of BER and SNR with respect to Δz

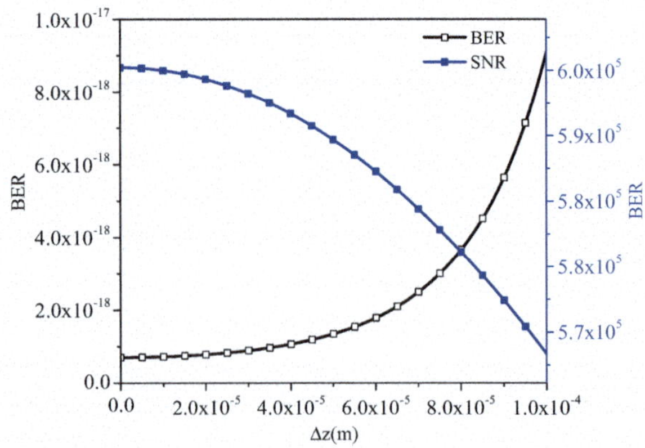

Comparing the mixing efficiency of Figs. 2.43, 2.45, and 2.47 with the system BER and SNRs in Figs. 2.44, 2.46, and 2.48, it can be seen that the relative aperture and mismatch angle are the most sensitive, followed by the radial error of the optical axis and finally the axial defocus; when the relative aperture value remains constant, the mixing efficiency produced by the mismatch angle decreases most significantly, and the axial defocus determines the mixing efficiency.

2.8 Signal-to-Noise Ratio, Bit Error Rate and Detection Sensitivity

2.8.1 Signal-to-Noise Ratio of Direct Detection and Heterodyne Detection

It is well known that the light field incident on the photodetector in the intensity modulation/direct detection (IM/DD) system is only the signal light field E_S(t). According to Eqs. (2.1) and (2.3), the output signal power of the IM/DD detection photodetector can be calculated as [2]

$$S_{PD} = i_{PD}^2 R_L = \beta^2 P_S^2 R_L \tag{2.83}$$

Comparing Eq. (2.5) (the photocurrent output from heterodyne detection) and Eq. (2.10) (which balances that output), the output signal power can be obtained as

$$S_{\mathrm{P}} = i_{\mathrm{P}}^2 R_L = 4\beta^2 R_L \left\{ \frac{P_S}{2} + \frac{P_L}{2} + \sqrt{P_S P_L} \cos[(\omega_S - \omega_L)t + (\varphi_S - \varphi_L)] \right\}^2 \tag{2.84}$$

$$S_{IF} = i_{IF}^2 R_L = 16\beta^2 P_S P_L R_L \tag{2.85}$$

In Eq. (2.84), $P_L \gg P_S$; thus, its influence on the output power can be neglected and the output power of heterodyne detection consists of a DC term and an AC term. The DC term $P_L/2$ does not help signal detection or demodulation but increases the DC intensity noise [2]. The AC term P_L amplifies the weak-signal optical power P_S to facilitate detection. Therefore, the optical power outputted by the detector during balanced heterodyne detection only retains the AC term and neglects the DC one. The intensity noise produced by the DC term forms part of the noise power and is its main component. Therefore, the noise of the heterodyne detection system comprises shot noise and thermal noise composed of signal light, local oscillator light, background light, and dark current. The IM/DD exhibits lower intensity noise and shot noise from the optical power of the local oscillator. Thermal noise can be expressed as [2]

$$\begin{cases} N_S = 2e\left[\frac{e\eta}{h\nu}(P_S + P_L + P_B) + I_D\right]\Delta f_{IF} R_L \\ N_J = 4kT\Delta f_{IF} \end{cases} \tag{2.86}$$

In Eq. (2.87), Δf_{IF} is the frequency difference bandwidth and N_S is shot noise, the first three terms on the right-hand side denote the shot noise produced by signal light, local oscillator light, and background radiation; the last term is attributable to the mixer dark current noise. This is the thermal noise produced by the internal resistance of the mixer and the load resistance. Therefore, from Eq. (2.83) to Eq. (2.86), the SNRs of direct detection, heterodyne detection, and balanced heterodyne detection systems are [2]

$$\left(\frac{S}{N}\right)_{PD} = \frac{\beta^2 P_S^2 R_L}{2e\left[\frac{e\eta}{h\nu}(P_S + P_B) + I_D\right]\Delta f R_L + 4kT\Delta f} \tag{2.87}$$

$$\left(\frac{S}{N}\right)_{\mathrm{P}} = \frac{2\beta^2 P_S P_L R_L}{2e\left[\frac{e\eta}{h\nu}(P_S + P_L + P_B) + I_D\right]\Delta f_{IF} R_L + 4kT\Delta f_{IF}} \tag{2.88}$$

$$\left(\frac{S}{N}\right)_{IF} = \frac{16\beta^2 P_S P_L R_L}{2e\left[\frac{e\eta}{h\nu}(P_S + P_L + P_B) + I_D\right]\Delta f_{IF} R_L + 4kT\Delta f_{IF}} \tag{2.89}$$

respectively. In Eqs. (2.87)–(2.88): β=0.9 is the responsivity of the detector, R_L = 50 Ω is the load resistance, $e = 1.6 \times 10^{-19}$C is the electronic charge, the quantum conversion efficiency η of the detector is 0.72, $h = 6.626 \times 10^{-34}$ J · s is the Planck's constant, ν=1.94 × 10^{14} Hz is the frequency of the light wave, T = 300 K is the thermodynamic temperature and k=1.38 × 10^{-23} J/K is the Boltzmann's constant.

In practice, the optical power P_L of a local oscillator light is significantly larger than the signal optical power P_S. In the simulation calculations and actual experiments, took P_L = 20 mW, P_L = 23μW, and Δf_{IF} = 10^8 Hz as the frequency difference signal bandwidth, furthermore, for convenience of calculation, we let in Eq. (2.84). From Eqs. (2.87)–(2.89), the SNRs in the three detection modes are simulated as shown in Fig. 2.49.

Figure 2.49 shows that the SNR of the heterodyne detection system is 9–10 orders of magnitude higher than that of the IM/DD system [2]; furthermore, when the signal light power increases, the output SNR exhibits a linearly increasing trend. When the signal light power is constant, the wider the bandwidth of the detector, the smaller the output SNR. This is because when the bandwidth of the detector is constant, the greater the signal light power, the greater the photocurrent outputted by the detector. An effective way to improve the SNR of a direct detection system is to increase the signal light power. However, in the optical wireless communication process, increasing the output power of the signal light will result in a series of technical difficulties and safety issues. Therefore, lower SNRs limit the development of IM/DD systems. In the optical heterodyne detection system, the local oscillator light power is used to effectively amplify the signal light power. The detector detects the frequency difference signal generated by the coherent mixing of the signal light

Fig. 2.49 SNR analysis curves for direct detection and heterodyne detection

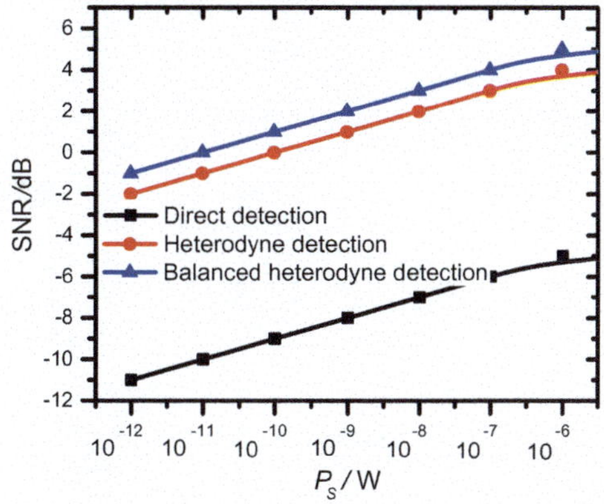

(a) Variation of SNR with respect to signal light.

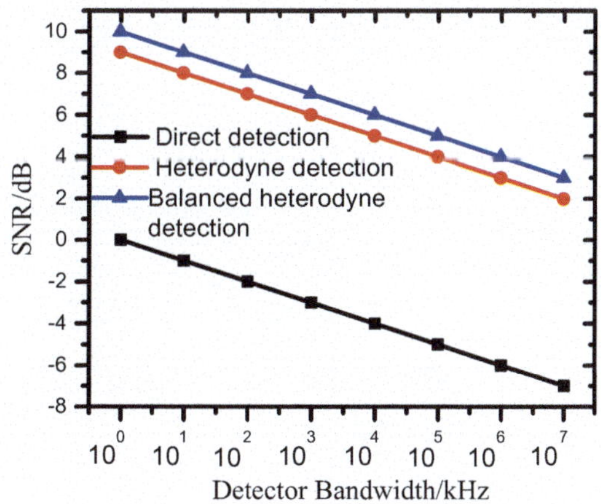

(b) Variation of SNR with respect to detector bandwidth.

and local oscillator light, which greatly improves the signal output power and SNR. In particular, the balanced heterodyne detection system effectively suppresses the influence of local oscillator light intensity noise on the SNR, and its output SNR is eight times that of the heterodyne detection system.

Comparing the output power of the direct detection system and heterodyne detection system, it can be seen that in the direct detection, the signal light power outputted by the detector; in the optical heterodyne detection system, the power of the frequency difference signal outputted by the mixer is $S_{IF} = 2\beta^2 P_L P_S R_L$. From this, the frequency difference conversion gain provided by optical heterodyne detection can be calculated as [2]

$$G_{IF} = \frac{S_{IF}}{S_P} = \frac{2\beta^2 P_L P_S R_L}{\beta^2 P_S^2 R_L} = \frac{2P_L}{P_S} \tag{2.90}$$

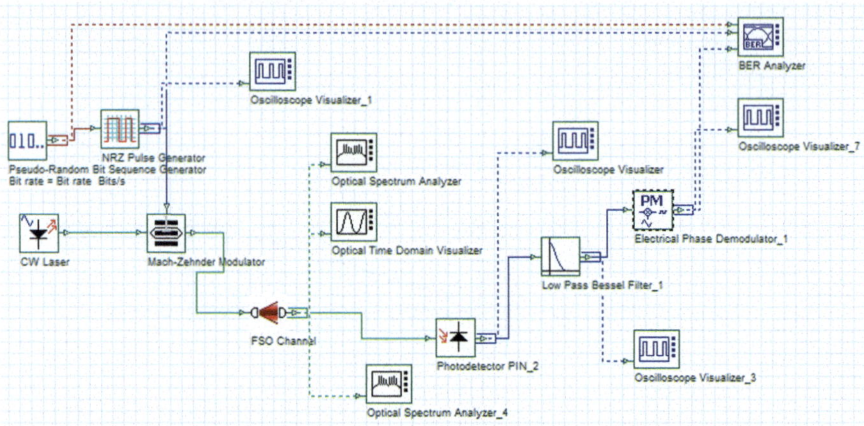

Fig. 2.50 Simulation diagram of direct detection system

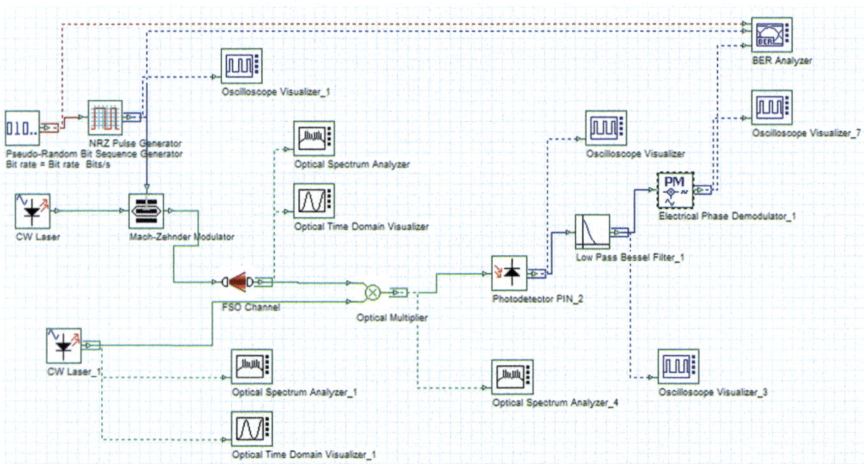

Fig. 2.51 Simulation diagram of heterodyne detection system

In actual experiments, the power of the local oscillator light P_L was considerably larger than the signal light power P_S. Typically, the maximum signal light power received by the detector is of the order of microwatts, whilst the local oscillator light power is of the order of milliwatts. According to Eq. (2.90), the frequency difference conversion gain of heterodyne detection is typically 10^3–10^8, and the frequency difference conversion gain of visible-light heterodyne detection is very high.

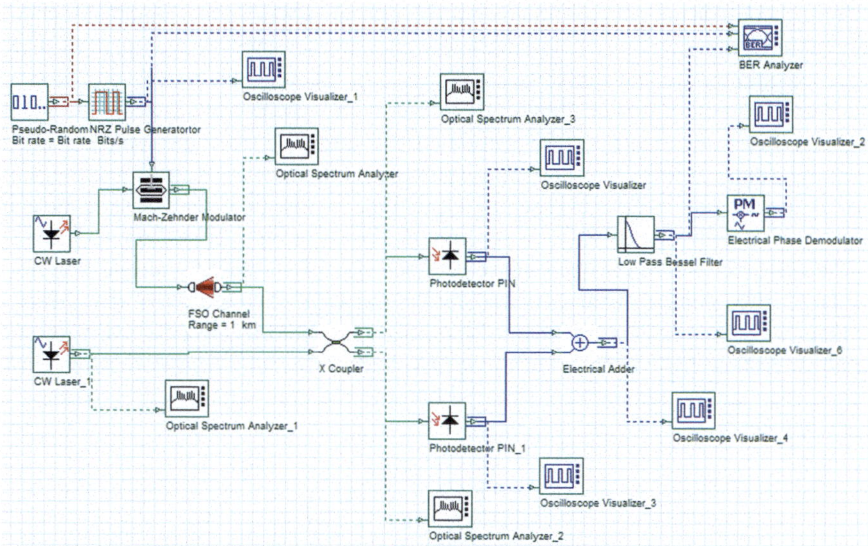

Fig. 2.52 Simulation diagram of balanced heterodyne detection system

2.8.2 *Bit Error Rate of Direct Detection and Heterodyne Detection*

Both the SNR and BER are important indicators for describing the performance of a communication system. Normally, the SNR is used to judge analog systems, and the BER is used to judge digital ones [36]. Direct and heterodyne detection systems can detect both analog and digital signals. In digital systems, the detection system no longer uses the SNR as the basis for evaluating system performance. This is because digital systems apply a "regeneration" function to the signal; this has no meaning for the SNR index of the output signal. Instead, the BER is used to evaluate system performance. The BER refers to the minimum bit loss rate and error rate of the pulse signal transmitted by the binary symbol after transmission attenuation and distortion; this expresses is the bit rate of the misjudgment. The ratio of the bit rate to the transmitted bit rate is determined according to the usage requirements [2], and it is generally in the range of 10^{-6} – 10^{-15}.

Different detection systems and methods exhibit different BERs. The unified detection system exhibits different BERs in different modulation directions. The statistical characteristics of the thermal, shot, and background noise of the amplifier in the system also differ. Thus, we obtain different BERs. Figures 2.50, 2.51 and 2.52 show the simulation models for the direct, heterodyne, and balanced heterodyne detection systems.

During simulation, it is assumed that the output power of the signal laser, the inherent linewidth of the laser, and the stable optical frequency are identical; furthermore, the attenuation and transmission distance of the atmospheric channel are the same, and the output power, inherent linewidth, and stable frequency of the local oscillator laser are identical. We simulated the direct, heterodyne, and balanced heterodyne detection systems under the assumption that the dark current and thermal noise of the detector and the bandwidth of the filter are identical, and we used the eye diagram to describe the BER of the system, as shown in Figs. 2.53, 2.54 and 2.55.

The graph marked by the black curve in Figs. 2.53, 2.54 and 2.55 is an eye diagram. The larger the opening of the eye, the smaller the BER. The red curve denotes the Q value, which represents the ratio of the noise threshold to the effective value of the noise voltage (or current). Different Q values correspond to different bit error rates; these can be obtained from the probability integral, which is approximately expressed as [2]

$$P_e = \frac{1}{Q\sqrt{2\pi}} \exp\left(-\frac{Q^2}{2}\right) \tag{2.91}$$

From Eq. (2.91), we can see that the higher the Q value, the lower the BER of the system. The relationship curve between the BER and Q is shown in Fig. 2.56.

Comparing Figs. 2.53, 2.54, 2.55 and 2.56, we can see that under the same simulation conditions as the transmission distance and noise model, the eye pattern obtained by the direct detection system tends to be closed, the minimum bit error rate is 1, the maximum Q value is 0, and the amplitude of the eye opening in the diagram is 0.

The eye pattern produced by the heterodyne detection system is open; however, numerous noise lines are visible around the eyes, which affect its final opening amplitude. Under the same conditions, the minimum BER of this system was 8.2066 $\times$ 10^{-10}, the maximum Q value was 6.0298, and the eye opening amplitude was 1.331 $\times$ 10^{-4}. Compared with the direct detection system, the BER, Q value, and eye-opening amplitude were considerably improved, the BER was reduced by ten orders of magnitude, and the Q value increased from 0 to 6 in the direct detection system.

The eye diagram generated by the balanced heterodyne detection system was in a fully open state, the various curves constituting it were compactly concentrated to form a clear eye diagram, and the noise lines around the eyes were also concentrated on the upper and lower sides. Meanwhile, the minimum BER of the system was 3.1016 $\times$ 10^{-15}, the Q value was 7.79973, and the eye height was 3.1103 $\times$ 10^{-4}. Compared with the heterodyne detection system, the minimum BER of the balanced heterodyne detection system was reduced by five orders of magnitude, the Q value was increased by a factor of ~ 1.3, and the eye height was increased by a factor of 2.34. Under the same simulation conditions, the BER of the digital transmission system was significantly affected by the detection method, and the BER of the balanced heterodyne detection system was minimized.

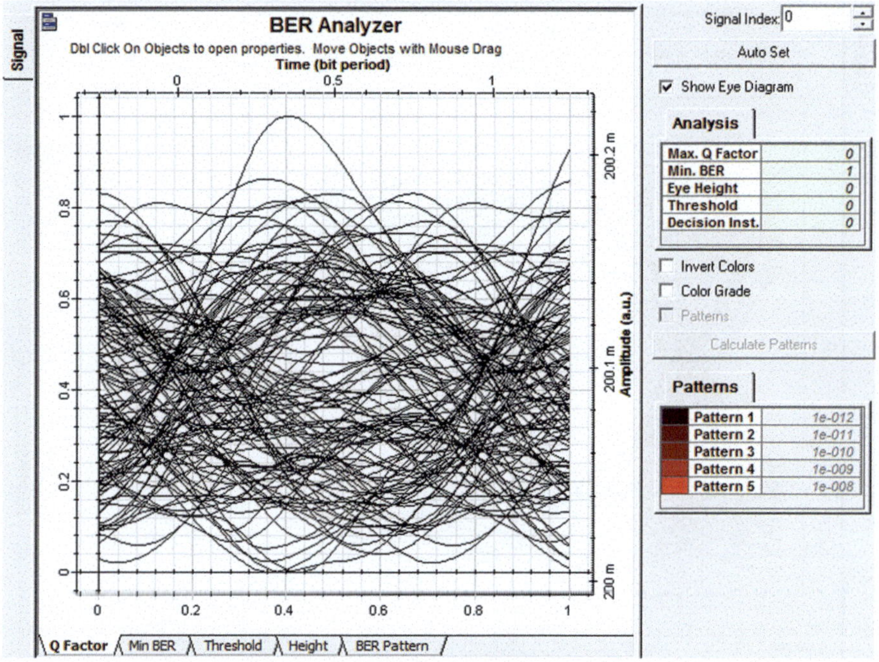

Fig. 2.53 Eye diagram of the direct detection system

2.8.3 Analysis of Detection Sensitivity of Direct and Heterodyne Detection

The sensitivity of the detection system refers to the minimum optical power required by the photoelectric detection system to ensure the SNR of communication quality. In general, the photocurrent generated by the signal light power exceeds the noise current. When the photocurrent is equal to the noise current, the photodetector can marginally detect the presence of signal light. Therefore, the limit sensitivity of the photodetector is defined as SNR = 1. The minimum optical power that can be detected—according to Eqs. (2.87)–(2.89) and assuming that the system is in the quantum noise time limit (i.e., the system noise only considers the shot noise generated by the signal light and local oscillator light)—and the system's SNR can be expressed (when other noises are neglected) as [2]

$$\left(\frac{S}{N}\right)_{PD} = \frac{\eta P_S}{2h\nu\Delta f} \tag{2.92}$$

$$\left(\frac{S}{N}\right)_{\mathrm{P}} = \frac{\eta P_S}{h\nu\Delta f_{IF}} \tag{2.93}$$

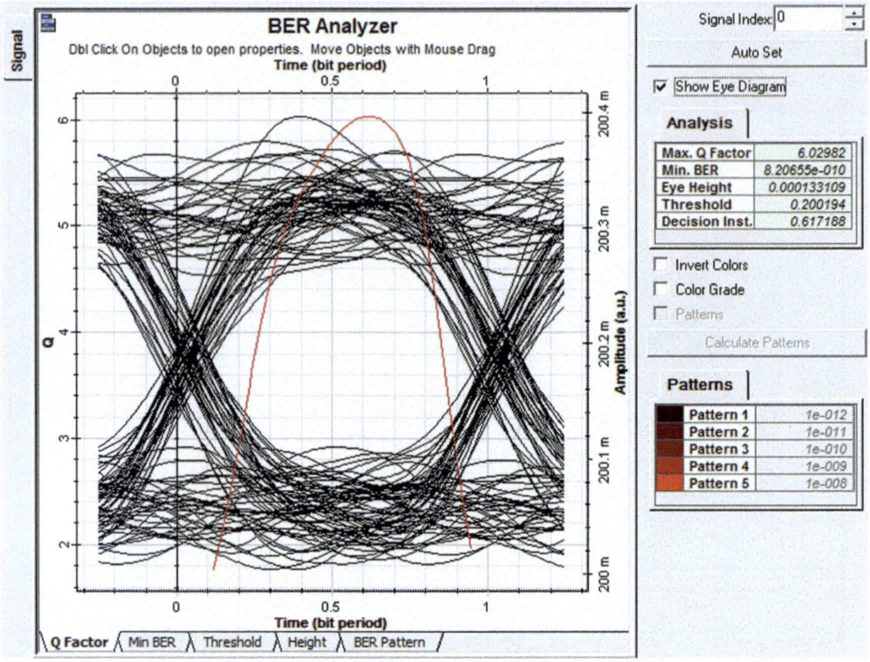

Fig. 2.54 Eye diagram of heterodyne detection system

$$\left(\frac{S}{N}\right)_{IF} = \frac{8\eta P_S}{h\nu \Delta f_{IF}} \tag{2.94}$$

According to Eqs. (2.92)–(2.94), the theoretical minimum optical power that the photodetector can detect under different detection methods is expressed as [2]

$$(P_S)_{PD} = 2h\nu\Delta f \big/ \eta \tag{2.95}$$

$$(P_S)_P = h\nu\Delta f_{IF} \big/ \eta \tag{2.96}$$

$$(P_S)_{IF} = h\nu\Delta f_{IF} \big/ 8\eta \tag{2.97}$$

When the response bandwidth of the detector is constant, the minimum optical power that the detector can detect is expressed as the sensitivity of the detection system. From Eqs. (2.95)–(2.97), the theoretical limit sensitivity of the direct detection system can be expressed as $P_S = 2h\nu\Delta f/\eta$, that of the classical optical heterodyne detection system can be expressed as $P_S = h\nu\Delta f_{IF}/\eta$ [36], and that of the balanced heterodyne detection system can be expressed as $P_S = h\nu\Delta f_{IF}/8\eta$. Numerical simulations of Eqs. (2.95)–(2.97) were carried out, and the simulation results are as shown in Fig. 2.57. This figure shows that when the detector bandwidth

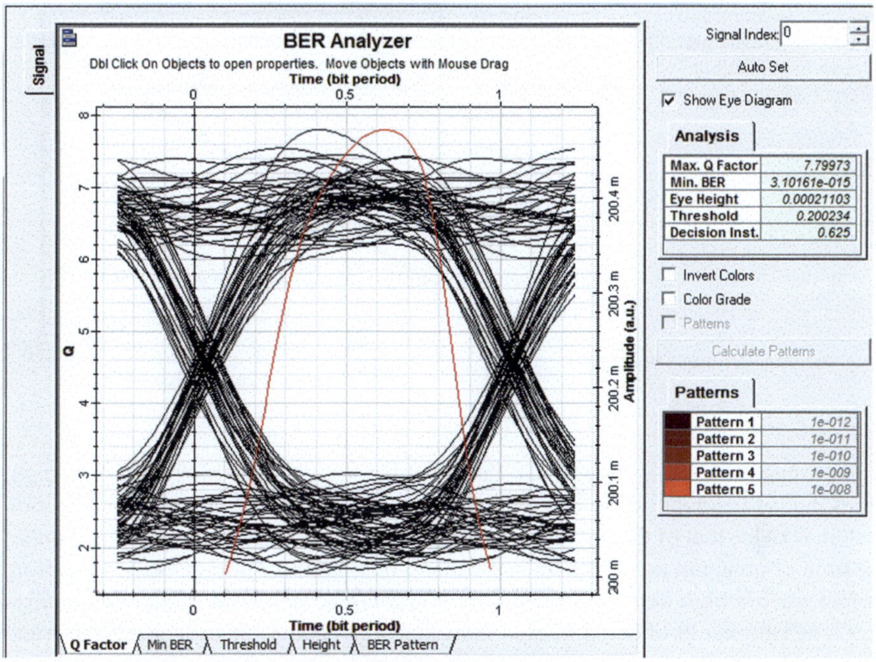

Fig. 2.55 Eye diagram of balanced heterodyne detection system

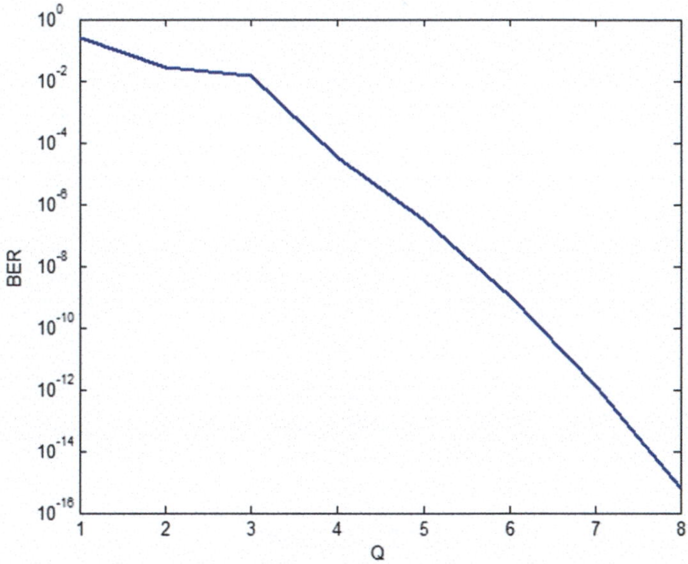

Fig. 2.56 Relationship between BER and Q

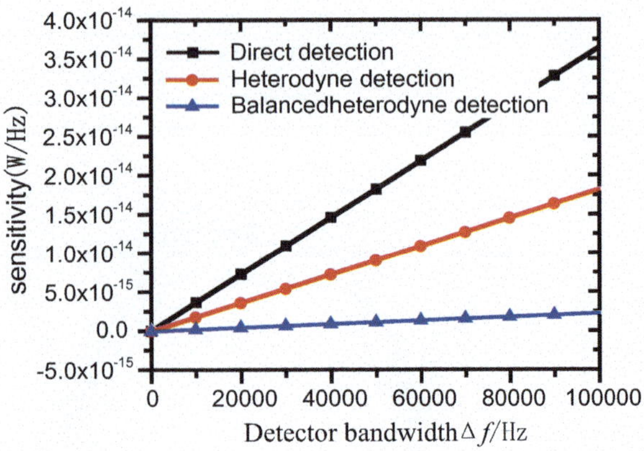

Fig. 2.57 Relationship between sensitivity and detector bandwidth in different detection modes

is unchanged, the sensitivity of the balanced heterodyne detection system is eight times that of the heterodyne detection system, and that of the heterodyne detection system is twice that of the direct detection system. This conclusion is obtained under the limit of quantum noise, which is the theoretical limit for photoelectric detection. In fact, such a high detection sensitivity could not be achieved. Equations (2.95)–(2.97) and the results of Fig. 2.57 also show that both optical heterodyne detection and balanced heterodyne detection can detect extremely weak optical signals; however, direct detection does not offer such advantages.

The discussion of the SNR, BER, and sensitivity of the space coherent optical communication system under different detection methods was conducted without considering the influence of atmospheric turbulence. Actual space optical communications are significantly affected by atmospheric turbulence. The wavefront distortion produced by atmospheric turbulence changes the transmission path of the signal light to varying degrees, eventually causing the signal light received on the photosensitive surface of the receiving end to lose its intensity, producing a powerful light pulse spread over a diffused spot with greatly attenuated energy. The larger the diameter of the light spot, the smaller the light energy node received by the detector; hence, the detector cannot respond to the correct signal photocurrent, which leads to a communication link interruption.

2.9 Influence of Wavefront Distortion on Spatial Coherent Optical Communication

2.9.1 Principle of Wavefront Distortion

The wavefront refers to the surface formed by points of the same phase in the vibration direction of the light waves. In physical optics, this surface is referred to as a wavefront. The principle of wavefront distortion states that the optical path from the light source to each point on the wave surface is no longer equal but exhibits a certain optical path difference. The size of the optical path difference reflects the change in the wavefront distortion. Figure 2.58 presents an optical schematic diagram of the signal light reaching the receiving-end optical antenna via space transmission (when no wavefront distortion is present) and then being received by the optical detector. As shown in the figure, when the signal light is transmitted over a long distance in a space without turbulence, the refractive index of each point on the spatial transmission path is equal everywhere, and the distance from any point on the isophase surface of the light wave to the light source is equal and transmitted in parallel. The light wave front is approximately planar. The plane wave is converged by the optical receiving system and then inputted to the photodetector. Because it is equal to the optical path transmission without wavefront distortion, the light at any point on the wave surface is concentrated in the center of the detector target surface, and the signal light energy is measured by the detector. The concentrated time-domain waveform of the light intensity is shown in Fig. 2.59. It can be seen from the figure that when there is no wavefront distortion, the signal light energy received by the detector is highly concentrated, which produces a very strong light pulse. The time Δt_1 required for the detector to receive the signal light at each point on the wavefront is called the "transit time" of the photoelectron.

When the signal light is transmitted in a turbulent atmosphere, the distances from each point on the cross section of the beam to the light source will no longer be equal (unequal optical path transmission). At this time, the wavefront formed by the phase surface of the light wave will no longer be planar but a distorted wavefront with certain fluctuations. Such a wavefront is called a distorted wavefront. When the wavefront of a parallel beam is distorted, the transmission path of each point on the distorted wave surface changes with the change in the refractive index of the turbulent atmosphere. Figure 2.60 shows that when wavefront distortion occurs, the signal light travels through space to reach the receiving-end optics; it is then received by the optical detector, as indicated in the schematic diagram. As shown in the figure, ΔL is the optical path difference for non-equal optical path transmission. After the wavefront of a parallel beam is distorted, the transmission direction of the light at each point on the wavefront changes irregularly. After this change, the rays of light transmitted to the photosensitive surface of the detector are no longer

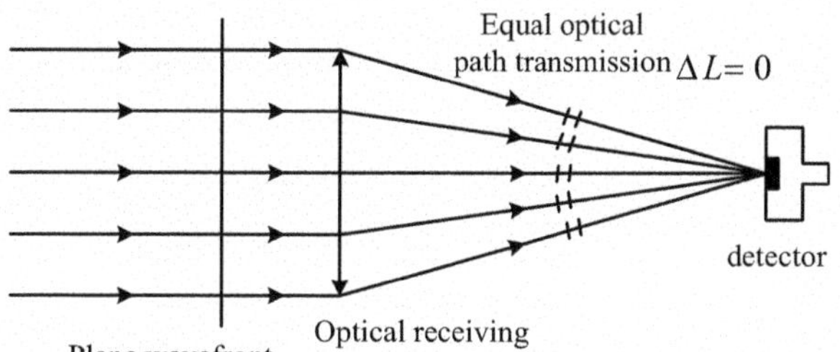

Fig. 2.58 Schematic diagram of plane-wave optical receiving system

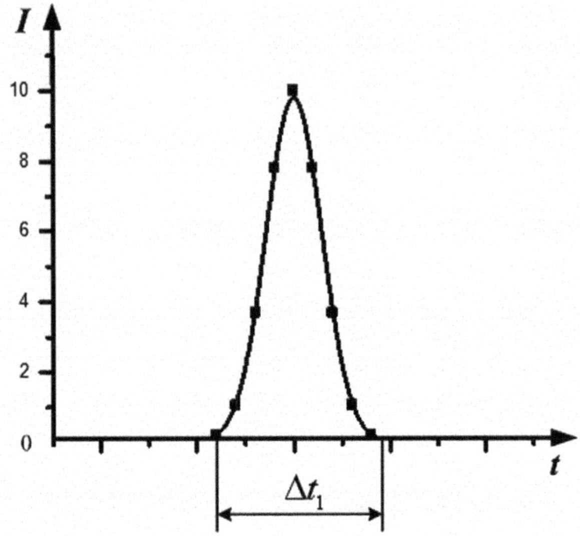

Fig. 2.59 Signal intensity waveform received by the detector in the absence of wavefront distortion

transmitted along equal optical paths, and a portion of the light deviates from the transmission direction of the beam; this exceeds the receiving range of the receiving optical system, resulting in the loss of light energy and greater wavefront distortion. The larger the deviation, the more serious the loss of light energy. At the same time, the time taken for each path of light transmitted by a non-equal optical path to reach the photosensitive surface of the detector is not the same. The longer the time, the longer the transit time of photoelectrons and the more serious the loss of light energy. Thus, the domain waveform will no longer be a light pulse with concentrated energy but a wide pulse with lower energy. This pulse width is called the "transition time dispersion" of the photoelectron, as shown in Fig. 2.61.

Figure 2.61 shows the influence of wavefront distortion produced by atmospheric turbulence on the time-domain response of the detector receiving the light energy. It

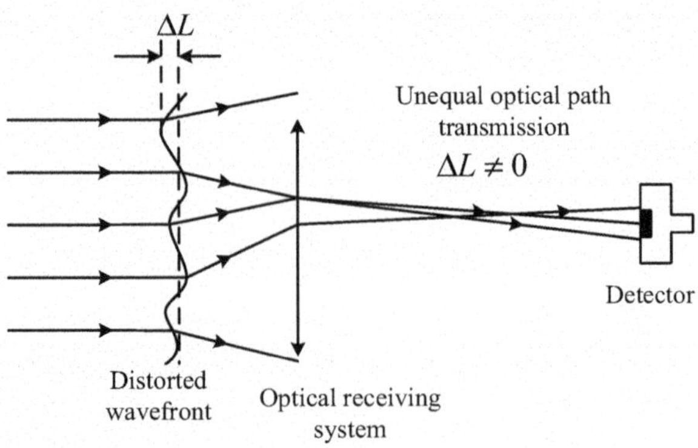

Fig. 2.60 Schematic diagram of the distorted wavefront optical receiving system

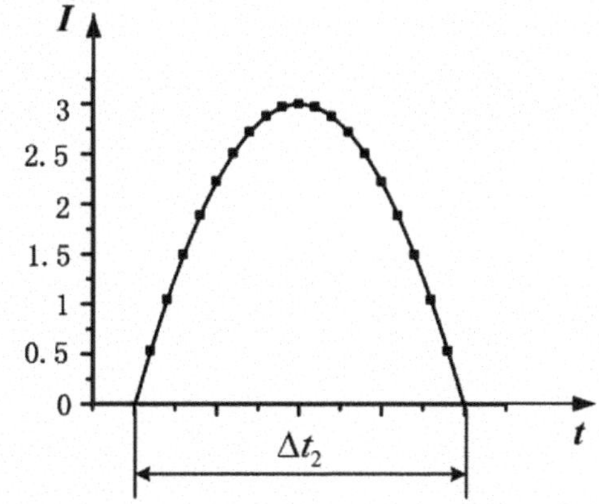

Fig. 2.61 Signal intensity waveform received by the detector under wavefront distortion

can be seen from the figure that when the received beam exhibits wavefront distortion, the time-domain response bandwidth of the received light energy is much wider than the time-domain response bandwidth without wavefront distortion, and the energy loss is very serious. The greater the dispersion of the transit time Δt_2, the more serious the wavefront distortion. The magnitude of the wavefront distortion can be expressed by the optical path difference ΔL between each path of light during non-equal optical path transmission. Assuming that the photoelectron transit time of the detector is Δt and the speed of light is c, then the optical path difference is

$$\Delta L = c \cdot \Delta t. \tag{2.98}$$

It can be seen from Eq. (2.98) that the magnitude of the wavefront distortion changes with the change in the photoelectron transit time Δt of the detector. The larger the value of Δt, the greater the optical path difference ΔL, and the more serious the wavefront distortion produced by atmospheric turbulence. The response frequency of the detector can be obtained via the Fourier transform of the photoelectron transit time Δt. Figures 2.62 and 2.63 show the frequency response diagrams obtained after the Fourier transforms of Δt_1 and Δt_2, respectively.

Comparing Figs. 2.62 and 2.63, we can see that the narrower the time-domain response bandwidth of the signal, the wider the frequency domain response bandwidth. Conversely, the wider its time-domain response, the narrower the frequency-domain response bandwidth. The bandwidth of the signal corresponds to the response bandwidth of the detector and the channel width. Generally, when the former matches the latter, the signal efficiency of the detector is maximized, and the BER is lowest. When the channel width exceeds the response bandwidth of the detector, it is difficult to receive complete transmission information; when the channel width is less than the response bandwidth of the detector, unnecessary resource wastage is entailed and the unnecessary cost waste is increased. The one-to-one correspondence between the time-domain bandwidth Δt and frequency-domain bandwidth Δf of the detector (i.e., $\Delta t = 1/\Delta f$) mean that the maximum wavefront distortion produced

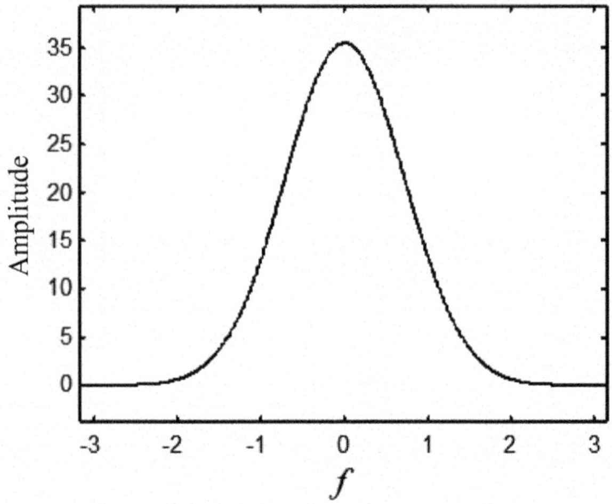

Fig. 2.62 Fourier transform spectrum

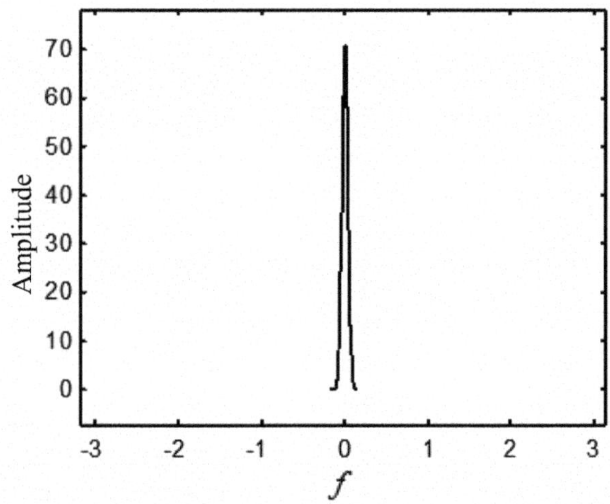

Fig. 2.63 Fourier transform spectrum of Δt_2

by the turbulent atmosphere can be obtained from Δt or Δf. Figure 2.64 shows the relationship between the wavefront distortion and the response bandwidth of the detector.

2.9.2 The Effect of Wavefront Distortion

The atmospheric refractive index fluctuations generated by atmospheric turbulence at each point on the cross section of the beam are not identical, resulting in a different phase shift at each point; thus, the isophase plane of the light wave received at the

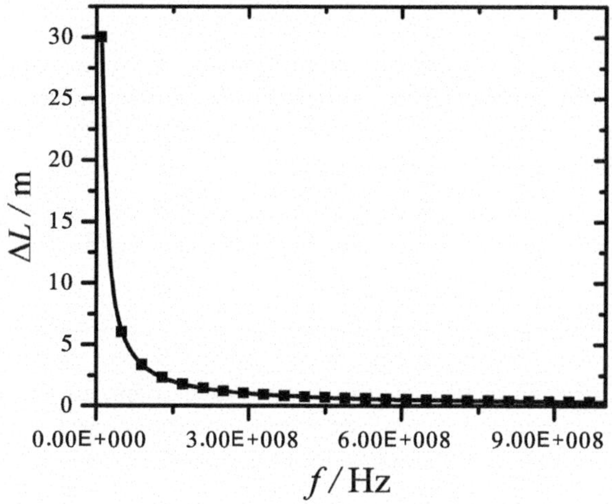

Fig. 2.64 Relationship between wavefront distortion and bandwidth

receiving end is a random fluctuation, and the curved surface causes the incident angle of each point on the focused spot of the optical receiving surface to differ. If the signal light received at this time and the local oscillator light are used for coherent mixing, the spatial mismatch angle θ between the wavefronts of the two beams will reduce the mixing efficiency of spatial coherent optical communication; the greater the mismatch angle, the lower the mixing efficiency. The spatial relationship between the local oscillator light wavefront and signal light wavefront on the photosensitive surface of the photodetector is shown in Fig. 2.65.

To simplify analysis, it was assumed that the photosensitive surface of the photodetector was a square of side length d. The local oscillation light was incident perpendicular to the photosensitive surface of the photodetector. Owing to the influence of atmospheric turbulence, a mismatch angle θ was formed between the signal light wavefront and local oscillation light wavefront. Therefore, when the signal light was incident on the photosensitive surface of the photodetector at a certain angle θ, the

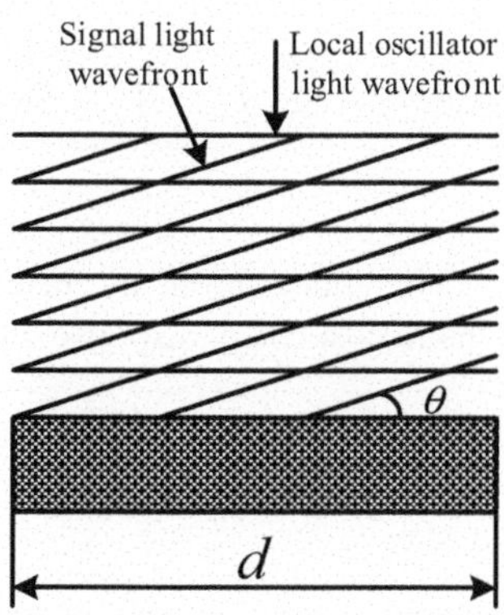

Fig. 2.65 Spatial relationships of coherent mixing

time taken for each point on the signal light cross section to reach the photosensitive surface of the photodetector differed, which can be equivalent to traveling at a speed v_x in direction x; thus, different points on the photosensitive surface form a wavefront phase difference (i.e., wavefront distortion). Therefore, the light field expression of the signal light can be written as [2]

$$\boldsymbol{E}_{\mathrm{S}}(t) = A_{\mathrm{S}} E_{\mathrm{S}} \cos(\omega_{\mathrm{S}} t + \phi_{\mathrm{S}} - \frac{\omega_{\mathrm{S}}}{v_x} \cdot x). \tag{2.99}$$

The expression of the light field of the local oscillator is given by Eq. (2.2). In Eq. (2.99), $k_x = \omega_{\mathrm{S}}/v_x$ is the component of the signal light wave vector $\boldsymbol{k}$ in direction $\boldsymbol{x}$. We set $\boldsymbol{c}$ as the speed of light, as shown in Fig. 2.65; furthermore, $k_x = k \sin\theta = (\omega_s/c)\sin\theta$; thus, the total light field incident on the photosensitive surface of the photodetector along the direction $\boldsymbol{x}$ is

$$E_x(t) = E_{\mathrm{S}}(t) + E_{\mathrm{L}}(t). \tag{2.100}$$

Thus, the instantaneous photocurrent outputted by the photodetector is

$$\begin{aligned} I(t) &= \frac{G\beta}{d^2}\int_{-d/2}^{d/2}\int_{-d/2}^{d/2}\left[E_{\mathrm{S}}\cos(\omega_{\mathrm{S}}t + \varphi_{\mathrm{S}} - \frac{2\pi\sin\theta}{\lambda_{\mathrm{S}}}x) + E_{\mathrm{L}}\cos(\omega_{\mathrm{L}}t + \varphi_{\mathrm{L}})\right]^2 dxdy. \\ &= \frac{G\beta}{d^2}\int_{-d/2}^{d/2}\int_{-d/2}^{d/2}\left[E_{\mathrm{S}}^2\cos^2(\omega_{\mathrm{S}}t + \varphi_{\mathrm{S}} - \frac{2\pi\sin\theta}{\lambda_{\mathrm{S}}}x) + E_L^2\cos^2(\omega_{\mathrm{L}}t + \varphi_{\mathrm{L}})\right. \\ &+ E_{\mathrm{S}}E_{\mathrm{L}}\cos((\omega_{\mathrm{L}} - \omega_{\mathrm{S}})t + (\varphi_{\mathrm{L}} - \varphi_{\mathrm{S}}) + \frac{2\pi\sin\theta}{\lambda_{\mathrm{S}}}x) \\ &\left. + E_{\mathrm{S}}E_{\mathrm{L}}\cos((\omega_{\mathrm{L}} + \omega_{\mathrm{S}})t - (\varphi_{\mathrm{L}} + \varphi_{\mathrm{S}}) + \frac{2\pi\sin\theta}{\lambda_{\mathrm{S}}}x)\right]dxdy \end{aligned} \tag{2.101}$$

In Eq. (2.101), G and β are the internal gain and responsivity of the photodetector, respectively, and d is the side length of the detector's photosensitive surface. The instantaneous difference in frequency photocurrent outputted by the photodetector in response is

$$i_{\mathrm{IF}}(t) = \frac{G\beta}{d^2}\int_{-d/2}^{d/2}\int_{-d/2}^{d/2} E_{\mathrm{S}}E_{\mathrm{L}}\cos((\omega_{\mathrm{L}} - \omega_{\mathrm{S}})t + (\varphi_{\mathrm{L}} - \varphi_{\mathrm{S}}) + \frac{2\pi\sin\theta}{\lambda_{\mathrm{S}}}x)dxdy. \tag{2.102}$$

After integration, the instantaneous difference frequency current of the photosensitive surface response output for the entire photodetector can be expressed as

$$i_{\mathrm{IF}}(t) = G\beta E_{\mathrm{S}}E_{\mathrm{L}}\cos[(\omega_{\mathrm{L}} - \omega_{\mathrm{S}})t + (\varphi_{\mathrm{L}} - \varphi_{\mathrm{S}})]\frac{\sin(\omega_{\mathrm{S}}d/2v_x)}{\omega_{\mathrm{S}}d/2v_x}. \tag{2.103}$$

Because $v_x = c/\sin\theta$, the difference frequency current i_{IF} is related to the mismatch angle θ. Clearly, the factor $\sin(\omega_s d/2v_x)/(\omega_s d/2v_x)$ in the heterodyne current in Eq. (2.103) is the mixing efficiency of the heterodyne detection. When its value is 1, the output difference frequency current value of heterodyne detection is maximal. At this time, we require that $\omega_S d/2v_x = 0$; that is, we require $\sin\theta = 0$. However, it is difficult to adjust the angle θ to zero because of the turbulence effect present in the transmission medium; therefore, the current value of the output difference frequency signal is generally below the maximum value of 1.

In a coherent optical communication system, the greater the spatial mismatch angle between the signal light and local oscillator light wavefront, the lower the mixing efficiency, and the smaller the difference frequency current outputted by the photodetector. If the difference frequency current is too small, it will cause signal detection and demodulation. Therefore, to maximize the difference frequency output, it is desirable that the two light beams be incident on the photosensitive surface of the detector in parallel and coincident, which requires that the signal light be incident in a form without wavefront distortion. Owing to the existence of atmospheric turbulence, the signal light inevitably undergoes wavefront distortion when it is transmitted through a turbulent atmosphere. To improve the mixing efficiency, the wavefront of the signal light must be corrected.

The problems that must be solved in coherent optical communication include the following: ① wavefront correction in atmospheric laser communication, ② narrow linewidth light sources, ③ transformation of the atmosphere-induced light polarization, and ④ consideration of the impact of Doppler frequency shifts on coherent detection for the high-speed movement of communication terminals. The solution to these problems will help in the practical application of coherent optical communication technology. The future development directions should include the following: ① low-cost narrow-band light sources; these have a great promotive effect on coherent optical communication, which has attracted people's attention. ② New coherence principles, such as subcarrier coherent detection and homodyne detection. ③ Simple and effective wavefront correction, as an effective method to suppress turbulence in atmospheric laser communication.

References

1. Xiao X, Chen MS (2011) Design of external step demodulation receiver for coherent optical communication. Popular Sci Technol 4:21–22
2. Wang QZ, Hu Y, Lin CJ (1994) Photoelectric detection technology. Electronic Industry Press (1994).
3. Lin CF(2010) High-frequency electronic circuits. Electronic Industry Press
4. Zhao F (2011) Research on receiving performance of intersatellite laser communication systems based on single-mode fiber coupled self-homodying detection. Harbin Institute of Technology.
5. Ke XZ, Chen JN (2014) Experimental investigation on non-optical heterodyne detection technology of 1 km atmospheric laser communication system. J Appl Sci 32(4):379–384
6. Fan CX Cao LN (2006) Principles of communication (6th edn). National Defense Industry Press

7. Ke XZ, Chen JN (2012) Research on factors and key techniques of space optical heterodyne detection. Semicond Optoelectron 33(4):548–557
8. Yu JJ, Chi N, Chen L (2013) Coherent optical communication technology based on digital signal processing. Posts and Telecom Press
9. Liu ZJ, Zhou YY, Hu LL et al (2008) Fiber optic communication (2nd edn). Xidian University Press
10. Qin YZ (2012) Research on the technology of the coherent optical communication link. University of Electronic Science and Technology of China
11. Zhu Y, Wang JP, Lu L (2011) Optical communication principles and technology (2nd edn). Science Press
12. An YY, Zeng XD, Feng ZJ (2010) Photoelectric detection and signal processing. Science Press
13. Liu HZ, Ji YF, Xu N et al (2011) Effect of amplitude profile difference of signal and local oscillator wave on heterodyne efficiency in the inter-satellite coherent optical communication system. J Opt 31(10):71–76
14. Wang LZ (2011) Study on optical heterodyne detection in coherent optical communication. Xi'an University of Electronic Science and Technology
15. Lv XG, Wang L, Xu DM et al (2013) Analysis of optical phase-locked loop for coherent optical communication based on QPSK modulation. J Changchun Univ Sci Technol 36(3):49–52
16. Bi GG (1990) Optical communication system with intensity modulation and nonlinear heterodyne detection. J Electron 2:69–75
17. Jacobaen G, Garrets I (1986) Optical ASK heterodyne receiver: comparison of a theoretical Model with Experiment 22(3):170
18. Liu C, Chen S, Li XY et al (2014) Performance evaluation of adaptive optics for atmospheric coherent laser communications. Opt Express 22(13):15554–15563
19. Barry JR, Lee EA (1990) Performance of coherent optical receivers. Proc IEEE 78(8):1369–1394
20. Liu Y, Tong SF, Chang SA et al (2018) Design of a phase sensor applied in the optical phase-loceked loop based on a high speed coherent laser communication system. IEEE Access 6:22131–22139
21. Feng F, White IH, Wilkinson TD (2014) Aberration correction for free space optical communications using rectangular Zernike modal wavefront sensing. J Lightwave Technol 32(6):1239–1245
22. Li ZK (2016) Research of adaptive optics for wave front distortion correction in free space optics communication. Jilin University
23. Noll RJ (1976) Zernike polynomials and atmospheric turbulence. J Opt Soc Am 66(3):207–211
24. Huang L, Zhou CL, Zhao WC et al (2017) Close-loop performance of a high precision deflectometry controlled deformable mirror unit for wavefront correction in adaptive optics system. Opt Commun 393:83–88
25. Rao RZ (2005) Light propagation in turbulent atmospheres. Anhui Science and Technology Press
26. Wang MJ, Jia WG, Zhang SY, et al (2015) Effect of Raman gain on the state of polarization evolution in a low-birefringence fiber. J Phys 64(3):034212(6)
27. Segura M, Vukovic N, White N et al (2015) Low birefringence measurement and temperature dependence in metre-long optical fibers. J Lightwave Technol 33(12):2697–2702
28. Waddy DS, Lu P, Chen L et al (2001) Fast state of polarization changes in aerial fiber under different climatic conditions. IEEE Photonics Technol Lett 13(9):1035–1037
29. Walker NG, Walker GR (1990) Polarization control for coherent communications. J Lightwave Technol 8(3):438–458
30. NoÉ R, Heidrich H, Hoffmann D (1988) Endless polarization control systems for coherent optics. J Lightwave Technol 6(7):1199–1208
31. Udayakumar R, Khanaa V, Kaliyamurthie KP (2013) High data rate for coherent optical wired communication using DSP. Indian J Sci Technol 6(6):4772–4776
32. Zhi Y, Feng L, Lei M (2014) Delay self-heterodyne measurement of narrow linewidth laser frequency drift characteristic. Optik-Int J Light Electron Opt 125(13):3124–3126

33. Chang S, Tong SF, Jiang HL et al (2017) Optical phase-locked loop technology in interplanetary high-speed coherent laser communication system. J Opt 2:68–76
34. Kong YX, Ke XZ, Yang Y (2016) Influence of local oscillation optical power on signal-to-noise ratio in spatially coherent optical communication. Infrared Laser Eng 45(2):0222002(6)
35. Candel SM (1981) An algorithm for the fourier-bessel transform. Comput Phys Commun 23(4):343–353
36. Yang YJ (2004) Fiber optic communication technology. Guangzhou: South China University of Technology Press, pp 240–261

Chapter 3
Spatial Light to Fiber Coupling and Beam Control

Contents

The space optical-fiber coupling receiving technology is used to couple optical signals into an optical fiber and perform optical amplification and photoelectric detection at the other end. This technology benefits the free space optical (FSO) communication system by improving the FSO communication rate, regulating the characteristics of the signal beam in the optical fiber [1, 2], improving system interchangeability, and so on. From the point of view of transmission, when the signal beam is coupled into the fiber before transmission, it can be modulated using an advanced high-speed fiber modulator, thus preventing the problems associated with high-power and high-speed laser modulation. Optical amplification technology can also be used to amplify the transmitted signal power without frequency bias.

X. Ke and J. Wu, *Coherent Optical Wireless Communication Principle and Application*, Optical Wireless Communication Theory and Technology,
https://doi.org/10.1007/978-981-19-4823-7_3

3.1 Space Optical-Fiber Coupling Technology

Optical-fiber coupling receiving technology [3] is already in use. A typical example is the demonstration of lunar laser communication (LLCD) project, which is an FSO communication platform developed by NASA that successfully realized optical communication between the Moon and the Earth in October 2013. This optical communication link is 400,000 km long—the longest one till date. Its terminal ground receiving antenna at Wrightwood (California) is a telescope with an aperture of 1 m. The core receiving technology used in this platform is the spatial light–multimode fiber coupling technology. The core receiving device is a tungsten silicide superconducting single-photon detector [4]. In 2010, Takahashi [5] proposed an FSO system capable of realizing optical wavelength division multiplexing based on an optical-fiber transceiver. In 2016, Abadi et al. [6] designed an RF-FSO dual-band antenna that employs optical receiving technology based on fiber coupling. Theoretical research on spatial light–fiber coupling receiving technology under atmospheric turbulence originated from lidar [7–9], and these theories are not directly applicable to the FSO system. In 1998, Ruilier [10] defined the efficiency of spatial light coupled into single mode fiber under Kolmogorov turbulence in the FSO system. In the same year, Belmonte [11] experimentally verified the feasibility of array arrangement and reception of telescope system array coupling system in satellite ground coherent optical communication system. In 2005, Dikmelik and Davidson [12] simplified the expression for coupling efficiency from a triple product to a double integral using the Gaussian approximation of the cross-correlation function of a plane light wave incident light field under Kolmogorov turbulence. Dikmelik proposed that the lens array could suppress the decrease in coupling efficiency caused by the turbulence. In 2010, the Institute of Mechanics, Chinese Academy of Sciences, conducted an in-depth theoretical analysis on the coupling efficiency of the lens array based on the literature [13] and proved that the coupling performance of the array receiver is better than that of a single lens. In 2009, Ma et al. [14] analyzed the impact of the angle of arrival fluctuation caused by atmospheric turbulence on the decline of the coupling efficiency. In 2012, Takenaka et al. [15] published a set of experimental data and theoretical analysis methods on the coupling efficiency fading of the optical inter-orbit communications engineering test downlink spatial light coupled through a single lens into a single-mode fiber. The experimental results indicated that the coupling efficiency faded between 10 and 19 dB. The theoretical calculation method presented in Ref. [15] is similar to that described in Ref. [12], except that the horizontal transmission calculation is changed into an oblique transmission calculation. In 2014, Tan [16] extended the coupling efficiency to the non-Kolmogorov turbulence spectrum and introduced the internal and external scales of turbulence as a factor affecting the coupling efficiency of spatial plane waves coupled into a single-mode fiber through a single lens.

3.1.1 Ideal Lens–Single-Mode Fiber Coupling

An optical fiber is a dielectric waveguide in the light wave band, and its cross section is usually circular. Optical fibers use the principle of total reflection to confine the energy of light waves within the interface and guide them along the axis of the fiber. The transmission characteristics are determined by the structure (core diameter D_f) and the material (core and cladding refractive indices n_1 and n_2). Owing to the limitation of the core boundary, its electromagnetic field solution is not continuous, and such a discontinuous field solution is referred to as a mode [17].

Optical fibers can be divided into single-mode and multimode optical fibers based on the number of transmission modes. In this study, the coupling efficiency of a single-mode fiber is examined, in which the lens is the phase converter of a light wave [18]. Under ideal conditions, the plane wave passes through an ideal thin lens (without considering the effect of the reflection difference) and becomes a convergent spherical wave, which converges at the focal point of the lens. As depicted in Fig. 3.1, if the axis of the optical fiber overlaps with the optical axis of the lens and the end face of the optical fiber is placed at the focal plane of the lens, the total reflection principle can be used to restrict the optical energy at the interface of the optical fiber and make the light wave travel along the axis of the optical fiber. If the fiber placed at the focal point is a single-mode fiber, this process is known as the "spatial plane wave lens–single-mode fiber coupling" process. As indicated in Eq. (3.1), the coupling efficiency η is defined as the ratio of the optical power P coupled into the optical fiber to the optical power P_0 incident on the receiving plane of the optical system. The coupling efficiency is a significant index used to measure the performance of coupled systems:

$$\eta = P/P_0 \tag{3.1}$$

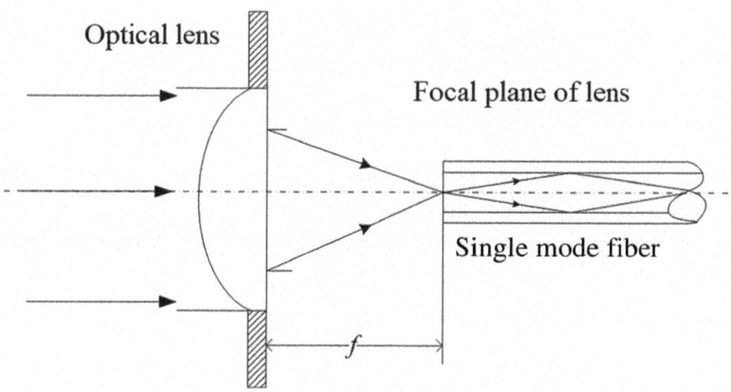

Fig. 3.1 Schematic of spatial plane wave single lens–single-mode fiber coupling under ideal conditions

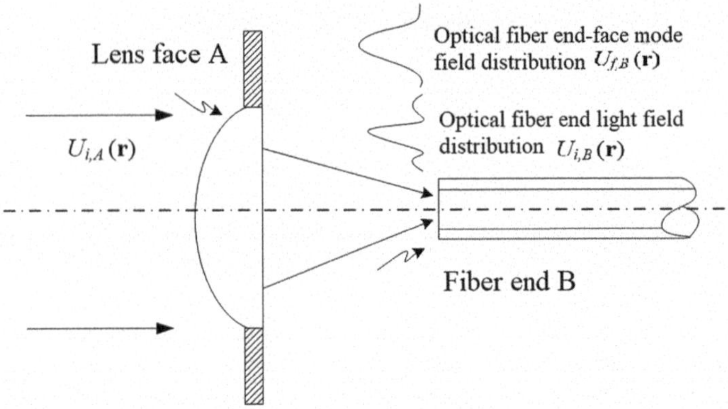

Fig. 3.2 Schematic of mode-field analysis of spatial plane wave lens–single-mode fiber coupling under ideal conditions

(1) **Mode-field analysis of coupling efficiency**

As depicted in Fig. 3.2, the diameter of the diffraction effect should be considered for the spatial light to fiber coupling optical system. In particular, when the wavelength of the incident light and the core diameter of the optical fiber (which is lower than that for single-mode fiber coupling) or the optical system diameter are the same, the diffraction effect is determined to be larger using the method of the mode-field matching analysis of the results, more in line with the actual situation.

In mode-field analysis, the coupling efficiency η is defined as follows:

$$\eta = \frac{\left|\int\int U_i^*(r)U_f(r)rdrd\theta\right|^2}{\int\int U_i(r)U_i^*(r)rdrd\theta \times \int\int U_f(r)U_f^*(r)rdrd\theta} \tag{3.2}$$

where $U_i(r)$ is the amplitude distribution of the signal light on the end face of the fiber $U_{i,B}(r)$ or lens $U_{i,A}(r)$ and $U_f(r)$ is the distribution of the electromagnetic field on the end face of the single-mode optical fiber $U_{f,B}(r)$ or lens $U_{f,A}(r)$.

(2) **Electromagnetic field distribution on optical fiber,** $U_{f,B}(r)$

As depicted in Fig. 3.3, single-mode optical fibers can transmit only the fundamental mode LP_{01}. The electromagnetic field distribution at the fiber cross section B of the fundamental mode LP_{01} is a zeroth order Bessel function, and the field distribution can be approximated using a Gaussian distribution:

$$U_{f,B}(r) = \sqrt{\frac{2}{\pi}}\frac{1}{W_m}\exp\left[-\left(\frac{r}{W_m}\right)^2\right] \tag{3.3}$$

where $r = \sqrt{x^2 + y^2}$ is the radial distance from any point on the cross section of the optical fiber to the center and W_m is the mode-field radius of the single-mode optical

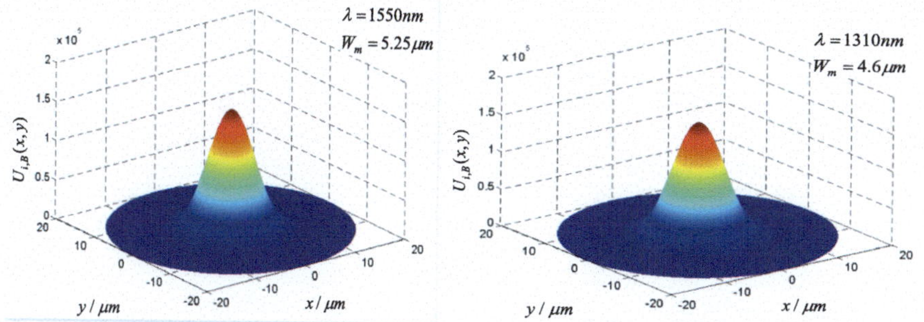

(a) λ = 1550 (nm) (b) λ = 1310 (nm)

Fig. 3.3 Electromagnetic field distribution on the end face of a single-mode optical fiber

fiber. The typical values are as follows: at $\lambda = 1150$ nm, $W_m = 5.25 \pm 0.5$ μm; at $\lambda = 1310$ nm, $W_m = 4.60 \pm 0.25$ μm; at $\lambda = 632.8$ nm, $W_m = 2.20 \pm 0.25$ μm [19].

(3) **Electromagnetic field distribution of plane light wave incident light field at focal point of single lens (end face of fiber)** $U_{i,B}(r)$

A laser light source is often used as a signal light source in optical wireless communication systems. Because of the effective directivity and monochromatic properties of the laser, its wavefront can be regarded as a monochromatic plane wave after long-distance transmission. Considering the limitation of the aperture D_A of the receiving lens and ignoring the effect of the lens phase difference, the Fresnel diffraction formula can be used to express the light field distribution of the incident plane wave on the focal plane of the receiving lens as follows:

$$U_{i,B}(r) = \frac{1}{i\lambda f}\exp(ikf)\exp(\frac{ikr^2}{2f})\pi\left(\frac{D_A}{2}\right)^2\left[\frac{2J_1\left(\frac{\pi D_A r}{\lambda f}\right)}{\frac{\pi D_A r}{\lambda f}}\right] \quad (3.4)$$

where f is the focal length of the coupled lens, D_A is the diameter of the coupled lens, $J_1(\cdot)$ is the first-order Bessel function of the first kind, and $k = 2\pi/\lambda$ is the spatial angular frequency of the wave (also known as wave loss). It can be seen that the electromagnetic field distribution of the plane light at the focus is related to the relative aperture of the lens D_A/f, and the wavelength of the incident light.

As depicted in Fig. 3.4, $D_A = 10$ mm and $\lambda = 1550$ nm, and Eq. (3.4) is used to simulate the light field distribution of plane waves at the focal plane through a single lens at various relative apertures. It can be seen that the lens diameter is unchanged, and with the increase in the focal length of the lens, the light spot is increasingly affected significantly by the diffraction effect and the light field distribution becomes increasingly diffuse.

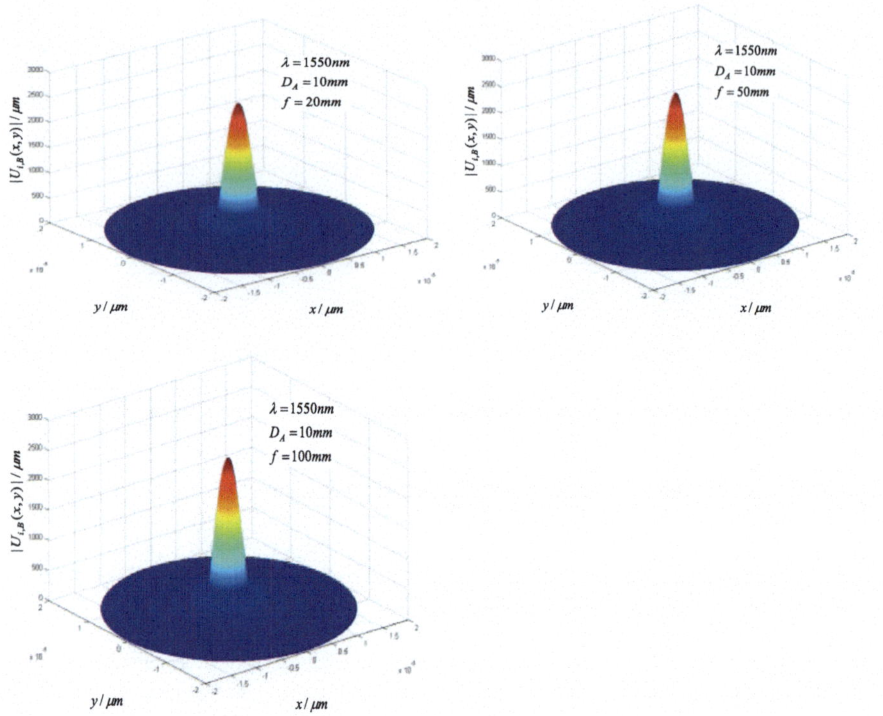

(a) $\lambda = 1550$ nm, $D_A = 10$ mm, $f = 20$ mm (b) $\lambda = 1550$ nm, $D_A = 10$ mm, $f = 50$ mm

(c) $\lambda = 1550$ nm, $D_A = 10$ mm, $f = 100$ mm

Fig. 3.4 Electromagnetic field distribution of plane wave incident light field at focal plane of single lens

(4) **Monochromatic plane wave single lens–single-mode fiber coupling efficiency**

The coupling efficiency was calculated at the cross section of the fiber, ignoring the phase difference, reflection, and absorption loss. Assuming that a monochromatic plane wave is incident and $\int\int U_{f,B}(r)U^*_{f,B}(r)rdrd\theta = 1$, the coupling efficiency is calculated by substituting the electromagnetic field distribution (3.3) at the end surface of the optical fiber and the plane wave light field distribution (3.4) at the focal plane in Eq. (3.2):

$$\eta_B = \frac{\left|\int\int U_{i,B}^*(r)U_{f,B}(r)rdrd\theta\right|^2}{\int\int U_{i,B}(r)U_{i,B}^*(r)rdrd\theta}$$

$$= \frac{4\left|\int_0^\infty \exp(-ikf)\exp\left[-ikr^2/(2f)\right]J_1\left(\frac{\pi D_A r}{\lambda f}\right)\exp\left[-\left(\frac{r}{W_m}\right)^2\right]dr\right|^2}{W_m^2\int_0^\infty\left[J_1\left(\frac{\pi D_A r}{\lambda f}\right)\right]^2/rdr} \tag{3.5}$$

As presented in Fig. 3.5, Eq. (3.5) is used to simulate the effect of the relative aperture of the lens and the wavelength of the beam on the coupling efficiency. During the simulation, we consider $D_A = 10$ mm. When the relative aperture increases from low to high, the focal length of the lens decreases from high to low. As can be seen in Fig. 3.5, the coupling efficiency first increases and then decreases as the relative aperture of the lens increases from small to large. This is because the relative aperture is low, the focal length of the lens is high, and the energy distribution of the spot is diffused by the diffraction effect. From the perspective of geometric optics, although the divergence angle of the beam is within the numerical aperture of the optical fiber, the area of the incident light spot is mismatched with the area of the end face of the optical fiber, leading to a decline in the coupling efficiency. As the relative aperture gradually increases and the focal length decreases, the effect of diffraction decreases and the coupling efficiency gradually increases. However, if the relative aperture continues to increase and the focal length continues to decrease beyond a certain value, the incident angle of the beam will not match the numerical aperture of the optical fiber, causing the coupling efficiency to decline again. Incident beams of different wavelengths require different coupling structures. When the wavelength of the incident beam changes, a lens with a relative aperture of 0.211 should be selected for the 1550-nm light wave to obtain the maximum coupling efficiency of 81.4%. For $\lambda = 1310$ nm, a lens with a relative aperture of 0.203 is required.

(5) **Coupling efficiency at end face of lens**

As depicted in Fig. 3.2, the light field distribution at the lens face and the focal plane of the optical field distribution are related to each other through the Fourier transform [21]. Based on the Parseval theorem, the calculation of coupling efficiency by converting the incident light field distribution to the end face of optical fiber through Fourier transform is equal to the convert the electromagnetic field distribution of optical fiber end face to the end face of lens through Fourier transform [22]. Therefore, the integration represented by Eq. (3.2) can be performed at the receiving aperture plane as follows:

$$\eta_A = \frac{\left|\int\int U_{i,A}^*(r)U_{f,A}(r)rdrd\theta\right|^2}{\int\int U_{i,A}(r)U_{i,A}^*(r)rdrd\theta \times \int\int U_{f,A}(r)U_{f,A}^*(r)rdrd\theta} \tag{3.6}$$

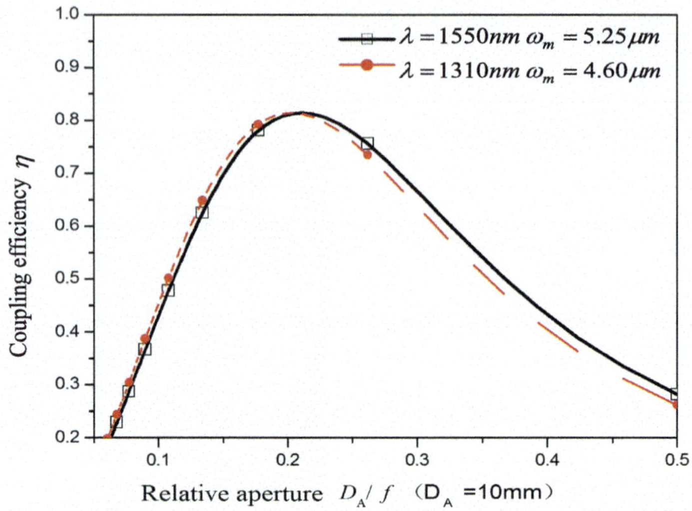

Fig. 3.5 Relationship curve between relative aperture and coupling efficiency [20]

where $U_{i,A}(r)$ is the optical field distribution on the plane of the receiving lens and $U_{f,A}(r)$ is the electromagnetic field distribution of the single-mode fiber on the plane of the receiving lens. The cross section of the optical fiber is assumed to be perpendicular to the incident light field and located at the center of the focal plane of the lens. Because the Fourier transform of the Gaussian function is still a Gaussian function, the normalized mode-field distribution of the optical fiber is converted into the mode-field distribution $U_{f,A}(r)$ of the receiving lens surface, which is expressed as follows [23, 24]:

$$U_{f,A}(r) = \frac{kW_m}{\sqrt{2\pi}f}\exp\left[-\left(\frac{kW_m}{2f}\right)^2 r^2\right] \tag{3.7}$$

where $k = 2\pi/\lambda$ is the wave vector, W_m is the radius of the fiber mode field, f is the focal length of the lens, and r is the distance from any radial point on the surface of the lens to the center of the lens. Assuming a completely coherent monochromatic plane light incident wave, that is, $U_{i,A} = 1$, the coupling efficiency (Eq. 3.6) can be expressed as follows:

$$\eta_A = \frac{\left|\int_0^{2\pi}\int_0^{D_A/2}\frac{kW_m}{\sqrt{2\pi}f}\exp\left[-\left(\frac{kW_m}{2f}\right)^2 r^2\right] r dr d\theta\right|^2}{\int_0^{2\pi}\int_0^{D_A/2} r dr d\theta} \tag{3.8}$$

Let $a = \pi W_m D_A/(2\lambda f)$ be the coupling parameter, where D_A is the diameter of the receiving aperture. Therefore, Formula (3.8) can be simplified as

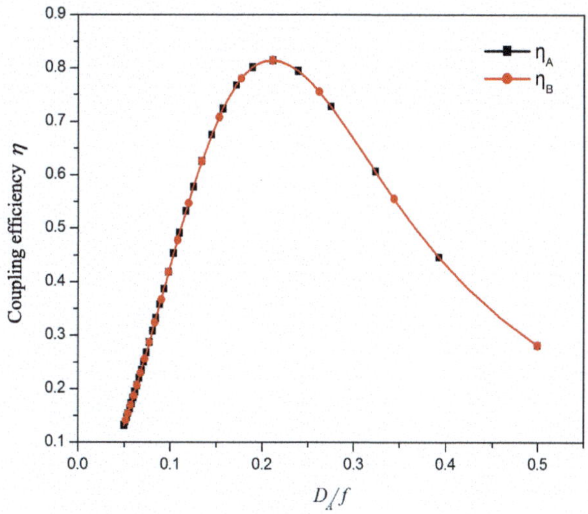

Fig. 3.6 Coupling efficiency curve at the fiber end (η_B) and the receiving lens end (η_A) [20]

$$\eta_A = \frac{2\left[1 - \exp(-a^2)\right]^2}{a^2} \tag{3.9}$$

The coupling efficiency expression (Eq. 3.9) is simpler and more intuitive than Eq. (3.5). It can be seen from Eq. (3.9) that, under ideal conditions, the coupling efficiency η_A is only a function of the coupling parameter a.

Figure 3.6 depicts the coupling efficiency curve calculated using Eqs. (3.5) and (3.9), respectively, when $\lambda = 1550$ nm and $W_m = 5.25$ μm. The coupling efficiency calculated using the two methods is identical, and the expression for calculating it on the lens plane is simple. When $D_A/f = 0.211$, $a = \pi W_m D_A / (2\pi f) \approx 1.12$ and the maximum coupling efficiency is 81.4%.

3.1.2 Gaussian Beam Coupling

The distance of laser transmission in space is relatively short. When the laser beam emitted by the semiconductor laser is directly coupled into the fiber, it is still represented by a Gaussian distribution [25, 26]. In this section, we discuss mainly the coupling efficiency of Gaussian beams.

As depicted in Fig. 3.7, assuming that the collimating Gaussian beam is incident perpendicular to the lens, the electromagnetic field distribution of the Gaussian light field on the end face of the lens is expressed as follows [27, 28]:

$$U_{G,i}(\boldsymbol{r}) = \exp(-\frac{\boldsymbol{r}^2}{W_s^2}) \tag{3.10}$$

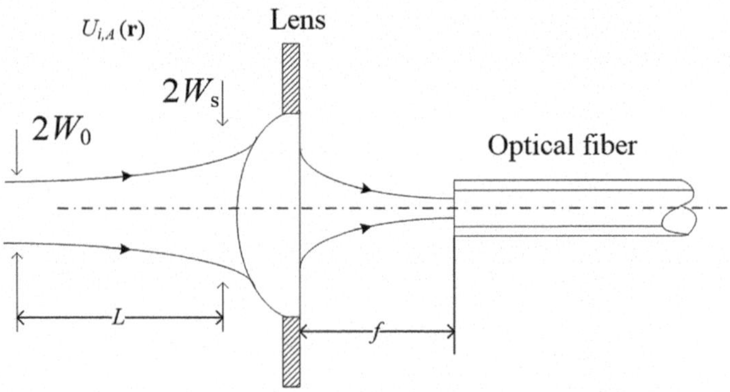

Fig. 3.7 Schematic of collimating a Gaussian beam coupled through a single lens into a single-mode fiber

where W_s is the radius of the lumbar spot of the Gaussian light field on the end face of the lens. The electromagnetic field distribution of the optical fiber is converted into a distribution at the surface of the receiving lens to calculate the coupling efficiency. By substituting Eqs. (3.10) and (3.7) in Eq. (3.6), the coupling efficiency of the Gaussian beam coupling through a single lens into a single-mode fiber can be expressed as follows:

$$\eta_G = \frac{\left|\int_0^{2\pi}\int_0^{D_A/2}\exp(\frac{r^2}{\omega_s^2})\frac{kW_m}{\sqrt{2\pi}f}\exp\left[-(\frac{kW_m}{2f})^2r^2\right]rdrd\theta\right|^2}{\int_0^{2\pi}\int_0^{D_A/2}\exp(\frac{2r^2}{\omega_s^2})rdrd\theta} \tag{3.11}$$

The integral in Formula (3.11) can be expressed as follows:

$$\eta_G = \frac{\left[\exp\left(\frac{D_A^2}{4W_s^2} - a^2\right) - 1\right]^2}{\left(\frac{f}{kW_mW_s} - \frac{kW_mW_s}{4f}\right)\left[\exp\left(\frac{D_A^2}{2W_s^2}\right) - 1\right]} \tag{3.12}$$

where $a = \pi W_m D_A/(2\lambda f)$ is the coupling parameter, D_A is the diameter of the lens, f is the focal length of the lens, and W_m is the mode-field radius of the optical fiber. As depicted in Fig. 3.8, Eq. (3.12) is used to simulate the coupling efficiency of a spatial collimating Gaussian beam coupled into a single-mode fiber through a single lens. To obtain the coupling efficiency curve for various values of the relative aperture, we consider $\lambda = 1550$ nm, $W_m = 5.25$ μm, and $D_A = 10$ mm and vary the focal length of the lens. The square curve depicted in Fig. 3.8 represents the coupling efficiency curve of the incident plane beam, and the others curves similarly represent Gaussian beams with various spot radii. It can be seen that the larger the radius of the beam waist, W_s, of the incident Gaussian beam, the higher the coupling efficiency and the closer it is to the coupling efficiency of the incident plane light. As the radius, the coupling efficiency also decreases gradually.

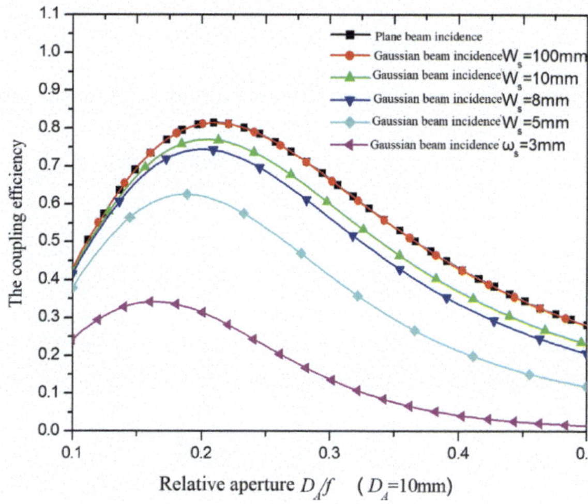

Fig. 3.8 Curves representing changes in the Gaussian beam coupling efficiency with respect to the relative aperture of the lens [20]

3.2 Spatial Plane Wave-Lens-Single Mode Fiber Coupling Under Weakly Turbulent Atmosphere

3.2.1 Light Field Distribution and Refractive Index Power Spectrum Under Atmospheric Turbulence

The diffraction, attenuation, atmospheric turbulence, and thermal blooming effects of the atmospheric channel affect the signal beam, and the impact mechanisms are different. The random characteristics of the light field distribution of the laser beam under atmospheric turbulence and its statistical description method are discussed in this section.

3.2.1.1 Rytov Approximation of Distribution of Light Field Under Atmospheric Turbulence

Weak turbulence affects the phase of the light wave, and the electric field distribution in this case can be expressed as follows [29]:

$$U(\boldsymbol{R}) \equiv U(\boldsymbol{r}, L) = U_0(\boldsymbol{r}, L)\exp([\psi(\boldsymbol{r}, L)] \tag{3.13}$$

where $\boldsymbol{r}$ is the radial vector on the observation plane, L is the transmission distance, and $\psi(\boldsymbol{r}, L)$ is the total complex phase disturbance at point $\boldsymbol{r}$. The objective of the Rytov approximation is to express the phase disturbances as the sum of nth-order complex phase disturbances as follows [29]:

$$\psi(\boldsymbol{r}, L) = \psi_1(\boldsymbol{r}, L) + \psi_2(\boldsymbol{r}, L) + \cdots \tag{3.14}$$

where $\psi_1(\boldsymbol{r}, L)$ and $\psi_2(\boldsymbol{r}, L)$ represent the first-order and second-order phase disturbances, respectively. The results of the Born perturbation are used to simplify the calculation. The normalized Born disturbance of order m is defined as follows:

$$\Phi(\boldsymbol{r}, L) = \frac{U_m(\boldsymbol{r}, L)}{U_0(\boldsymbol{r}, L)} \quad \mathrm{m} = 1, 2, 3 \ldots \tag{3.15}$$

where $U_m(\boldsymbol{r}, L)$ represents the mth-order scattering field distribution in the Born perturbation method. Let the Rytov first-order disturbance be equal to the Born first-order disturbance; therefore,

$$\begin{aligned} U_0(\mathbf{r}, L)\exp[\psi_1(\mathbf{r}, L)] &= U_0(\mathbf{r}, L) + U_1(\mathbf{r}, L) \\ &= U_0(\mathbf{r}, L)[1 + \Phi_{u,1}(\mathbf{r}, L)] \end{aligned} \tag{3.16}$$

Dividing both sides of Eq. (3.16) by $U_0(\mathbf{r}, L)$ and calculating the natural logarithm, we obtain the following expression:

$$\psi_1(\mathbf{r}, L) = \ln[1 + \Phi_{u,1}(\mathbf{r}, L)] \cong \Phi_{u,1}(\mathbf{r}, L) \quad \left|\Phi_{u,1}(\mathbf{r}, L)\right| \ll 1 \tag{3.17}$$

Similarly, the Rytov second-order disturbance is equivalent to the Born second-order disturbance; therefore,

$$\psi_2(\mathbf{r}, L) = \Phi_{u,2}(\mathbf{r}, L) - \frac{1}{2}\Phi^2_{u,1}(\mathbf{r}, L) \quad \left|\Phi_{u,1}(\mathbf{r}, L)\right| \ll 1 \quad \left|\Phi_{u,2}(\mathbf{r}, L)\right| \ll 1 \tag{3.18}$$

Three basic second moments are defined and associated with the refractive index power spectrum $\Phi_n(\kappa, z)$:

$$\begin{aligned} E_1(0,0) &= \langle\Phi_{u,2}(\mathbf{r}, L)\rangle = \langle\psi_2(\mathbf{r}, L)\rangle + \frac{1}{2}\langle\psi_1^2(\mathbf{r}, L)\rangle \\ &= 4\pi^2k^2 \int_0^L dz \int_0^\infty d\kappa\, \Phi_n(\kappa, z) J_0(\kappa|\gamma r_1 - \gamma^* r_2|) \exp[-\frac{i\kappa^2}{2k}(\gamma - \gamma^*)(L - z)] \\ &= 4\pi^2k^2 \int_0^L \int_0^\infty \kappa\Phi_n(\kappa, z) J_0(\kappa|[1 - \overline{\Theta}(1 - z/L)]\mathbf{p} - 2i\Lambda(1 - z/L)\mathbf{r}|) \\ &\quad \times \exp[-\frac{\Lambda L\kappa^2(1 - z/L)^2}{k}] d\kappa\, dz \end{aligned} \tag{3.19}$$

$$E_2(\mathbf{r}_1,\mathbf{r}_2)=\left\langle \Phi_{u,1}(\mathbf{r}_1,L)\Phi^*_{u,1}(\mathbf{r}_2,L)\right\rangle=\left\langle \psi_1(\mathbf{r}_1,L)\psi_1^*(\mathbf{r}_2,L)\right\rangle$$

$$=4\pi^2k^2\int_0^L dz\int_0^\infty d\kappa\,\Phi_n(\kappa,z)J_0(\kappa|\gamma r_1-\gamma^* r_2|)$$

$$\exp[-\frac{i\kappa^2}{2k}(\gamma-\gamma^*)(L-z)]$$

$$=4\pi^2k^2\int_0^L\int_0^\infty \kappa\Phi_n(\kappa,z)J_0(\kappa|[1-\overline{\Theta}(1-z/L)]\mathbf{p}-2i\Lambda(1-z/L)\mathbf{r}|)$$

$$\times\exp[-\frac{\Lambda L\kappa^2(1-z/L)^2}{k}]d\kappa dz \tag{3.20}$$

$$E_3(\mathbf{r}_1,\mathbf{r}_2)=\left\langle \Phi_{u,1}(\mathbf{r}_1,L)\Phi_{u,1}(\mathbf{r}_2,L)\right\rangle=\langle \psi_1(\mathbf{r}_1,L)\psi_1(\mathbf{r}_2,L)\rangle$$

$$=-4\pi^2k^2\int_0^L dz\int_0^\infty d\kappa\,\Phi_n(\kappa,z)J_0(\gamma\kappa|r_1-r_2|)\exp[-\frac{i\kappa^2\gamma}{k}(L-z)]$$

$$=-4\pi^2k^2\int_0^L\int_0^\infty \kappa\Phi_n(\kappa,z)J_0\left\{\kappa\rho[1-(\overline{\Theta}+i\Lambda)(1-z/L)]\right\}$$

$$\times\exp[-\frac{\Lambda L\kappa^2(1-z/L)^2}{k}]$$

$$\exp\left\{-\frac{iL\kappa^2}{k}(1-z/L)[1-\overline{\Theta}(1-z/L)]\right\}d\kappa dz \tag{3.21}$$

where $J_0(\cdot)$ represents the zeroth order Bessel function,

$$\begin{array}{ll}\mathbf{r}=\frac{1}{2}(\mathbf{r}_1+\mathbf{r}_2), & \mathbf{p}=\mathbf{r}_1-\mathbf{r}_2\\ r=|\mathbf{r}|, & \rho=|\mathbf{p}|\end{array} \tag{3.22}$$

$$\gamma=1-(\overline{\Theta}+i\Lambda)(1-z/L),\quad 0\le z\le L \tag{3.23}$$

In Eq. (3.23), $\overline{\Theta}$ and Λ are related parameters describing the Gaussian beam on the receiving plane. They have the following relationship with the parameters Θ_0 and Λ_0 of the Gaussian beam on the incident plane (F_0 and w_0 are the radii of curvature of the Gaussian beam on the incident plane and the optical waist radius on the incident plane, respectively, as depicted in Fig. 3.9):

Gaussian beam parameters on the input plane:

$$\Theta_0=1-\frac{L}{F_0},\quad \Lambda_0=\frac{2L}{kw_0^2} \tag{3.24}$$

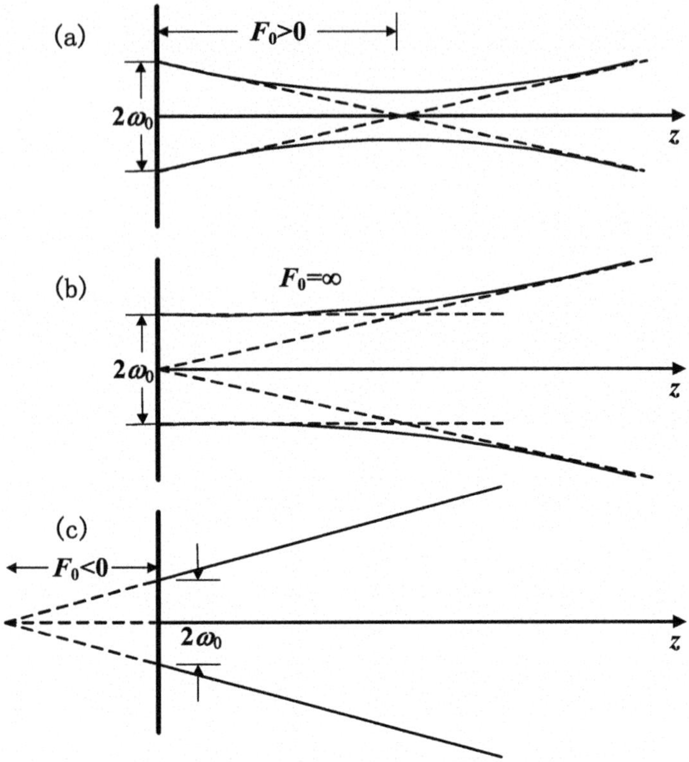

Fig. 3.9 Beam models with different wavefront curvature radiuses [30]: **a** focusing beams, **b** collimating beams, **c** divergent beams

Gaussian beam parameters on the receiving plane:

$$\Theta = \frac{\Theta_0}{\Theta_0^2 + \Lambda_0^2} = 1 + \frac{L}{F}, \quad \overline{\Theta} = 1 - \Theta = -\frac{L}{F}$$
$$\Lambda = \frac{\Lambda_0}{\Theta_0^2 + \Lambda_0^2} = \frac{2L}{kw^2} \tag{3.25}$$

The radius of curvature of the Gaussian beam and the optical waist radius on the receiving plane are represented by F and w, respectively. Special cases: for plane light waves, $\overline{\Theta} = \Lambda = 0$; for spherical light waves, $\overline{\Theta} = 1$, $\Lambda = 0$. In general, ensemble averaging of the first-, second-, and fourth-order moments of the complex phase perturbation is often used to calculate the propagation properties of light in random media. Hence, Eq. (3.14) can be expressed as follows:

First-order moment:

$$\langle \exp[\psi(\mathbf{r}, L)] \rangle = \langle \exp[\psi_1(\mathbf{r}, L) + \psi_2(\mathbf{r}, L)] \rangle \tag{3.26}$$

Second-order moment:

$$\left\langle \exp[\psi(\mathbf{r}_1, L) + \psi^*(\mathbf{r}_2, L)] \right\rangle = \langle \exp[\psi_1(\mathbf{r}_1, L) + \psi_2(\mathbf{r}_1, L) + \psi_1^*(\mathbf{r}_2, L) + \psi_2^*(\mathbf{r}_2, L)] \rangle \tag{3.27}$$

Fourth-order moment:

$$\begin{aligned} &\left\langle \exp[\psi(\mathbf{r}_1, L) + \psi^*(\mathbf{r}_2, L) + \psi(\mathbf{r}_3, L) + \psi^*(\mathbf{r}_4, L)] \right\rangle \\ &= \langle \exp[\psi_1(\mathbf{r}_1, L) + \psi_2(\mathbf{r}_1, L) \\ &+ \psi_1^*(\mathbf{r}_2, L) + \psi_2^*(\mathbf{r}_2, L) \\ &+ \psi_1(r_3, L) + \psi_2(\mathbf{r}_3, L) + \psi_1^*(\mathbf{r}_4, L) + \psi_2^*(\mathbf{r}_4, L)] \rangle \end{aligned} \tag{3.28}$$

$\psi(\mathbf{r}, L)$ fulfils the conditions for the Gaussian distribution. The cumulant method [31] can be used to calculate the three basic moments; only the first and second orders are considered here:

$$\begin{aligned} \langle \exp[\psi(\mathbf{r}, L)] \rangle &= \lim_{t \to -i} \langle \exp(it\psi(\mathbf{r}, L)) \rangle \\ &= \exp\left(K_1 + \frac{1}{2} K_2 + \cdots \right) \end{aligned} \tag{3.29}$$

where

$$\begin{aligned} K_1 &= \langle \psi(\mathbf{r}, L) \rangle \\ K_2 &= \left\langle \psi^2(\mathbf{r}, L) \right\rangle - \langle \psi(\mathbf{r}, L) \rangle^2 \end{aligned} \tag{3.30}$$

Equations (3.29) and (3.30) are used to evaluate Eqs. (3.26)–(3.28). Thereafter, the ensemble average of the first-, second-, and fourth-order moments of the complex phase disturbance of the optical field can be related to the three basic second moments represented by Eqs. (3.19)–(3.21). A further functional relationship can be established using the refractive index spectral density function, which is convenient for the subsequent calculation of coupling efficiency and relative variance in coupling power.

First-order moment:

$$\langle \exp[\psi(\mathbf{r}, L)] \rangle = \exp[E_1(0, 0)] \tag{3.31}$$

Second-order moment:

$$\left\langle \exp[\psi(\mathbf{r_1}, L) + \psi^*(\mathbf{r_2}, L)] \right\rangle = \exp[2E_1(0, 0) + E_2(\mathbf{r_1}, \mathbf{r_2})] \tag{3.32}$$

Fourth-order moment:

$$\begin{aligned}&\left\langle \exp[\psi(\mathbf{r_1}, L) + \psi^*(\mathbf{r_2}, L) + \psi(\mathbf{r_3}, L) + \psi^*(\mathbf{r_4}, L)]\right\rangle \\ &= \exp[4E_1(0, 0) + E_2(\mathbf{r_1}, \mathbf{r_2}) \\ &+ E_2(\mathbf{r_1}, \mathbf{r_4}) + E_2(\mathbf{r_3}, \mathbf{r_2}) \\ &+ E_2(\mathbf{r_3}, \mathbf{r_4}) + E_3(\mathbf{r_1}, \mathbf{r_3}) + E_3^*(\mathbf{r_2}, \mathbf{r_4})]\end{aligned} \tag{3.33}$$

where $E_1(0, 0)$, $E_2(\mathbf{r}_1, \mathbf{r_2})$, and $E_3(\mathbf{r}_1, \mathbf{r_3})$ are represented by Eqs. (3.19), (3.20), and (3.21), respectively.

3.2.1.2 Refractive Index Power Spectrum Model

Atmospheric turbulence can be considered a vortex hierarchy system. The movement of the wind transfers energy to a large-scale whirlpool, and the large-scale whirlpool splits into smaller ones. Energy is transferred from the large-scale whirlpool to the smaller ones and gradually dissipated.

The power spectrum of the atmospheric refractive index random field $n_1(\mathrm{R})$ is defined as the three-dimensional Fourier transform of its cross-correlation function, denoted as $\Phi_n(k)$:

$$\Phi_n(k) = \frac{1}{(2\pi)^3} \int\int\int_{-\infty}^{\infty} B(r)\exp(-ik \cdot r)dr \tag{3.34}$$

where $k = (k_x, k_y, k_z)$ is the space wave vector. If $n_1(\mathrm{R})$ represents an isotropic random field, then its power spectrum is a one-dimensional power spectrum:

$$\Phi_n(k) = \frac{1}{2\pi^2 k} \int_{-\infty}^{\infty} B(r)r\sin(kr)dr \tag{3.35}$$

A variety of atmospheric refractive index power spectrum models, referred to as turbulence spectrum models, have been proposed to accurately describe the distribution of random changes caused by turbulence in the refractive index.

1. **Kolmogorov turbulence spectrum**

The Kolmogorov turbulence hypothesis and refractive index fluctuation structure function conform to the "two-thirds" law. The Kolmogorov turbulence spectrum for the fluctuation of refractive index in the inertia region can be obtained as follows:

$$\Phi_n(k) = 0.033C_n^2 k^{-11/3} \quad (2\pi/L_0 \ll k \ll 2\pi/l_0) \tag{3.36}$$

This function is depicted in Fig. 3.10. In the figure, l represents the turbulence scale parameter. It should be noted that the turbulence spectrum represented by Formula (3.36) is only a spectral model of the inertia region. The Kolmogorov turbulence

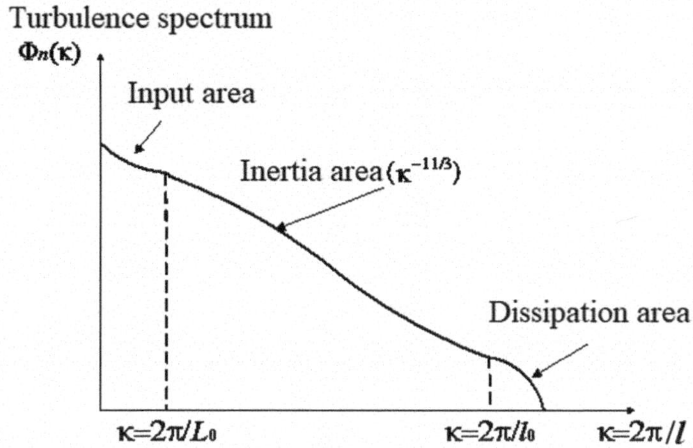

Fig. 3.10 Kolmogorov spectrum model of turbulence [32]

spectrum does not describe the input and dissipation regions mathematically because these are usually anisotropic and consume energy rapidly.

2. **Tatarskii turbulence spectrum**

In dissipation zone, $k > k_m$, the energy dissipation exceeds the kinetic energy, and the energy is very small. Therefore, turbulence spectrum $\Phi_n(k)$ decreases rapidly, where $k_m \approx 2\pi / l_0$ and l_0 are the internal scales of turbulence, and the turbulence in this region can be summarized using the Tatarskii turbulence spectrum, as follows [33]:

$$\Phi_n(k) = 0.033 C_n^2 k^{-11/3} \exp(-k^2/k_m^2) \quad k \gg 1/L_0 \tag{3.37}$$

The Tatarskii turbulence spectrum considers the internal scale of turbulence, and an attenuation factor attributed to the Gaussian function in the region with a large wave number k addresses the deficiency in the Kolmogorov turbulence spectrum. However, these two spectra still do not describe the region with small wave number k, and, in practice, these two spectra are still turbulent spectra with certain limitations.

3. **von Karman turbulence spectrum**

In 1948, von Karman [34] proposed a turbulence spectrum that addresses the problem that when the wave number k is smaller ($k \to 0$), $\Phi_n(k)$ cannot be arbitrarily large. The von Karman turbulence spectrum model can be expressed as follows:

$$\Phi_n(k) = 0.033 C_n^2 (k^2 + k_0^2)^{-11/6} \tag{3.38}$$

where $k_0 = 2\pi / L_0$ and L_0 are the outer scales of turbulence. The inner scale of turbulence is also introduced into this model. Thus, the modified von Karman turbulence spectrum is proposed, which can be expressed as follows:

$$\Phi_n(k) = 0.033C_n^2 \frac{\exp(-k^2/k_m^2)}{(k^2 + k_0^2)^{-11/6}} \quad 0 < k < \infty \quad k_0 = \frac{2\pi}{L_0} \quad k_m = \frac{5.92}{l_0} \tag{3.39}$$

When the inner scale is l_0 and the outer scale is $L_0 = \infty$, the modified von Karman turbulence spectrum can be reduced to the Kolmogorov turbulence spectrum. When compared with the Kolmogorov and Tatarskii turbulence spectra, the modified von Karman turbulence spectrum model is most widely used in the theoretical study of atmospheric turbulence. It can best reflect the real turbulence condition, addressing the problem of singularities at the $k = 0$ points in the Kolmogorov integration.

3.2.2 Lens Coupling Under Atmospheric Turbulence

As depicted in Fig. 3.11, the laser propagates along a distance L through the collimating beam of the optical system under atmospheric turbulence and is incident on the surface of the receiving aperture A. The receiving system converges and shapes the distorted beam and couples it into the fiber.

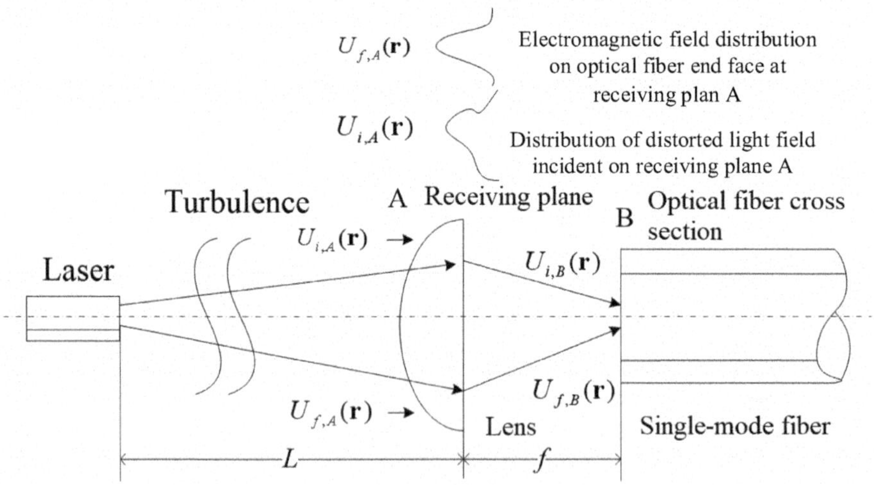

Fig. 3.11 Spatial light–single-lens–single-mode fiber coupling

3.2.2.1 Coupling Efficiency Model Under Kolmogorov Turbulence Spectrum

Owing to the random fluctuation of the atmospheric refractive index, the coupling efficiency changes randomly under turbulence, which can be described using a statistical method. It is assumed that the incident beam is a monochromatic unit plane wave, and the fiber cross section is perpendicular to the direction of propagation of the incident light field and located at the center of the optical axis of the coupling system. Using the cross-correlation function $\Gamma_2(r_1, r_2, L)$ of the incident light field, the turbulence-free coupling efficiency η_A calculated on the lens plane using Eq. (3.2) can be extended to the ensemble average of the coupling efficiency $\langle\eta_A\rangle$ under atmospheric turbulence:

$$\langle\eta_A\rangle = \frac{1}{\pi D_A^2/4}\iint_A \Gamma_2(r_1, r_2)U_{f,A}^*(r_1)U_{f,A}(r_2)dr_1dr_2 \tag{3.40}$$

where $U_{f,A}(r)$ is the electromagnetic field distribution on the end face of the optical fiber converted to a distribution on the receiving plane A (as represented by Eq. (3.7)), D_A is the diameter of the coupling lens, r_1 and r_2 are the radial vectors from the center of the receiving plane to the receiving plane.

If the plane wave is transmitted along a distance L on the optical axis, the distribution of the light field on the receiving plane is represented by $U(r, L)$, r, which denotes the radial vector in the receiving plane. The second moment $\Gamma_2(\mathbf{r}_1, \mathbf{r}_2, L)$ of the light field on the receiving plane is defined by Eq. (3.41), which represents the spatial coherence of the coherent light field, also known as the mutual coherence function (MCF):

$$\begin{aligned}\Gamma_2(\mathbf{r}_1, \mathbf{r}_2, L) &= \langle U(\mathbf{r}_1, L)U^*(\mathbf{r}_2, L)\rangle \\ &= U_0(\mathbf{r}_1, L)U_0^*(\mathbf{r}_2, L)\langle \exp[\psi(\mathbf{r}_1, \mathrm{L}) + \psi^*(\mathbf{r}_2, \mathrm{L})\rangle \\ &= U_0(\mathbf{r}_1, L)U_0^*(\mathbf{r}_2, L)\exp[2E_1(0,0) + E_2(\mathbf{r}_1, \mathbf{r}_2)]\end{aligned} \tag{3.41}$$

Assuming a unit monochromatic incident plane wave, $U_0(\mathbf{r}_1, L) = U_0^*(\mathbf{r}_2, L) = 1$ and $\overline{\Theta} = \Lambda = 0$. Using Eqs. (3.19) and (3.20), Eq. (3.41) can be simplified as follows:

$$\Gamma_2(\rho, L) = \exp\{-4\pi^2 kL\int_0^\infty k\Phi_n(k)[1 - J_0(k\rho)]dk\} \tag{3.42}$$

where $\rho = |\mathbf{r}_1 - \mathbf{r}_2|$ and $\Phi_n(k)$ represents the turbulence spectrum models. Equation (3.38) is substituted in Eq. (3.42) and simplified to obtain the MCF ($\Gamma_{2,k}(\rho, L)$) for the Kolmogorov turbulence spectrum:

$$\Gamma_{2,k}(\rho, L) = \exp(1.46C_n^2k^2L\rho^{5/3}) \quad l_0 \ll \rho \ll L_0 \tag{3.43}$$

By substituting Eqs. (3.43) and (3.7) in Eq. (3.40), the plane vector distance can be expanded as $\rho^2 = |\mathbf{r_1} - \mathbf{r_2}|^2 = r_1^2 + r_2^2 - 2r_1r_2\cos(\theta_1 - \theta_2)$ using the cosine theorem and the vector integral can be converted into a scalar integral. Finally, the coupling efficiency $\langle\eta_{A,K}\rangle$ for the Kolmogorov turbulence spectrum is expressed as follows:

$$\langle\eta_{A,K}\rangle = \frac{8a^2}{\pi}\int_0^1\int_0^1 \exp[-a^2(x_1^2 + x_2^2)] \times F\left(\frac{A_R}{A_{C,K}}(x_1^1 + x_2^2), \frac{2x_1x_2}{x_1^2 + x_2^2}\right)x_1x_2dx_1dx_2 \tag{3.44}$$

where $a = \pi W_m D_A/(2\lambda f)$ is the coupling parameter, $A_R = \pi D_A^2/4$ is the aperture area of the lens, $A_{C,K} = \pi\rho_c^2$ is the spatial coherent region of the incident light field (also known as the speckle size), $\rho_c = \left(1.46C_n^2k^2L\right)^{-3/5}$ is the coherence radius, A_R/A_C is the number of speckles in the receiving aperture, $x_1 = 2r_1/D_A$ and $x_2 = 2r_2/D_A$, D_A is the lens diameter, and F is a function with two parameters, expressed as follows:

$$F(v, u) = \int_0^\pi \exp\{-v^{5/6}[1 - u\cos\theta]^{5/6}\}d\theta \tag{3.45}$$

If the MCF of the Kolmogorov turbulence spectrum plane light incident on the light field is represented by a Gaussian approximation:

$$\Gamma'_{2,K}(r_1, r_2) = \exp(-|r_1 - r_2|^2/\rho_c^2) \tag{3.46}$$

Then the coupling efficiency can be simplified as follows:

$$\langle\eta'_{A,K}\rangle = 8a^2\int_0^1\int_0^1 \exp\left[-\left(a^2 + \frac{A_R}{A_{C,K}}\right)(x_1^2 + x_2^2)\right] \times I_0\left(2\frac{A_R}{A_C}x_1x_2\right)x_1x_2dx_1dx_2 \tag{3.47}$$

where I_0 is the modified Bessel function of the first class of zeroth order and the other parameters are consistent with Eq. (3.44). By comparing Eqs. (3.44) and (3.47), the Gaussian approximation can be used to simplify the expression for coupling efficiency from the triple product to the double integral, thus simplifying the calculation.

3.2.2.2 Coupling Efficiency Model for Von Karman Turbulence Spectrum

The incident light MCF is extended to the von Karman turbulence spectrum, and Eq. (3.39) is substituted in Eq. (3.42) using the first type of Bezier function expansion and changing the order of integration and summation. Thereafter, the MCF ($\Gamma_{2,V}(\rho, L)$) of the plane wave under the von Karman turbulence spectrum can be expressed as follows:

$$\Gamma_{2,V}(\rho, L) = \exp\left\{-4\pi^2 k^2 L(0.033)C_n^2 \sum_{n-1}^{\infty} \frac{(-1)^{n-1}}{2^{2n}(n!)^2} \int_0^{\infty} \kappa^{2n+1} \exp\left(-\frac{\kappa^2}{\kappa_m^2}\right)\left(\kappa^2+\kappa_0^2\right)^{-11/6} d\kappa\right\} \quad (3.48)$$

As represented by Eq. (3.49), using the integral form of the confluent hypergeometric function of the second kind [28], Eq. (3.65) can be simplified to the form represented by Eq. (3.50) as follows [28]:

$$U(a;c;x) = (1/\Gamma(a)) \int_0^{\infty} t^{a-1} e^{-xt} (1+t)^{c-a-1} dt \quad a > 0,\ \ \mathrm{Re}(z) > 0 \quad (3.49)$$

$$\Gamma_{2,V}(\rho, L) = \exp\left\{-2\pi^2 k^2 L(0.033)C_n^2 \kappa^{-5/3} \sum_{n=1}^{\infty} \frac{(-1)^{n-1}(\rho\kappa_0)^{2n}}{2^{2n} n!} U\left(\mathrm{n}+1; n+\frac{1}{6}; \frac{\kappa_0^2}{\kappa_m^2}\right)\right\} \quad (3.50)$$

When $k_0^2/k_m^2 \ll 1$, that is, $l_0^2/L_0^2 \ll 1$, Eq. (3.50) can be simplified using the second kind of confluent hypergeometric function approximation, as represented by Eqs. (3.51) and (3.52):

$$U(a;c;z) = \frac{\Gamma(1-c)}{\Gamma(1+a-c)} + \frac{\Gamma(c-1)}{\Gamma(a)} z^{1-c} \quad |z| \ll 1 \quad (3.51)$$

$$F_1(a,c,z) = \sum_{n=0}^{\infty} \frac{\frac{\Gamma(a+n)}{\Gamma(a)}}{\frac{\Gamma(c+n)}{\Gamma(c)}} \frac{z^n}{n!} \quad (3.52)$$

where $\Gamma(\cdot)$ is the gamma function. Therefore, Eq. (3.50) can be further simplified as follows:

$$\Gamma_{2,V}(\rho, L) = \exp\left\{-2\pi^2 k^2 L(0.033)C_n^2 \left\{\Gamma(-5/6)\kappa_m^{-5/3}\left[1 - \mathrm{F}_1\left(-\frac{5}{6}; 1; -\frac{\kappa_m^2 \rho^2}{4}\right)\right] - \frac{9}{5}\kappa_0^{1/3}\rho^2\right\}\right\} \quad (3.53)$$

Using the approximate and difference formulas, the MCF of a plane light wave at an arbitrary radial distance ρ under the modified von Karman spectrum is expressed in the range of error less than 2% as follows [35]:

$$\Gamma_{2,V}(\rho, L) = \exp\left(-1.64C_n^2k^2Ll_0^{-1/3}\rho^2\left\{\left[1+2.03(\rho/l_0)^2\right]^{-1/6} - 1.32(l_0/L_0)^{1/3}\right\}\right) \tag{3.54}$$

The MCF under the von Karman turbulence spectrum (Eq. 3.54) and the electromagnetic field distribution expression on the end face of the single-mode fiber converted to a distribution on the receiving plane A (Eq. 3.7) are substituted in Eq. (3.40). Therefore, the coupling efficiency $\langle\eta_{A,V}\rangle$ under the von Karman turbulence spectrum can be simplified as follows:

$$\langle\eta_{A,V}\rangle = \frac{8a^2}{\pi}\int_0^1\int_0^1\int_0^\pi \exp\left[-a^2(x_1^2+x_2^2)\right] \exp\left(-\frac{A_R}{A_{C,V}}N(x_1,x_2,\theta)\left\{[1+GN(x_1,x_2,\theta)]^{-1/6} - C\right\}\right)x_1x_2dx_1dx_2 \tag{3.55}$$

where $N(x_1, x_2, \theta) = (x_1^2 + x_2^2 - 2x_1x_2\cos\theta)$, $G = 0.51D_A^2/l_0^2$, $C = 1.32(l_0/L_0)^{1/3}$, l_0 and L_0 represent the inner and outer scales of turbulence, respectively, $A_{C,V} = \pi\rho_s^2$, and $\rho_s = \left(1.64C_n^2k^2Ll_0^{-1/3}\right)^{-1/2}$ can only represent the spatial coherence radius to a certain extent and is not the exact expression for the coherence radius under the von Karman turbulence spectrum; the other parameters are consistent with Formula (3.44).

By comparing Eq. (3.9) with Eqs. (3.44), (3.47), and (3.55), it can be seen that when atmospheric turbulence is considered, the coupling efficiency is no longer a function of the coupling parameter A but a function of the ratio of the diameter of the receiving lens to the atmospheric coherence length (D_A/ρ_c). For the von Karman turbulence spectrum, the internal and external scales of turbulence also affect the coupling efficiency.

3.2.2.3 Coupling Efficiency of Slant Transmission Under Von Karman Turbulence Spectrum

For slant transmission, the Hufnagel–Valley model is the widely used model $C_n^2(h)$, as depicted in Fig. 3.12. Here, we introduce the concept of the zenith angle ψ, which is the acute angle formed between the connection between two communication terminals and the vertical surface of the receiving terminal. Therefore, the uplink ρ_s is expressed as follows:

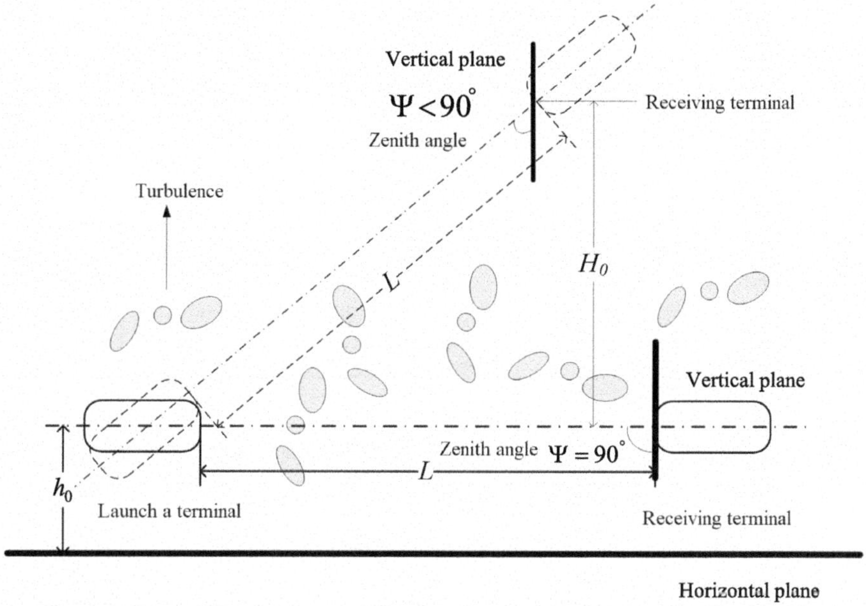

Fig. 3.12 Schematic of slant transmission [20]

$$\rho_s = \left(1.64k^2 l_0^{-1/3} \int_0^L C_n^2(h_0 + z\cos\psi)dz\right)^{-1/2} \tag{3.56}$$

In Fig. 3.12, h_0 indicates the distance between the lower terminal of the uplink and the horizontal plane. By substituting Eq. (3.56) into Eq. (3.55), the coupling efficiency of the two communication end machines with specific horizontal drops can be analyzed.

Figure 3.13 depicts the curves representing the communication distance and coupling efficiency at various zenith angles when the communication terminal has a certain horizontal drop (slope). These curves are obtained by substituting Eq. (3.56) in Eq. (3.55). Assuming that the lower end of the communication terminal is installed at a distance of 2 m from the horizontal plane H_0, it can be seen that the smaller the zenith angle, the higher the uplink coupling efficiency. This can be attributed to the fact that the smaller the zenith angle, the shorter the transmission distance of the beam in the strong turbulence area near the ground and the shorter the distance affected by the turbulence and, thus, the higher the coupling efficiency.

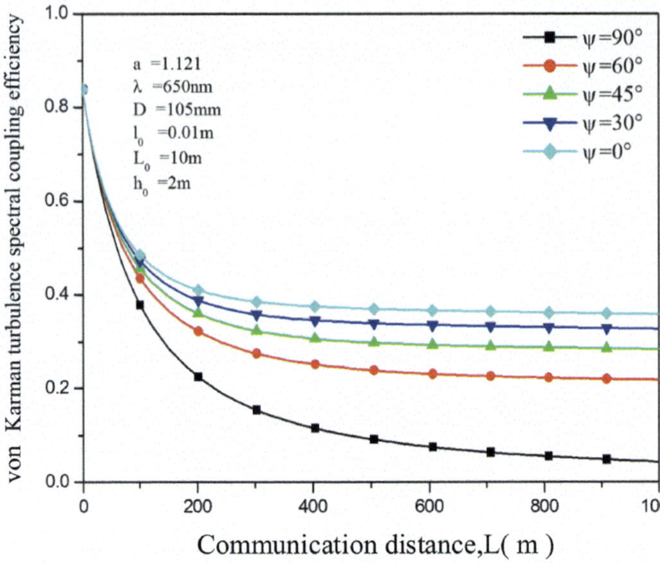

Fig. 3.13 Coupling efficiency curves of communication distance and von Karman turbulence spectrum at various zenith angles [20]

3.2.3 *Relative Variance in Fluctuation of Lens Coupled Optical Power Under Atmospheric Turbulence*

Under atmospheric turbulence, another significant parameter characterizing the performance of the coupling system is the relative variance in fluctuation of the optical power coupled into the optical fiber. The aim is to have as high a coupling efficiency and as stable an optical power as possible.

3.2.3.1 Relative Variance in Fluctuation of Lens–Single Mode Fiber Coupling Power Under Atmospheric Turbulence

The relative variance in fluctuation of optical power coupled into single-mode fiber is expressed as follows:

$$\sigma_{A,V}^2 = \langle P_{A,V}^2 \rangle / \langle P_{A,V} \rangle^2 - 1 \tag{3.57}$$

$$\langle P_{A,V} \rangle^2 = \left(\langle \eta_{A,V} \rangle \pi D_A^2 / 4 \right)^2 \tag{3.58}$$

where $\langle \eta_{A,V} \rangle$ is the ensemble average of the coupling efficiency under the von Karman turbulence spectrum (as represented by Eq. (3.55)). Therefore,

$$\left\langle P_{A,V}^2\right\rangle = \int\int\int\int_A \Gamma_{4,V}(r_1, r_2, r_3, r_4, L)U_{f,A}^*(r_1)U_{f,A}^*(r_2)$$
$$U_{f,A}^*(r_3)U_{f,A}^*(r_4)dr_1dr_2dr_3dr_4 \tag{3.59}$$

where $U_{f,A}(r)$ is the mode-field distribution in the optical fiber on the receiving lens surface (as represented by Eq. (3.7)) and $\Gamma_4(r_1, r_2, r_3, r_4, L)$ is the fourth moment of the monochromatic plane light field under atmospheric turbulence (as represented by Eq. (3.60)). The fourth moment is also known as the cross-coherence function (CCF), which represents the jitter in the optical power:

$$\begin{aligned}\Gamma_4(r_1, r_2, r_3, r_4, L) &= \left\langle U(r_1, L)U^*(r_2, L)U(r_3, L)U^*(r_4, L)\right\rangle \\ &= U_0(r_1, L)U_0^*(r_2, L)U_0(r_3, L)U_0^*(r_4, L) \\ &\quad \times \left\langle \exp\left[\psi(r_1, L) + \psi^*(r_2, L) + \psi(r_3, L) + \psi^*(r_4, L)\right]\right\rangle \\ &= \Gamma_2(r_1, r_2, L)\Gamma_2(r_3, r_4, L) \\ &\quad \times \exp\left[E_2(r_1, r_4) + E_2(r_3, r_2) + E_3(r_1, r_3) + E_3^*(r_2, r_4)\right]\end{aligned} \tag{3.60}$$

where $\Gamma_2(r_1, r_2, L)$ is the second moment of the light field, as represented by Eq. (3.42). Assuming a unit monochromatic incident plane wave, $U_0(r_1, L) = U_0^*(r_2, L) = U_0(r_3, L) = U_0^*(r_4, L) = 1$ and $\overline{\Theta} = \Lambda = 0$. Therefore, $E_2(r_1, r_2)$ and $E_3(r_1, r_2)$ can be simplified as follows:

$$E_2(r_1, r_2) = 4\pi^2k^2L\int_0^\infty \kappa\Phi_n(\kappa)J_0(\kappa\rho_{1,2})d\kappa \tag{3.61}$$

$$\begin{aligned}E_3(r_1, r_2) &= -4\pi^2k^2L\int_0^1\int_0^\infty \kappa\Phi_n(\kappa)J_0(\kappa\rho_{1,2})\exp\left(-\frac{i\kappa^2L\xi}{k}\right)d\kappa d\xi \\ &= 4\pi^2k^2L\int_0^\infty \kappa\Phi_n(\kappa)J_0(\kappa\rho_{1,2})\frac{k}{i\kappa^2L}\left[\exp\left(-\frac{i\kappa^2L\xi}{k} - 1\right)\right]d\kappa\end{aligned} \tag{3.62}$$

Thereafter, the von Karman turbulence spectrum is substituted and denoted as $\Gamma_{4,V}(r_1, r_2, r_3, r_4, L)$. The vector integrals are transformed into scalar integrals and simplified. Thus, Eq. (3.57) can be expressed as follows:

$$\sigma_{A,V}^2 = \frac{\int_0^1\int_0^1\int_0^1\int_0^1\int_0^{2\pi}\int_0^{2\pi}\int_0^{2\pi}\int_0^{2\pi}\exp\left[-a^2\left(x_1^2+x_2^2+x_3^2+x_4^2\right)\right]\times\Gamma_{4,V}(x_1, x_2, x_3, x_4, L)x_1x_2x_3x_4dx_1dx_2dx_3dx_4d\theta_1d\theta_2d\theta_3d\theta_4}{\left(\int_0^1\int_0^1\int_0^{2\pi}\int_0^{2\pi}\exp\left[-a^2\left(x_1^2+x_2^2\right)\right]\times\Gamma(x_1, x_2, L)x_1x_2dx_1dx_2d\theta_1d\theta_2\right)^2} - 1 \tag{3.63}$$

where $x_i = 2r_i/D_A (i = 1, 2, 3, 4)$, A is the coupling parameter, $\Gamma_{2,V}(r_1, r_2, L)$ is the MCF of the plane wave under the von Karman turbulence spectrum (as represented by Eq. (3.59)), and $\Gamma_{4,V}(r_1, r_2, r_3, r_4, L)$ is the CCF under identical conditions (as represented by Eq. (3.60)).

The Monte Carlo sample mean value method [36] (with 2 million iterations) was used to simulate Eq. (3.63), and the results are presented in Fig. 3.14. It can be seen that in the case of medium–strong turbulence, at the same relative aperture and communication wavelength, the relative variance in fluctuation of the coupling power decreases first, then increases, and finally decreases with an increase in the lens diameter. Figure 3.15 depicts the curve representing the relative variance in fluctuation of the coupling power and the internal scale of medium–strong turbulence with a communication distance of 1 km. It can be seen that the internal scale of turbulence has a weak effect on the relative variance in fluctuation of the coupled light power, and the trend of the impact is inconsistent. The wavelength is short ($\lambda = 650$ nm), and the relative variance in fluctuation of the coupled power increases with an increase in the internal scale of turbulence. At long wavelengths ($\lambda = 1550$ nm), the turbulence decreases with an increase in the internal scale.

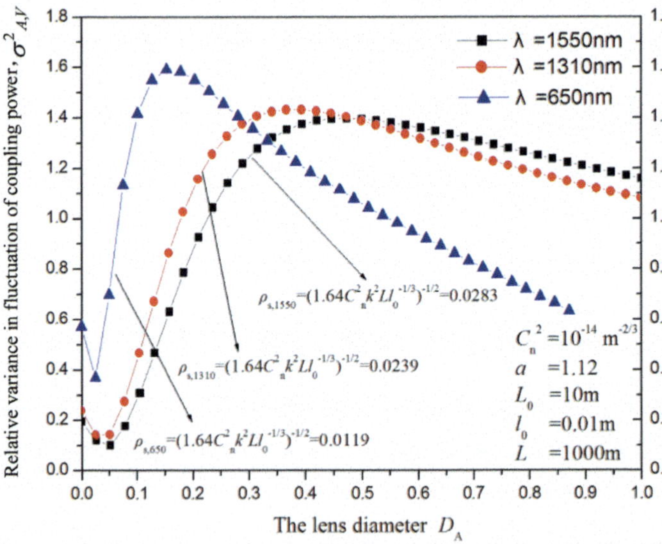

Fig. 3.14 Relation between lens diameter and relative variance in fluctuation of the coupling power [20]

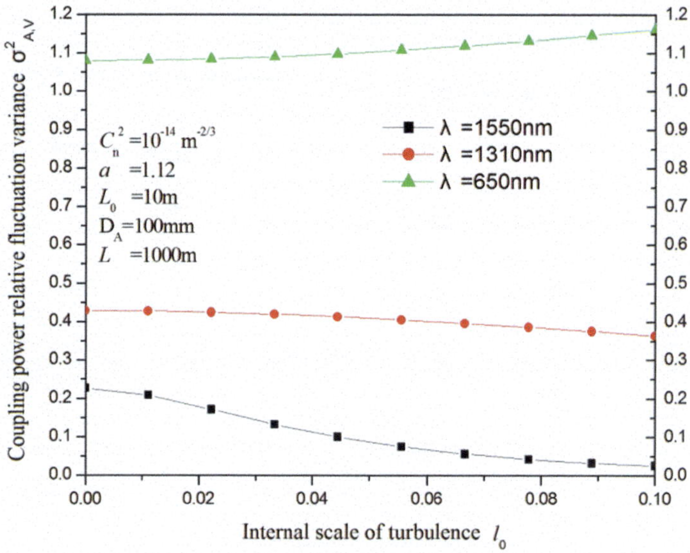

Fig. 3.15 Relation between internal scale of turbulence and relative variance in fluctuation of the coupling power [20]

3.2.3.2 Effect of Coupling Efficiency and Jitter Variance in Coupling Power on Bit Error Rate of FSO Communication System

The bit error rate (BER) is an indicator used to measure the accuracy of data transmission within a specified time, that is, the probability of code error in the transmission system. Ignoring the attenuation of optical power on the path, the BER of a system with on–off keying (OOK) modulation can be expressed as follows [37, 38]:

$$\mathrm{BER}_{\mathrm{ook}} = \int_0^\infty p_I(u) erfc(\eta' s P u (\frac{1}{2\sigma_n^2})^{1/2}) du \tag{3.64}$$

where η' is the transmitting efficiency of the receiving antenna, s is the photoelectric responsivity, P is the average transmitting power, σ_n^2 is the variance in the Gaussian white noise of the system, $erfc(\cdot)$ is the complementary error function, and $p_I(u)$ is the probability density distribution function of the optical power fluctuation. Therefore, considering that the coupling system adopts OOK modulation, the BER of the system can be expressed as follows:

$$\mathrm{BER}_{\mathrm{ook}} = \int_0^\infty p_{I,f}(u) \frac{1}{2} erfc(\eta' \eta_c s P u (\frac{1}{2\sigma_n^2})^{1/2}) du \tag{3.65}$$

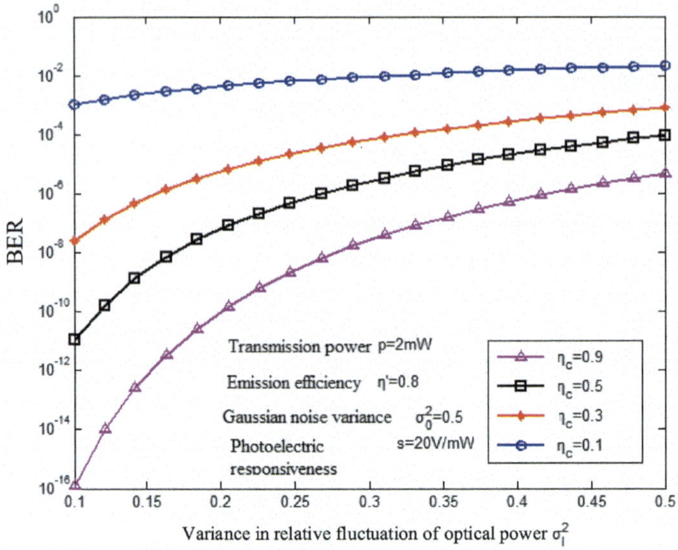

Fig. 3.16 Impact of coupling system on the BER of the FSO system under weak turbulence [20]

where η_c is the coupling efficiency and $p_{I,f}(u)$ is the probability density distribution function representing the optical power in the optical fiber, which is approximated using a lognormal distribution under weak turbulence. Figure 3.16 depicts the impact of the coupling system simulated by Eq. (3.65) on the BER of the OOK-modulated FSO system under weak turbulence. It can be seen that with an increase in the relative variance in fluctuation of the coupled optical power and a decrease in the coupling efficiency, the BER of the system continues to increase. Therefore, it is necessary to continuously explore engineering techniques to reduce this variance and improve the coupling efficiency.

3.2.4 Spatial Optical Coupling of Lens Array Under Atmospheric Turbulence

The main difference between the lens and array coupling systems is that lens coupling couples the light energy incident on the surface of the lens into an optical fiber (as illustrated in Fig. 3.17a), which means that the light incident on N effective array elements is coupled into N equal-length and mutually independent optical fibers [39] (as illustrated in Fig. 3.17b). The effective element in the array is also a lens, but its radius and focal length are different from those of the lens-coupled system. To verify that the lens array can effectively suppress turbulence, an equal-area lens and lens array structure, as depicted in Fig. 3.18, are designed for comparison. Figure 3.18a illustrates a single lens with diameter D_A and focal length F. Figure 3.18b illustrates a single-lens array with total diameter D_A, lens array element diameter $d_A = D_A/3$, and a focal length $F/3$. It can be seen that the total areas of the two structures are the same, the relative apertures (D_A/f) of the single lens and the lens array elements are the same, and the coupling parameter $a = \pi W_m D_A/(2\lambda f)$ can be guaranteed to be consistent for the coupling structure with the same incident wavelength and coupled into the same type of fiber. For the array illustrated in Fig. 3.18b, the effective area occupancy rate η_s (nonshaded area indicated in Fig. 3.18b) should be considered when the spot completely covers the surface of the array. For the array structure presented in Fig. 3.18b,

$$\eta_s = \frac{7 \times \pi (D_A/3)^2/4}{\pi D_A^2/4} = \frac{7}{9} \approx 0.778 \tag{3.66}$$

Equation (3.60) can be used to analyze the coupling efficiency of the single lens depicted in Fig. 3.18a for the von Karman turbulence spectrum, ignoring the mirror-through difference, reflection loss, and absorption loss. As for the lens array structure depicted in Fig. 3.18b, because the MCF of the planar light field is related only to the distance between the radial vectors, the coupling efficiency of each array element can be calculated separately. Considering the effective area occupancy rate of the array element, $\eta_s \approx 0.778$, when monochromatic plane waves are incident, the ensemble average of the coupling efficiency of the lens array for the von Karman turbulence spectrum is denoted as $\langle \eta_{Z,V} \rangle$:

$$\langle \eta_{Z,V} \rangle = \eta_s \left| \frac{8a^2}{\pi} \int_0^1 \int_0^1 \int_0^\pi \exp[-a^2(x_1^2 + x_2^2)] \times \exp\left(-\frac{A_R}{A_{C,V}} N(x_1, x_2, \theta) \right. \right.$$
$$\left. \left. \left\{ [1 + GN(x_1, x_2, \theta)]^{-\frac{1}{6}} - C \right\} \right) x_1 x_2 d\theta dx_1 dx_2 \right|_{D_A = d_A} \tag{3.67}$$

where $\eta_s \approx 0.778$ is the effective area occupancy rate of the array elements and $|\cdot|_{D_A=d_A}$ represents D_A from Formula (3.60) and is substituted to calculate the lens diameter d_A of the array elements. For the lens array structure illustrated in Fig. 3.18b, $d_A = D_A/3$ and the other parameters are consistent with Formula (3.55).

Figure 3.19 presents the coupling efficiency curve of the isometric lens and lens array under medium–strong turbulence simulated using Eqs. (3.67) and (3.55), respectively. The following conclusions can be drawn:

(1) When the communication distance is small, the coupling efficiency of the array is lower than that of a single lens with equal area. This is because there is a certain gap in the lens array, and the occupancy rate of the effective area is only 0.778, and beam incident on the aperture plane cannot be fully coupled into the fiber.
(2) The turbulence intensity increases with an increase in the communication distance. Because the diameter of the lens array element is equal to one-third the diameter of the single lens, the number of speckles $A_R/A_{C,V}$ on the surface

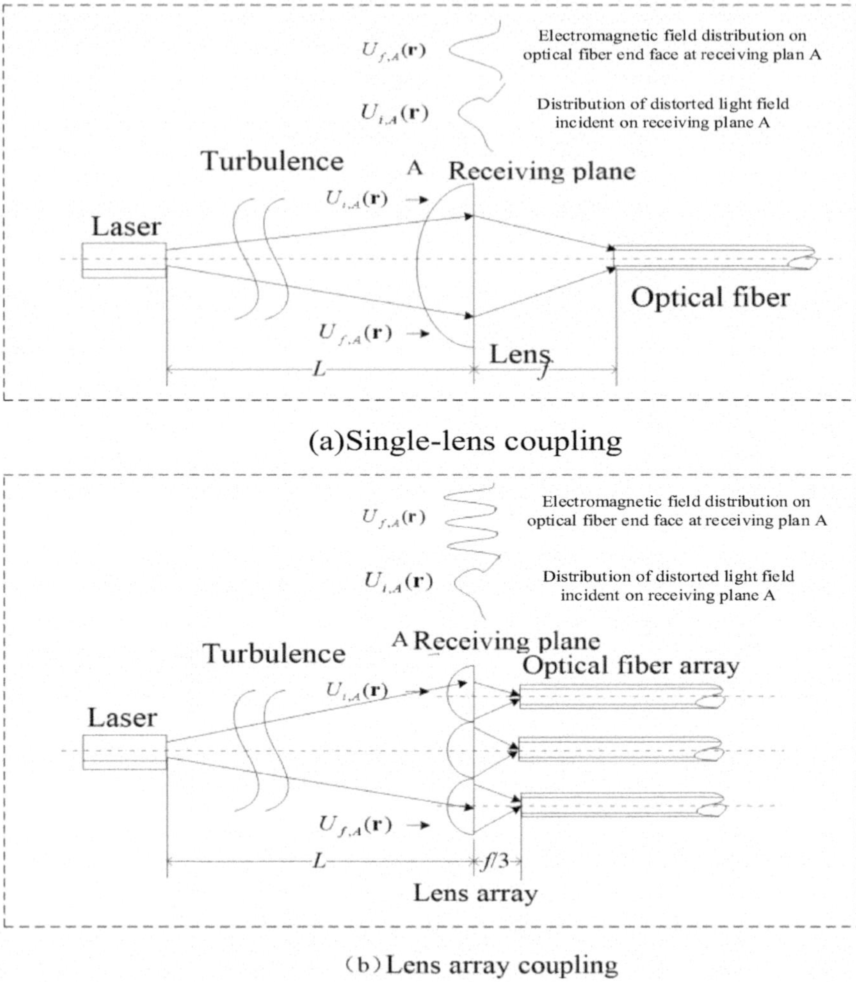

Fig. 3.17 Configurations of lens and array coupling systems under turbulence

of the lens is equal to only one-ninth of that of the single lens, which causes the coupling efficiency of the array to decrease slowly with an increase in the communication distance. Finally, over a certain distance under turbulence, the coupling efficiency of the array as a whole is higher than that of a single lens with equal area. When the diameters of both the lens and lens array are 30 mm, the diameter of the lens array element is 10 mm. In particular, $\lambda = 650$, 1310, and 1550 nm at communication distances of $L > 400$ m, $L > 1724$ m, and $L > 2413$ m, respectively, and the coupling efficiency of the lens array is higher than that of a single lens.

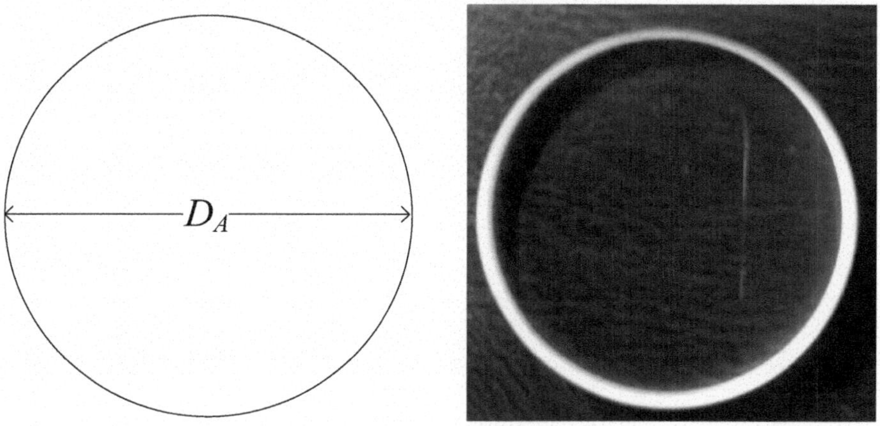

(a) Lens

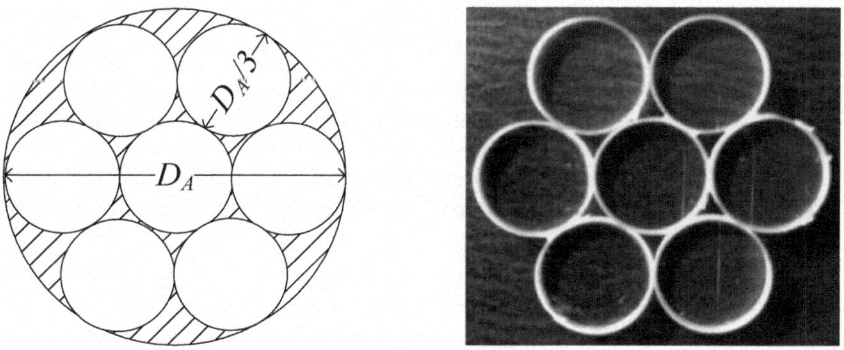

(b) Lens array

Fig. 3.18 Flat plans of lens and lens array

(3) Because the areas of the two structures are the same, the values of the average optical power incident on the surface of the lens and the lens array are the same and the power improvement coefficient K of the array relative to the lens is defined as

$$K = \eta_Z / \eta_A = \langle P_Z \rangle / \langle P_A \rangle \tag{3.68}$$

(4) $\langle P_Z \rangle$ is the average optical power coupled into the array, and $\langle P_A \rangle$ is the average optical power coupled into the lens. In the FSO system, owing to the limited capability of the collimating system and the large spot area at the receiving end, it is difficult to directly measure the optical power reaching the surface of the receiving aperture and there are reflection losses on each end face of the

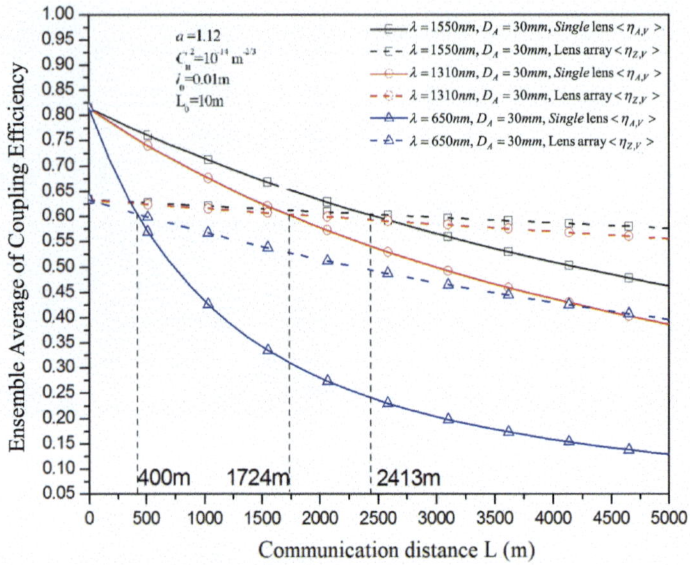

Fig. 3.19 Coupling efficiency curve of lens and lens array under moderate–strong turbulence [20]

optical system. Therefore, it is difficult to measure the actual coupling efficiency. Hence, in the experimental system, the ratio of the mean power coupled into the optical fiber, namely, the power improvement coefficient, can be used to measure the trend of variation in the coupling efficiencies of the lens and array. As depicted in Fig. 3.20, Eqs. (3.67) and (3.55) are substituted in Eq. (3.68) to vary the communication distance, and a numerical simulation is carried out to determine the relationship curve between the power improvement coefficient and the communication distance. When $K > 1$, the optical power coupled into the array exceeds that into the lens, and a conclusion consistent with the curves presented in Fig. 3.19 is reached.

(5) As depicted in Fig. 3.21, the communication distance is $L = 5$ km, the turbulence intensity is changed, and a numerical simulation is conducted to determine the relationship curve between the power improvement coefficient and the communication distance. As can be seen, when the communication wavelengths are 650, 1310, and 1550 nm and the turbulence intensities are greater than 1.14×10^{-15}, 3.75×10^{-15} m$^{-2/3}$, and 5.3×10^{-15} m$^{-2/3}$, respectively, the coupling efficiency of the array will be higher when compared with an equal-area single lens.

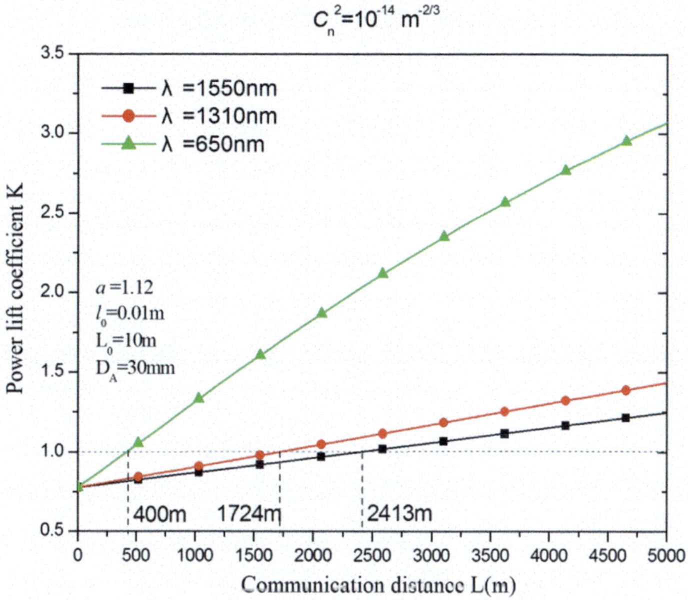

Fig. 3.20 Power improvement coefficient K of array relative to lens varies with the communication distance under moderate–strong turbulence [20]

Fig. 3.21 Curve representing variation in array relative to lens power improvement coefficient K with turbulence intensity when the communication distance $L = 5$ km [20]

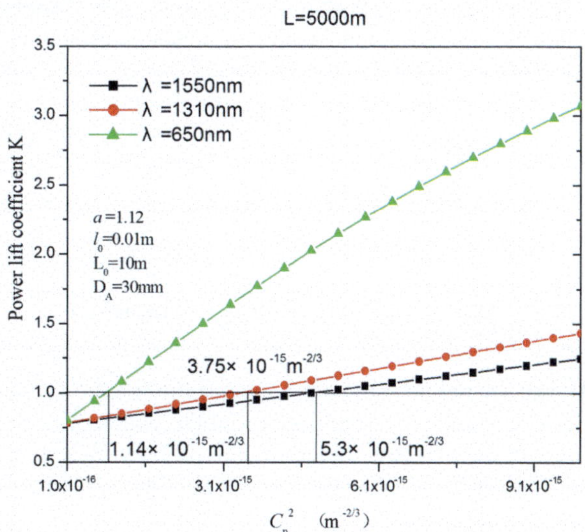

3.3 Automatic Alignment Algorithm for Spatial Light–Optical-Fiber Coupling

The coupling efficiency of space light to the optical fiber is closely related to the front optical system of the coupling system, optical-fiber parameters, optical fiber end processing technology, phase distribution of space light, and alignment attitude of optical fiber and space light. When the coupling system is identified, the key factor in determining the coupling efficiency is to adjust the alignment attitude of the space light and optical fiber adaptively to obtain a higher coupling efficiency. Because the diameter of the core of the optical fiber measures only a few micrometers, it is difficult to realize manual alignment. This problem can be addressed by designing an automatic coupling alignment system. Assuming that the beam is transmitted along the Z-axis in the Cartesian coordinate system, the alignment posture of the space light and optical fiber can have at most five meaningful degrees of freedom: three translational degrees of freedom in the X, Y, and Z directions and two rotational degrees of freedom in the X and Y directions. The five input variables control the output variables. The automatic alignment of spatial light and the optical fiber is realized using an intelligent algorithm.

As illustrated in Fig. 3.22, the automatic alignment system for spatial light and optical-fiber coupling mainly includes a light source, an optical system, an optical fiber, a mechanical system for alignment (also known as a control system, such as a high-precision stepper motor or piezoelectric ceramics), a feedback system, and a control algorithm. Signal acquisition can be implemented through an optical signal acquisition system such as an optical power meter with an acquisition function (as illustrated in Fig. 3.22a). An electrical signal acquisition system can also be used. If such a system is used, a photoelectric detector must first be used to carry out photoelectric conversion before the acquisition of electrical signals (as illustrated in Fig. 3.22b). In this section, we focus on the control algorithm.

3.3.1 Simulated Annealing Algorithm

The simulated annealing (SA) algorithm is used for determining the global optimal value of a function of multiple variables, which theoretically converges to the optimal value of the objective function with probability 1. In the following sections, we use the algorithm to realize the alignment of the optical axes between the space light and the optical fiber.

3.3.1.1 Flow of SA Algorithm

Several parameter conditions can be used as stopping parameters in the SA algorithm, such as the objective function reaching a specific value, j-times the outer cycle

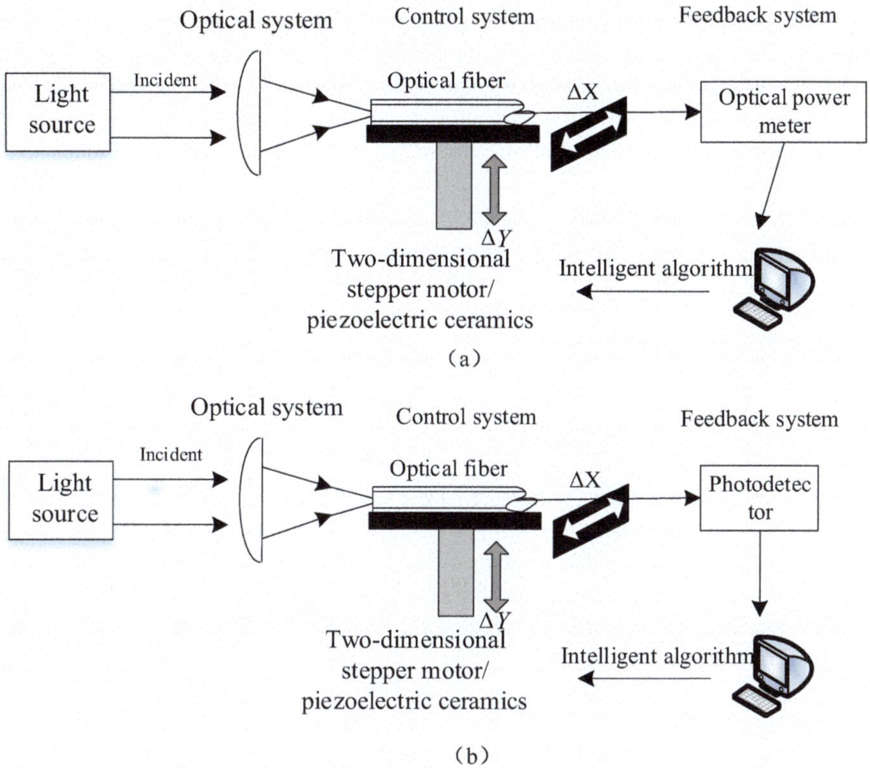

Fig. 3.22 Automatic alignment system for spatial light–optical-fiber coupling

reaching a specific value, and the annealing temperature T_j decreasing to a specific value. These conditions can be used as the basis for determining the end of the algorithm. The performance of the SA algorithm (including the speed of the algorithm and the quality of the final solution) is closely related to all parameters of the SA cooling schedule (such as, the search region of space Ω_0, initial temperature T_0, inner loop time I, number of outer cycles j, annealing temperature T_j, and temperature drop factor Q_T). It is concluded that in the construction of the most efficient schedule, only the size of and interaction between the various parameters impact the performance of the algorithm. The impact of the parameters on the quality of the final solution in decreasing order is as follows: search area, temperature drop function, and initial temperature. The order of impact on the time required for the solution is as follows: search area scope, internal cycle number, and temperature drop function [40]. Theoretically, as long as the initial temperature T_0 is sufficiently high, the cooling process is sufficiently slow, and the sampling durations I at each temperature are sufficiently large, the algorithm will converge to the optimal solution with probability 1. However, in practical applications, it is difficult to search the solution space globally, which makes it difficult for the algorithm to converge to the global optimal solution. Further, to improve the search efficiency and shorten the amount of time required for the search, the algorithm must reduce the initial temperature and increase the amplitude of each cooling process. In this way, although the algorithm can help improve the efficiency, it cannot guarantee theoretical accuracy. Therefore, it is necessary to combine empirical rules with the theoretical analysis to obtain a cooling schedule with a fast convergence rate and good fitting effect. An SA algorithm whose termination condition is "the number of external cycles reaching a specific value" is described in the following along with a flowchart [41]:

Let the value of the objective function in the first iteration be $f(x_i)$ and the value of the $(i+1)$th objective function after random perturbation be $f(x_{i+1})$. Therefore, based on the Metropolis criterion:

$$p = \exp\left[-\frac{|f(x_{i+1}) - f(x_i)|}{kT_j}\right] \tag{3.69}$$

where k is a constant (usually equal to 1), T_j is the temperature of the jth state (i is the number of internal cycles and j is the number of external cycles), and x_i and x_{i+1} are points in the feasible domain of the objective function. Let the difference between the two objective functions be $f(x_{i+1}) - f(x_i) = \Delta f$.

As presented in Fig. 3.23, the process for obtaining the maximum value of the objective function is as follows:

If $\Delta f > 0$, the $(i+1)$th state is received as the current state. If $\Delta f < 0$, it is checked whether the $(i+1)$th state can be accepted as the current state based on the probability. A random number q (between 0 and 1) subject to uniform distribution is generated, and compared with the random number p and q calculated using the Metropolis criterion at the current temperature T_j. When $p > q$, the receiving state $i+1$ is considered as the current state. Otherwise, the ith state is retained as the current state. Following several comparisons and state transitions, an equilibrium state is reached at temperature T_j. Thereafter, the temperature T_{j+1} is continuously lowered to determine the equilibrium state at this temperature. By repeating the cycle, the temperature approaches zero, at which point the maximum objective function can be obtained.

3.3.1.2 SA Algorithm Applied to Single Lens–Spatial Light Coupling

The theoretical model of the SA algorithm applied to the single lens–spatial light coupling alignment mainly involves the elements described in the following sections.

a. **Objective function for alignment of single lens**

When space light is coupled, the radial alignment error has the most significant effect on the coupling efficiency. Theoretically (without turbulence), the radial alignment error of the single-mode fiber exceeds 3.6 μm and the coupling efficiency decreases by −3 dB. In the Cartesian coordinate system, the radial alignment coordinates are

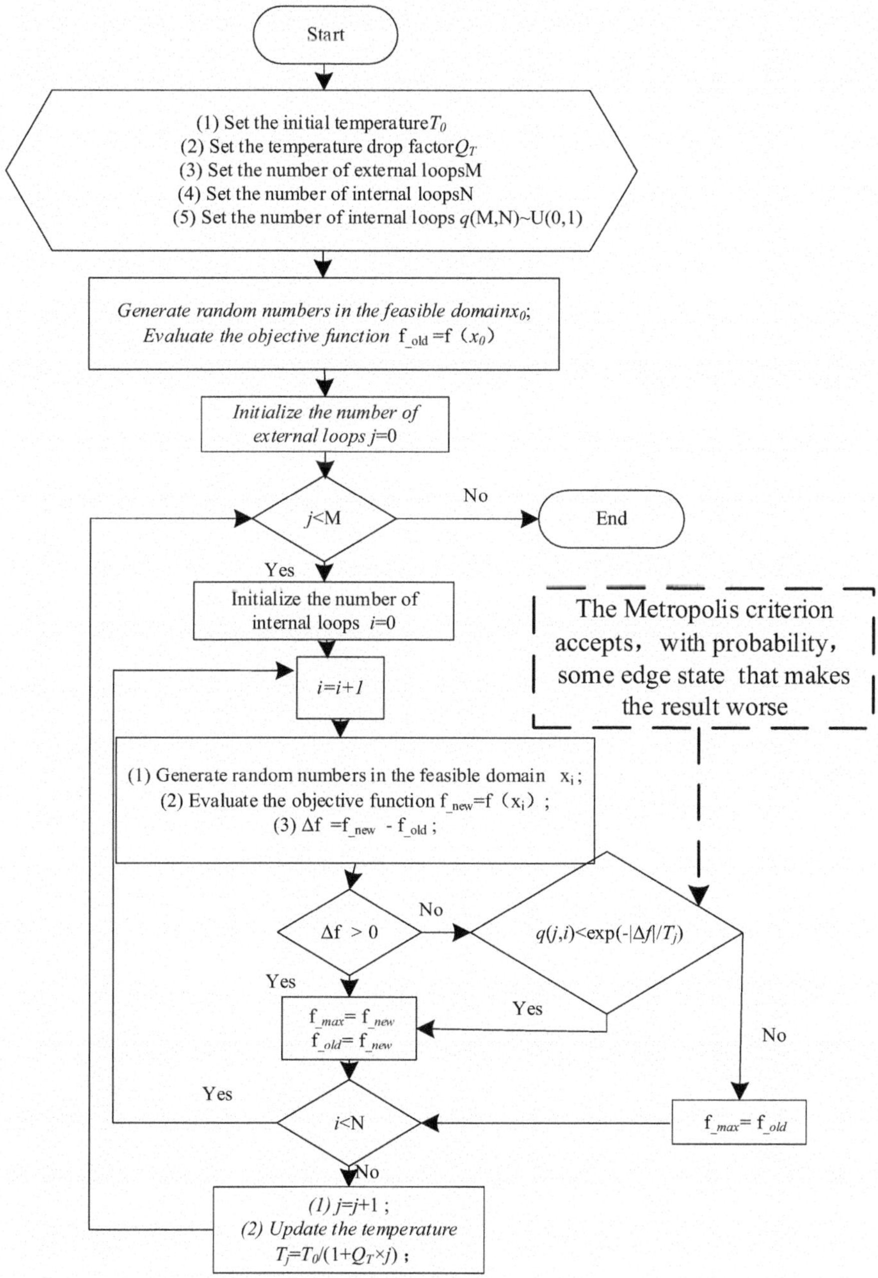

Fig. 3.23 Flowchart for maximum SA algorithm [20]

decomposed into X and Y alignments. In the SA algorithm, for the alignment of single lens–spatial light coupling, the power coupled into a single-mode fiber is considered as the objective function P_B, which can be expressed as follows:

$$P_B(\Delta x, \Delta y) = \frac{\pi D_A^2}{4} \times \frac{\left| \int_0^\infty \frac{1}{-i} \exp(-ikf) \exp\left(\frac{-ikr^2}{2f}\right) J_1\left(\frac{\pi D_A r}{\lambda f}\right) \exp\left[-\left(\frac{r-\sqrt{\Delta x^2+\Delta y^2}}{W_m}\right)^2\right] r dr \right|^2}{\int_0^\infty J_1\left(\frac{\pi D_A r}{\lambda f}\right)^2 / r dr \times \int_0^\infty \exp\left[-2\left(\frac{r-\sqrt{\Delta x^2+\Delta y^2}}{W_m}\right)^2\right] r dr} \quad (3.70)$$

where Δx and Δy are the alignment errors in the X and Y directions, respectively. Theoretically, when $\Delta x = 0$ and $\Delta y = 0$, the coupling power has a maximum value.

b. **Feasible region Ω_0**

The solution (x, y) satisfying the constraints of the optimization problem is known as the feasible solution, and the set composed of all the feasible solutions is known as the feasible region. Owing to the limited precision of the optical fiber and lens, there will always be a certain initial alignment error. Assuming that the initial installation error is less than 10 μm, the feasible region of the objective function P_B can be considered as $\Delta x \in (-10^{-5}, 10^{-5})$ m, $\Delta y \in (-10^{-5}, 10^{-5})$ m. Assuming that the installation error is within 5 μm, the feasible region of P_B can be considered as the following: $\Delta x \in (-5 \times 10^{-6}, 5 \times 10^{-6})$ m, $\Delta y \in (-5 \times 10^{-6}, 5 \times 10^{-6})$ m.

c. **Cooling schedule**

The relevant parameters in the cooling schedule need to be optimized, and the empirical rules and theoretical analysis must be combined to obtain a cooling schedule with a high convergence rate and good fitting effect. The primary cooling schedule parameters include the following.

1. Initial temperature T_0

   In general, the higher the initial temperature, the better. However, considering the efficiency of the algorithm, the initial temperature of the SA algorithm is generally obtained as follows:

   (1) From a uniform sampling of a group of states, the variance of the objective function under each state is considered as the initial temperature.
   (2) A group of states is randomly generated to determine the maximum difference in the target value between the two states $\Delta_{\max}$. This difference and certain functions are used to determine the initial temperature; for example, $T_0 = |\Delta_{\max}|/p_r$, where p_r represents the initial reception probability.
   (3) The empirical formula is used.
   (4) Multiple experiments are conducted.

2. Inner loop time I (Markov chain length)

   In the SA algorithm, whether the new state is received depends only on the new and current states, and it is controlled by temperature. If, at each temperature, the algorithm calculates the change in the Markov chain until it is stable and then drops the temperature, it is referred to as a time symmetry algorithm. If the algorithm does not need to reach a stationary distribution at each temperature but drops the temperature at a certain rate, it is known as an asynchronous algorithm.
3. Number of external cycles (j)

   This parameter controls the cooling function.
4. Annealing function

   For example, the temperature drop factor Q_T in the annealing function.

3.3.1.3 Alignment Effect of SA Algorithm

Assuming a parallel light incident at 1550 nm, optical power density $P_s = 1\ \mathrm{W/m^2}$, and diameter (of the coupled lens) $D_A = 0.1$ m, the optical power incident on the aperture of the single lens is calculated as $P_s \pi D_A/4 = 0.0079\ \mathrm{W} = 7.9$ mW. The maximum coupling efficiency is 0.814 when the mode-field radius is 5.25 μm. The value of the maximum objective function for single lens–spatial light coupling on punctuality is approximately 6.4 mW. At this point, when the initial alignment error is less than 10 μm, the theoretical model of the objective function P_B is depicted in Fig. 3.24, and Eq. (3.72) represents its specific expression. The effects of the relevant parameters in the SA algorithm on its convergence are analyzed in the following.

a. Initial temperature T_0

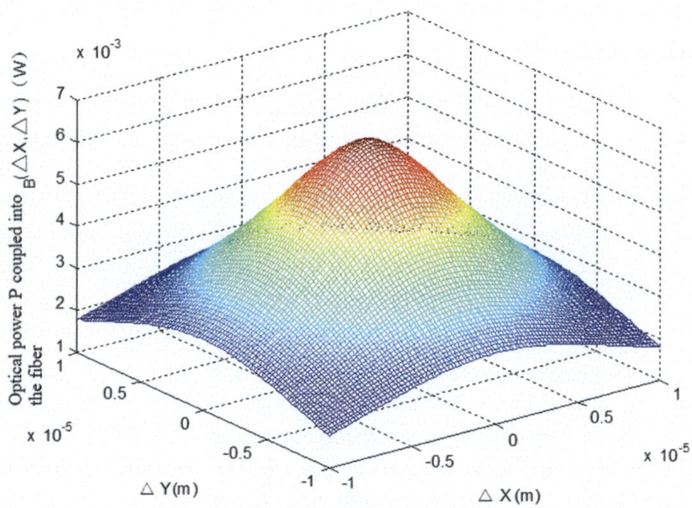

Fig. 3.24 Target function model of single lens–spatial light coupling [20]. ($\lambda = 1550$ nm, $P_s = 1\ \mathrm{W/m^2}$, $D_A = 0.1$ m, $\Delta x \in (-10^{-5}, 10^{-5})$ m, $\Delta y \in (-10^{-5}, 10^{-5})$ m)

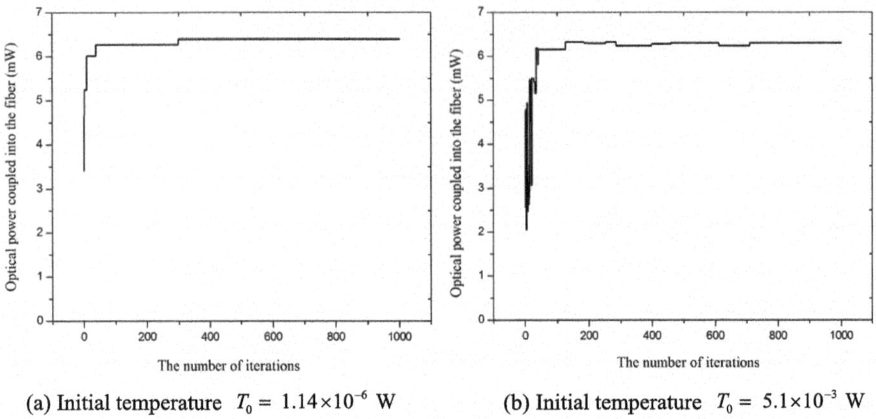

Fig. 3.25 Effect of various initial temperatures on SA algorithm [20]

The number of internal cycles, $i = 5$; number of external cycles, $j = 200$; and temperature drop factor, $Q_T = 1$, were used to generate the initial temperature, as described in the following.

Method 1: Within the range $\Delta x \in (-10^{-5}, 10^{-5})$ m, $\Delta y \in (-10^{-5}, 10^{-5})$ m, 100 groups of states were uniformly sampled, and the variance of the objective function for each state was considered as the initial temperature, yielding $T_0 = 1.14\times 10^{-6}$ W. The results of the optimization are presented in Fig. 3.25a.

Method 2: The objective function is convex in the range $\Delta x \in (-10^{-5}, 10^{-5})$ m, $\Delta y \in (-10^{-5}, 10^{-5})$ m. The difference between the maximum and minimum values of the objective function was calculated and divided by the reception probability of 0.9 to obtain the initial temperature $T_0 = 5.1 \times 10^{-3}$ W. The results of the optimization are presented in Fig. 3.25b.

As can be seen from Fig. 3.25, the convergence effect of the algorithm evidently varies based on the initial temperature. Because the objective function of single lens–spatial light coupling on punctuality is convex, the initial temperature has little effect on the final optimization result of the algorithm. However, it is evident that the lower the initial temperature, the faster the algorithm converges to the maximum value. However, if the objective function is not a convex function, the initial temperature is too low. Therefore, the algorithm is likely to converge to an extreme value of the objective function rather than to the maximum value.

b. Feasible region Ω_0

Internal cycle number $i = 5$, external cycle number $j = 200$, and temperature drop factor $Q_T = 1$. When the feasible region Ω_0 is reduced from $\Delta x \in (-10^{-5}, 10^{-5})$ m and $\Delta y \in (-10^{-5}, 10^{-5})$ m to $\Delta x \in \left(-5 \times 10^{-6}, 5 \times 10^{-6}\right)$ m and $\Delta y \in \left(-5 \times 10^{-6}, 5 \times 10^{-6}\right)$ m, that is, the initial alignment error is reduced from 10 to 5 μm, the objective function model is as depicted in Fig. 3.26. Method 1 is used to calculate the initial temperature: $T_0 = 3.89 \times 10^{-7}$ W; using method 2, the initial temperature is calculated as $T_0 = 3 \times 10^{-3}$ W. Figure 3.27 presents the results of the optimization at various initial temperatures after the feasible region is narrowed.

By comparing Figs. 3.25a and 3.27a, it can be seen that for the initial temperature obtained using the same method, reducing the range of the solution space can significantly accelerate the convergence speed of the algorithm. The same conclusion can be drawn by comparing Figs. 3.25b and 3.27b. That is, if the initial alignment error is reduced from 10 to 5 μm, the convergence speed of the algorithm can be accelerated

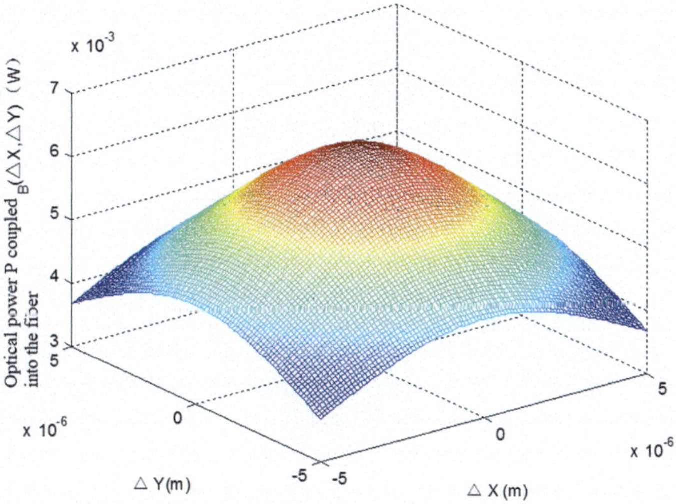

Fig. 3.26 Target function model of single lens–spatial light coupling [20]. ($\lambda = 1550$ nm, $P_s = 1$ W/m^2, $D_A = 0.1$ m, $\Delta x \in \left(-5 \times 10^{-6}, 5 \times 10^{-6}\right)$ m, $\Delta y \in \left(-5 \times 10^{-6}, 5 \times 10^{-6}\right)$ m)

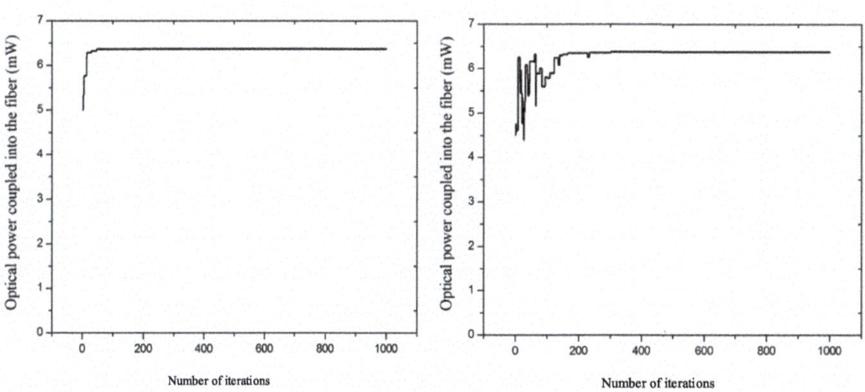

Fig. 3.27 Effect of narrowing feasible region on SA algorithm at various initial temperatures [20]

and the time required for automatic alignment can be reduced; this is convenient for automatic alignment.

c. Effect of method of generating candidate solution

The candidate solution of the traditional SA algorithm is generally obtained by randomly sampling the feasible region Ω_0 based on a probability density function. The probability distribution can be uniform, normal, exponential, and so on. In the above simulation, the candidate solutions Δx_i and Δy_i of Δx and Δy satisfy a uniform distribution in the feasible region. The method for generating candidate solutions is improved as follows. Let Δx_i still satisfy the uniform distribution $\Delta x_i \sim U(-5 \times 10^{-6}, 5 \times 10^{-6})$ m in the feasible region. The value of Δy_i is a floating value in a random small range near Δx_i, that is, $\Delta y_i = \Delta x_i + \text{rand}(-0.01 \times 10^{-6}, 0.01 \times 10^{-6})$ m. Under identical conditions, that is, internal cycle number $i = 5$, external cycle number $j = 200$, temperature drop factor $Q_T = 1$; and initial temperature $T_0 = 3 \times 10^{-3}$ W (the simulation parameter values in here are consistent with the values depicted in Fig. 3.27b). Figure 3.28 illustrates the effect of the improved candidate solution generation method on the SA algorithm.

By comparing Figs. 3.27b and 3.28, it can be seen that after improving the generation method of the candidate solution, the optical power jitter degree coupled into the optical fiber decreases significantly during the convergence process of the algorithm, and the power jitter variance decreases from 5.8×10^{-8} to 9.5×10^{-9} mW. This is because, theoretically, by comparing Figs. 3.27b and 3.28, it can be seen

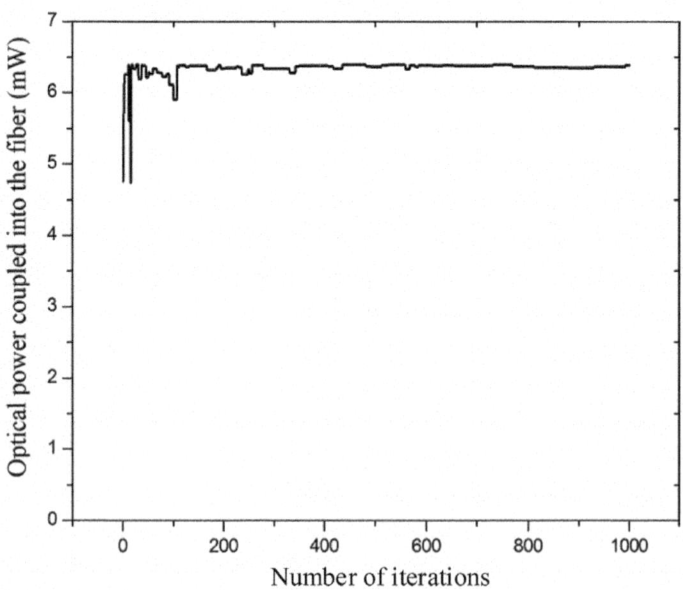

Fig. 3.28 Effect of improved candidate solution generation method on the results of the optimization of the SA algorithm [20]

that after improving the method of generation of the candidate solution, the optical power jitter degree coupled into the optical fiber decreases significantly during the convergence process of the algorithm, and the variance in power jitter decreases from 5.8×10^{-8} to 9.5×10^{-9} mW. This is because, theoretically, the objective function has a maximum value and the randomness of the alignment error is limited by improving the method of generation of the candidate solutions. Therefore, by modifying this method and further combining it with a motor with a high response speed, the algorithm can not only be used for automatic alignment but also for the automatic tracking of atmospheric turbulence coupling.

After the optimization of the SA algorithm, the optical power coupled into the fiber can reach a theoretical maximum of 6.4 mW, and $i = 5$ and $j = 200$. Theoretically, the probability of successful optimization using the SA algorithm can reach 100%.

3.3.2 *Particle Swarm Optimization*

Particle swarm optimization (PSO) is an intelligent optimization algorithm based on parallel operation inspired by birds searching for food in nature [42]. The parameters used in PSO are simpler than those used in the SA algorithm. In this section, the use of the PSO algorithm to simulate the alignment of the optical axes between the space light and the optical fiber is described.

3.3.2.1 Flow of PSO Algorithm

Suppose that in the d-dimensional search space, N particles form a particle swarm and the position of the ith particle is expressed as $X_i = (X_{i1}, X_{i2}, L, X_{id})$ using a d-dimensional vector. The position of each particle represents a potential solution in the solution space, and the quality of the solutions is generally determined by the objective function. The displacement velocity of the ith particle is expressed as $V_i = (V_{i1}, V_{i2}, L, V_{id})$ using the d-dimensional vector. The optimal position obtained by the ith particle in the search process is expressed as $Pbest_{id} = (P_{i1}, P_{i2}, L, P_{id})$, and the optimal position obtained by all particles during the entire search process is expressed as $Gbest_{id} = (P_{g1}, P_{g2}, L, P_{gd})$.

The next search speed and position of the particle ith are expressed as follows [43]:

$$V_{id}^{k'+1} = w \cdot V_{id}^{k'} + c_1 \cdot \text{rand}_1 \cdot (Pbest_{id}^{k'} - X_{id}^{k'}) + c_2 \cdot \text{rand}_2 \cdot (Gbest_{d}^{k'} - X_{id}^{k'}) \tag{3.71}$$

$$X_{id}^{k'+1} = X_{id}^{k'} + V_{id}^{k'+1} \tag{3.72}$$

In Eqs. (3.71) and (3.72), $i(1, 2, \ldots, N)$ represents the ith particle in the particle swarm, V_{id} represents the velocity vector of the d-dimensional component of the ith particle, $Pdest_{id}^{k'}$ represents the d-dimensional component of the optimal position of the ith particle, $Gdest_{gd}^{k'}$ represents the d-d component of the optimal position in the entire population, c_1 and c_1 denote the weight factors, rand() is a function that generates an evenly distributed random number within [0,1], w denotes the inertial weight, and k' represents the required number of iterations of the algorithm. Figure 3.29 depicts the process of PSO.

The main parameters involved in PSO and their meanings are as follows:

(1) Particle length d: Dimension of the solution space of the problem.
(2) Particle population size N: The selection of the particle population size depends on the specific problem, but the number of particles is generally set to 20–50. For most problems, 10 particles are good enough, but for more difficult problems or certain special types of problems, the number of particles can reach up to 100 or 200. In addition, the larger the number of particles, the larger the search space of the algorithm and the easier it is to find the global optimal solution; however, the algorithm also runs for a long time.

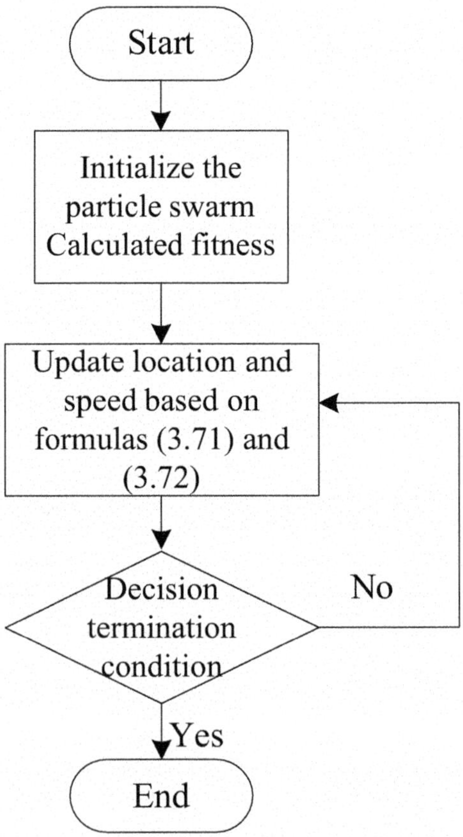

Fig. 3.29 Basic flowchart representing the process of PSO [20]

(3) Acceleration constants C_1 and C_2: The maximum stride length of flight is adjusted in the direction of Pbest and Gbest to determine the effects of the individual particle and group experiences on the particle trajectory, respectively, and reflect the information exchange between the particle swarms. When $C_1 = 0$, the particle has only a group experience and its convergence speed is high but easily falls into the local optimum. When $C_2 = 0$, there is no group of particles that share information. A group of M particles is equivalent to M particles running independently, and the probability of obtaining a solution is very small. Therefore, in general, C_1 is set equal to C_2. In this way, the individual and group experiences have the same significant impacts, making the final optimal solution more accurate. Varying these constants alters the "tension" in the system, with the lower values of C_1 and C_2 causing the particles to hover far from the target and the higher values of C_1 and C_2 producing steep movements or overtaking the target. Shi and Eberhart [44] suggested that, in general, $C_1 = C_2$.

(4) Maximum speed of particle, v_{max}: The particle speed in each dimension of space has a maximum speed limit value v_{max}, which is used to clamp the particle speed so that it is controlled within the range $[-v_{max}, +v_{max}]$. This range determines the strength of the problem space search and is generally set by the user. v_{max} is a highly significant parameter. If it is too high, the particles may fly through the good region. However, if it is too small, the particles may not be able to detect the region beyond the local optimum sufficiently. In fact, they may fall into local optima and would not be able to move far enough out of the local optima to reach a better position in space.

(5) rand_1 and rand_2 are random numbers between (0,1), which increases the randomness of the particle's flight.

(6) Iteration termination conditions: These are generally set as the maximum iteration number T_{max}, calculation accuracy, or maximum stagnation step Δt of the optimal solution.

3.3.2.2 Theoretical Model of PSO for Single Lens–Spatial Light Coupling Alignment

The PSO algorithm is applied to the spatial optical coupling alignment system, which only considers radial alignment, and the optimization process of optical power coupled into the optical fiber is completed in the two-dimensional (X- and Y-axis) search space Ω_0. The search space is divided into a population $X = (X_1, X_2, X_3, \ldots, X_n)$ composed of n particles, where X_i represents the distance between the location of the ith "particle" and the center of the single lens, $X_i = (\Delta x_i, \Delta y_i)$, and its corresponding adaptive value $P_i^{k'} = (\Delta x_i, \Delta y_i)$ can be calculated using Eq. (3.72). It is assumed that the rate of the ith "particle" is $V_i^{k'}$ and k' denotes the number of iterations of the algorithm. If the adaptive value of the lth ($l \leq n$) particle in the population is the largest, it is referred to as the extreme value of the population and denoted as $P_l^{k'} = (\Delta x_l, \Delta y_l)$. During each iteration, the "particle" updates its position and velocity based on Eqs. (3.73) and (3.74), and their corresponding adaptive values are $P_i^{k'} = (\Delta x_i, \Delta y_i)$ and extremum values of the population are $P_l^{k'} = (\Delta x_l, \Delta y_l)$. That is,

$$V_i^{k'+1} = \omega \cdot V_i^{k'} + c_1 \cdot \text{rand} \cdot (P_i^{k'} \times 10^{-6} - X_i^{k'}) + c_2 \cdot \text{rand} \cdot (P_i^{k'} \times 10^{-6} - X_i^{k'}) \tag{3.73}$$

$$X_i^{k'+1} = X_i^{k'} + V_i^{k'+1} \tag{3.74}$$

Equation (3.73) improves the traditional particle swarm updating function, primarily because the maximum moving position of the fiber line on the focal plane of the lens is measured in micrometers, whereas the value of the coupling power is measured in milliwatts. To match the dimensions and improve the precision of the spatial search, the position-updating speed is controlled to be measurable in milliwatts. Therefore, a dimensional constraint is applied to the adaptive value $P_i^{k'} = (\Delta x_i, \Delta y_i)$ and the population extremum value $P_l^{k'} = (\Delta x_l, \Delta y_l)$.

3.3.2.3 Effect of Parameters of PSO Algorithm on Single Lens–Spatial Light Coupling Alignment

a. **Fixed inertia weight**

The PSO algorithm is used to complete the coupling alignment process of the single lens and spatial light; the objective function is represented by Eq. (3.70). The values of the acceleration constants are set to $C_1 = 1$ and $C_2 = 1$, and the particle number is 10. Figure 3.30 depicts the effect of single-lens coupling on punctuality and various inertia weights w on the PSO algorithm. As can be seen from Fig. 3.30a and c, when $w = 0.4$ or $w = 0.8$, the PSO algorithm converges to the maximum value of the objective function approaching 6.4 MW. By comparing Fig. 3.30b and d, it can be seen from the particle motion track that the particle motion range at $w = 0.8$ is evidently larger. As depicted in Fig. 3.30e, when $w = 1.2$, the PSO algorithm can no longer converge to the maximum value of the objective function, because with the increase in the inertia weight, the length of a particle “leap” is large, which is not conducive to local convergence. As can be seen from Fig. 3.30 and Eq. (3.73), the inertia weight w directly affects the step size of the solution space search. If the step size is too small, the global convergence ability of the algorithm decreases, and it easily converges to the local extremum. When the step size is too large, the global convergence ability of the algorithm increases, but local convergence is not easy. Theoretically, the PSO algorithm is applied to single-lens coupling pair punctuality. Because the objective function is a convex function with a single extreme value, the inertia weight can be restricted within the range $0.4 < w < 0.8$ to narrow down the range of particle motion relatively and ensure the local convergence ability of the algorithm.

b. **Linearly adjusted inertia weight**

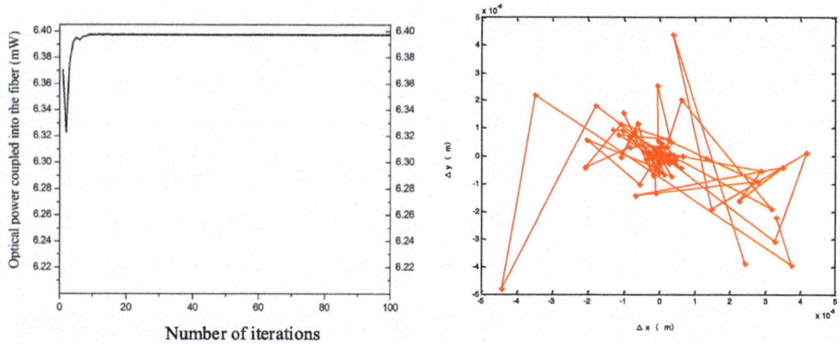

(a) Results of optimization of PSO algorithm and (b) particle movement trajectory for w=0.4

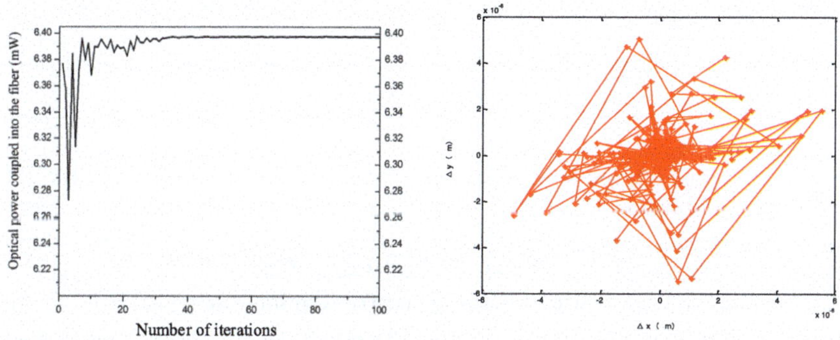

(c) Results of optimization of PSO algorithm and (d) particle movement trajectory for w=0.8

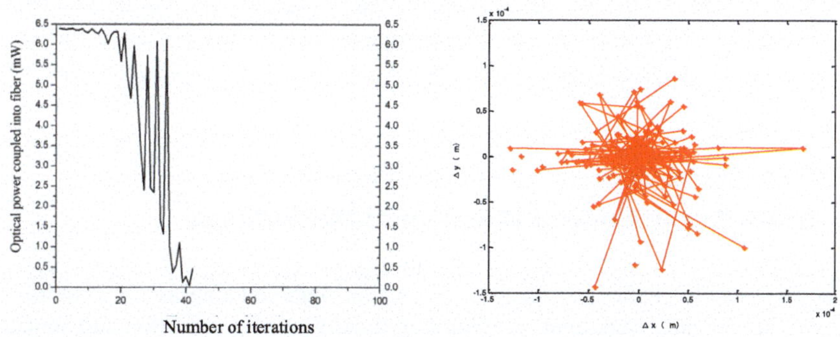

(e) Results of optimization of PSO algorithm and (f) particle movement trajectory for w=1.2

Fig. 3.30 Effect of single-lens coupling on punctuality and inertia weight w on PSO algorithm [20]

The concept of decreasing inertia weight was proposed by Shi and Eberhart [45]. The formula for calculating the inertia weight is as follows:

$$w = w_{\max} - \frac{w_{\max} - w_{\min}}{k_{\max}} \times k' \tag{3.75}$$

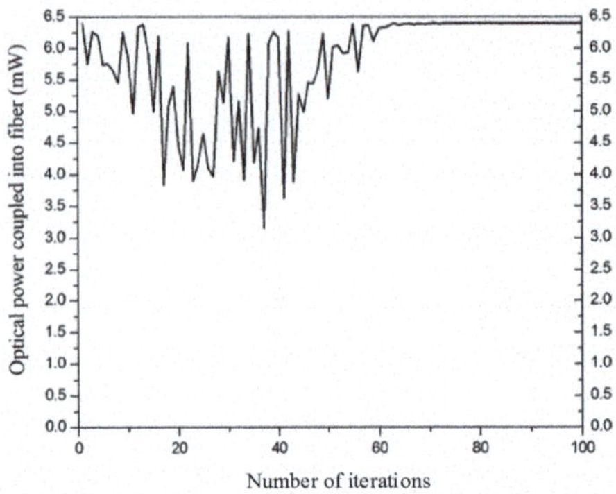

(a) Results of optimization for w=1.2

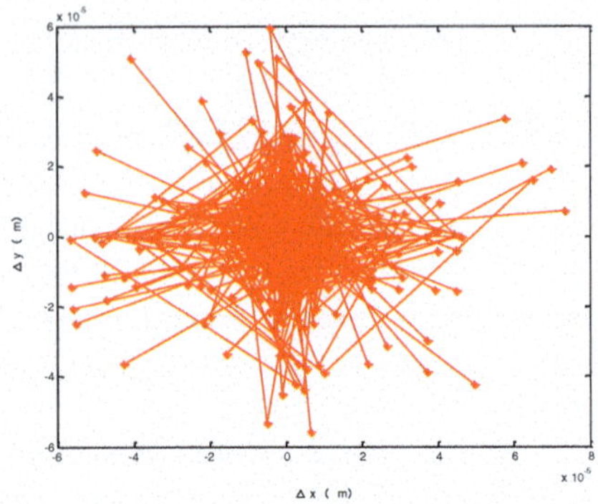

(b) particle movement trajectory for w=1.2

Fig. 3.31 Effect of single-lens coupling on punctuality and linear adjustment of inertia weight on PSO algorithm [20]

where $w_{\max}$ and $w_{\min}$ represent the maximum and minimum values of inertia weight, k' is the current iteration number, and $k'_{\max}$ represents the maximum number of iterations. In this study, a linear decline from 1.2 to 0.4 is set for w (the values of the other parameters are consistent with the results presented in Fig. 3.30), as depicted in Fig. 3.31, so that PSO can explore a larger area in the beginning and quickly locate the approximate position of the optimal solution. As w gradually decreases, the particle velocity slows down and fine local search begins. By comparing Figs. 3.30e and 3.31a and Figs. 3.30f and 3.31b, it can be seen that the PSO algorithm with linear decreasing inertia weight can still converge to the maximum value of the objective function while expanding the search space, thus improving the convergence performance of the algorithm.

3.4 Beam Array Control Based on Maka Antenna

In this section, the challenges faced in using the Maka antenna as a transmitting and receiving antenna were analyzed. Further, the off-axis transmission and control technology of the Maka antenna array Gaussian beam and the theoretical modeling of the Gaussian beam array transmission were studied under atmospheric turbulence. In addition, the spatial optical coupling performance of the Maka antenna as a receiving antenna was analyzed.

3.4.1 Maka Antenna and Existing Problems

3.4.1.1 Structural Overview of Maka Antenna

The Cassegrain system is widely used in FSO systems [46]. As early as 1994, Dong and Ren [47] modeled the collimation emission of a Gaussian light source in a Cassegrain antenna system. Subsequently, researchers have conducted extensive studies on various aspects of its performance. Zhang et al. [48] analyzed the coupling performance of the Cassegrain telescope and conical catheters in detail. Yang et al. [49] performed a theoretical analysis of laser beam transmission through a Cassegrain antenna under turbulent atmosphere. Cao et al. [50] understood that temperature would cause the Cassegrain system to deform, leading to the degradation of the performance of optical signal–single-mode fiber coupling. Ran et al. [51] analyzed the performance of the off-axis timing of a Cassegrain system and derived the expressions for the received light spot area and power attenuation curve corresponding to various off-axis angles. Yang and Su [52] analyzed the stray light problem of a Cassegrain telescope system against the background of daytime strong light.

The Cassegrain system is a typical double-reflection telescopic objective lens with no chromatic aberration. The primary mirror is usually parabolic, and the secondary mirror is a convex hyperboloid. The Maksutov–Cassegrain [53] telescope is a telescopic objective lens that improves the structure of the Cassegrain system. A spherical reflector is used as the basis, refraction elements are added for the correction of aberrations, and a refraction structure is constructed, referred to as the Maka antenna. When compared with the Cassegrain system, the manufacturing cost of the Maka antenna can be reduced because the components are spherical. The typical feature of the Maka antenna is that its secondary mirror is very small, usually an aluminized dot on the correction mirror. Figure 3.32 illustrates a schematic of the Maka antenna, which is mainly composed of a second shield (also known as a conical shield), first shield, primary mirror, secondary mirror, and meniscus correction mirror. The refraction and reflection structures make the Maka antenna compact and can form a lens system with a large aperture and focal length. When the antenna is used as a space optical coupling system, large-aperture coupling receiving technology can be realized; it also has a strong ability to gather optical power. The large focal length satisfies the requirement of the numerical aperture of an optical fiber. For the same relative aperture (D/f), the length of the lens system can be reduced by nearly half.

3.4.1.2 Impact of Secondary Reflector on Transmission Efficiency

Certain studies [54] have reported that for uniformly distributed light sources, the energy loss is more than 30% because of the secondary reflector shielding of the Maka antenna. For Gaussian-distributed laser sources, the energy loss can reach 50% or higher. In an FSO communication system, the signal light source is usually

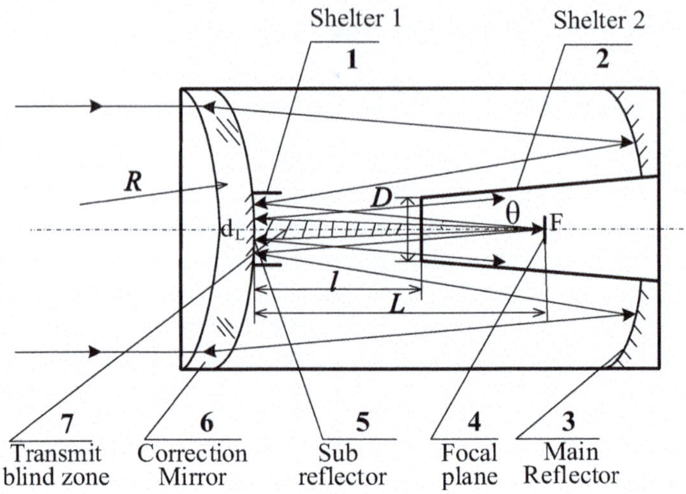

Fig. 3.32 Schematic of the structure of the Maka antenna

a Gaussian laser light source. Although the area of the secondary mirror is small, the loss of light energy is large. The parameters of the existing Maka antenna are substituted in the nonsequence of ZEMAX [55] for simulation. As illustrated in Fig. 3.33, an exit pupil light power detector is placed at the front end of the Maka antenna. A loss detector is placed at the back to detect the light power reflected back to the conical aperture by the secondary mirror. A collimating Gaussian beam with a wavelength of 650 nm, power of 100 mW, and beam width of 2 mm is placed 25 mm in front of the tapered diaphragm.

Using the additional optical device method, the circular light source with a dark spot in the center is simulated using the circular optical-fiber array. Each fiber is a Gaussian light source, and each light source adopts off-axis emission. The problem

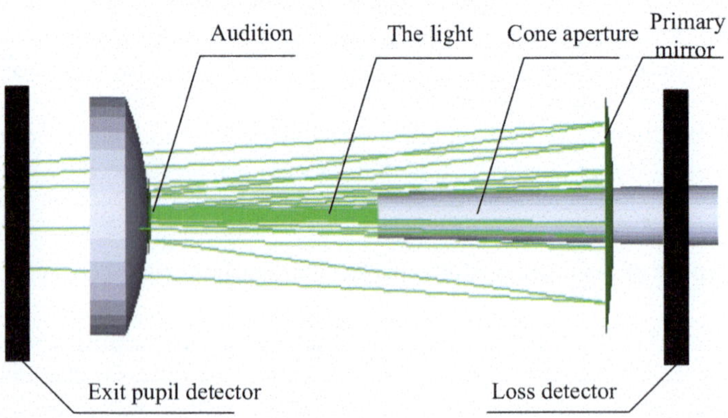

Fig. 3.33 Normal transmission in a Maka antenna

of secondary mirror occlusion can be addressed by combining the two methods [56]. This method can also be used to construct the signal source array. The array of the signal light source plays a significant role in the FSO system. Certain studies have reported that a multibeam transmitting system can improve the fluctuation in the optical power on the receiving surface and reduce the system error code [57]. When the wavelength of the array light source is consistent, the power of a single light source can be reduced by replacing it with the array light source. When the wavelength of each light source in the array is different, wavelength-division multiplexing and multiple-input multiple-output communication can also be introduced into the system.

3.4.2 Array Gaussian Beam Control Based on Maka Antenna

The Gaussian beam emitted by a laser light source is very different from an ordinary beam in the following ways [58]:

(1) The Gaussian beam has good directivity, but it also constantly diverges, and its divergence law is different from that of a spherical wave.
(2) The wavefront curvature of a Gaussian beam changes continuously during its propagation but never becomes zero.
(3) Gaussian beams do not travel in straight lines except for at the beam center.

3.4.2.1 Off-Axis Gaussian Beam Emission Control

a. Transformation of Gaussian beam by ideal optical system

For a light beam in an axisymmetric ideal optical system being transmitted along the optical axis *Z*, it is assumed that the distance between the initial ray and the axis and the included angle with the axes are r_0 and θ_0, respectively. After passing through any optical system, these parameters change to r_1 and θ_1. If θ_0 and θ_1 are small, transmission through the optical fiber can be described using the fiber pursuit equation:

$$\begin{pmatrix} r_1 \\ \theta_1 \end{pmatrix} = \begin{pmatrix} A & B \\ C & D \end{pmatrix} \begin{pmatrix} r_0 \\ \theta_0 \end{pmatrix} \tag{3.76}$$

Let $T = \begin{pmatrix} A & B \\ C & D \end{pmatrix}$, *T* is referred to as the ABCD matrix of optical fiber change [59], and Eq. (3.76) is known as the ABCD law of geometric ray transformation. When an optical system is composed of *n* basic optical units, light travels in the order 1, 2, 3, …, *n* through the *n* optical units. Thus, the total transformation matrix is the inverse product of the transformation matrix for each element:

$$\begin{pmatrix} A & B \\ C & D \end{pmatrix} = \begin{pmatrix} A_n & B_n \\ C_n & D_n \end{pmatrix} \cdots \begin{pmatrix} A_2 & B_2 \\ C_2 & D_2 \end{pmatrix} \begin{pmatrix} A_1 & B_1 \\ C_1 & D_1 \end{pmatrix} \tag{3.77}$$

The propagation of Gaussian beams can also be described in terms of the ABCD matrix [60, 61]. When the Gaussian beam is transmitted along the optical axis Z, the expression for the Gaussian beam on the incident plane ($Z = 0$) is as follows:

$$E_{in}(r, 0) = \exp(\frac{1}{2}\alpha_0 k r^2) \tag{3.78}$$

where $\alpha_0 = \frac{2}{k\omega_0^2} + i\frac{1}{F_0}$ (measured in m^{-1}) and r represent the radial vectors.

After passing through the optical system consisting of n basic optical units, the Gaussian light field at the exit plane ($Z = L$) can be expressed as follows:

$$E_{out}(r, L) = \frac{1}{p(L)} \exp(ikL) \exp\left[-\frac{1}{2}\alpha(L) k r^2\right] \tag{3.79}$$

$$\begin{cases} p(L) = A + i\alpha_0 B = \frac{2}{kW^2} + i\frac{1}{F} \\ \alpha(L) = \frac{\alpha_0 D - iC}{A + i\alpha_0 B} \end{cases} \tag{3.80}$$

where A, B, C, and D are parameters in the total transformation matrix. Table 3.1 provides the details of the ABCD matrices of several commonly used optical elements.

b. Transformation of eccentric Gaussian beam by an ideal optical system

Generally, the transmission axis of a Gaussian beam is coincident with its peak intensity axis, but when off-axis transmission is adopted, as illustrated in Fig. 3.34, the transmission axis Z deviates from its peak intensity axis Z', and the beam is referred to as an eccentric Gaussian beam [62]. In [63], the variation law of the optical system with an eccentric Gaussian beam has been discussed in detail. In this study, the field distribution of an eccentric Gaussian beam transformed by the optical system has been expressed as follows:

$$\begin{aligned} E_{out}(x, y, L) = \frac{1}{p(L)} \exp(ikL) \exp\{ & -\frac{1}{2}\alpha(L)k\left[(x - x_d(L))^2 + (y - y_d(L))^2\right] \\ & + ik\left[\varepsilon_x(L)x + \varepsilon_y(L)y\right] + i\left[\phi_x(L) + \phi_y(L)\right]\} \end{aligned} \tag{3.81}$$

where x_d and y_d are the offsets of the central position of the beam waist, ε_x and ε_y are the tangent values on the planes xoz and yoz of the angle between the axes Z' and Z, and $i[\phi_x(L)x + \phi_y(L)y]$ is the additional phase factor. Here,

$$\begin{pmatrix} x_d(L) \\ \varepsilon_x(L) \end{pmatrix} = \begin{pmatrix} A & B \\ C & D \end{pmatrix} \begin{pmatrix} x_d(0) \\ \varepsilon_x(0) \end{pmatrix}; \quad \begin{pmatrix} y_d(L) \\ \varepsilon_y(L) \end{pmatrix} = \begin{pmatrix} A & B \\ C & D \end{pmatrix} \begin{pmatrix} y_d(0) \\ \varepsilon_y(0) \end{pmatrix} \sqrt{2} \tag{3.82}$$

Table 3.1 ABCD matrix of common optical elements

Name	ABCD matrix	Schematic
Homogeneous medium	$\begin{pmatrix} 1 & l \\ 0 & 1 \end{pmatrix}$	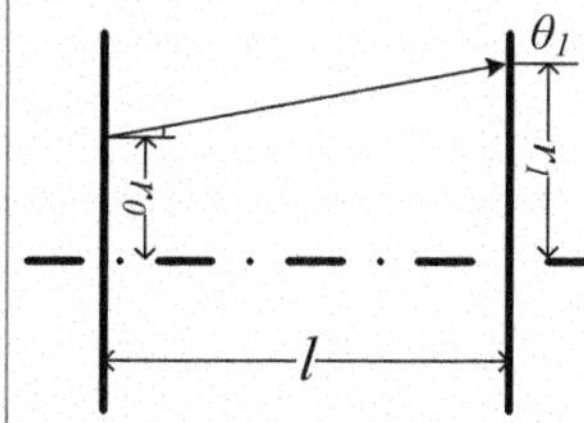
Thin lens	$\begin{pmatrix} 1 & 0 \\ -\frac{1}{f} & 1 \end{pmatrix}$	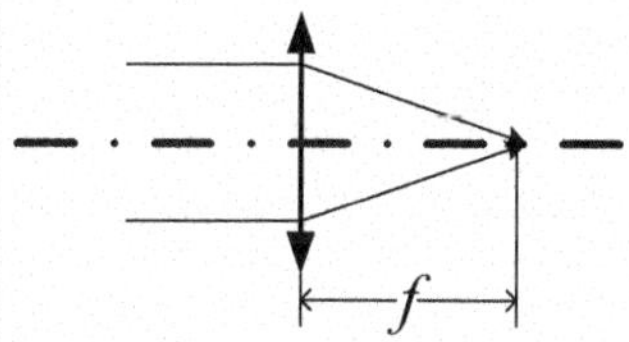
Spherical mirror	$\begin{pmatrix} 1 & 0 \\ -\frac{2}{R} & 1 \end{pmatrix}$	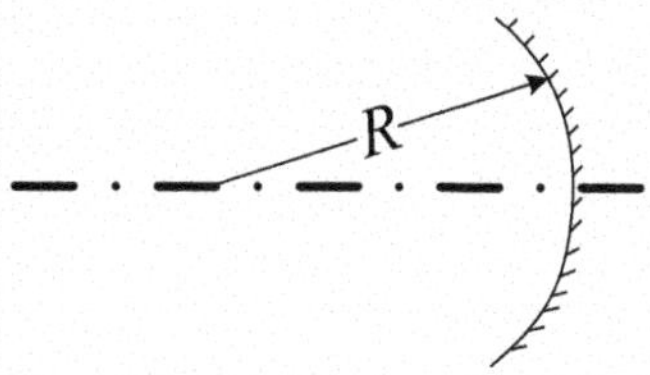

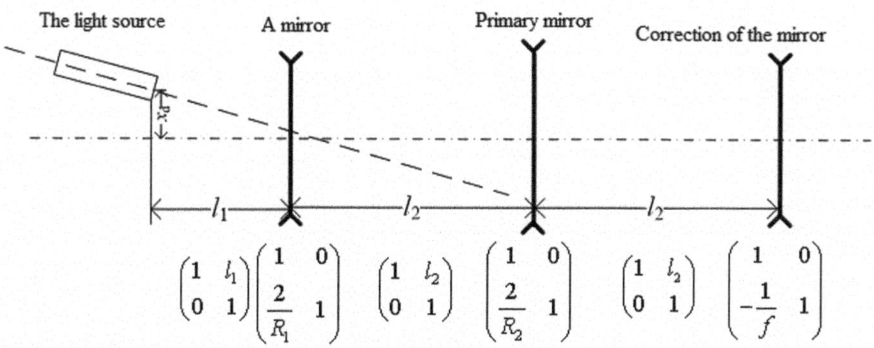

Fig. 3.34 Equivalent structure depicting Maka antenna transmitting principle

$$\phi_x(L) = \frac{k}{2}[Cx_d x_d(L) + B\varepsilon_x \varepsilon_x(L)];\quad \phi_y(L) = \frac{k}{2}\left[Cy_d y_d(L) + B\varepsilon_y \varepsilon_y(L)\right] \tag{3.83}$$

Therefore, the following can be concluded:

(1) The parameters of the eccentric beam $\alpha(L)$ satisfy the ABCD law, as indicated in Eq. (3.80), and are not related to the eccentric parameters.
(2) The parameters x_d, y_d, ε_x, and ε_y, representing the deviation between the transmission axis z and peak light intensity axis z', satisfy the ABCD law of geometric light transformation, as represented by Eq. (3.82).
(3) Following the transformation, the distribution of the eccentric Gaussian beam increases the phase factor $i[\phi_x(L) + \phi_y(L)]$, which is related not only to the eccentricity parameters $(x_d, y_d, \varepsilon_x, \varepsilon_y)$ but also to the transformation matrix of the optical system represented by Eq. (3.83).

c. ABCD matrix of Maka antenna

Starting from the position of the light source, the primary parameters of the Maka antenna are as follows: radius of secondary mirror, R_1; radius of primary mirror, R_2; distance from light source to secondary mirror, l_1; distance from secondary to primary mirror, l_2; and focal length of orthoscope, f. The structure of the Maka antenna (Fig. 3.32) and the principle of beam transmission indicate that the Maka antenna could be understood as being equivalent to the structure presented in Fig. 3.34. The most significant difference is that the primary and secondary mirrors are equivalent to divergent lenses.

Therefore, the ABCD matrix for the Maka antenna can be expressed as follows:

$$T_m = \begin{pmatrix} A & B \\ C & D \end{pmatrix} = \begin{pmatrix} 1 & 0 \\ -\frac{1}{f} & 1 \end{pmatrix}\begin{pmatrix} 1 & l_2 \\ 0 & 1 \end{pmatrix}\begin{pmatrix} 1 & 0 \\ \frac{2}{R_2} & 1 \end{pmatrix}\begin{pmatrix} 1 & l_2 \\ 0 & 1 \end{pmatrix}\begin{pmatrix} 1 & 0 \\ \frac{2}{R_1} & 1 \end{pmatrix}\begin{pmatrix} 1 & l_1 \\ 0 & 1 \end{pmatrix} \tag{3.84}$$

where

$$A = F(R_1, R_2, l_2) = \left(\frac{4}{R_1 R_2}\right) l_2^2 + \left(\frac{4}{R_1} + \frac{2}{R_2}\right) l_2 + 1 \tag{3.85}$$

$$B = F(R_1, R_2, l_1, l_2) = \left(\frac{4l_1}{R_1 R_2} + \frac{2}{R_2}\right) l_2^2 + \left(2 + \frac{4l_1}{R_1} + \frac{2l_1}{R_2}\right) l_2 + l_1 \tag{3.86}$$

$$C = F(R_1, R_2, l_2, f) = \left(-\frac{4}{f R_1 R_2}\right) l_2^2 + \left(-\frac{2}{f R_2} - \frac{4}{f R_1} + \frac{4}{R_1 R_2}\right) l_2 + \left(\frac{2}{R_1} - \frac{1}{f} + \frac{2}{R_2}\right) \tag{3.87}$$

$$D = F(R_1, R_2, l_1, l_2, f) = -\left(\frac{2}{fR_2} + \frac{4l_1}{fR_1R_2}\right)l_2^2 + \left(-\frac{2l_1}{fR_2} - \frac{2}{R_2} + \frac{4l_1}{R_1R_2}\right)l_2 + \left(\frac{2l_1}{R_1} - \frac{3l_1}{f} + \frac{2l_1}{R_2} - \frac{4l_1^2}{fR_1} + 1\right) \quad (3.88)$$

Therefore, the following can be concluded:

(1) If the light source is only transversally and longitudinally eccentric $(x_d(0), y_d(0))$, its transmission axis Z is parallel to its peak intensity axis Z', $(\varepsilon_x(0) = 0, \varepsilon_y(0) = 0)$. Using Eq. (3.82), the following relations can be obtained:

$$x_d(L) = Ax_d(0), \varepsilon_x(L) = Cx_d(0); y_d(L) = Ay_d(0), \varepsilon_y(L) = Cy_d(0) \quad (3.89)$$

Because the parameters A and C in the ABCD matrix of the Maka antenna are independent of the distance between the light source and the secondary mirror l_1, as indicated in Eqs. (3.85) and (3.87), the eccentricity parameters of the outgoing beam $(x_d(L), y_d(L), \varepsilon_x(L), \varepsilon_y(L))$ are independent of the distance between the light source and the secondary mirror l_1 and related only to the structural parameters of the Maka antenna and the offset of the center of the initial waist of the Gaussian beam $(x_d(0), y_d(0))$.

(2) If the light source is only transversally and longitudinally off-axis $(x_d(0), y_d(0))$, its transmission axis Z is parallel to its peak intensity axis Z', $(\varepsilon_x(0) = 0, \varepsilon_y(0) = 0)$. Using Eq. (3.83), the following relations can be obtained:

$$\phi_x(L) = \frac{kAC}{2}x_d^2(0); \phi_y(L) = \frac{kAC}{2}y_d^2(0) \quad (3.90)$$

The additional phase factor is not related to the distance l_1 from the light source to the secondary mirror but only to the structural parameters of the antenna, the wavelength of the light source, and the offset $(x_d(0), y_d(0))$ of the initial center of the Gaussian beam waist.

(3) The parameters of the emanating eccentric beam $\alpha(L)$ satisfy the ABCD law represented by Eq. (3.80), which is unrelated to the eccentric parameters but closely related to the structural parameters of the Maka antenna, distance between the light source and the secondary mirror l_1, and initial parameters of the Gaussian light source $\alpha_0 = \frac{2}{k\omega_0^2} + i\frac{1}{F_0}$.

3.4.2.2 Physical Model of Gaussian Beam Propagation in Space Array Under Ideal Conditions

As illustrated in Fig. 3.35, in the Cartesian coordinate system, it is assumed that N Gaussian beams is evenly distributed on a circle with radius r_0 and the angular

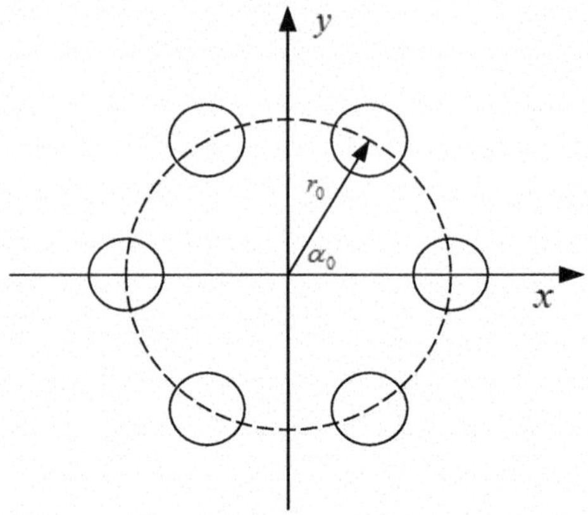

Fig. 3.35 Schematic of radial distributed beam array

spacing between adjacent sub-beams is $\alpha_0 = 2\pi/N$. These N sub-beams constitute a radially distributed beam array. The electric field of the array source field ($Z = 0$) can be expressed as follows:

$$E(x, y, 0) = \sum_{i=1}^{N} E_i(x, y, 0) \tag{3.91}$$

where $E(x, y, 0)$ represents the electric field of the ith Gaussian beam in the beam array. The electric field of a single paraxial Gaussian beam with unit amplitude on the plane $Z = 0$ can be expressed as follows:

$$E_i(\rho, 0) = \exp\left[-\frac{(\rho - r_i)^2}{\omega_0^2} - \frac{ik}{2F_0}(\rho - r_i)^2\right] \tag{3.92}$$

where $\rho = (x, y)$ represents the coordinate vector of a point on the source plane, ω_0 represents the waist spot radius of the fundamental-mode Gaussian beam array, $k = 2\pi/\lambda$ (λ is the wavelength) denotes the wave number in free space, r_i represents the central vector coordinates of the ith beam in the beam array, and F_0 represents the radius of the curvature of the wavefront. When $F_0 > 0$, the transmitting beam is a focused one; when $F_0 < 0$, it is a divergent one; and when $F_0 \to \infty$, it is a quasi-direct beam.

Combining Eqs. (3.91) and (3.92), the electric field of the radially distributed Gaussian beam array on the $Z = 0$ plane can be expressed as follows:

$$\begin{aligned} E_{im}(x, y, z = 0) &= \sum_{i=1}^{N} \exp\left[-\frac{(\rho - r_i)^2}{\omega_0^2} - \frac{ik}{2F_0}(\rho - r_i)^2\right] \\ &= \sum_{i=1}^{N} \exp\left[-\frac{1}{2}\alpha_0 k(\rho - r_i)^2\right] \end{aligned} \tag{3.93}$$

Under ideal conditions, the Gaussian beam array transmits a distance L along the z-axis of the optical axis, and the electric field of the radially distributed Gaussian beam array on the $Z = L$ plane can be expressed as follows:

$$E_{out}(x, y, z = L) = \sum_{i=1}^{N} \frac{1}{1 + i\alpha_0 L} \exp\left[ikL - \frac{1}{2}\left(\frac{\alpha_0 k}{1 + i\alpha_0 L}\right)(\rho - r_i)^2\right] \quad (3.94)$$

To study the uniform radial distribution of four Gaussian beams in the array, where each Gaussian beam waist is 0.01 m and beam center is 0.02 m away from the center of the observation plane, N is set to 4 and Eq. (3.94) is used. Table 3.2 indicates the changes in light intensity when the pound axis of the Gaussian beam in this array is transmitted to various distances under ideal conditions. It describes the divergence of the array of the Gaussian beam ($F_0 < 0$), the collimating array Gaussian beam ($F_0 = \infty$), and the convergent array of the Gaussian beam ($F_0 > 0$) when it transmission horizontally along the optical axis. The wave curvature of the Gaussian beam may change by itself with the increase of the transmission distance. Finally, the light beam can evolve to a discrete Gaussian beam; the farther the distance, the better the combination effect and the higher the attenuation of the central energy.

3.4.2.3 Physical Model of Gaussian Beam Propagation in Space Array Under Atmospheric Turbulence

The expression for the cross-spectral density of a Gaussian beam array on plane $Z = 0$ is as follows:

$$W^{(0)}(\rho_{s1}, \rho_{s2}; z = 0) = \langle E^*(\rho_{s1}, z = 0)E(\rho_{s2}, z = 0)\rangle \quad (3.95)$$

When the radially distributed Gaussian beam arrays are independent of each other, referred to as incoherent synthesis, the expression for the cross-spectral density on the $Z = 0$ plane is as follows [64, 65]:

$$W^{(0)}(\rho_{s1}, \rho_{s2}; 0) = \sum_{i=1}^{N} \exp\left[-\frac{(\rho_{s1}-r_i)^2 + (\rho_{s2}-r_i)^2}{\omega_0^2}\right] \times \exp\left[-\frac{ik}{2F_0}\left((\rho_{s1} - r_i)^2 - (\rho_{s2} - r_i)^2\right)\right] \quad (3.96)$$

where ρ_{s1} and ρ_{s2} are the vector coordinates of the two points on the source plane and r_i are the vector coordinates of the center of the circle of the first beam in the beam array. Based on the generalized Huygens–Fresnel principle, the cross-spectral density function of the two points (ρ_1, z) and (ρ_2, z) on the receiving plane under atmospheric turbulent transport i can be expressed as follows [66]:

Table 3.2 Pound-axis transmission intensity variation in array Gaussian beam under ideal conditions ($N = 4$) [20]

Simulation parameters: $r_0 = 0.02$ m, $\omega_0 = 0.01$ m		
$F_0 = -700$	$F_0 = \infty$	$F_0 = 700$
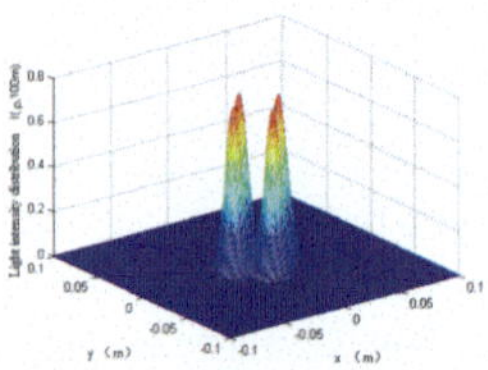	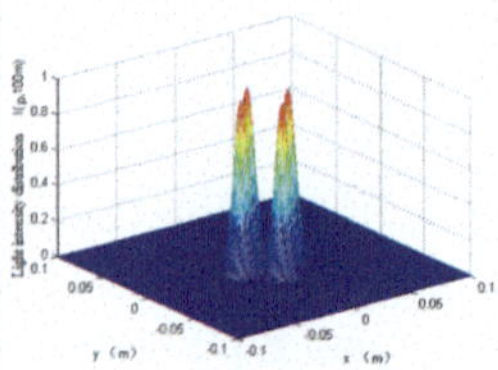	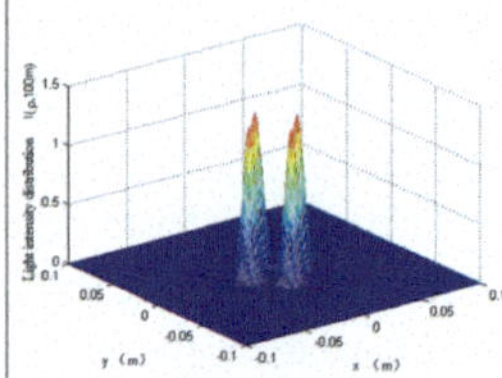
(a) $L = 100$ m		
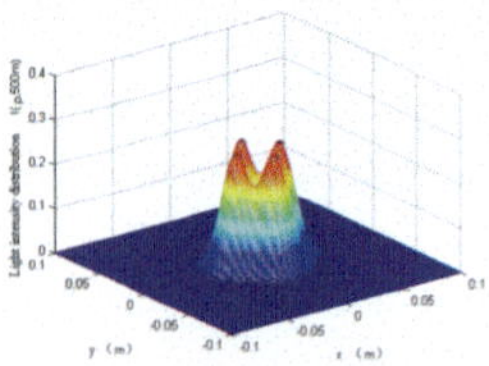	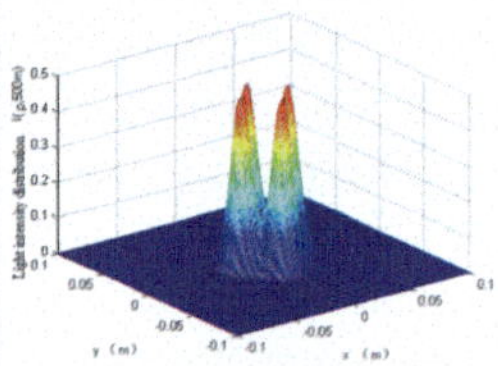	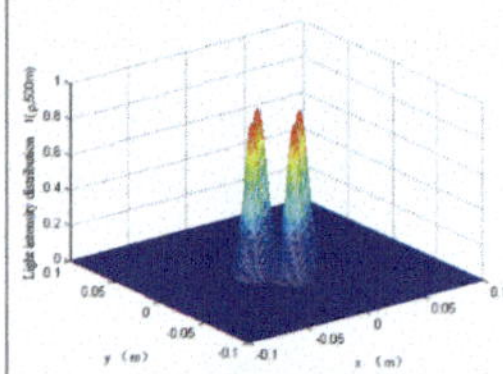
(b) $L = 500$ m		
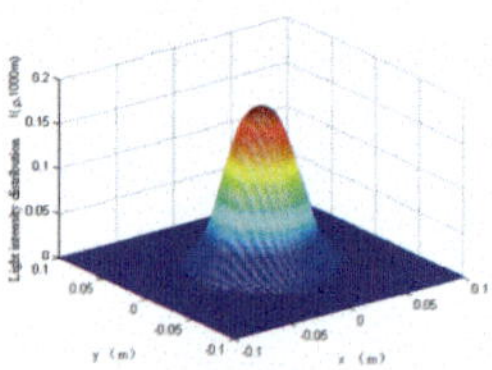	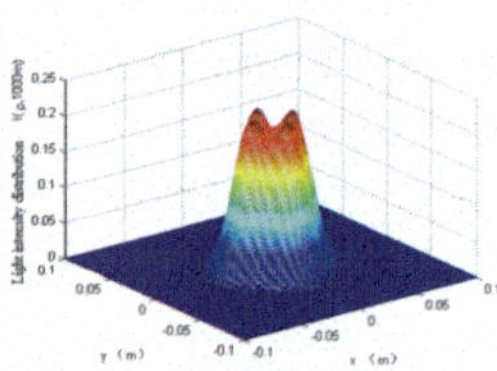	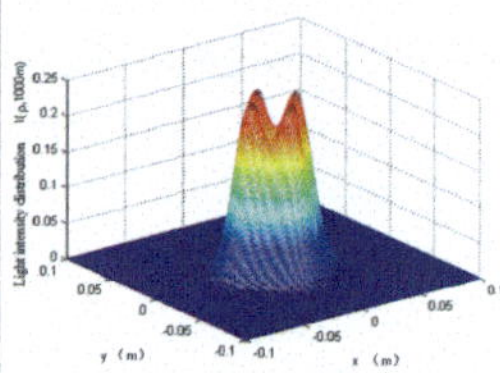
(c) $L = 1000$ m		
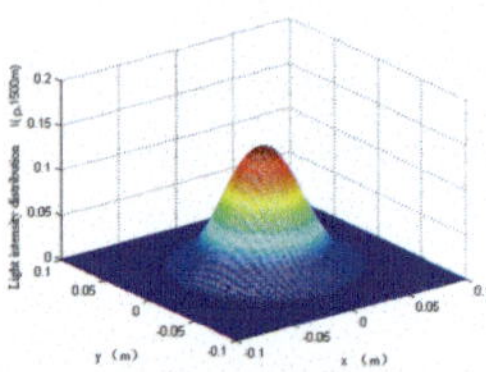	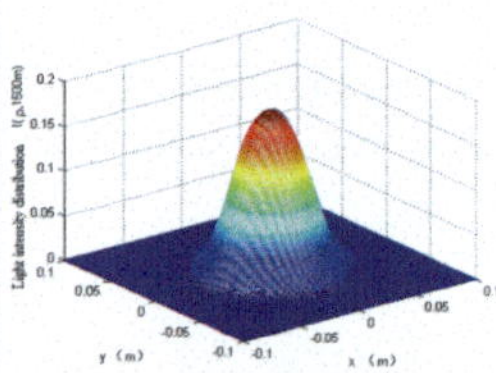	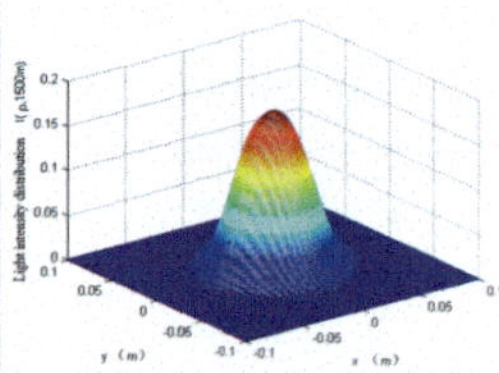
(d) $L = 1500$ m		

$$W(\boldsymbol{\rho}_1, \boldsymbol{\rho}_2; z) = (\frac{k}{2\pi z})^2 \int\int d^2\boldsymbol{\rho}_{s1} \int\int d^2\boldsymbol{\rho}_{s2} W^{(0)}(\boldsymbol{\rho}_{s1}, \boldsymbol{\rho}_{s2}; 0) \times \left\{\exp\left[\frac{-ik}{2z}(\boldsymbol{\rho}_1 - \boldsymbol{\rho}_{s1})^2 - (\boldsymbol{\rho}_2 - \boldsymbol{\rho}_{s2})\right]\right\} \times \left\langle \exp[\psi(\boldsymbol{\rho}_1, \boldsymbol{\rho}_{s1}, z) + \psi^*(\boldsymbol{\rho}_2, \boldsymbol{\rho}_{s2}, z)]\right\rangle \quad (3.97)$$

where $\psi(\rho, \rho_s)$ is the complex random perturbation of spherical waves propagating from the emission aperture $(\rho_s; 0)$ to the target point $(\rho; z)$ in the turbulent medium, * denotes the complex conjugate, and $\langle\cdot\rangle$ represents the ensemble average of the turbulent medium.

It is assumed that the fluctuations in the light wave and the atmospheric turbulence are statistically independent, and the mean values of these parameters can be calculated separately. The angular brackets in Eq. (3.97) describing the impact of turbulence can be expressed as follows [28]:

$$\begin{aligned}&\left\langle \exp[\psi^*(\rho_1, \rho_{s1}) + \psi(\rho_2, \rho_{s2}, z)]\right\rangle \cong \\ &\exp\left[\frac{-\pi^2k^2z}{3}[((\rho_1 - \rho_2)^2 + (\rho_1 - \rho_2)\cdot(\rho_{s1} - \rho_{s2}) \right. \\ &\left. +(\rho_{s1} - \rho_{s2})^2)]\int_0^\infty \kappa^3\Phi_n(\kappa)d\kappa\right]\end{aligned} \tag{3.98}$$

where $\int_0^\infty \kappa^3\Phi_n(\kappa)d\kappa$ represents the impact of turbulence and $\Phi_n(\kappa)$ represents the power spectrum of the fluctuation in the refractive index. In this section, the modified von Karman spectrum (represented by Eq. (3.44)), considering both the internal and external scales, is used for the simulation analysis. Let $\rho_1 = \rho_2 = \rho$ represent an integral solution. Therefore, we obtain the light intensity distribution expression for a Gaussian beam array with radial distribution as follows:

$$I(\rho, z) = W(\rho, \rho, z) = \frac{1}{T}\sum_{i=1}^{N}\exp\left[-\frac{2}{\omega_0^2T}(\rho - r_i)^2\right] \tag{3.99}$$

where

$$T = \frac{4z^2}{k^2\omega_0^4} + \frac{8Mz^2}{k^2\omega_0^2} + \frac{z^2}{F_0^2} + \frac{2z}{F_0} + 1 \tag{3.100}$$

$$M = \frac{1}{3}\pi^2k^2z\int_0^\infty \kappa^3\Phi_n(\kappa)d\kappa \tag{3.101}$$

It can be seen from Eq. (3.99) that the light intensity distribution of the radial Gaussian beam array is related to the wavelength, intensity of turbulence, internal and external scales, central position of the beam, transmission distance, and other parameters.

Considering a collimating beam as an example ($F_0 = \infty$), Eq. (3.99) is used to simulate the transmission of an array Gaussian beam at various transmission distances under atmospheric turbulence. Table 3.3 summarizes the changes in the pound-axis transmission intensity of the collimating-array Gaussian beam under atmospheric turbulence. It can be seen that when $C_n^2 = 10^{-16}$ m$^{-2/3}$ (weak turbulence) and $C_n^2 = 10^{-14}$ m$^{-2/3}$ (medium–strong turbulence), the combination of the four collimating Gaussian beams at 1000 m is more apparent (as described in (c) in Table 3.3), and the light intensity decreases significantly, leading to a larger spot diffusion area after the combination (see Fig. 3.36). This theory indicates that atmospheric turbulence can cause the beam array to coalesce in advance.

Table 3.3 Pound-axis transmission intensity changes in the collimating array Gaussian beam under atmospheric turbulence ($N = 4$) [20]

Simulation parameters: $r_0 = 0.02$ m, $\omega_0 = 0.01$ m, $F_0 = \infty$		
Under ideal conditions	$C_n^2 = 10^{-16}$ m$^{-2/3}$ (weak turbulence)	$C_n^2 = 10^{-14}$ m$^{-2/3}$ (strong turbulence)
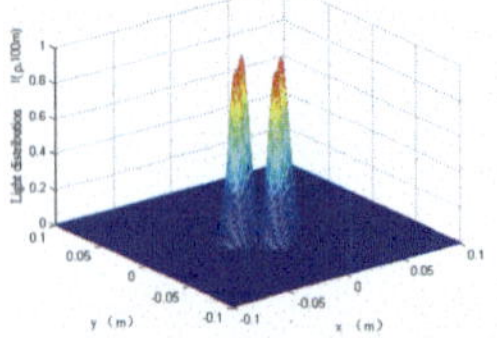	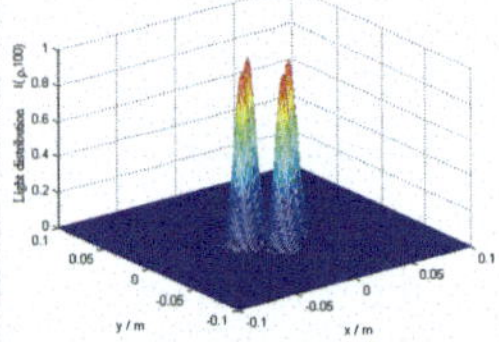	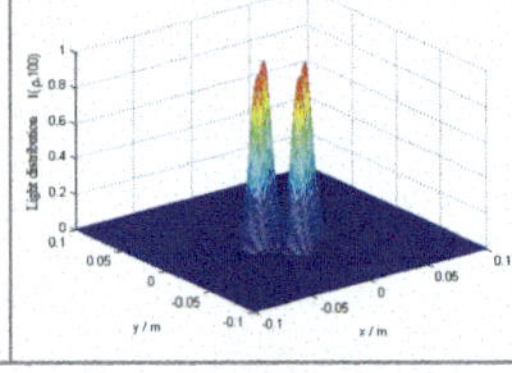
(a) $L = 100$ m		
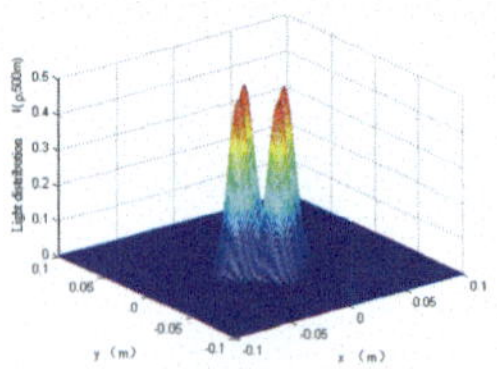	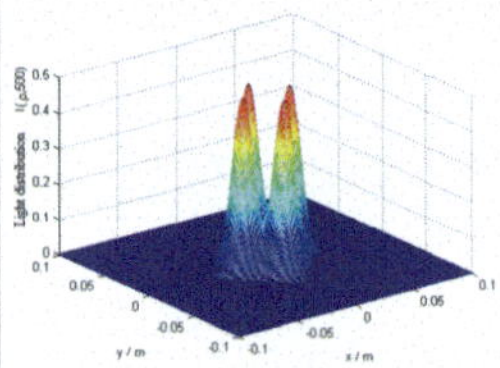	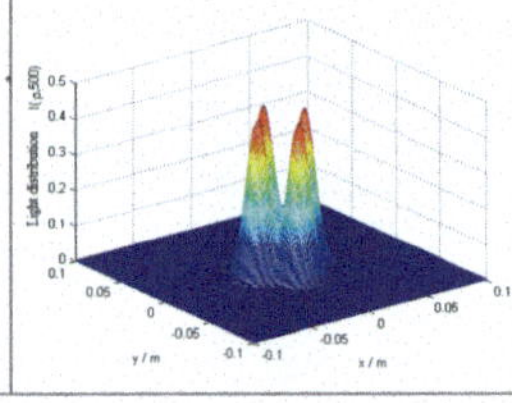
(b) $L = 500$ m		
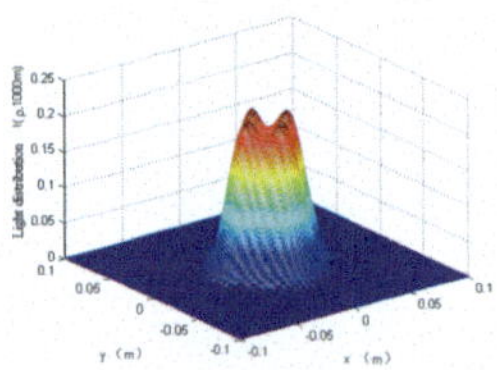	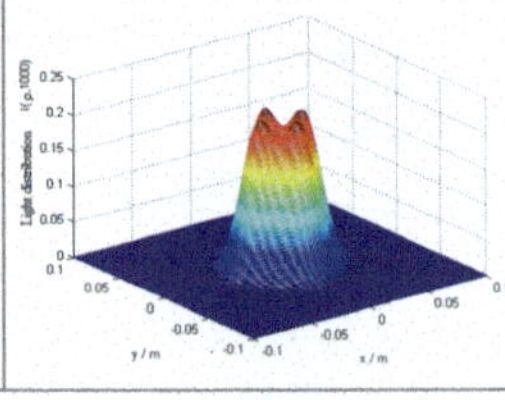	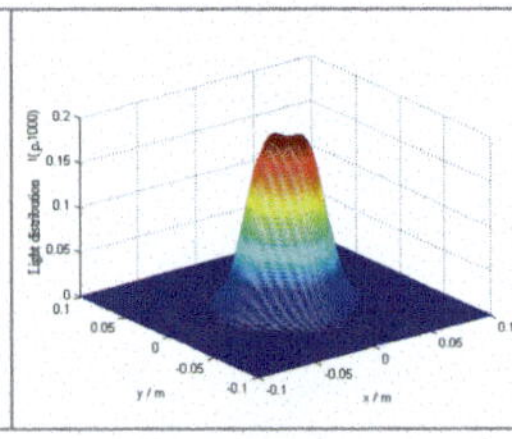
(c) $L = 1000$ m		
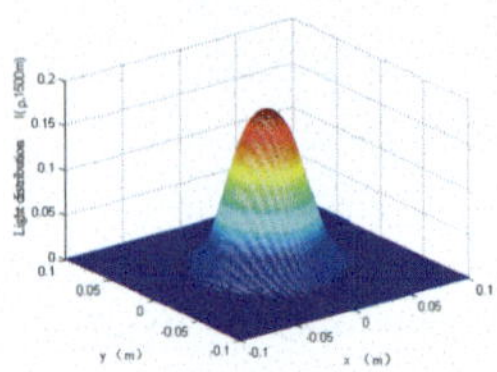	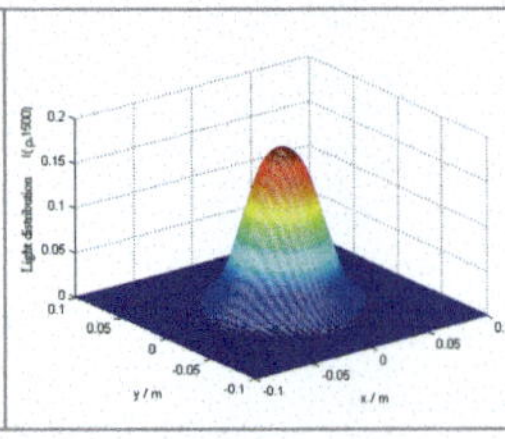	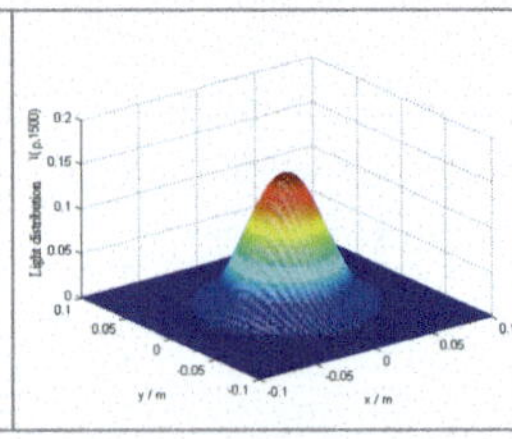
(d) $L = 1500$ m		

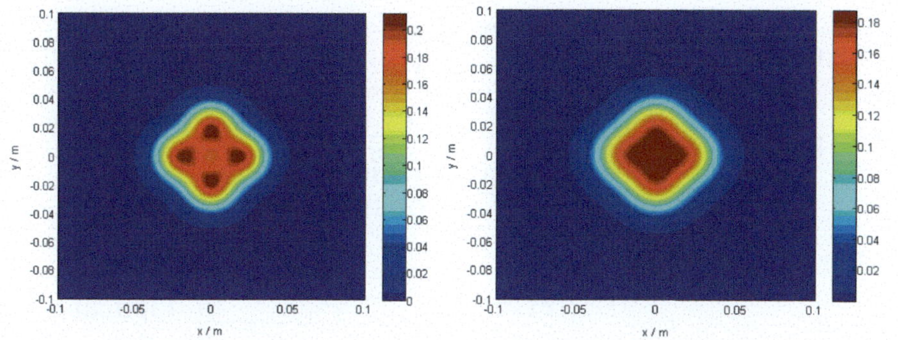

(a)Weak turbulence (b) Strong turbulence

Fig. 3.36 Average intensity distribution of Gaussian beam transmitted for 1 km under turbulence [20]

3.4.3 Coupling Efficiency of Maka Antenna Under Atmospheric Turbulence

Unlike ordinary lenses, owing to the existence of a secondary reflector, the existence of a secondary reflector shielding ratio should be considered when the Maka antenna is equivalent to a lens system [67]. As illustrated in Fig. 3.37, the Maka antenna system is equivalent to a long-focus single-lens system, and its incident aperture plane A is a ring-shaped structure with a certain shielding ratio. The shielding ratio $b = d_M / D_M$, where d_M and D_M are the diameters of the secondary reflector and meniscus correction mirror, respectively.

As illustrated in Fig. 3.38, spatial light is coupled into the single-mode fiber through the Maka antenna. The laser propagates for a certain distance through the optical system collimation under atmospheric turbulence and is incident on the surface of the receiving aperture A. The Maka antenna concentrates and shapes the distorted beam and is incident on the surface of the optical fiber. Using the definition

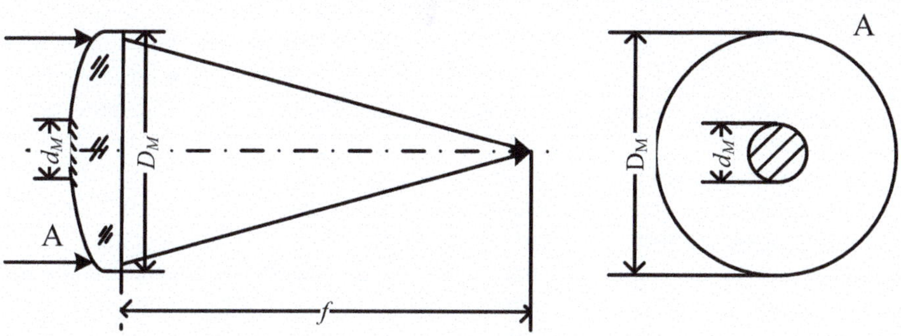

Fig. 3.37 Equivalent structure intention of Maka antenna

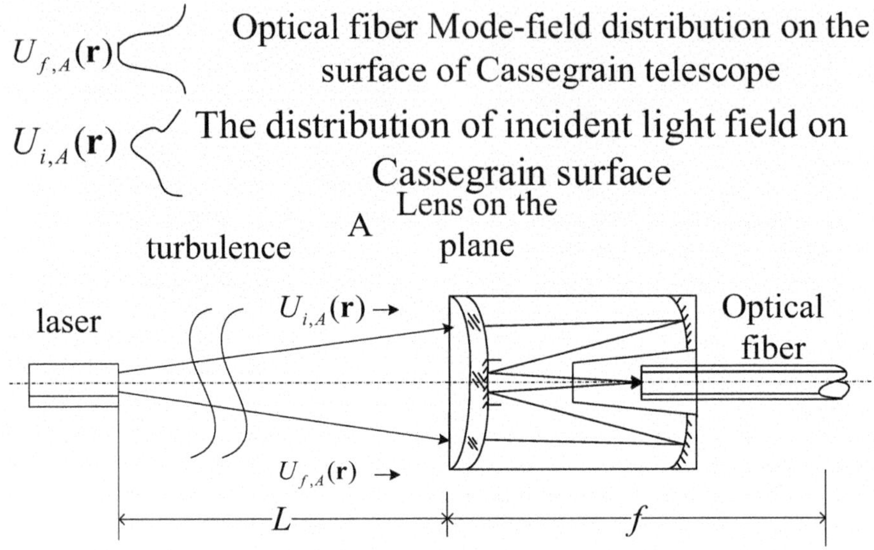

Fig. 3.38 Spatial optical–Maka antenna–fiber coupling diagram

of coupling efficiency and the equivalent structure of the Maka antenna, the ensemble average of the coupling efficiency of the Maka antenna in the lens array under the von Karman turbulence spectrum can be expressed as follows:

$$\begin{aligned}\langle\eta_{M,V}\rangle = \Bigg|\frac{8a^2}{\pi}\int\limits_b^1\int\limits_b^1\int\limits_0^\pi \exp[-a^2(x_1^2+x_2^2)] \\ \times \exp\bigg(-\frac{A_R}{A_{C,V}}N(x_1,x_2,\theta) \\ \Big\{[1+GN(x_1,x_2,\theta)]^{-\frac{1}{6}} - C\Big\}\bigg)x_1x_2d\theta dx_1dx_2\Bigg|_{D_A=D_M}\end{aligned} \tag{3.102}$$

where b is the blocking ratio of the card antenna, D_M is the diameter of the Maka antenna, and $|\cdot|_{D_A=D_M}$ indicates the replacement of D_A in the representative Eq. (3.102) with the Maka antenna diameter D_M for calculation. The other parameters are consistent with Eq. (3.60).

The curves presented in Fig. 3.39 are plotted using Eq. (3.102) for the numerical simulation of the horizontal transmission under weak turbulence ($C_n^2 = 10^{-16}$ m$^{-2/3}$) with different obscuration ratio b (the simulation parameters are specified in the figure). It can be seen that with an increasing obscuration ratio, the maximum coupling efficiency declines and the theoretical optimum coupling parameters match. When the shielding ratio increases by 0.2, the coupling efficiency decreases by at least 10%. In particular, the maximum coupling efficiency ($\eta_c \approx 80\%$) can be obtained

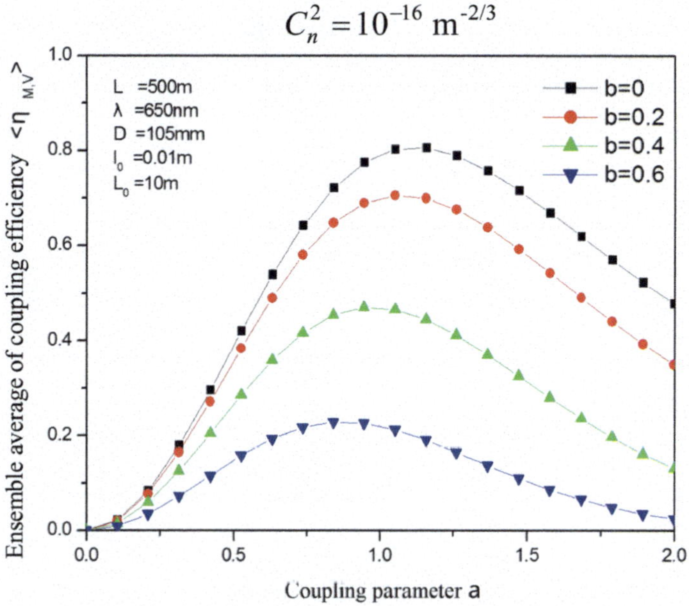

Fig. 3.39 Coupling parameter versus coupling efficiency under weak turbulence [20]

when $b = 0$ and $A = 1.121$; when $b = 0.2$ and $a = 1.074$, the maximum coupling efficiency $\eta_c \approx 70\%$. Figure 3.40 presents the relationship between the beam transmission distance and coupling efficiency for horizontal transmission and under weak turbulence $C_n^2 = 10^{-16}$ m$^{-2/3}$ (the simulation parameters are specified in the figure). In Fig. 3.42, the coupling efficiency at $b = 0$ (solid line) is significantly better than that at $b = 0.2$ (dashed line). It can be seen that under weak turbulence, the increase in the shielding ratio b of the Maka antenna is the primary cause for the decrease in the coupling efficiency.

The curves presented in Fig. 3.41 are plotted using Eq. (3.102) for the numerical simulation of the horizontal transmission under strong turbulence ($C_n^2 = 10^{-14}$ m$^{-2/3}$) with different obscuration ration b (the simulation parameters are specified in the figure). Compared with the weak turbulence shown in Fig. 3.39, the maximum coupling efficiency under different obscuration ratio decreases rapidly in medium turbulence. When $b = 0$, the coupling efficiency drops to $\eta_c \approx 16\%$; when $b = 0.2$, $\eta_c \approx 13\%$. Figure 3.42 presents the curves plotted between the beam transmission distance and coupling efficiency for horizontal transmission (the simulation parameters are specified in the figure). The figure indicates that the coupling efficiency at 1550 nm (line with squares) is the highest, followed by that at 850 nm (line with circles) and 650 nm (line with triangles). The following conclusions can be drawn:

(1) Under medium–strong turbulence, the coupling efficiency of long wavelengths is clearly better than that of short wavelengths. When $b = 0$ and $A = 1.121$, the

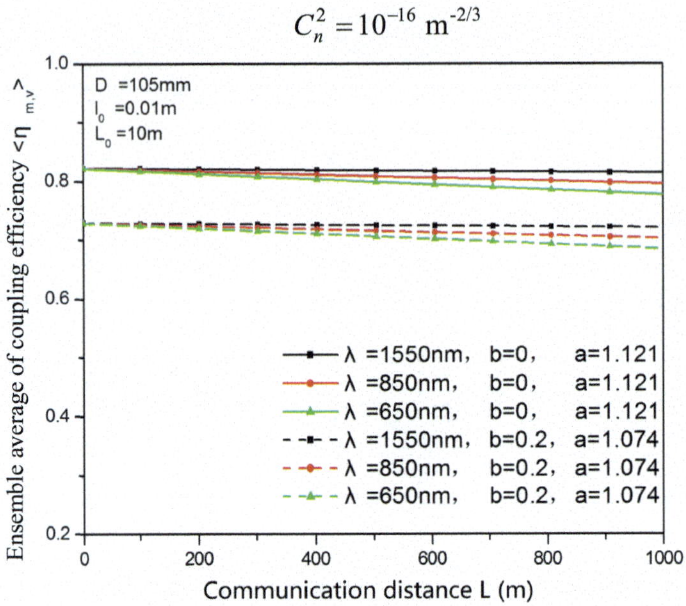

Fig. 3.40 Communication distance versus coupling efficiency under weak turbulence [20]

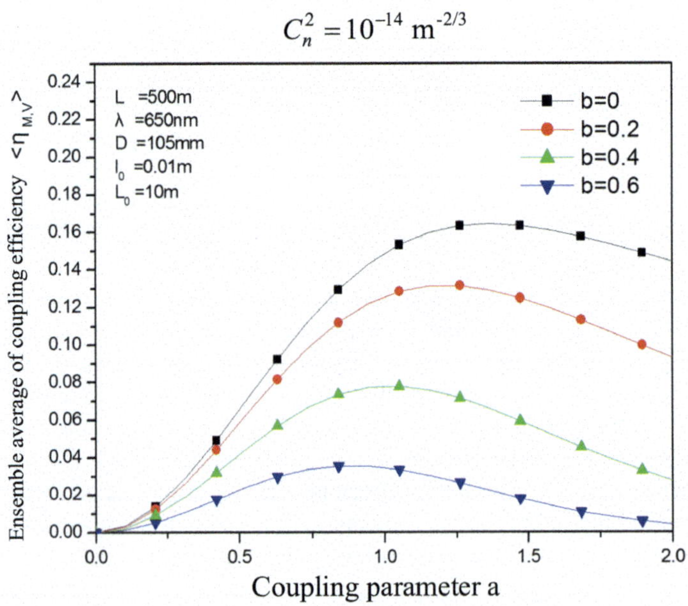

Fig. 3.41 Coupling parameter versus coupling efficiency under medium–strong turbulence [20]

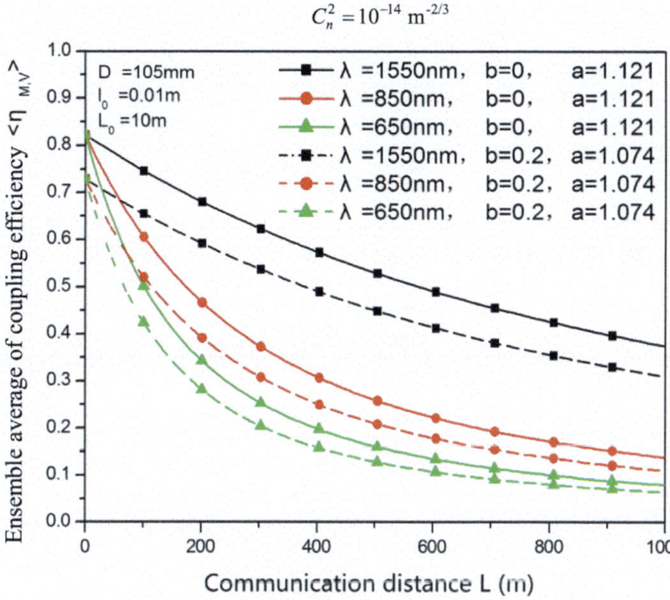

Fig. 3.42 Communication distance versus coupling efficiency under medium–strong turbulence [20]

coupling efficiency of the 1550-nm light wave can still be higher than 40% at 1 km, whereas the coupling efficiency of the 650-nm light wave is reduced to approximately 10%.

(2) Under medium–strong turbulence, an increase in the light transmission distance is the primary factor that causes the coupling efficiency to reduce. The trend of the solid ($b = 0$) and dotted ($b = 0.2$) lines getting closer indicates that the effect of the shielding ratio $b = 0.2$ of the Maka antenna on the coupling efficiency can be ignored when the transmission distance of the beam under strong turbulence continues to increase to more than a kilometer (its effect on short wavelength is evident). Therefore, the shielding ratio of the Maka antenna should not exceed 0.2; otherwise, although the birefringence structure of the antenna can cause the size of the antenna to reduce, its shielding has too significant an impact on the coupling efficiency.

References

1. Lv BD (2000) Developments in the propagation, transformation and beam control of high-power lasers. Infrared Laser Eng 29(1):40–45
2. Guan XL (2010) Technical of liquid crystal application in controlling system of laser beam. Microcomput Inf 26(34):88–89

3. Yoshida K, Tanaka K, Tsujimura T (2013) Assisted focus adjustment for free space optics system coupling single-mode optical fibers. IEEE Trans Industr Electron 60(11):5306–5314
4. Shaw MD, Birnbaum K, Cheng M et al (2014) A receiver for the lunar laser communication demonstration using the optical communications telescope laboratory. Sci Innov 1–2
5. Takahashi K, Nakajima N (2010) Design and evaluation of fiber direct coupling optical antennas for next generation optical wireless communication systems. In: 2010 International conference on broadband, wireless computing, communication and applications, Victoria, British Columbia, Canada, pp 129–136
6. Abadi MM, Ghassemlooy Z, Zvanovec S et al (2016) Dual purpose antenna for hybrid free space optics/RF communication systems. J Lightwave Technol 34(14):3432–3439
7. Leeb WR, Winzer PJ, Kudielka KH (1998) Aperture dependence of the mixing efficiency, the signal-to-noise ratio, and the speckle number in coherent lidar receivers. Appl Opt 37:3143–3148
8. Li SC, Hua DX, Song YH (2011) Research on micro-lens coupling system of all-fiber Raman Lidar. Acta Opt Sin 31(6):0601001
9. Boroson DM (1993) Overview of Lincoln Laboratory development of lasercom technologies for space. Proc SPIE 18(6):30–39
10. Ruilier C (1998) A study of degraded light coupling into single mode fibers. Proc SPIE 3350:319–329
11. Belmonte A, Kahn KM (2011) Field conjugation adaptive arrays in free-space coherent laser communications. J Opt Commun Network 11(3):830–838
12. Dikmelik Y, Davidson FM (2005) Fiber-coupling efficiency for free-space optical communication through atmospheric turbulence. Appl Opt 44(23):4946–4952
13. Wu H, Yan H, Li X (2010) Modal correction for fiber-coupling efficiency in free-space optical communication systems through atmospheric turbulence. Opt Int J Light Electorn Opt 121:1789–1793
14. Ma J, Zhao F, Tan LY et al (2009) Plane wave coupling into single-mode fiber in the presence of random angular jitter. Appl Opt 48(27):5184
15. Takenaka H, Toyoshima M, Takayama Y (2012) Experimental verification of fiber-coupling efficiency for satellite-to-ground atmospheric laser downlinks. Opt Expr 20(14):15301
16. Tan L, Zhai C, Yu S et al (2014) Fiber-coupling efficiency for optical wave propagating through non-Kolmogorov turbulence. Opt Commun 331:291–296
17. Liao YB (2007) Fiber optics. Tsinghua University Press, Beijing
18. Lv NG (2006) Fourier optics. China Machine Press, Beijing
19. Shao XF, Zhang X, Wu JF et al (1986) Study on mode field radius of single mode fiber. J Commun 3:49–54
20. Lei SC (2016) Research on optical coupling and beam control in free space optical communication. Xi'an University of Technology, Xi'an
21. Steward EG (1983) Fourier optics. Halsted Press, New South Wales
22. Li ZQ (1986) Application of Parseval's theorem in Fourier optics. J Sichuan Normal Univ (Nat Sci) 1:75–80
23. Li SC (2014) Research on key technology of all-fiber optical fiber rotational Raman temperature measurement LiDAR system. Xi'an University of Technology, Xi'an
24. Gisin N, Passy R, Perny B (1993) Optical fiber characterization by simultaneous measurement of the transmitted and refracted near field. J Lightwave Technol 11(11):1875–1883
25. Chen JB, Peng RL (2008) Laser principle and application. Publishing House of Electronics Industry, Beijing
26. Xiao ZG, Li BC (2008) Single-lens coupling efficiency of a fundamental Gaussian beam to an optical fiber. Opto-Electron Eng 35(8):29–34
27. Lu YX, Yang YP, Chen SF (1999) Laser beam transmission and transformation technology. University of Electronic Science and Technology of China Press, Chengdu
28. Sakai J, Kimura T (1980) Design of a miniature lens for semiconductor laser to single-mode fiber coupling. IEEE J Quantum Electron 16(10):1059–1066
29. Andrews LC (1992) Special functions of mathematics for engineers. McGraw-Hill, New York

30. Andrews LC, Phillips RL, Miller WB (1997) Mutual coherence function for a double-passage retroreflected optical wave in atmospheric turbulence. Appl Opt 36(3):698–708
31. Kendall M, Stuart A, Ord JK (1987) Kendall's advanced theory of statistics. Griffin, London
32. Zhang YX (2002) Light propagation and imaging in random media. National Defense Industry Press, Beijing
33. Tatarskii VI (1961) Wave propagation in a turbulent medium. Dover Publication, New York
34. Miller WB, Richlin JC, Andrews LC (1994) Effects of the refractive index spectral model on the irradiance variance of a Gaussian beam. J Opt Soc Am A 11(10):2719–2726
35. Andrews LC, Vester S (1993) Analytic expressions for the wave structure function based on a bump spectral model for refractive index fluctuations. J Mod Opt 40(5):931–938
36. Wang M, Heng Y (2001) The sample mean Monte Carlo method for approximate integral using measure theory. J South China Normal Univ (Nat Sci) 01:50–52
37. Ji Y, Yue P, Yan RQ et al (2016) BER performance analysis of the atmospheric laser communication system on the slant path in weak turbulence. J Xidian Univ (Nat Sci) 1:66–70
38. Du W, Yao Z, Ai R et al (2013) Influence of non-Kolmogorov turbulence on bit-error rates in laser satellite communications. J Russ Laser Res 34(4):351–355
39. Lei SC, Ke XZ (2015) Coupling efficiency of lens array spatial light in turbulence. Chin J Lasers 42(6):179–186
40. Liu Y (2012) Research on wavefront correction method based on improved simulated annealing algorithm. Huazhong University of Science and Technology, Wuhan
41. Li SY (2001) Parallel realization of simulated annealing algorithm: modifications and applications. Acta Phys Sin 50(7):1260–1263
42. Ning WG (2012) Study on polarization control in coherent optical communication system. Xi'an University of Technology, Xi'an
43. Jiang JW, Wu WF (2016) An improved particle swarm optimization algorithm. Comput Program Skills Maintenance 6:12–13, 15
44. Shi Y, Eberhart RC (1998) A modified particle swarm optimizer. In: IEEE International conference on evolutionary computation proceedings, Anchorage, pp 69–73
45. Shi Y, Eberhart RC (1999) Empirical study of particle swarm optimization, vol 3. Congress on evolutionary computation, Washington, DC, USA, pp 1945–1950
46. Li XH, Chen ZL, Gu LL et al (2009) Design of Cassegrain optical receiving antenna in free-space optical communication. Commun Technol 42(3):38–39, 42
47. Dong HF, Ren KX (1994) A quasi-optical analysis of Cassegrain antenna systems. J Xidian Univ 1:60–63
48. Zhang PF, Wang ZB, Li X (2014) Simulation and tolerance analysis of axicon duct for coupling Cassegrain telescope into multimode fiber. J Appl Opt 35(3):432–437
49. Yang YJ, Ma JX, Deng K et al (2012) Propagation of laser beams in the turbulent atmosphere passing through Cassegrain antenna. Laser Infrared 42(7):739–743
50. Cao GY, Yang YQ, Yu JJ et al (2014) Influence of temperature on the coupling efficiency of spatial light into single model fiber. Acta Opt Sin 34(10):65–70
51. Ran YH, Yang HJ, Xu Q et al (2009) Properties of Cassegrain optical antenna with inclined optical axis. Acta Phys Sin 58(2):946–951
52. Yang R, Su Y (2003) Analysis for stray light of a Cassegrain telescope. Opto-Electron Eng 30(5):48–50, 65
53. Marasco A, Romano A (2007) Maksutov's cameras and telescopes. Int J Eng Sci 45(1):34–40
54. Zhang Y, Ai Y (2005) Drawback of Cassegrain antenna system based on space optical communication. Infrared Laser Eng 34(5):560
55. Almeida J, Liang D (2012) Design of a high brightness solar-pumped laser by light-guides. Opt Commun 285(24):5327–5333
56. Ke XZ, Lei SC, Wang TY (2015) Cassegrain transceiver antenna based on fiber array. Laser Optoelectron Progr 52(12):120602
57. Ke XZ, Zhan J, Pei GQ (2012) Multiple-beam transmission techniques for wireless laser communication. Opto-Electron Eng 39(7):1–7
58. Chen HM (2009) Laser principle and application. Electronic Industry Press, Beijing

59. Casperson LW (1981) Synthesis of Gaussian beam optical systems. Appl Opt 20(13):2243–2249
60. Qiu JH, Mei RH (1991) Discussion of ABCD matrix and Gaussian beam optical systems. Opt Instrum 3:25–32
61. Chen K, Cao JQ, Lu QS et al (2011) Polarization properties of Gaussian Schell-model beams through ABCD optical system. Electro-Opt Technol Appl 26(4):28–31, 48
62. Al-Rashed AAR, Saleh BEA (1995) Decentered Gaussian beams. Appl Opt 34(30):6819–6825
63. Lu YX, Yang YP, Chen SF (1999) Transmission and transformation of laser beams. University of Electronic Science and Technology of China Press, Chengdu
64. Li CQ, Zhang HY, Wang TF et al (2013) Investigation on coherence characteristics of Gauss-Schell model beam propagating in atmospheric turbulence. Acta Phys Sin 62(22):224203-7
65. Ke XZ, Wang W (2015) Intensity and expansion of partially coherent beam propagating in slant and horizontal atmospheric turbulence. J Appl Sci 33(2):142–154
66. Cai Y, Lin Q (2004) Partially coherent flat-topped multi-Gaussian–Schell-model beam and its propagation. Opt Commun 239(1):33–41
67. Ke X, Lei S (2016) Spatial light coupled into a single-mode fiber by a Maksutov-Cassegrain antenna through atmospheric turbulence. Appl Opt 55(15):3897–3902

Chapter 4
Beam Polarization Control Technology

Contents

Starting from the basic theory of light polarization, the basic representation of light polarization state is introduced, the principles of polarization controller and mixer are analyzed in detail, the influence of signal light polarization state on IF signal amplitude and mixing efficiency in coherent optical communication system is deduced, the actual situation of signal light at the receiver and several common polarization control schemes are analyzed, and finally the polarization control scheme applicable to coherent optical communication system is given.

X. Ke and J. Wu, *Coherent Optical Wireless Communication Principle and Application*, Optical Wireless Communication Theory and Technology,
https://doi.org/10.1007/978-981-19-4823-7_4

4.1 Advances in Beam Polarization Control

(1) Research status of polarization controller

In free-space optical communication systems, it is generally assumed that the polarization state of the beam is constant when transmitted in the atmospheric channel [1], and this assumption is based on the assumption that the light transmitted in the atmosphere is a coherent plane wave, but in the actual system the signal light is partially coherent, and the polarization state of the signal light changes randomly when the partially coherent beam is transmitted in the atmosphere [1], and the speed of change is related to the turbulence size [2], and the temperature also causes a slow drift of the polarization state of the signal light [3], and in order to avoid signal damage caused by the mismatch between the polarization state of the signal light and the native light in coherent optical communication systems, a polarization control system must be incorporated. Based on the principle that the magnitude and direction of pressure determines the direction of the fast and slow axes and the magnitude of birefringence when an optical fiber is subjected to pressure, Ulrich first proposed an optical fiber squeezed polarization controller in 1979 [4], which stabilizes the polarization state of the signal light at 10^{-2} to 10^{-3} rad using the light intensity value as feedback. Granestrand is able to track polarization changes of about 0.3 rad/s using this type of polarization controller [5]. In 1980 Lefevre proposed fiber ring polarization controller using the birefringence phenomenon introduced by single-mode fiber bending [6], this polarization controller is larger and has a slow control rate. The polarization controller was first fabricated by Kidoh in 1981 using two LiNbO3 electro-optic crystal materials with 45° Z-tangent to each other to enable the output of stable linearly polarized light with an insertion loss of 6 dB. The LiNbO3-type polarization controller has a fast response but a large insertion loss [7]. The polarization control in the outlier detection system was first accomplished by Imai in 1985 using a polarization controller made of two wave pieces, reducing the loss caused by polarization mismatch to less than 0.2 dB with a feedback power of less than −70 dBm [8]. 1988 Barnes made polarization controller using electro-optical effect of liquid crystal, the electro-optical coefficient of liquid crystal wave sheet polarization controller is low, the driving voltage is lower than PLZT, and the fiber coupling efficiency is higher than LN, but its response speed is slow in the order of ms or more [9]. In 2001, Tie Wang reported an all-optical polarization controller [10], which uses a polarizing beam splitter to divide the beam into two orthogonally polarized beams, selects one of the beams to pass through the polarizing device to rotate the polarization direction by 90°, and then couples the two beams together to obtain polarized light with a constant polarization state. The polarization controller has a fast response speed and no external driving voltage, but can only output a single polarization state, and the insertion loss is relatively large. In the Ref. [10], its insertion loss is less than 0.7 dB. In 2014, Shen et al. introduced a polarization controller made of metamaterials [11], which is controlled by designing a microscopic unit structure on the surface of the metamaterial to transmit light in one polarization direction losslessly, while the beam orthogonal to it passes through the metamaterial after The polarization direction is rotated by 90° so as to obtain a stable polarization state of the beam, which has a transmission efficiency of about 74% for polarized light of wavelength 1550 nm in the Ref. [11], and the size of a single cell of this device is only 4 × 4 μm, and the polarization controller integrated by it is able to generate The integrated polarization controller is able to generate more complex vector beams, but this polarization controller is wavelength sensitive, and it has a large insertion loss and can only output a single polarization beam.

The polarization controller of rotatable waveplate and fiber ring does not need to design complex reset algorithm, but the response speed of the whole device is slow and cannot be used for high-speed polarization control; the polarization controller composed of fiber extrusion, electro-optical crystal and lithium niobate, although the response speed is faster, but it needs to design complex reset algorithm to realize the endless polarization control; the all-optical polarization controller and the polarization controller made of metamaterials The advantages are fast response and no mechanical parts, the disadvantage is that it is more sensitive to wavelength, insertion loss is large, and can only output a single polarization state of the beam. A good polarization controller should have low insertion loss, low drive voltage, stable linear delay, no mechanical displacement parts, and fast response speed, but the above polarization controller does not fully meet the above requirements of the polarization controller, the rapid development of metamaterials in recent years to provide new ideas for the production of polarization controllers, polarization controllers are developing in the direction of miniaturization, low loss, and fast response speed. The polarization controller is developing towards miniaturization, low loss and fast response.

(2) Research status of polarization control algorithm

In 2004, Shuo Chen used a 3-cell fiber squeezed polarization controller to improve the search and tracking speed by using variable step algorithm and prediction technique, which can track polarization changes larger than 1000 rad/s with insertion loss less than 0.6 dB [12]. In 2005 and 2006, Weiwen Li successively used Simulated Annealing (SA) SA) and Genetic Algorithm (GA) for polarization control, and achieved stable polarization control with light intensity fluctuation less than 1%. However, the authors only performed theoretical analysis [13]. In 2005, Zhang proposed to use the Particle Swarm Optimization (PSO) algorithm for adaptive polarization mode dispersion (PMD) compensation, and demonstrated experimentally that the PSO algorithm can successfully track the time-varying PMD [14]. In 2008, Yang Liu proposed a fast localization algorithm for polarization control, the basic principle of which is an algorithm designed by combining the advantages of both the gradient algorithm and the simulated annealing algorithm, and then outputting a stable polarization state [15]. In 2016, Yuyue Liu applied the Stochastic Parallel Gradient Descent (SPGD) algorithm applied to the scheme of polarization compensation in quantum key distribution, and experimentally verified that the polarization compensation scheme of SPGD algorithm can adjust the polarization extinction ratio of the system to a more ideal state [16]. In 2012, Li Tie analyzed the effect of polarization on the efficiency of outlier detection in coherent optical communication systems and showed through simulations that the simulated annealing algorithm can achieve automatic control of polarization state in coherent optical communication [17]. In 2017, Chen Kode used a combination of coarse tuning of divisional approximation algorithm and fine tuning of proportional-derivative algorithm to reduce the complexity of polarization control algorithm and improve the speed of polarization control, and through experimentally proved that the response speed of the polarization controller is in the order of 10 ms and can output the purpose of stabilizing the polarization state [18]. In 2018, Yin Ming used the polarization control of SPGD for theoretical simulation and analyzed the effect of the algorithm performance evaluation function, perturbation voltage type, gain step, perturbation amplitude and other parameters on the polarization control effect to obtain the laser output with the extinction ratio greater than 11 dB [19].

In the high-speed polarization control system, the response speed of polarization control is not only related to the polarization controller but also has a greater relationship with the control algorithm. In addition to the emergence of new material polarization controller, the integration and innovation of algorithm is also an effective way to improve the speed of polarization control.

Polarization controllers can be generally divided into three types of polarization controllers: fixed direction-phase variable, variable direction-phase fixed, and variable direction-phase variable. For the phase adjustable polarization controller, there is a fixed range of phase change. When the polarization controller works outside the phase shift range, the polarization controller needs to be readjusted so that it can work within the normal specified range of change, and the adjustment process should ensure that the polarization state will not change drastically, which is the endless reset of the polarization controller. The polarization control has been studied from 1979 to 1986, and various polarization controller materials and polarization controller composition schemes have appeared, but the problem of endless reset of the polarization controller has not been solved, and the Ref. [8] is the first to realize the endless polarization control, because it is a polarization controller composed of two rotatable waveplates, so the reset is only through the direct reset of the voltage, and does not fundamentally solve the problem.

In 1986 Rysdale showed theoretically that unterminated polarization control can be achieved with a sufficient number of waveplates in series [20], and in 1986 Noe proposed an unterminated polarization control which requires only three fiber squeezers but can only control the output of a single horizontal line of polarized light [21]. In 1990 Walker analyzed two unterminated polarization control schemes for fiber squeezers and lithium niobate polarization control schemes, while pointing out that in order to satisfy the unterminated operation of the polarization controller, at least three fiber squeezers are required [22]. 2003 Martinelli year used quaternion theory to analyze the combination of linearly rotatable half-wave and quarter-wave equivalent that can be used with variable amount of delay waveplates to achieve unterminated control of the polarization controller [23]. 2005 Weiwen Li introduced a method of forced endless reset [24], but the authors only conducted theoretical simulation studies, after which Rong also studied the reset problem of polarization control in depth and proposed a variety of reset methods [25]. The reset problem has been a difficult point of polarization control, and despite the emergence of various reset algorithms, the reset process in the actual system will inevitably be due to the limitations of the reset algorithm and the introduction of various errors cause signal jitter, so the study of a better reset algorithm is essential for the improvement of the performance of the communication system, and the reset algorithm applicable to the system should be selected according to the different systems.

Depending on the environment, internal structure and length of the optical fiber, the polarization state changes at a very different rate when light is transmitted in the fiber, and the high-speed changes in polarization state seriously degrade the performance of optical communication systems. In China, represented by Zhang Xiaoguang's group at Beihang University, the polarization stabilizer velocity of 12.6 krad/s was reported in 2008 [26]. In 2014, Professor Noe from the University of Paderborn (Germany) increased the speed of polarization stabilization to >100 krad/s [27], and the combination of a well-performing polarization controller and a control algorithm is what constitutes a high-speed polarization control system.

4.2 Coherent Optical Communication System with Polarization Control

The polarization state of the signal light fluctuates randomly in coherent optical communication systems due to atmospheric turbulence, and the optical mixer is sensitive to the polarization state of the signal light and the fundamental light, so the polarization state matching of the signal light and the fundamental light must be controlled to improve the mixing efficiency.

4.2.1 Representation of Light Polarization

The representation of optical polarization has basic mathematical expressions, Jones vector, Stokes vector, and bunga sphere [29], and the Jones vector and Stokes vector are more frequently used in the study of polarization variations in complex optical systems.

(1) Jones vector and Jones matrix

A light wave is an electromagnetic wave with an extremely short wavelength, which can be characterized by the electric vector E, the potential shift D, the magnetic induction strength B and the magnetic vector H. The relationship between them is determined by Maxwell's set of equations. Since the force of the electric field on the electrons is much greater than the force of the magnetic field on the electrons when light interacts with matter, the light wave is generally characterized by the electric field strength E. A light wave is a transverse wave whose propagation direction and vibration direction are perpendicular to each other, as shown in Fig. 4.1, if light is propagated along the z-axis direction, the nature of the curve depicted by the electric vector endpoints at a representative point in space is the polarization of light, in terms of polarization, light can generally be divided into polarized light, natural light and partially polarized light [28], which describes the vector nature of the electric field oscillation, if the direction of rotation of E and the propagation direction constitutes a right-handed helix is called right-handed polarization, and constitutes a left-handed helix is called left-handed polarization, and from the z-axis direction according to the trajectory of the curve can be divided into elliptically polarized waves, circularly polarized waves, and linearly polarized waves.

Assuming that the light is transmitted along the z-direction, the polarized photoelectric vector in the fully polarized state can be divided into two orthogonal components varying in the x–y plane, which can be written in the form of the following matrix as in [29]:

$$\begin{aligned} E &= \begin{bmatrix} E_x \\ E_y \end{bmatrix} = \begin{bmatrix} a_x \exp[i(wt - kz + \varphi_x)] \\ a_y \exp\big[i\big(wt - kz + \varphi_y\big)\big] \end{bmatrix} \\ &= \exp[i(wt - kz + \varphi_x)] \begin{bmatrix} a_x \\ a_y \exp(i\delta) \end{bmatrix} \\ &= \exp[i(wt - kz + \varphi_x)] \begin{bmatrix} A_x \\ A_y \end{bmatrix} \end{aligned} \tag{4.1}$$

where Ax and Ay denote the complex amplitudes of the x- and y-direction components, respectively, w indicates the center frequency of the optical signal, k is the wave number, φ_x, φ_y denotes the phase of the two split amplitudes, respectively, the phase difference between the x and y components $\delta = \varphi_y - \varphi_x$, then, Eq. 4.1 is called the Jones vector.

Since the absolute phase of light waves is not often used, after omitting the absolute phase and optical frequency parts of Eq. 4.1, the Jones vector of light can be written as

$$E = \begin{bmatrix} A_x \\ A_y \end{bmatrix} = \begin{bmatrix} a_x \\ a_y \exp(i\delta) \end{bmatrix} \tag{4.2}$$

The Jones vector of polarized light after normalization can be expressed as

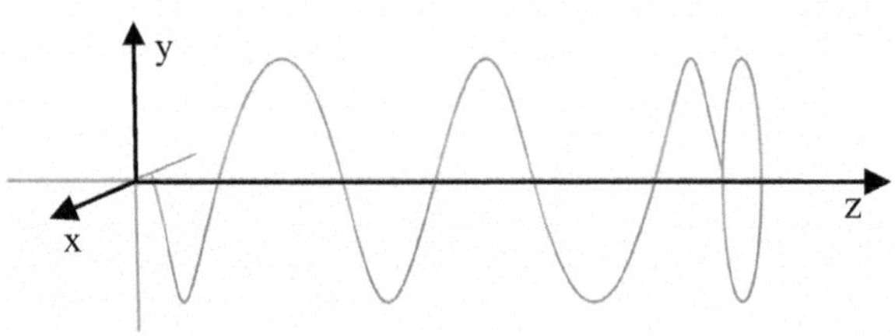

Fig. 4.1 Polarization properties of light

Table 4.1 Jones vectors for several common types of polarized light

Polarization state	Normalized Jones vector	Polarization state	Normalized Jones vector
Horizontal linear polarization light	$\begin{bmatrix} 1 \\ 0 \end{bmatrix}$	−45°linearly polarized light	$\frac{1}{\sqrt{2}}\begin{bmatrix} 1 \\ -1 \end{bmatrix}$
Vertical linearly polarized light	$\begin{bmatrix} 0 \\ 1 \end{bmatrix}$	Right circularly polarized light	$\frac{1}{\sqrt{2}}\begin{bmatrix} 1 \\ -i \end{bmatrix}$
45° linearly polarized light	$\frac{2}{\sqrt{2}}\begin{bmatrix} 1 \\ 1 \end{bmatrix}$	Left circularly polarized light	$\frac{1}{\sqrt{2}}\begin{bmatrix} 1 \\ i \end{bmatrix}$

$$E(\alpha, \delta) = \begin{pmatrix} \cos\alpha \\ \exp(i\delta)\sin\alpha \end{pmatrix} \tag{4.3}$$

where $\alpha = \arctan(\frac{a_y}{a_x})$, Jones vector through parameters A_y, A_x, δ in turn, the polarization state of the light is determined, Table 4.1 gives the Jones vectors after normalization of several common polarized light.

When polarized light passes through various polarization devices, such as polarizers, phase delayers, and polarization controllers, the polarization state of the beam will change, and the optical components with birefringence properties are collectively referred to as waveplates. Suppose the Jones vector of the polarization state of the input signal light is E_{in}, Then the Jones vector of the polarization state of the output signal light after passing through the polarization device $E_{out} = J \cdot E_{in}$ is expressed, where J is a 2×2 matrix, this matrix is called Jones matrix, which is used to describe the conversion relationship between the input light and the output light polarization state after the polarization light passes through the polarization device, when the polarization light passes through multiple polarization devices, the Jones vector of the output light polarization state can be obtained by Eq. 4.4.

$$E_{out} = J_N J_{N-1} \ldots E_{in} \tag{4.4}$$

where $J_N J_{N-1} \cdots$ denotes the cascaded multiplication of the Jones matrix of individual polarization devices.

(2) Stokes parametric and Muller matrix

Jones vector cannot represent partially polarized light, but Stokes parameters can represent partially polarized and unpolarized light, Stokes Parameters are four parameters introduced by Stokes in 1852 when he studied partially coherent light [30], all four parameters are about the magnitude of light intensity, which can be obtained by simple experiments and then, the ordinary Stokes parameters are the following four quantities.

$$\begin{cases} S_0 = \langle |E_x|^2 \rangle + \langle |E_y|^2 \rangle \\ S_1 = \langle |E_x|^2 \rangle - \langle |E_y|^2 \rangle \\ S_2 = 2\langle E_x E_y \cos\delta \rangle \\ S_3 = 2\langle E_x E_y \sin\delta \rangle \end{cases} \tag{4.5}$$

The Stokes covariate after taking its normalization can be expressed as

$$S = \begin{bmatrix} S_0/S_0 \\ S_1/S_0 \\ S_2/S_0 \\ S_3/S_0 \end{bmatrix} \tag{4.6}$$

The polarization P of an optical wave can be expressed as:

$$P = \frac{\sqrt{S_1^2 + S_2^2 + S_3^2}}{S_0} \tag{4.7}$$

The polarization P represents the ratio of the polarized light to the total light intensity in a partially polarized beam.

The Stokes vector after normalization can be represented by a unit ball Σ of radius 1, This ball Σ is called a bunga ball, As shown in Fig. 4.2, When the intensity of a plane monochromatic wave is given (s_0 = constant), each polarization state of the monochromatic light corresponds to a point on the sphere Σ. When $\chi \in [0, \pi/4]$,$(S_1/S_0, S_2/S_0, S_3/S_0)$ Corresponding ball Σ Point on the northern hemisphere, That is, all the points on the northern hemisphere of ball Σ represent left-rotating polarized light, and the north pole is left-rotating circularly polarized light, and all the points on the corresponding southern hemisphere represent right-rotating polarized light, and the south pole is right-rotating circularly polarized light. When $\chi = 0$, $S_3 = 0$, at this time, the polarization state is in the equatorial plane, and the difference between the orthogonal components of the polarized light is either zero or an integer multiple of the polarization state, which is linearly polarized light. After the light passes through the crystal medium, the trajectory of the polarization state can be directly traced on the bunga sphere to visualize the specific changes of the polarization state, and the Stokes vectors of several common polarized light are given in Table 4.2.

The polarization state of the input light is represented by the Stokes vector S_{in}, After the light passes through various polarization devices, the change of light polarization by the polarization devices can be represented by a 3×3 Mille matrix (Mueller) M, Then when the light passes through multiple polarization devices, the polarization state S_{out} of the output light can be expressed as $S_{in} = MS_{out}$, where $M = M_N M_{N-1} \ldots M_1$ denotes the inverse sequential product of individual component Mille matrices to represent.

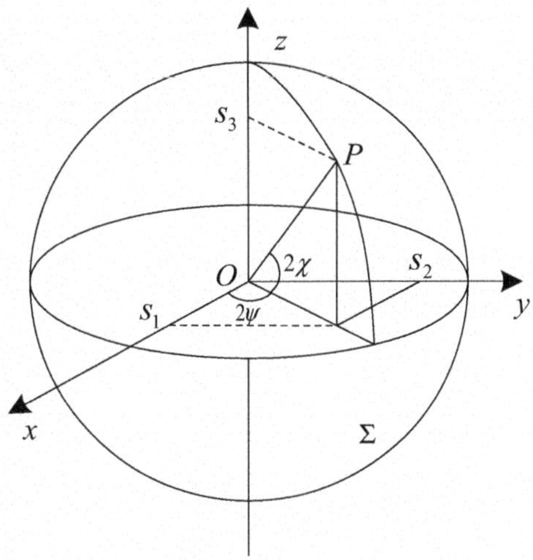

Fig. 4.2 Bonga sphere representation of polarization state

Table 4.2 Stokes vectors for several common types of polarized light

Natural Light	Horizontal linear polarization light	Vertical linearly polarized light	45° linearly polarized light	−45°linearly polarized light	Right circularly polarized light	Right circularly polarized light
$\begin{bmatrix}1\\0\\0\\0\end{bmatrix}$	$\begin{bmatrix}1\\1\\0\\0\end{bmatrix}$	$\begin{bmatrix}1\\-1\\0\\0\end{bmatrix}$	$\begin{bmatrix}1\\0\\1\\0\end{bmatrix}$	$\begin{bmatrix}1\\0\\-1\\0\end{bmatrix}$	$\begin{bmatrix}1\\0\\0\\1\end{bmatrix}$	$\begin{bmatrix}1\\0\\0\\-1\end{bmatrix}$

4.2.2 *Polarization Control of Coherent Optical Communication Systems*

(1) The basic principle of polarization controller

In this paper, an optical fiber squeezed type of polarization controller is used, as shown in Fig. 4.3, when the fiber is subjected to pressure F in a determined direction, the delay between the fast axis f and the slow axis s of the fiber will change, and the direction of its slow axis is the same as the direction of the force F. At this time, the squeezed part of the fiber is equivalent to a birefringent waveplate, and its polarization state changes when the signal light passes through it, and the relationship between the force F and the resulting birefringent phase difference satisfies [31].

$$\gamma = KF/d \tag{4.8}$$

Fig. 4.3 The relationship between the direction of pressure and the axis of production

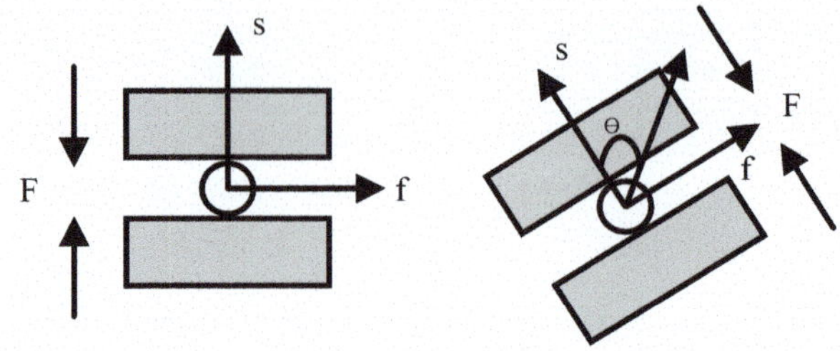

Where F is the pressure on the fiber cross-section and d is the fiber diameter, constants $K = 9.5 \times 10^{-5}$ rad/m.

If the Jones matrix is used to describe the principle of the polarization controller, the Jones matrix for the change in the optical polarization state of the fiber to the input signal when the fiber is subjected to a pressure at an angle of θ with respect to the x-axis direction can be expressed as [29]:

$$J(\theta,\gamma) = \begin{bmatrix} \cos\left(\frac{\gamma}{2}\right) + i\cos(2\theta)\sin\left(\frac{\gamma}{2}\right) & i\sin(2\theta)\sin\left(\frac{\gamma}{2}\right) \\ i\sin(2\theta)\sin\left(\frac{\gamma}{2}\right) & \cos\left(\frac{\gamma}{2}\right) - i\cos(2\theta)\sin\left(\frac{\gamma}{2}\right) \end{bmatrix} \tag{4.9}$$

The Jones matrices corresponding to the fiber optic extruders shown in Fig. 4.4, respectively, are

$$J(0^\circ,\gamma) = \begin{bmatrix} \exp\left(i\frac{\gamma}{2}\right) & 0 \\ 0 & \exp\left(-i\frac{\gamma}{2}\right) \end{bmatrix} \quad J(45^\circ,\gamma) = \begin{bmatrix} \cos\left(\frac{\gamma}{2}\right) & i\sin\left(\frac{\gamma}{2}\right) \\ i\sin\left(\frac{\gamma}{2}\right) & \cos\left(\frac{\gamma}{2}\right) \end{bmatrix} \tag{4.10}$$

As shown in Fig. 4.4, a schematic diagram of this paper using a polarization controller is shown. It is made up of 0°, 45°, 0° three fiber optic extruders at 45° to each other. By controlling the voltage of the three channels of the polarization controller U_1, U_2, U_3 Generates force F_1, F_2, F_3, which in turn changes the polarization state of the beam. It is capable of converting between any input polarization state to any output polarization state.

As in Fig. 4.5, the polarization state on the bunga ball will rotate clockwise around the OX axis when the first or third fiber squeezer is added, and counterclockwise

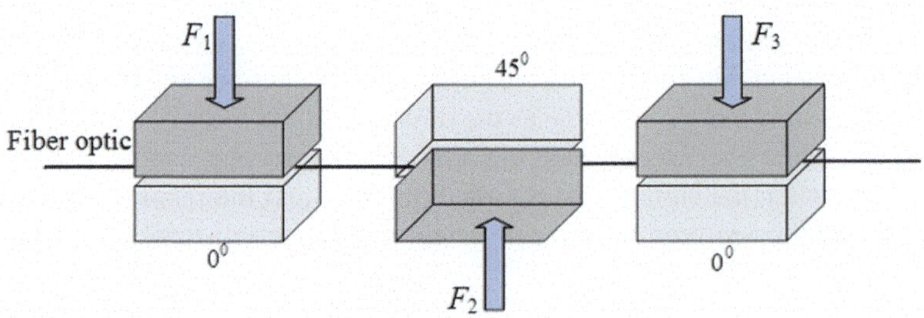

Fig. 4.4 Schematic diagram of the internal structure of the polarization controller

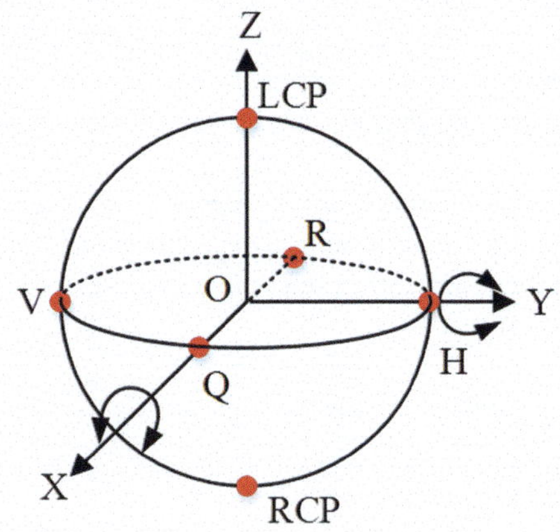

H: Horizontal linear polarization light LCP: Left circularly polarized light
V: Vertical linearly polarized light RCP: Right circularly polarized light

Fig. 4.5 The change of polarization state on the Bonga sphere

when the voltage is reduced, and when the pressure F_2 of the second fiber squeezer is added, the polarization state will rotate clockwise around the OY axis. The same reduced voltage rotates counterclockwise around the OY axis, and it can be seen that the polarization controller composed of three fiber optic squeezers is able to satisfy the change from any input polarization state to any output polarization state, and the polarization states are spread throughout the bunga ball.

Let the Jones vectors of the input and output polarization states be E_S, E_{out}, then the polarization state E_{out} of the output signal light can be calculated by the following equation after the signal light has passed through the polarization controller:

$$E_{in} = J(0^\circ, \gamma_3)J(45^\circ, \gamma_2)J(0^\circ, \gamma_1)E_{out} \tag{4.11}$$

where γ_3, γ_2, γ_1 indicates the phase shift produced by the action of pressure F_3, F_2, and F_1, respectively.

When using the Mille matrix to describe the principle of the polarization controller, the Mille matrix of the change in the polarization state of the fiber to the signal light when the fiber is subjected to a pressure at an angle of θ with the x-axis direction can be expressed as

$$M = \begin{bmatrix} \cos^2(\frac{\gamma}{2}) + \sin^2(\frac{\gamma}{2})\cos(4\theta) & \sin^2(\frac{\gamma}{2})\sin(4\theta) & -\sin(\gamma)\sin(2\theta) \\ \sin^2(\frac{\gamma}{2})\sin(4\theta) & \cos^2(\frac{\gamma}{2}) - \sin^2(\frac{\gamma}{2})\cos(4\theta) & \sin(\gamma)\cos(2\theta) \\ \sin(\gamma)\sin(2\theta) & -\sin(\gamma)\cos(2\theta) & \cos(\gamma) \end{bmatrix} \tag{4.12}$$

Then the Mille matrices of the 0° and 45° fiber extruders, respectively, are

$$M(0^\circ,\gamma)=\begin{bmatrix}1 & 0 & 0\\ 0 & \cos(\gamma) & \sin(\gamma)\\ 0 & -\sin(\gamma) & \cos(\gamma)\end{bmatrix}\quad M(45^\circ,\gamma)=\begin{bmatrix}\cos(\gamma) & 0 & -\sin(\gamma)\\ 0 & 1 & 0\\ \sin(\gamma) & 0 & \cos(\gamma)\end{bmatrix}\tag{4.13}$$

Let the polarization state of the input signal light be expressed by the Stokes vector as S_{in}, Then, after the signal light passes through this polarization controller, the Stokes vector S_{out} of the polarization state of the output signal light can be derived from the following equation:

$$S_{out}=M(0^\circ,\gamma_3)M(45^\circ,\gamma_2)M(0^\circ,\gamma_1)S_{in}\tag{4.14}$$

(2) The basic principle of optical mixer

In a coherent optical communication system, the signal light received at the receiving end is coupled into a single-mode fiber for coherent mixing with the local oscillation light through wavefront correction, as shown in Fig. 4.6 for the optical path of the mixer, we can see that the local oscillation light and the signal light enter the mixer first through two polarizers with horizontal transmission direction P1 and P2, the light generated by the local oscillation laser is a beam with good and stable polarization characteristics. We can control the incident direction of the local light is to enter the mixer with horizontal line polarization light, the signal light at the transmitting end through modulation, after amplification by the EDFA (Erbium Doped Fiber Amplifier, EDFA) generated by the signal light is also a good polarization characteristics of stable polarized light. However, after the signal light passes through the transmitting antenna, atmosphere, receiving antenna, various optical devices of AO system and then coupled into the fiber, the signal light arriving at the front of the mixer is a signal light with unknown polarization state, the change of atmospheric turbulence will make the polarization state of the signal light fluctuate randomly, and the change of temperature will make the polarization state of the signal light drift slowly. If the polarization state of the signal light is not controlled at the receiving end, the change of polarization state will be converted into signal light intensity fluctuation after passing through the polarizer P2, when the polarization direction of the signal light is perpendicular to the transmission direction of P2, that is, when the signal light is 90° line polarized light incident to the mixer, the signal light will be filtered out by the polarizer P2, resulting in signal interruption, so in order to avoid this situation in the coherent optical communication system The polarization state of the signal light must be controlled.

We use the Jones vector and Jones matrix to analyze the operating principle of the optical mixer. In coherent optical communication systems, the polarization state of the fundamental light emitted by the local oscillation laser is generally stable and constant, assuming that the fundamental light entering the mixer is a beam of stable linearly polarized light vibrating along the *x*-axis, light intensity is I_{LO}, Then the Jones matrix of the local vibrational light can be written as

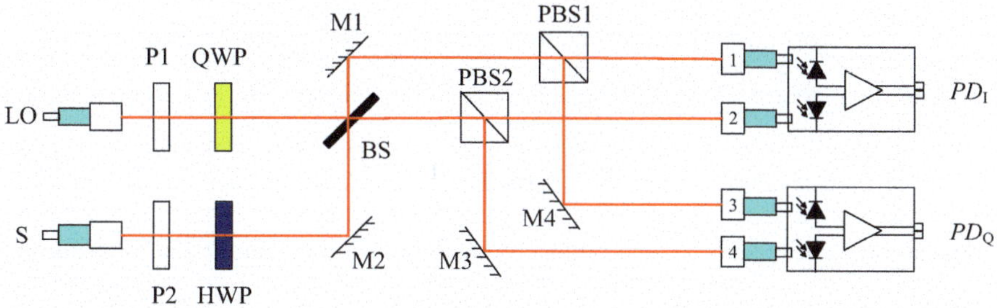

QWP:1/4Wave Sheet HWP:1/2Wave Sheet BS:50/50 Beam splitter

PBS:Polarization beam splitter M:Reflector P:Polarizer PD：Photoelectric detection

Fig. 4.6 Optical diagram of the mixer

$$E_{LO} = \sqrt{I_{LO}}\exp(iw_{LO}t)\begin{pmatrix}1\\0\end{pmatrix} \tag{4.15}$$

where w_{LO} is the angular frequency of the local oscillation light.

While the signal light is a polarized light with uncertain polarization characteristics when it reaches the front end of the mixer, the Jones vector of the signal light can be written as:

$$E_S = \sqrt{I_S}\exp[iw_St + i\phi(t)]\begin{pmatrix}\cos\alpha\\ \sin\alpha\cdot\exp(i\varphi_S)\end{pmatrix} \tag{4.16}$$

where I_S is the light intensity of the signal light; w_S is the angular frequency of the signal light; ϕ is the modulation phase of the signal light; $\tan\alpha = E_{Sy}/E_{Sx}$, E_{Sy}, E_{Sx} are the sub-amplitudes of the electric vectors of the signal light in the *y*-direction and *x*-direction, respectively. $\alpha \in [0, \pi/2]$; $\varphi_S = \varphi_{Sy} - \varphi_{Sx}$ is the phase difference between the electric vector of the signal light in the *y*-direction and the *x*-direction. $\varphi_S \in [-\pi, \pi]$.

As shown in Fig. 4.6 when the principal and signal light pass through P1 and P2 respectively, the Jones vector of the signal light and the principal light can be written as

$$E_{LO} = \sqrt{I_{LO}}\exp(iw_{LO}t)\begin{pmatrix}1\\0\end{pmatrix} \tag{4.17}$$

$$E_S = \sqrt{I_S}\exp[iw_St + i\phi(t)]\begin{pmatrix}\cos\alpha\\0\end{pmatrix} \tag{4.18}$$

When the beam passes through the 1/4 and 1/2 waveplates, the change of polarization state by the waveplates can be replaced by a Jones matrix, then the Jones matrix of a delayer with arbitrary azimuth θ and delay amount δ can be expressed as:

$$J(\theta,\delta) = R(-\theta)W(\delta)R(\theta) = \begin{pmatrix} \cos\theta & \sin\theta \\ -\sin\theta & \cos\theta \end{pmatrix} \begin{pmatrix} \exp(j\delta/2) & 0 \\ 0 & \exp(-j\delta/2) \end{pmatrix} \begin{pmatrix} \cos\theta & -\sin\theta \\ \sin\theta & \cos\theta \end{pmatrix} \tag{4.19}$$

The Jones matrices corresponding to the 1/4 and 1/2 wavelets in the mixer are

$$J\left(\frac{\pi}{4},\frac{\pi}{4}\right) = \frac{\sqrt{2}}{2}\begin{pmatrix} 1 & -j \\ -j & 1 \end{pmatrix} \quad J\left(\frac{\pi}{8},\pi\right) = \frac{1}{\sqrt{2}}\begin{pmatrix} 1 & 1 \\ -1 & 1 \end{pmatrix} \tag{4.20}$$

Then the Jones vectors of the principal and signal light after passing through the 1/4 and 1/2 waveplates, respectively, are

$$E_{LO} = \sqrt{\frac{I_{LO}}{2}}\exp(iw_{LO}t)\begin{pmatrix} 1 \\ \exp(-i\pi/2) \end{pmatrix} \tag{4.21}$$

$$E_S = \sqrt{\frac{I_S}{2}}\exp[iw_S(t) + i\phi(t)]\begin{pmatrix} \cos\alpha \\ \cos\alpha \end{pmatrix} \tag{4.22}$$

Then the Jones vectors of the four optical signals obtained from the polarization beam splitter with 90° difference are

$$E_{out1} = \frac{1}{2}\begin{pmatrix} \overline{S}\cdot\cos\alpha + \overline{LO}\cdot\exp(-i\pi/2) \\ 0 \end{pmatrix} \tag{4.23}$$

$$E_{out2} = \frac{1}{2}\begin{pmatrix} \overline{S}\cdot\cos\alpha\cdot\exp(-i\pi/2) + \overline{LO} \\ 0 \end{pmatrix} \tag{4.24}$$

$$E_{out3} = \frac{1}{2}\begin{pmatrix} 0 \\ \overline{S}\cdot\cos\alpha + \overline{LO}\cdot\exp(-i\pi) \end{pmatrix} \tag{4.25}$$

$$E_{out4} = \frac{1}{2}\begin{pmatrix} 0 \\ (\overline{S}\cdot\cos\alpha + \overline{LO})\cdot\exp(-i\pi/2) \end{pmatrix} \tag{4.26}$$

The two electrical signals output I/Q by the double balanced detector are

$$E_I = \sqrt{I_{LO}I_S/2}\cdot\cos\alpha\cdot\sin[(w_S - w_{LO})t + \phi(t)] \tag{4.27}$$

$$E_Q = \sqrt{I_{LO}I_S/2}\cdot\cos\alpha\cdot\cos[(w_S - w_{LO})t + \phi(t)] \tag{4.28}$$

Then the photocurrent sum can be expressed as [32]:

$$E = E_I + E_Q$$

$$= \sqrt{I_{LO} I_S} \cdot \cos\alpha \cdot \sin[(w_S - w_{LO})t + \phi(t) + \pi/4] \tag{4.29}$$

Then the mixing efficiency can be expressed as

$$\eta = E_{actual}^2 / E_{ideal}^2 = \cos^2\alpha \tag{4.30}$$

The above equation shows that only when $\alpha = 0$, the highest mixing efficiency $\eta = 100\%$, when the signal light is along the x-axis line polarization light, the signal light polarization direction and polarizer transmission direction is the same, the signal light all into the optical mixer for coherent mixing. When $\alpha \in (0, \pi/2)$, $0 < \eta < 1$, At this time, the signal light is elliptically polarized light, and only part of the signal light enters into the optical mixer. As α increases from 0 to $\pi/2$, the mixing efficiency gradually decreases, when $\alpha = \pi/2$, $\eta = 0$, The signal light is linearly polarized along the y-axis, and all the signal light is filtered out, when the mixing efficiency is the lowest. Therefore, in order to achieve high efficiency mixing in coherent optical communication system, the signal light must be polarization control.

(3) Design of polarization control scheme

The signal light at the receiving end of the coherent optical communication system after transmission through the atmosphere has the following main characteristics.

(1) The signal light received at the receiving end is very weak after the transmission attenuation of the atmosphere;
(2) The polarization state of the signal light is unknown and fluctuates randomly before entering the mixer because the signal light has to pass through the transmitting antenna, atmosphere, receiving antenna, AO, optical fiber, etc. when it reaches the receiving end;
(3) The light intensity value of the signal light received at the receiving end fluctuates randomly due to beam drift, change in wavefront and fiber coupling, etc.

The characteristics of the signal light limit the polarization control scheme of the coherent optical communication system, as shown in Fig. 4.7 for a commonly used polarization control scheme [31], in which the signal light from the laser first passes through the scrambler, where the scrambler simulates the random perturbation of the polarization state of the signal light, and the detector detects the polarization state of the signal light in real time, then construct the objective function $F_{\min}\left[(S_1 - S_1'), (S_2 - S_2'), (S_3 - S_3')\right]$, then through the data processing module and polarization controller to complete the real-time control of the signal light polarization state, this polarization control scheme must be divided into a way to complete the detection of the signal light polarization state. The POD-001 module produced by General Photonics is currently available on the market. The module should detect the polarization state of the input light and output at least 5mw of signal light, but the optical power of the signal light received in the coherent optical communication system is generally about 5mw, which is difficult to meet the detection requirements.

This polarization control scheme differs from the polarization control scheme in Fig. 4.7 in that the polarization beam splitter (PBS) is used instead of the polarizer

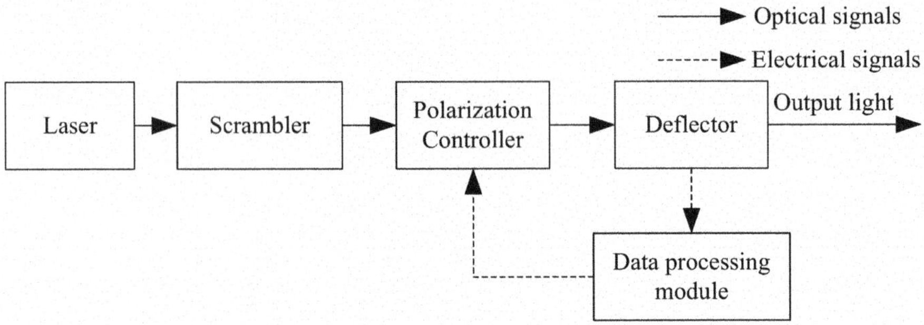

Fig. 4.7 Polarization control scheme with stokes parameters as feedback

module. The polarization control scheme shown in Fig. 4.7 can control the output of any polarization state, while the polarization control scheme shown in Fig. 4.8 can only output fixed polarization light, and the objective functions constructed by the two are different. For fully polarized light, the objective function shown in Fig. 4.8 is to minimize the photocurrent at the feedback end to zero, at which time a stable polarized light is output at the other end of the PBS. However, in the coherent optical communication system, after the signal light is transmitted through the atmosphere due to the atmospheric depolarization effect, the signal light is depolarized from a fully polarized beam to a partially polarized beam, and when the signal light passes through the PBS, the PBS decomposes the unpolarized light into two polarized beams with orthogonal polarization direction, and the spot jitter and light intensity flicker caused by atmospheric turbulence will make the objective function without a definite target value This will inevitably increase the complexity of the control algorithm, and the light passing through the PBS will also cause some attenuation of the light energy when coupled into the fiber.

The above two commonly used control schemes are not very suitable for coherent optical communication, in the coherent optical communication system for a stable beam of polarized light, from Eqs. 4.27 and 4.28 can be seen that the amplitude of the IF signal can reflect the signal light polarization information, so we design

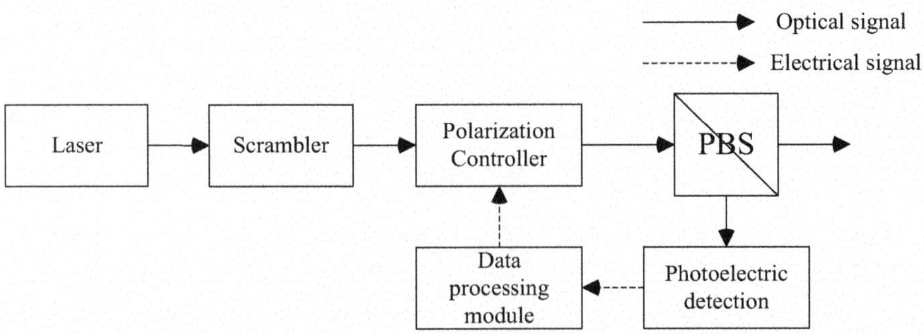

Fig. 4.8 Polarization control scheme with light intensity value as feedback

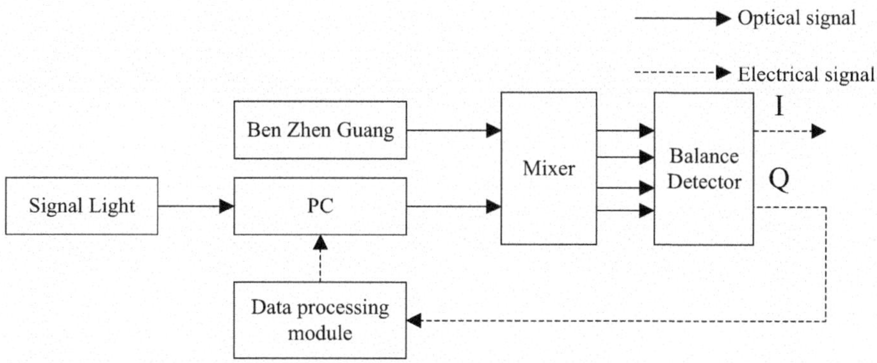

Fig. 4.9 Polarization control scheme with IF signal as feedback

the coherent optical communication system polarization control scheme as shown in Fig. 4.9.

The polarization control system shown in Fig. 4.9 contains only two parts, the data processing module and the polarization controller, where the Q-way signal is used as the feedback signal for the polarization control, then the data processing module completes the extraction of the IF signal amplitude information and the generation of the control signal through the algorithm, and the polarization controller completes the conversion of the control signal and the change of the signal light polarization state, this polarization control scheme reduces the loss of light energy in the coupling process compared to the polarization control scheme in Fig. 4.8.

4.3 Coherent Optical Communication Polarization Control Model and Control Algorithm

4.3.1 Polarization Control Model for Coherent Optical Communication Systems

Generally speaking, two fiber optic squeeze modules can meet any input polarization state to any output polarization state change, but for some special localization polarization state, using two units of polarization controller will have a control blind spot [33], In order to avoid the occurrence of control blindness, a three-unit polarization controller must be used. Each polarization control unit is driven by the applied voltage, the applied voltage has a certain control range, in the process of control, if the limit of the voltage range is reached, it is necessary to reset or reset (Reset) the voltage of the unit, this reset method is the first reset. If the reset process is to ensure that the signal cannot be interrupted, three or more control units are required to meet the continuous work of the polarization controller, too many control units, the polarization controller is larger and more expensive, and the control algorithm is more complex, too few control units can not complete the conversion from any input polarization state to any output polarization state, so generally choose the best polarization controller composed of three or four control units.

In coherent optical communication system, only the drive voltage of the polarization controller needs to be adjusted by the optimization algorithm to keep the amplitude of the IF signal to the maximum, when the polarization state of the signal light is fed into the mixer with horizontal linearly polarized light, and its mathematical model can be written as follows:

$$\max A_{IF} = f(\gamma_1, \gamma_2, \gamma_3) \tag{4.31}$$

where A_{IF} is the amplitude of the Intermediate Frequency (IF) signal., γ_1, γ_2, γ_3 is the phase change produced by the three waveplates of the polarization controller. When the polarization state of the input signal light is (0.5, −0.49, 0.71) stable and constant, as shown in Fig. 4.10a for the relationship between the phase change and the IF signal amplitude produced by the first and second squeezing unit squeezing in Figs. 4.4 and 4.10b for the relationship between the phase change and the IF signal amplitude produced by the second and third squeezing unit applying voltage, it can be seen from the figure Therefore, in the selection of the control algorithm, the simulated annealing algorithm, particle swarm algorithm or genetic algorithm are generally chosen to solve the multipole problem, which can better solve the problem of polarization control.

4.3.2 Simulated Annealing Algorithm in Polarization Control

The simulated annealing algorithm is a more commonly used control algorithm in polarization control algorithm, which can better solve the multipole problem of polarization control.

(1) Simulated annealing algorithm

The Simulated Annealing Algorithm (SA) idea was first proposed by Metropolis et al. in 1953, and Kirkpatrick used it for combinatorial optimization in 1983, proposing the Simulated Annealing Algorithm, which is a stochastic optimization-seeking algorithm based on the Monte Carlo iterative solving strategy [35]. Initially the method based on Monte Carlo simulation was simple but computationally intensive, in 1953 Metropolis proposed to receive the new state with probability, which is the Metropolis criterion, it significantly reduces the computation of SA and shortens the optimization time, the expression of the probability function of the Metropolis criterion is:

$$p_r = \exp\left(-\frac{E_j - E_i}{kT}\right) \tag{4.32}$$

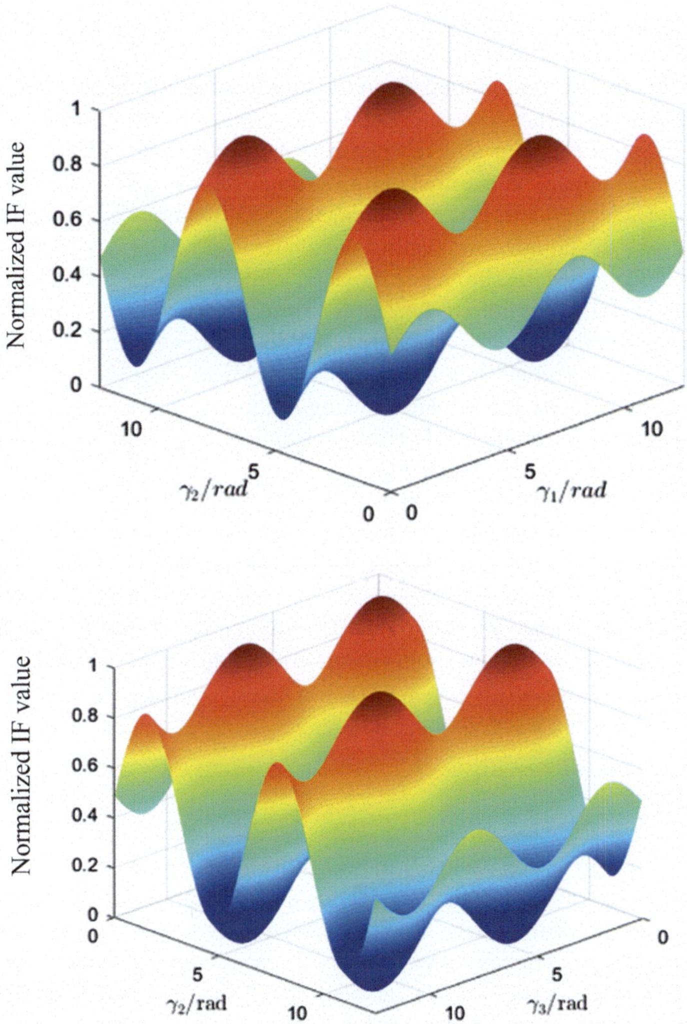

Fig. 4.10 The relationship between the phase change of the polarization controller waveplate and the amplitude of the IF signal [34], **a** the first waveplate and the second waveplate, **b** The second waveplate and the third waveplate

where E_i and E_j denote the internal energy of the system at state i and state j, respectively, k is Boltzmann constants, taking 1 in the simulated annealing algorithm, T is the temperature of the state i, Metropolis criterion indicates that at the time of $E_i < E_j$, receiving the solution for the new state j, when $E_i > E_j$, taking a random number p_n between [0, 1] and p_r for comparison, If p_r is greater than the random number p_n, the solution of the worse state j is received, when p_r is less than the random number p_n, the solution of the original state i is retained, and the internal energy of the system is minimized after continuous iterations, and the optimal solution of the objective function is achieved.

The main parameters and functions in the simulated annealing algorithm include the initial temperature, the temperature update function, the state generation function, the state reception function and the number of internal and external cycles, and they have a large impact on the final optimization results. For different optimization problems, the selection strategies of parameters and functions are different. In this section, we take the polarization control in coherent optical communication system as a model, and analyze the influence of each parameter on the polarization control results through simulation into to further determine the optimal parameters.

The selection of the initial temperature in SA determines the quality of the optimal solution and the processing time of the algorithm when the optimum is reached. A larger initial temperature can enhance the algorithm's ability to search the solution space and improve the quality of the optimal solution, but it will also directly affect the convergence speed of the algorithm.

The temperature update function is used to regulate the way the temperature drops in the outer loop, the speed of temperature drop is jointly determined by the initial temperature and the coefficient p. A slower temperature drop speed allows the algorithm to traverse more solutions in the solution space to get a better solution, but accordingly in order to reach the final termination temperature $T_{\min}$ leads to an increase in the number of outer loop iterations and an increase in the processing time of the algorithm, in this paper we select the temperature update function as exponential decreasing temperature, i.e.:

$$T_{i+1} = pT_i \tag{4.33}$$

when $p(0 < p < 1)$ is a constant, T_i indicates the temperature when the state is i.

The state generation function can generate a new solution for the next iteration, which is decided according to the specific problem. Equation 4.34 is the state generation function designed in this paper, and by adjusting s, the size of the neighborhood where the new solution is generated can be decided, so as to adjust the ability of the algorithm to explore the solution space.

$$U_n = \begin{cases} U_c + s \cdot r_1 \cdot (2\pi - U_c) & x_i \leq 0, x_i \geq 4\pi \\ U_c + s \cdot r_2 \cdot U_c & 0 < x_i < 4\pi \end{cases} \tag{4.34}$$

where r_1 is random number between 0 and 1, r_2 is the random number of $[-1\,,\,1]$, s generate function factors for states.

The final optimization result of simulated annealing algorithm has the advantage of not relying on the initial value, so in general the initial value of simulated annealing algorithm is generated randomly. However, in the polarization control system, if the range of the solution is beyond the solution space in the process of iteration, the first reset is required. The state generation function designed in this paper comes with the first reset capability, but this reset method is equivalent to the direct reset method in the reset mode, and the reset process will cause significant fluctuations in the signal, so a reasonable initial value setting can effectively reduce the first reset occurring in the iterative process of the algorithm Therefore, a reasonable initial value can effectively reduce the number of times the first reset occurs during the iteration of the algorithm, and Eq. (4.35) is the initial value of the simulated annealing algorithm design.

$$U_0^1 = U_0^2 = U_0^3 = U_{MAX}/2 \tag{4.35}$$

where U_0^1, U_0^2, U_0^3 indicates the initial voltage of the three channels of the polarization controller, respectively, U_{MAX} indicates the maximum voltage value of 140 V that can be applied by the polarization controller.

A flow chart of polarization control using the simulated annealing algorithm is shown in Fig. 4.11.

(2) Simulation study of polarization control based on simulated annealing algorithm

The polarization controller uses three fiber optic squeezers at 45° to each other to form the polarization controller, using the amplitude of the IF signal as the feedback signal, the phase shift range generated by the three fiber optic squeezers is $0 \sim 4\pi$. In order to study the effectiveness of the algorithm and the influence of the parameters on the optimization results, the effects of the initial temperature T_0, p in the temperature update function and the state generation function factor s in the simulated annealing algorithm on the performance of the simulated annealing algorithm are analyzed separately, taking the signal light with the input polarization state determined as an example. The polarization state of the signal light is assumed to be a linearly polarized light perpendicular to the x-axis, and the corresponding amplitude of the IF signal is zero at this time, and the polarization state of the signal light will be controlled to be a horizontal linearly polarized light when the amplitude of the IF signal is maximized by the simulated annealing algorithm.

The curves of the effect of different initial temperatures on the convergence of the algorithm are shown in Fig. 4.12, we set $p = 0.5$, $s = 0.5$, $i_0 = 100$, $l_0 = 3$, the horizontal coordinate is the number of iterations, and the vertical coordinate is the magnitude of IF after normalization. As can be seen from the figure, the SA algorithm converges only after 250 iterations when $T_0 = 10$, and has less influence on the convergence of the algorithm when T_0 is other values, and the algorithm converges after about 100 iterations, although a larger T_0 can improve the quality of the final solution, the corresponding time spent to reach the optimal solution is longer. In order to make the algorithm converge faster and have higher quality solutions, we can choose $T_0 = 1.0$ (Fig. 4.13).

Take $T_0 = 1.0$, $s = 0.015$, the curves of the effect of different temperature update function factors p on the algorithm are shown in Fig. 4.12. From the figure, we can see that the value of p is proportional to the number of iterations required for the SA algorithm to reach the optimum, the larger the value of p, the longer the convergence

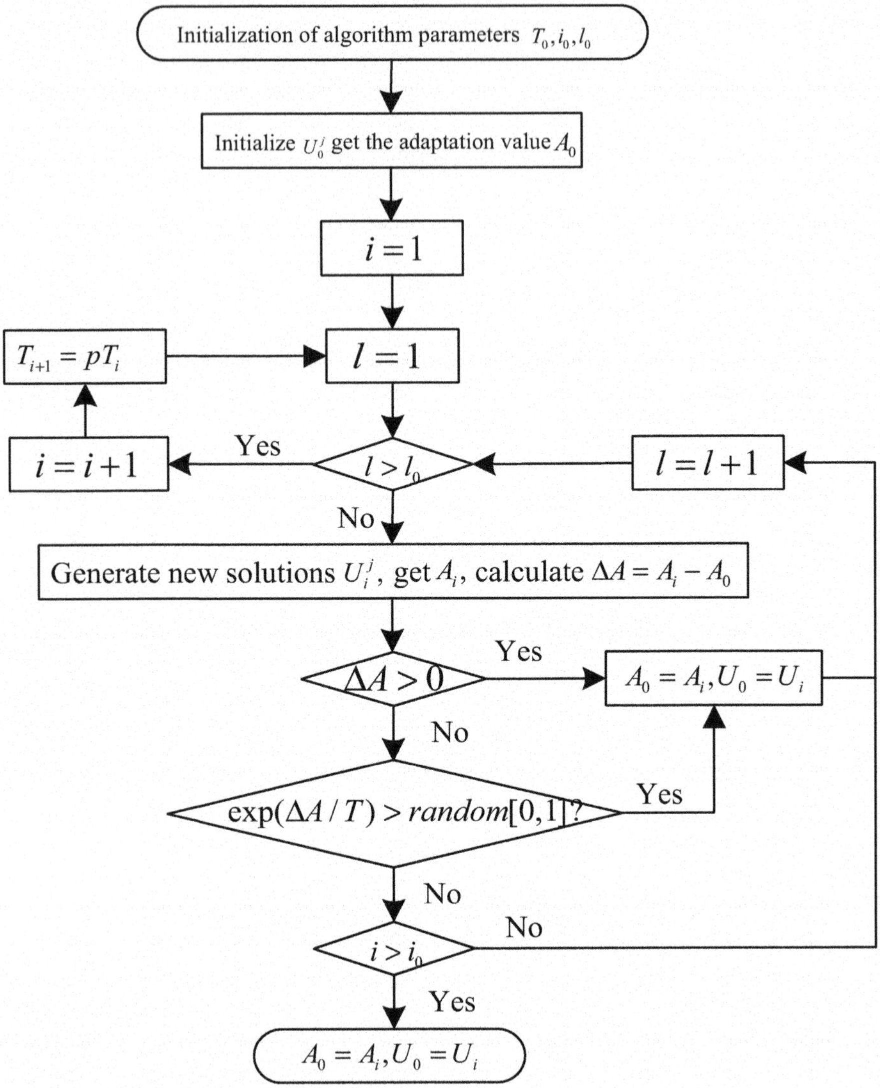

Fig. 4.11 Flow chart of simulated annealing algorithm for polarization control [34]

time of the algorithm, since $T_0 = 1.0$, experiencing the same number of iterations with different p will only affect the probability of the state acceptance function, the larger the p, the greater the probability of receiving the solution of the poor state, so the longer the iteration time, but it can enhance the search ability of the SA algorithm for the solution space. In order to be able to obtain the optimal solution in a short time can be chosen as $p = 0.5$.

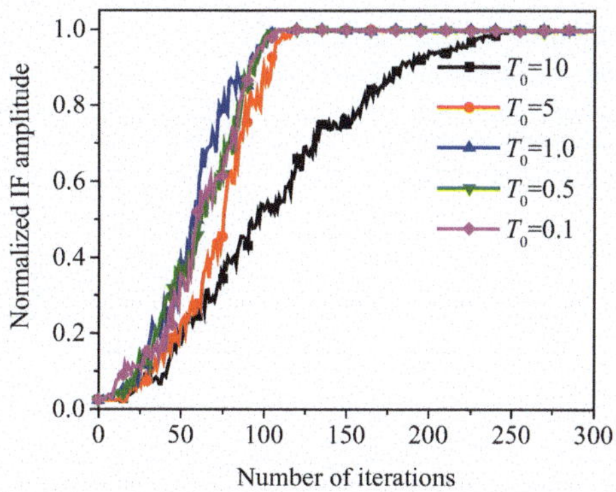

Fig. 4.12 Curve of the effect of T_0 on the performance of simulated annealing algorithm [34]

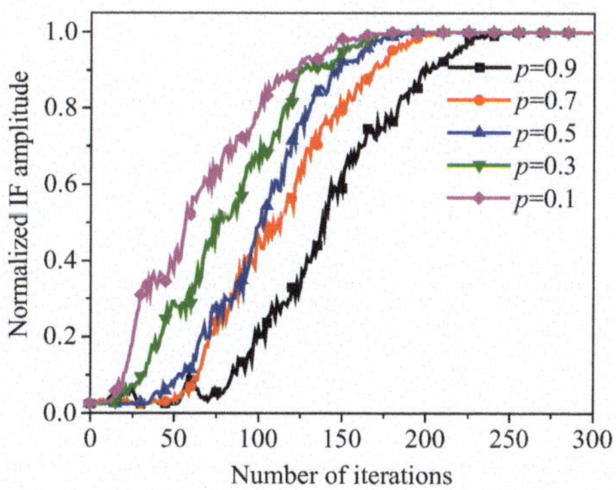

Fig. 4.13 Curve of the effect of p on the performance of the simulated annealing algorithm [34]

As Fig. 4.14 shows the influence curve of the state generating function factor s on the convergence performance of the algorithm, it can be seen from the figure: when $s = 0.09$, the maximum step length of the algorithm in the domain search is 32.4°, we can see that the algorithm can converge quickly, but the fluctuation of IF signal amplitude is larger after convergence, when $s = 0.01$, the maximum step length of the algorithm in the domain search is 3.6°, the IF amplitude of the algorithm after convergence is small, but the algorithm converges slower. When $s = 0.03$, the maximum step length of the algorithm in the domain search is 10.8°, and the fluctuation of IF signal amplitude is still larger, so we compromise by choosing $s = 0.015$, at which time the maximum step length of the algorithm in the domain search is 5.4°, and the fluctuation of IF signal amplitude is smaller under the premise of faster convergence of the algorithm.

The above analysis determined the value of the simulated annealing parameters, taking $p = 0.5$, $s = 0.015$ and $T_0 = 1.0$, there has been a large number of studies have shown [2] that the signal optical polarization state due to atmospheric turbulence is always randomly varying, but the rate of change is not fast, the Ref. [2] concluded that for a 1 km wireless optical communication system, when $C_n^2 = 2 \times 10^{-14} m^{-2/3}$, the RMS of polarization angle change is about 0.05 rad. Therefore, we assume that the

Fig. 4.14 Impact curve of the state generating function factor *s* on simulated annealing algorithm performance [34]

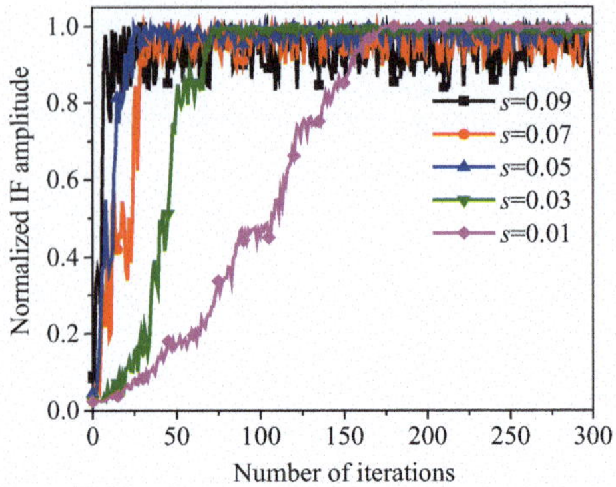

Fig. 4.15 The effect curve of simulated annealing algorithm on the amplitude of IF signal [34]

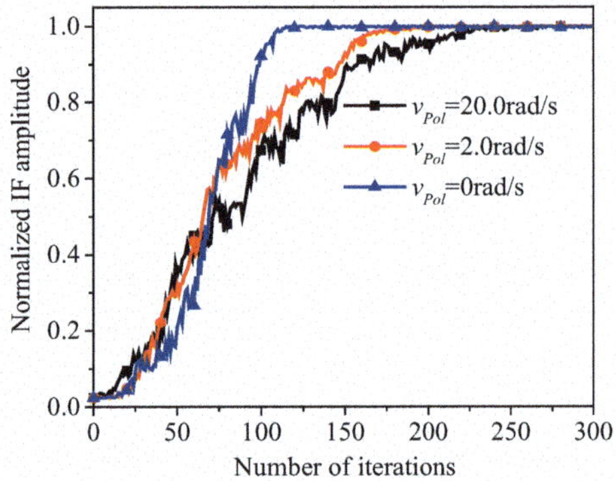

change rate of signal light polarization state v_{Pol} is 20.0, 2.0 and 0 rad/s, respectively, to study the optimization performance of the simulated annealing algorithm, as shown in Fig. 4.15: the larger the change rate of signal light polarization state, the slower the algorithm converges, and the fluctuation of the signal IF amplitude after convergence is larger, but the fluctuation is less than 1%, and overall the simulated annealing The final optimization performance of the algorithm is better even in the case of fluctuations in the signal polarization state.

4.3.3 Application of Particle Swarm Algorithm in Polarization Control

The particle swarm algorithm is a parallel swarm intelligence optimization algorithm, which can well solve the multipole optimization problem of polarization control. In this section, the particle swarm optimization algorithm is used for polarization control, and the optimal parameter combination of the particle swarm algorithm is analyzed, and the influence of the particle swarm optimization algorithm on the polarization control results is analyzed.

(1) Particle swarm algorithm

The Particle Swarm Optimization (PSO) algorithm is also one of the more commonly used algorithms for polarization control, PSO is a swarm intelligence optimization algorithm proposed by Kennedy and Eberhart in 1995 [36], due to its simple structure, the optimization process does not need to know the advantages of the objective function is widely used in many fields [37]. The main parameters involved in the PSO algorithm are population size N, speed limit $V_{\max}$, inertia weight w, acceleration factor c_1 and c_2. Where w, c_1 and c_2 have a greater impact on algorithm performance, therefore, there are more literatures on w, c_1, and c_2 to further improve the optimization performance of the algorithm through different selection strategies [38]. The existing research on $V_{\max}$ has shown that it will cause a certain limit to the convergence speed of the algorithm [39], and it is not necessary in the PSO algorithm. The parameters of the PSO algorithm interact with each other, and the strategy of taking each parameter and the mutual cooperation between them are necessary to make the algorithm have the best optimization performance.

In the PSO algorithm, each particle represents a solution in the solution space, and each particle has its own corresponding adaptation value. Equations 4.36 and 4.37 are the update formulas for the velocity and position of the standard PSO algorithm.

$$v_i^d(t+1) = wv_i^d(t) + c_1 \cdot r_1\left(p_{best}(t) - x_i^d(t)\right) + c_2 \cdot r_2\left(g_{best}(t) - x_i^d(t)\right) \quad (4.36)$$

$$x_i^d(t+1) = x_i^d(t) + v_i^d(t+1) \quad (4.37)$$

where t is the number of iterations of the algorithm, d is the dimension of the solution space, i is the number of particles $i = 1, 2, 3 \ldots, N$ v_{id} denotes the velocity of the i th particle in the d th dimension, x_{id} denotes the corresponding position, w denotes inertia weights and learning factors, c_1 and c_2 denote the cognitive and social parameters, respectively, also known as learning factor, p_{id} denotes the particle position corresponding to the individual optimum obtained by the i th particle in the traversal process, p_{gd} denotes the global optimal position obtained by all particles in the process of traversal, r_1, r_2 denotes two random numbers that obey [0, 1] uniform distribution independently of each other. The exploration ability of the particle swarm algorithm and the selection of the algorithm parameters have a great relationship. In this section, we use the classical PSO algorithm, i.e.,w, c_1 and are taken as constant values, and study the effect of different values of w, c_1 and c_2 on the polarization control results by taking the population number $N = 10$, $V_{\max} = 0.1 * X_{\max}$. The flow chart of polarization control using particle optimization algorithm is shown in Fig. 4.16.

(2) Simulation study of polarization control based on particle swarm algorithm

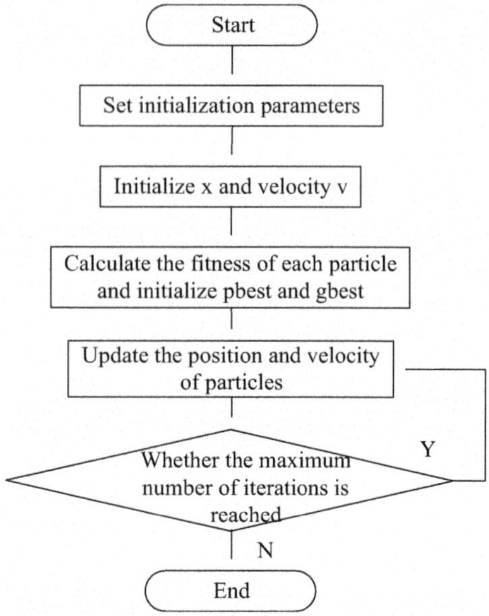

Fig. 4.16 Polarization control flow chart using particle swarm algorithm [34]

PSO algorithm particle position indicates the solution space of the solution, in the polarization control model indicates the voltage value applied to the three squeezers, the adaptation value of each particle corresponds to the amplitude of IF, as can be seen from Chap. 2 by the particle swarm algorithm to find the particle corresponding to the maximum amplitude, that is, the signal light polarization state control for the horizontal line polarized light applied to the fiber squeezer corresponding to a set of voltage values, that is, the optimal solution. In the PSO algorithm, w, c_1 and c_2 have a large impact on the final optimization results of the algorithm, and each parameter has many different strategies for taking values, but in this paper, the parameters are set according to the classical PSO algorithm, and w, c_1 and c_2 are taken as fixed values. Assuming that the polarization state of the input signal light is linearly polarized light perpendicular to the x -axis, the number of particles $N = 10$, and the number of iterations $T = 100$, a coherent optical communication system model is established, and the influence of the values of w, c_1 and c_2 on the convergence of the PSO algorithm is analyzed.

Inertia weights w generally take the number between [0.3, 0.9], when $c_1 = 1.0$, $c_2 = 2.0$, Fig. 4.17 analyzes the influence curve of w on the convergence of the algorithm for $w = 0.90$, 0.75, 0.6, 0.45, and 0.3. The vertical coordinates in the figure are the mean values of the adaptation values corresponding to 10 particles in each iteration, $w = 0.60$, 0.45 and 0.30, the POS algorithm converges and stabilizes in less than 30 iterations, while the PSO algorithm finds the maximum value only after 80 iterations when $w = 0.75$. The algorithm's ability to explore globally and locally can be balanced by adjusting the value of w. From the figure, it can be seen that the algorithm has better convergence results at $w = 0.60$, $w = 0.60$ can be chosen.

Take $w = 0.6$ and $c_2 = 2.0$, Fig. 4.18 shows the influence curve of learning factor c_1 on PSO algorithm, as seen from the figure: the algorithm converges only when $c_1 = 0.5$ and 1.0, and the number of iterations required for convergence is 20 and 40, respectively. In order to make the PSO algorithm have faster convergence speed and not easy to fall into local optimum, $c_1 = 1.0$, is taken in the next simulation. The influence curves of different learning factors c_2 on the PSO algorithm are analyzed by taking $w = 0.6$ and $c_1 = 1.0$ in Fig. 4.19. It can be seen from the figure that: the PSO algorithm does not converge throughout the iterative process when $c_2 = 0.5, 1.0$, and the number of iterations required for the PSO algorithm to reach the maximum is 32, 15, and 36 when $c_2 = 2.5,\ 2.0,\ 1.5$, respectively, and the number of iterations required from 4.36 equations can be seen that c_2 is to adjust the current degree of particle awareness of the global optimum, too large c_2 although convergence is faster but easy to fall into the local optimum, so $c_2 = 2.0$ is chosen.

Through the above analysis we determined the optimal values of the parameters in the PSO algorithm, and chose $w = 0.6$, $c_1 = 1.0$, $c_2 = 2.0$. Figure 4.20 analyzes the simulation graph of the optimization by the PSO algorithm when the change rate

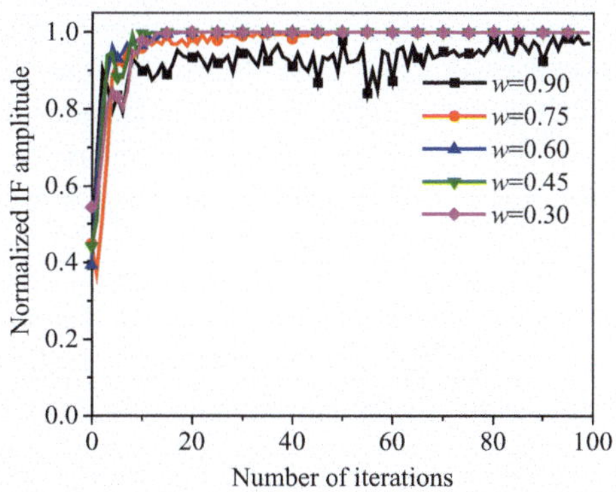

Fig. 4.17 Curve of the effect of w on the PSO algorithm [34]

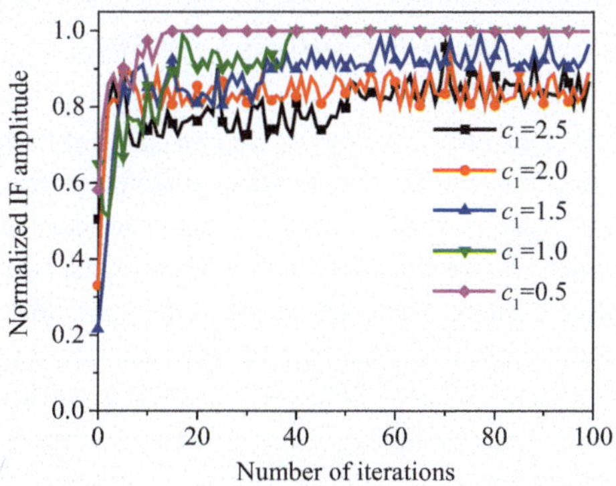

Fig. 4.18 Curve of the effect of c_1 on PSO algorithm [34]

Fig. 4.19 Curve of the effect of c_2 on PSO algorithm [34]

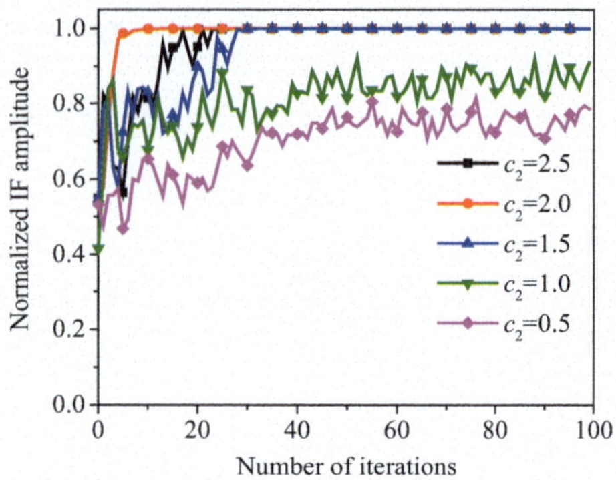

Fig. 4.20 Curve of the effect of PSO algorithm on the amplitude of IF signal [34]

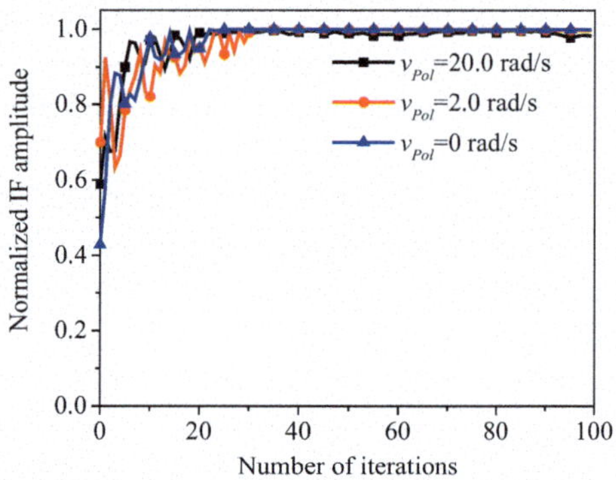

of the polarization state of the signal light υ_{Pol} is 20.0, 2.0, 0 rad/s, respectively. At $\upsilon_{Pol} = 20\,\text{rad/s}$, the fluctuation of IF signal amplitude is less than 2% after convergence of the algorithm.

4.3.4 Design of SPO Algorithm and Its Application in Polarization Control

From the above analysis, we can see that in the polarization control, both SA and PSO can complete the stable control of the polarization state, but in SA, in order to make the algorithm have a faster convergence speed, then the state generation function factor must be set larger, but at this time, the amplitude of the signal fluctuates more after convergence, which has a certain impact on the accuracy of the final solution, when s is small, the algorithm needs to spend a lot of time to iterate in order to reach Optimal, when we choose $s = 0.015$ algorithm convergence takes about 150 times, assuming that the response time of the polarization controller is t_r, then the time of SA convergence is $150t_r$. While the number of iterations required for the PSO algorithm to reach the optimum is 20, with a total of 10 particles, the convergence time of the PSO algorithm is 200 t_r. In practical engineering applications when we apply the first particle we record the adaptation value of the particle, and when the second particle is applied, the changed polarization state needs to be restored to the initial value to record the adaptation value of the second particle, but in the polarization control This process is redundant and almost wastes twice the time of the polarization control, in this paper we call this reset the second reset, so the total convergence time required by the PSO algorithm is 400 t_r, and the particles of the PSO algorithm are generated randomly, so the process of algorithm optimization also increases the possibility of the first reset accordingly. In addition to the SA and PSO algorithms, there is also a common algorithm Genetic Algorithm (GA) in polarization control, and since this algorithm and the PSO algorithm also belong to the group intelligence optimization parallel algorithm, the first reset and the second reset also exist. The second reset limits the optimization performance of the PSO algorithm and the GA algorithm, while the SA algorithm does not exist for the second algorithm, but the SA algorithm requires a large number of iterations to reach the optimum. In summary, this paper designs an optimization algorithm SPO (Single Particle Optimization, SPO) specifically for polarization control to complete the polarization control in coherent optical communication systems.

(1) Design of SPO algorithm

SPO is an optimization algorithm designed by combining the advantages of SA, PSO and GA algorithms, where each particle in the velocity update formula of PSO algorithm is in the direction of individual optimum and global optimum, which can provide faster convergence speed, but SA algorithm has the advantage of no second reset, this paper draws on the particle swarm velocity update formula, the concept of variation in GA algorithm and the advantage of SA algorithm, the SPO algorithm has a fast convergence speed and does not have the advantage of the second reset. In this paper, we use the SPO algorithm to implement the polarization control in coherent optical communication systems.

(1) Design of initial voltage and initial step size

The algorithm is initialized by designing the control voltage U_1, U_2, U_3 of the three channels of the polarization controller to be half of the maximum value of the voltage applied to each channel $U_{\max}$, i.e.,

$$U_1 = U_2 = U_3 = U_{\max}/2 \tag{4.38}$$

Since the phase shift range of each channel of the polarization controller used in this paper is $0 \sim 4\pi$, in this way the SPO algorithm is able to produce $0 \sim 2\pi$ phase change in each channel during the initial seeking process, assuming that the polarization state of the signal light is linearly polarized light perpendicular to the x-axis, and the polarization state of the sought target is linearly polarized light in the horizontal direction, with a difference of π between the two polarization states on the bunga ball, this design can reduce the possibility of the first reset occurring during the seeking process and reduce the jitter of the signal during the seeking process, the design formula for the initial step is

$$\Delta_1 = \Delta_2 = \Delta_3 = r_1 \times \Delta_{max} = r_1 \times 0.1 \times U_{max} \tag{4.39}$$

where r_1 is a random number between [−1,1] and the randomness of the step size is set to enhance the randomness of the algorithm in the initial search process.

(2) Design of step update formula

The step update formula we designed combines the PSO algorithm speed update formula with the concept of variation in the GA algorithm, which can be expressed as

$$\Delta_j^{i+1} = \begin{cases} w \cdot \Delta_j^i + c \cdot r_2 \cdot \left(U_{Gbest,j} - U_j^i\right) + r_1 \cdot \Delta_{max} \cdot p_2 \cdot [(N-i)/N]^3, & r_2 \le p \\ r_1 \cdot \Delta_{max}, & r_2 > p \end{cases} \tag{4.40}$$

where the first two are the same as the formula for the particle swarm velocity update.w and c are inertia weights and learning factors, respectively.r_2 is a uniformly distributed random number on [0, 1]; $U_{Gbest,j}$ is the voltage of channel j corresponding to $a_{allbest}$, $a_{allbest}$ is the best adaptation value obtained during the iteration of the algorithm.$i = 1, 2, \ldots, N$ is the number of iterations.$j = 1, 2, 3$ denotes the jth channel of the polarization controller; Δ_j^i is the step size of the jth channel during the ith iteration, the step size of the jth channel during the ith iteration. U_j^i is the value of the voltage applied to the jth channel of the ith iteration.p_2 is the dithering factor, was introduced to suppress premature convergence of the algorithm causing it to stop and to improve the control accuracy of the algorithm.p is the probability of variation of the step size, the expressions are

$$p = \exp\left\{-p_1 \cdot [(N-i)/N] \cdot \left(\left|\Delta_j^i\right|/a_{allbest}\right)\right\} \tag{4.41}$$

where p_1 is introduced to enhance the global search capability of the algorithm and the ability to jump out of the local optimum. The expression of the variation probability p involves variables Δ_j^i and $a_{allbest}$ in addition to the variation factor p_1 and variable i. The three variables together determine the variation probability of that step in the next step. A flow chart of polarization control using the SPO algorithm is shown in Fig. 4.21.

The specific steps for polarization control using the SPO algorithm are as follows.

(1) Algorithm initialization, initialization of algorithm parameters, set the maximum number of iterations N, variable factors p_1, jitter factor p_2, acceleration constants c, inertia weights w, initial voltage U_j and initial step size Δ_j.

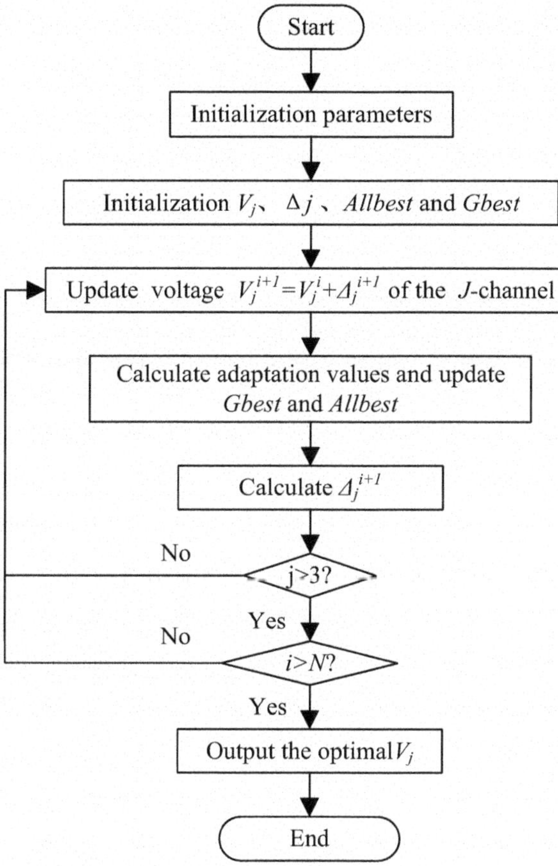

Fig. 4.21 Flow chart of polarization control using SPO algorithm [34]

(2) Apply the initial voltage value of the three channels to obtain the initial adaptation value, and use the initial adaptation value as the global optimal adaptation value, and update U_{Gbest} and $a_{allbest}$.

(3) Update the voltage of the jth channel according to Eq. (4.42), obtain the adaptation value after the update, determine whether the adaptation value is better than $a_{allbest}$, and update U_{Gbest} and $a_{allbest}$ if it is better.

$$U_j^{i+1} = U_j^i + \Delta_j^{i+1} \tag{4.42}$$

(4) generating the step size of the *j*th channel of the next iteration by Eqs. 4.40 and 4.41,

(5) determine whether the three channels are updated at one time, if the condition is true, execute the next step; if the condition is false, jump back to step (3),

(6) determine whether the maximum number of iterations is reached: if the condition is true, the optimal voltage value applied to the three channels is obtained and

the next step is executed; otherwise, jump back to step (4) to enter the next iteration.

(7) The maximum value of the IF signal amplitude $a_{allbest}$ and the corresponding U_j, are obtained, at which time the optimal voltage value U_j, for each channel of the PCD-M02 is input and the program enters a small range of perturbation tracking.

(2) Determination of parameters of SPO algorithm

The parameters involved in the SPO algorithm are the variation factor p_1, jitter factor p_2, inertia weights w and learning Factor c, Together they affect the convergence speed of the algorithm and the final optimization performance. In the simulation, we assume that the polarization state of the input signal light is (−1, 0, 0) and the light intensity values of the signal light and the local oscillation light are certain.

Set $p_2 = 0.6$, $w = 0.5$ and $c = 1.5$, As shown in Fig. 4.22, the variation factor p_1 is 0.1, 0.08, 0.06, 0.04 and 0.02, respectively, and the number of variants of the algorithm in the iterative process is 10, 14, 6, 2 and 2. The size of the variation factor determines the number of step variants of the algorithm in the iterative process, and when the value of is small, the number of step variants in the whole iterative process is small Therefore, in order to make the SPO algorithm have a faster convergence speed while having the ability to jump out of the local optimum, we choose $p_1 = 0.06$.

Set $p_1 = 0.6$, $w = 0.5$ and $c = 1.5$, as Fig. 4.23 shows the curve of the effect of the jitter factor p_2 on the performance of the algorithm, when $p_2 = 1$ the algorithm reaches the optimum through 166 iterations, and the IF amplitude has a large fluctuation at the number of 100 iterations. When $p_2 = 0.2$, the magnitude of IF reaches the maximum after 70 iterations of SPO, and the fluctuation of IF magnitude is less than 0.01% after 100 iterations. p_2 is too large to enhance the search range of the solution space at the early stage of the optimization search, but it also increases the convergence time of the algorithm and affects the variance of IF magnitude fluctuation after convergence. Therefore, we choose $p_2 = 0.6$ in order to have a fast convergence speed with a strong search capability.

The inertia weight w and learning factor c in the SPO algorithm have the same meaning as in the PSO algorithm, setting $p_1 = 0.06$, $p_2 = 0.6$ and $c = 1.5$.

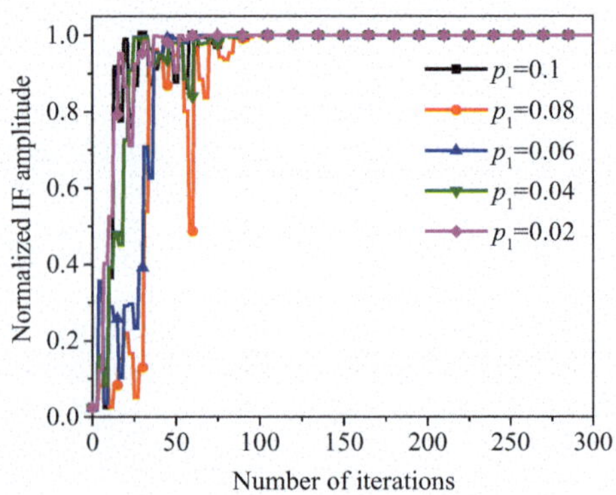

Fig. 4.22 Influence curve of p_1 on SPO algorithm [34]

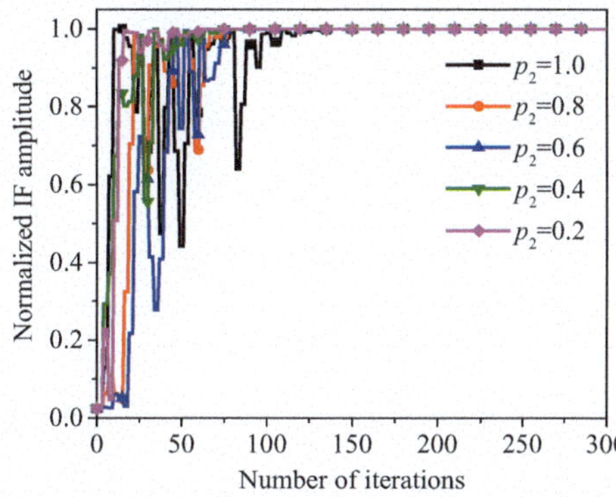

Fig. 4.23 Curve of the effect of p_2 on SPO algorithm [34]

Figure 4.24 shows the influence curve of the inertia weight w on the performance of the algorithm. w has a large influence on the convergence speed of the algorithm, when $w = 0.9$ the algorithm needs to pass 150 iterations to reach the optimum, fluctuates more in the process of seeking the optimum, and has a strong ability to explore the solution space. The exploration ability of the space is stronger. In the process of simulation, it is found that when $w < 0.8$, the influence of w on the algorithm becomes less obvious, but for smaller w, when $w = 0.1$, the algorithm is easy to fall into the local optimum, so in order to make the algorithm have better optimization performance, $w = 0.5$ is taken.

Set $p_1 = 0.06$, $p_2 = 0.6$ and $w = 0.5$, as shown in Fig. 4.25, the impact curves of different learning factors c on the algorithm are shown. c_1 and c_2 have a greater impact on the algorithm in the PSO algorithm, but in the SPO algorithm when other parameters are determined, as shown in the figure, the algorithm is able to converge after less than 100 iterations. c has less impact on the convergence speed of the algorithm and the control accuracy, and we take $c = 1.5$.

The influence of w, c, p_1 and p_2 on the performance of SPO algorithm is analyzed above, when taking the optimal values obtained from simulation, $w = 0.5$, $c = 1.5$,

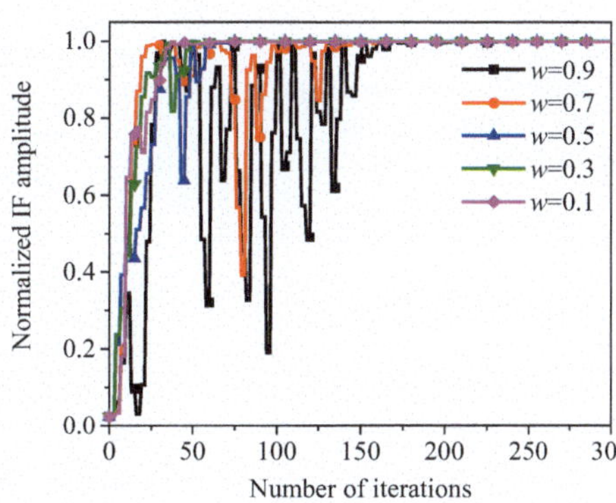

Fig. 4.24 Curve of the effect of w on the SPO algorithm [34]

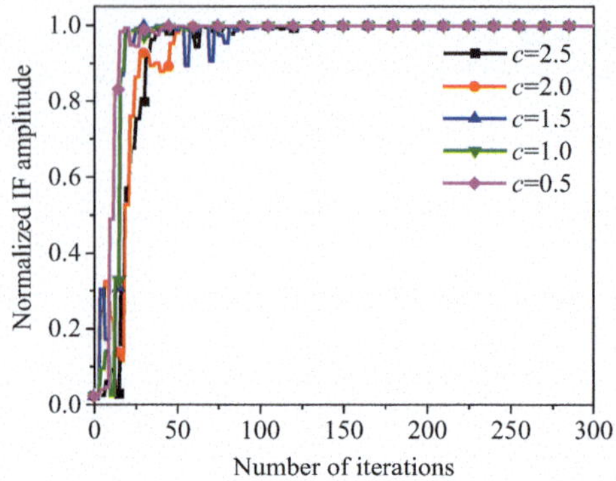

Fig. 4.25 Curve of the effect of c on the SPO algorithm [34]

$p_1 = 0.06$ and $p_2 = 0.6$. In the coherent optical communication system, the polarization state of the signal light is unknown when the signal light enters the mixer at the receiver side, and due to the atmospheric turbulence, fiber jitter, temperature changes will make the signal light polarization state change at different rates. Therefore, the polarization state of the signal light is not fixed in the initial optimization process, but is randomly perturbed at a certain rate, as shown in Fig. 4.26 for the SPO algorithm optimization performance when the polarization change rate v_{pol} is 20.0, 2.0 and 0 rad/s, respectively. It can be concluded that the control effect of the SPO algorithm is better even in the case of rapid changes in the polarization state of the signal light.

(3) Simulation study of polarization control based on SPO algorithm

Through the above analysis, the simulation parameters of SPO algorithm are determined: $w = 0.5$, $c = 1.5$, $p_1 = 0.06$ and $p_2 = 0.6$, and the polarization controller used in this paper is used to build the polarization control model of coherent optical communication system. When the polarization state of the signal light is controlled as linearly polarized light along the x-axis, the amplitude of the IF signal will reach the maximum and stabilize.

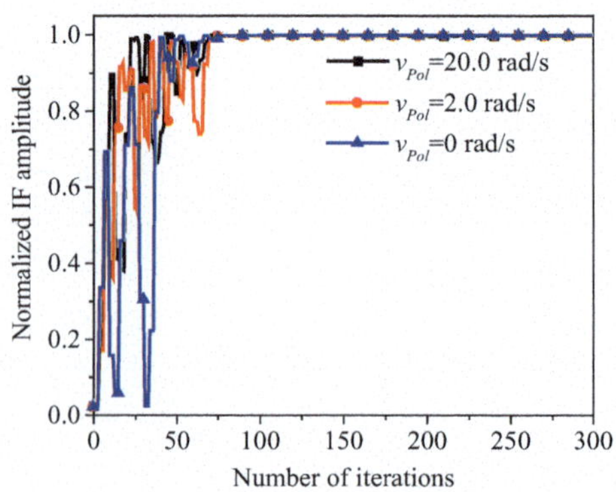

Fig. 4.26 Curve of the effect of SPO algorithm on the amplitude of IF signal [34]

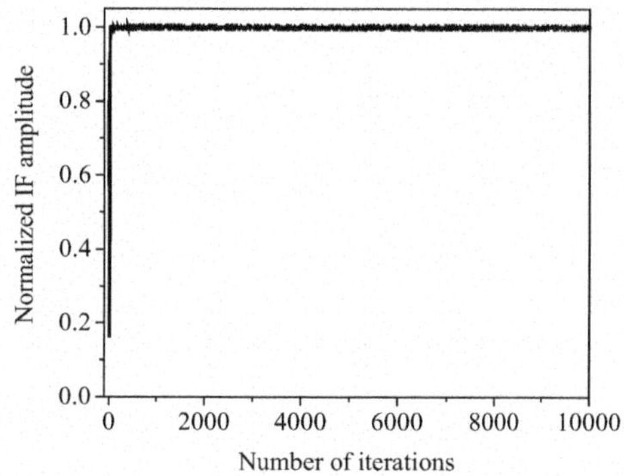

Fig. 4.27 Variation curve of IF signal amplitude [34]

Figure 4.27 The horizontal coordinate is the number of iterations, and the vertical coordinate is the change of the IF signal amplitude after normalization. The polarization state of the signal light is generated randomly, and its corresponding Stokes parameter is (0.5814, 0.6252, −0.5207), and the fluctuation rate satisfies $0 \leq v_{Pol} \leq 20$ rad/s. From Fig. 4.27, we can see that the amplitude of the IF signal of the SPO algorithm increases rapidly after simple iterations and stabilizes to the maximum value with fluctuations less than 1%, and the initial optimization process of the SPO algorithm reaches the maximum amplitude of the IF signal through about 75 iterations, and after reaching the maximum number of iterations, the algorithm will enter the tracking process

The red dot in Fig. 4.28 shows the change of the polarization state of the signal light on the bunga ball during the whole iteration of the algorithm, and Fig. 4.29 shows the curve corresponding to the change of the Stokes parameter. It can be seen that after the iteration of the SPO algorithm, the Stokes parameter of the signal light is stabilized at (1, 0, 0), and at this time the signal light is controlled as linearly polarized light along the x-axis, and the signal light will all enter the optical mixer to mix coherently with the local oscillation light, corresponding to $\alpha = 0$, $\eta = 100\%$, and the mixing efficiency reaches the maximum.

Figures 4.30, 4.31 and 4.32 show the phase shift changes of the three waves P_1, P_2 and P_3 during the control process. The phase shift change range of each wave inside the polarization controller is $[0, 4\pi]$.

4.3.5 Comparison of the Three Algorithms

For the same polarization controller, after the response time of the polarization controller is determined, the time of polarization control mainly depends on the optimization time of the algorithm, the response time of the polarization controller used in this paper is less than 400 μs, assuming that the signal light polarization state fluctuation rate satisfies $0 \leq v_{Pol} \leq 20$ rad/s. As shown in Fig. 4.33, the time required

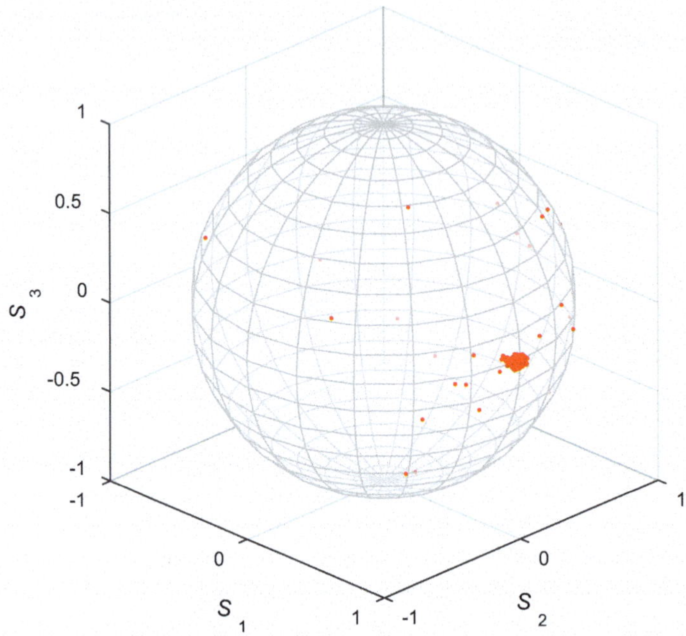

Fig. 4.28 Change of polarization state of signal light [34]

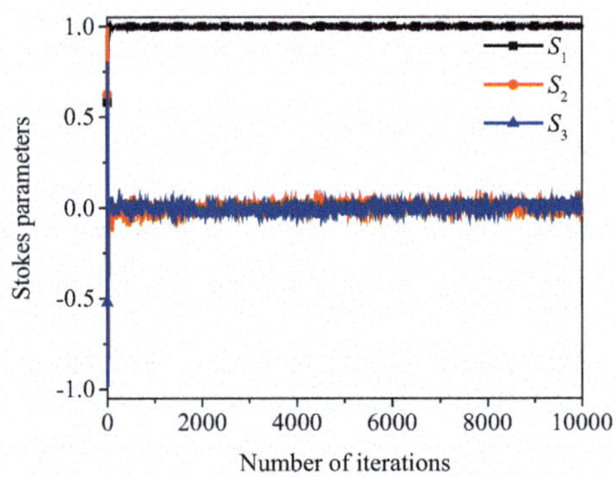

Fig. 4.29 The variation curve of signal light stokes parameter [34]

for the optimization process of SA, PSO and SPO are analyzed by simulation, and the final result is the average value of each algorithm after 10 independent runs.

From Fig. 4.33, it can be seen that the convergence speed of SPO algorithm is significantly better than that of PSO and SA. The SPO algorithm reaches the maximum amplitude of IF in only about 0.05 s, while the SA and PSO algorithms take about 0.2 s. In the actual engineering application, due to the existence of the second reset, the PSO algorithm takes a total of 0.4 s to reach the optimum, in terms of optimization time. The optimization time of SPO algorithm is 4 times shorter than

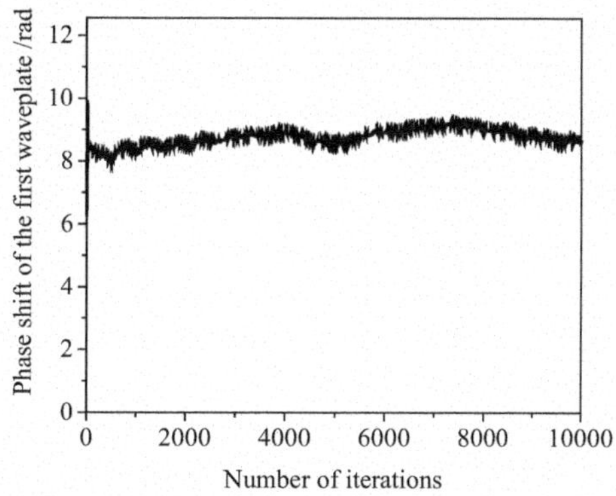

Fig. 4.30 Variation of phase shift of the first waveplate P_1 of the polarization controller with the SPO algorithm control [34]

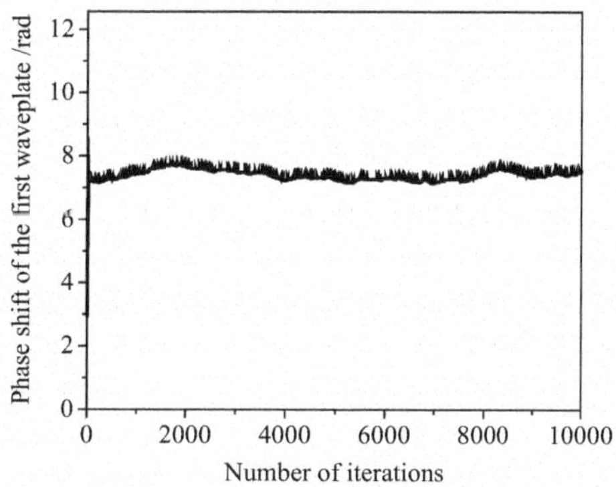

Fig. 4.31 Variation of the phase shift of the second waveplate P_2 of the polarization controller with the SPO algorithm control [34]

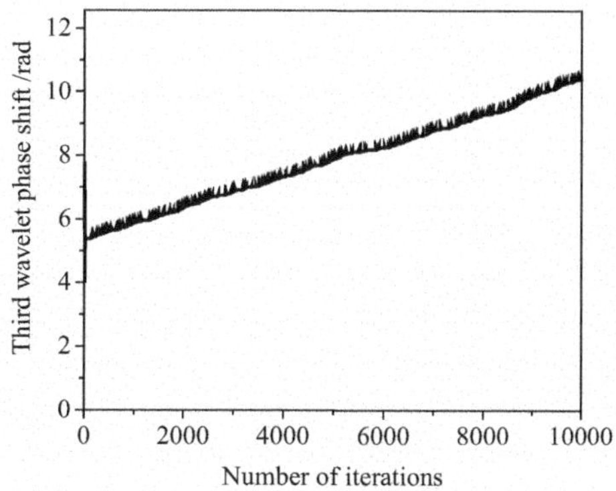

Fig. 4.32 Variation curve of the phase shift of the third waveplate P_3 of the polarization controller with the SPO algorithm control [34]

that of SA and 8 times shorter than that of PSO, which greatly shortens the response time of polarization control.

The SPO algorithm and SA algorithm belong to the same serial algorithm, but the SPO algorithm provides a faster optimization time for the SPO algorithm by borrowing the update formula of the PSO algorithm speed, provides the algorithm

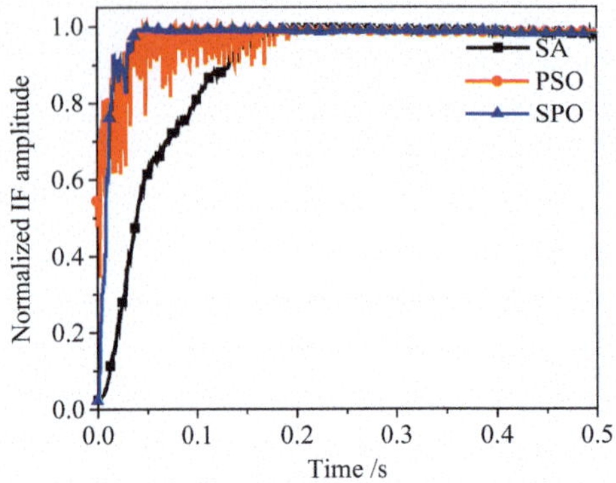

Fig. 4.33 Comparison of convergence time of three algorithms [34]

with the diversity of the selected solutions by using the concept of variation, so that the SPO algorithm is not easy to fall into the local optimum, increases the diversity of the algorithm's solution search at the initial time and improves the control accuracy of the algorithm by adding the perturbation term.

4.4 Endless Reset of the Polarization Controller

The polarization controller used in the experiment is a fiber-optic squeeze type polarization controller, and a reset algorithm must be designed to achieve continuous operation of the polarization controller. This chapter investigates the reset problem in the polarization control and finds a reset algorithm applicable to this polarization controller.

4.4.1 Small Step Backward Reset Method and Direct Reset Method

Fiber optic squeeze type polarization controller is a phase adjustable polarization controller, which is widely used in optical communication system due to small insertion loss and fast response, but the phase change of this type of polarization controller has a certain range limitation, which is also called the first reset method.

There are two general ways to solve the reset: one is to choose the polarization controller material without complex reset algorithm, such as the rotatable waveplate mentioned in the Ref. [8], which has no limitation on the rotation range. And the reset of external drive voltage does not cause fluctuation of the output optical polarization state; the second is to design the reset algorithm, such as fiber-optic squeezed polarization controller, which requires a complex reset algorithm to complete the endless reset of the polarization controller because the reset process will have an impact on the output polarization state. The design of the reset algorithm is a difficult point of polarization control and a research hotspot. There are a lot of studies on the reset algorithm of polarization control at home and abroad, mainly direct reset method, small step backward reset method, Noe reset method, and equivalent wavelet reset method [21, 23, 39, 40], among which Noe reset method is only applicable when the output polarization state is required to be horizontal linearly polarized light, and the coherent optical communication system and polarization controller involved in this paper use single-mode fiber, and the light output by the polarization controller is not necessarily horizontal linearly polarized light, so Noe reset is not applicable. The equivalent waveplate reset method is to use three delay-variable waveplates equivalent to 1/4 waveplate and four delay-variable waveplates equivalent to 1/2 waveplate to complete the polarization control without end by the combination of two waveplates, and the simplest combination to complete the polarization control without end requires four delay-variable waveplates [41], while the polarization controller used in this paper has only three delay-variable waveplates, this This resetting method is also not applicable. In this section, a small-step backward reset method is firstly designed and a theoretical analysis of the designed small-step backward reset method as well as the direct reset method is done to further obtain the reason why the third waveplate needs to be reset frequently.

(1) The basic principle and theoretical simulation of the small step backward reset method

In the Ref. [40], a reset method is mentioned as a small step backward method, and the reset principle is relatively simple. The reset method mentioned requires at least four waveplates with 45° difference in azimuth, two of which are used as the main control waveplates and the other two as the compensation waveplates. When one of the waveplates needs to be reset, the other three waveplates work at the same time for the reset of the waveplate, and after the reset is completed, the compensation waveplate returns to the median value for the next reset, this reset method has a longer reset time because the four waveplates work together during the reset process.

We have designed a reset method similar to the small step backward, which requires only three waveplates. The reset principle is that when the polarization controller starts to work, three waveplates work simultaneously to output a fixed polarization state. When one of the waveforms needs to be reset, the waveform stops working to reset, and the other two waveforms work. For each reset step, the signal jitter caused by the reset is compensated by the other two waveforms. Generally speaking, two waveplates can complete the work from any input polarization state to any output polarization state, the signal jitter caused by this reset method is related to the polarization state of the output signal and the reset step set by the reset, if the reset step set by the reset process is short, the signal jitter caused by each reset step of the reset waveplate is smaller and can be quickly compensated by the other two waveplates, but the reset If the reset step length is set larger, each reset step may cause large jitter on the signal, but the reset time is shorter, the time required for reset mainly depends on the sum of the step length set for reset and the time required for compensation, in order to avoid large fluctuations in the signal caused by the reset process should be reasonably designed reset step length.

Three fiber extruders with 0°, 45°, and 0° at 45° to each other are used as the polarization controller model, and the polarization control in the coherent optical communication system is completed using the SPO algorithm and the small-step backward reset method designed in this paper. Figures 4.34, 4.35 and 4.36 show the phase shift changes of the three waveplates P_1, P_2, and P_3 of the polarization controller when the input signal optical polarization state is (−0.1, −0.19, −0.98) and the polarization state change rate v_{Pol} satisfies $0 \leq v_{Pol} \leq 20$ rad/s. The reset method is the small step backward reset method designed in this paper, and the reset step size is set to $\pi/3$. From Fig. 4.36, it can be seen that only wavelet three needs to be reset frequently during the whole control process, and no reset occurs for wavelet one and wavelet two, similar to the phenomenon mentioned in the Ref. [28].

As shown in Fig. 4.37 for the control process of the intermediate frequency (IF) signal amplitude change, in the entire control process of the IF signal amplitude jitter less than 3%, although the third wave in Fig. 4.36 need to reset often, but from Fig. 4.37 can be seen that the reset process of wave three does not produce any IF signal amplitude, IF signal amplitude fluctuation size is only related to the rate of

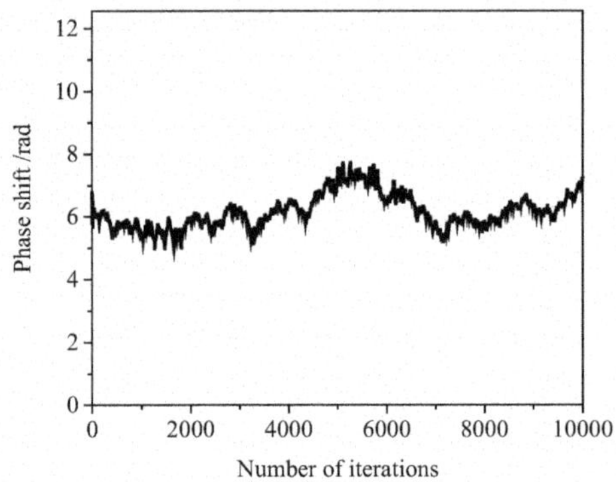

Fig. 4.34 Variation of phase shift of the first waveplate P_1 of the polarization controller with the small-step backward reset [34]

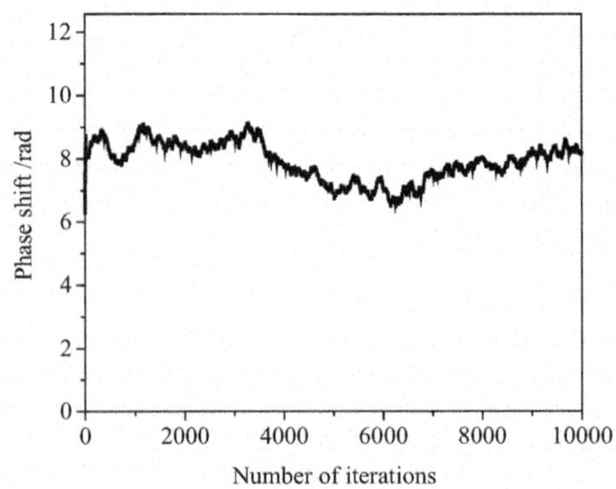

Fig. 4.35 Variation of phase shift of the second waveplate P_2 of the polarization controller with the small-step backward reset [34]

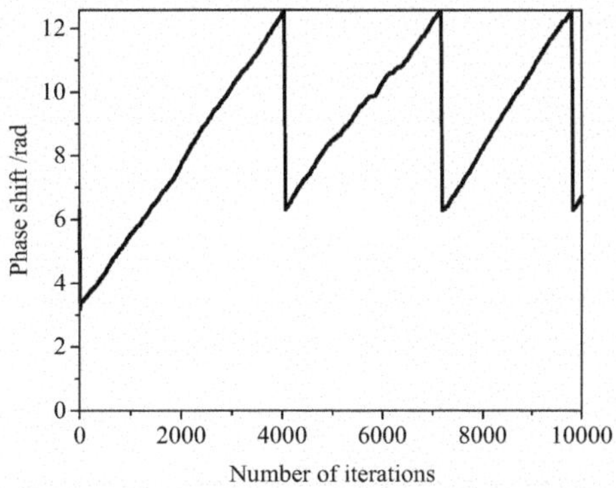

Fig. 4.36 Variation curve of the phase shift of the third waveplate P_3 of the polarization controller with the small-step backward reset [34]

change of the signal light polarization state, the entire reset process control effect is good.

(2) The basic principle and theoretical simulation of the direct reset method

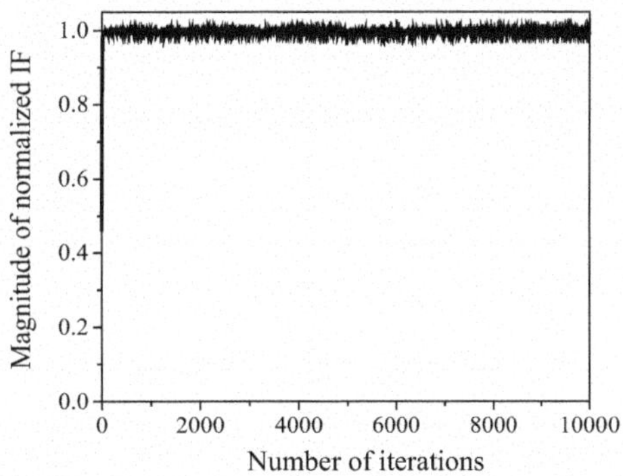

Fig. 4.37 Variation of IF signal amplitude with the small-step backward reset [34]

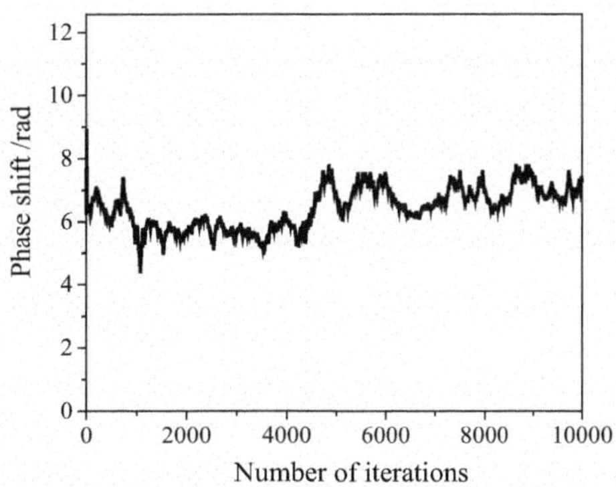

Fig. 4.38 Variation of phase shift of the first waveplate P_1 of the polarization controller with direct reset [34]

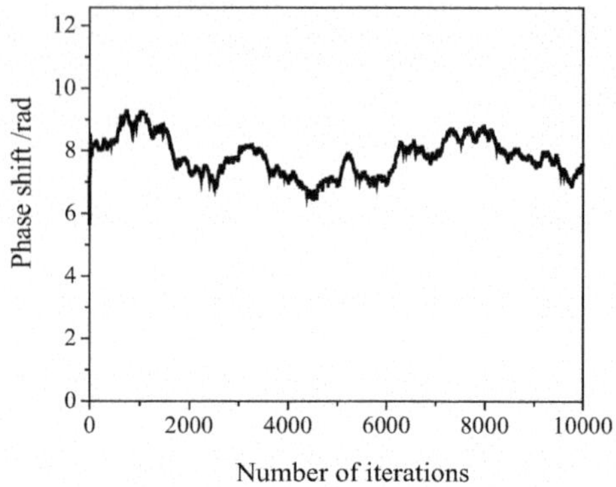

Fig. 4.39 Variation of phase shift of the second waveplate P_2 of the polarization controller with direct reset [34]

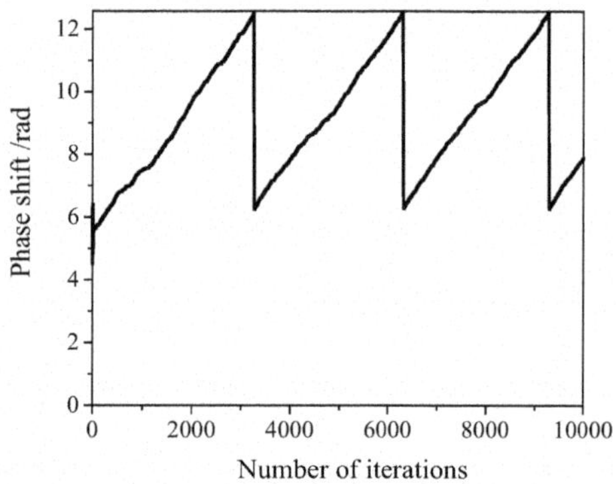

Fig. 4.40 Variation curve of the phase shift of the third waveplate P_3 of the polarization controller with direct reset [34]

For a controller consisting of three fast axes with horizontal reference direction at 0°, 45° and 0° waveplates, it is known from Chap. 2 that the relationship between the input signal optical polarization state E_{in} and the output signal optical polarization state E_{out} is satisfied by

$$E_{out}(\gamma_3, \gamma_2, \gamma_1) = J_3 \cdot J_2 \cdot J_1 \cdot E_{in} \tag{4.43}$$

Since $\gamma_3, \gamma_2, \gamma_1 \in [0, 4\pi]$, when one of the waveplates of the polarization controller reaches the limit of the phase shift range of 0 or 4π, the phase change of the waveplate by 2π will not change the polarization state of the output signal light, and its reset formula can be written as follows.

$$E_{out}(\gamma_i) = E_{out}(\gamma_i \pm 2\pi) \tag{4.44}$$

where $\gamma_i = 0, 4\pi; i = 1, 2, 3$, direct reset is based on this principle, through this reset makes the three waveplates always work within $0 \sim 4\pi$, this phase jump is achieved by the rapid rise and fall of the applied drive voltage, requiring a certain

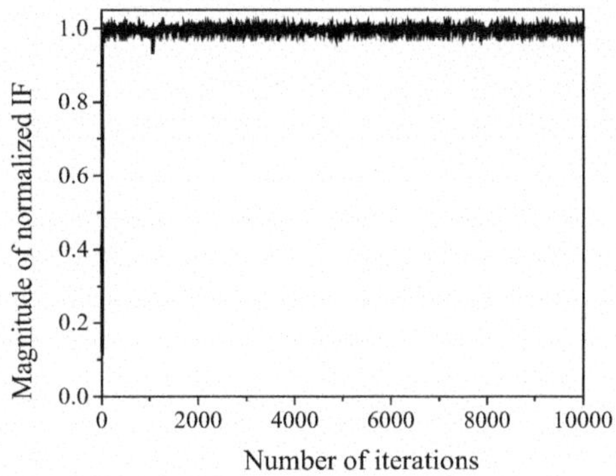

Fig. 4.41 Variation of IF signal amplitude with direct reset [34]

response time, although the reset will not affect the polarization state of the output signal light, but in the reset, the sudden voltage change process may cause significant signal jitter.

The polarization control in the coherent optical communication system is also modeled by using the SPO algorithm and direct reset method. Figures 4.38, 4.39 and 4.40 show the phase shift changes of the three waveplates P_1, P_2 and P_3 of the polarization controller when the polarization state of the randomly generated signal light is $(-0.22, 0.067, -0.97)$, and the polarization state of the control required output light is $(1, 0, 0)$. The direct reset method is the same as the small step backward reset, and only wavelet three needs to be reset frequently, while wavelet one and wavelet two generally do not need to be reset, and it can be seen from Fig. 4.41 that in the direct reset, the reset of wavelet three also does not have any effect on the amplitude of the IF signal, and the reset effect is good.

By comparing the small step backward reset and direct reset method, it is found that these two types of reset are the third waveplate need to reset often, waveplate one and waveplate two generally do not need to reset, the signal jitter caused by the reset is not obvious, through the analysis of its main reason is due to the coherent optical communication system in the three waveplates simultaneously involved in polarization control, waveplate one and waveplate two play a major role in the control, assuming that the polarization state of the input signal light is E_{in}, the polarization state of the output signal light after passing the first waveplate and the second waveplate can be written as follows.

$$E_{out} = J(45^\circ, \gamma_1)J(0^\circ, \gamma_2)E_{in} \tag{4.45}$$

The SPO algorithm can quickly control the randomly polarized state of the signal light into horizontal linearly polarized light, at this point $E_{out} \approx [1, 0]^T$, when the signal light then passes through the third waveplate after the final output signal light polarization state can be written as

$$E_{out}^{'} = \left[\exp(i\gamma_3/2), 0\right]^T \tag{4.46}$$

This phenomenon can also be seen in Fig. 4.21. Regardless of the small step backward reset method or direct reset method, only waveplate three needs to be reset frequently, so waveplate three will not affect the polarization state of the output signal light regardless of the way it is reset. Similarly, when the polarization state of the output signal light is vertical line polarization light, $E_{out}^{'} = \left[0, \exp(-i\gamma_3/2)\right]^T$, at this time any change γ_3 will not affect the polarization state of the output light, so for the output polarization state of horizontal or vertical line polarization light, you can use small step backward reset or direct reset is simpler.

4.4.2 Experiment of Direct Reset Method

As can be seen from Chap. 2, the theoretical model of coherent optical communication is to control the polarization state of the signal light after passing through the atmosphere as a horizontal line polarized light input to the mixer, theoretically the Noe reset method, small step backward reset and direct reset method can be used in this system, but in the actual system the fibers used are single-mode fibers, assuming that the polarization state of the signal light output from the polarization controller is E_{P_out}, the polarization state of the signal light input to the mixer is E_{H_in}, the transmission matrix formed by the single-mode fiber between the polarization controller and the mixer is J_c, and the relationship between them satisfies.

$$E_{H\_in} = J_c E_{P\_out} \tag{4.47}$$

In order to achieve the maximum mixing efficiency needs $E_{H\_in} = [1, 0]^T$, so the final polarization state of the signal light output by the polarization controller is not necessarily horizontal line polarized light, it is obvious that the Noe reset method is not applicable. Small step backward reset method only when the polarization state of the output signal light is horizontal line polarization light or vertical line polarization light, the third wave chip needs to be reset frequently, and at this time the reset of the third wave chip will not have an impact on the polarization state of the output polarization light, so the reset effect is better, but in the output of other polarization states of the signal light, it is possible to encounter two of the wave chips reset at the same time or the second wave chip reset, which requires a combination of small step backward reset and direct reset method, the reset effect is poor, so in this paper we choose the simplest direct reset method to complete the endless reset of the polarization controller.

As shown in Fig. 4.42 is the polarization control system, the manual polarization controller behind the laser is the line polarization light incident mixer of the local vibration light control level, the manual polarization controller behind the signal light is used to simulate the random changes of the signal light polarization state, the I-way signal from the balance detector is used as the monitoring signal, the Q-way is

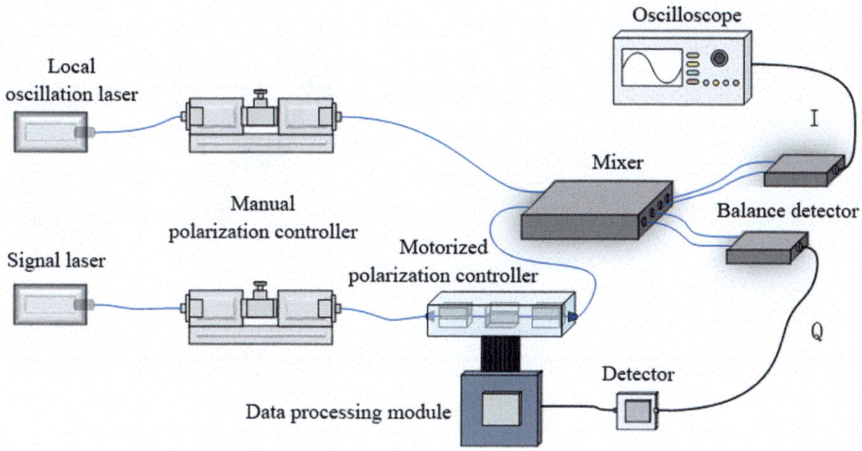

Fig. 4.42 Polarization control system [34]

used as the feedback signal of the polarization control, and the polarization control is completed by using the direct reset method and SPO algorithm.

Figures 4.43, 4.44 and 4.45 show the changes of the three phases of the polarization controller in the working process, the horizontal coordinate is the sampling time, the vertical coordinate is the phase shift, and it can be seen from Fig. 4.43 that in the control process, wavelet one reaches the minimum limit of P_1 at about 2 s and a reset occurs, wavelet two reaches the maximum limit of P_2 at about 13 s in Fig. 4.43 and a reset occurs. In Fig. 4.45, wavelet three is not reset within 30 s, as shown in Fig. 4.46, the change of the corresponding IF signal amplitude, when wavelet one and wavelet two are reset, it can be seen from Fig. 4.46 that the IF signal amplitude decreases instantly and then recovers, so it can be obtained that the direct reset method used in this paper will inevitably cause signal amplitude jitter in the process of reset. The direct reset method used in this paper will inevitably cause signal amplitude jitter during the reset process, which will have a greater impact on the signal quality.

The design of reset algorithm in polarization control system is a difficult point of polarization control, especially for the actual application system, there are more influencing factors, which limit the use of some reset algorithms. In the next work, we will focus on researching and designing better reset algorithms for polarization control in coherent optical communication system to improve the system performance.

4.5 Experiment of Polarization Control

In practice, the feedback signal is affected by many factors together, which will affect the control effect to some extent. The single-particle optimization algorithm and direct reset method are used in coherent optical communication system

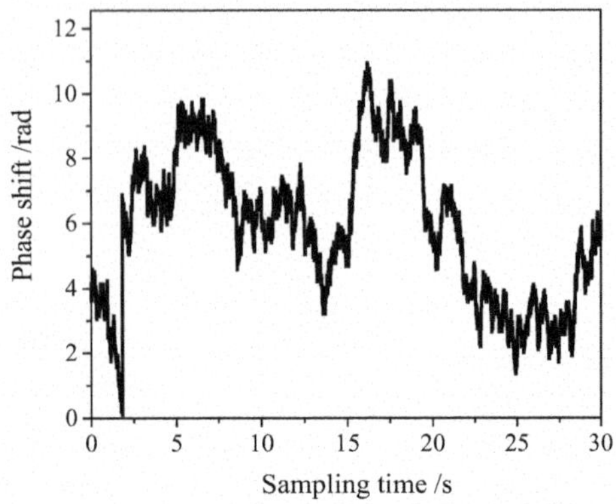

Fig. 4.43 Variation of phase shift of the first waveplate P_1 of the polarization controllercurve with direct reset in practical system [34]

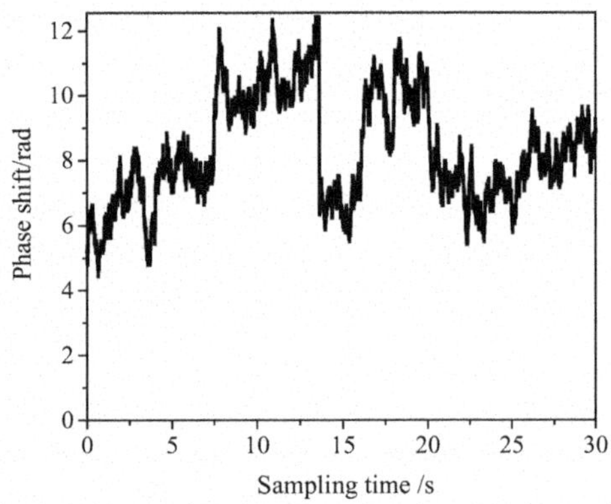

Fig. 4.44 Variation curve of the phase shift of the second waveplate P_2 of the polarization controller with direct reset in practical system [34]

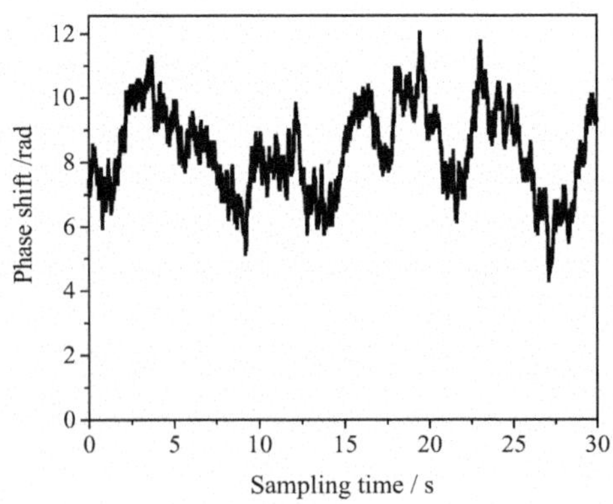

Fig. 4.45 Variation curve of the phase shift of the third waveplate P_3 of the polarization controller with direct reset in practical system [34]

to complete the experiment of measuring and controlling polarization in coherent optical communication.

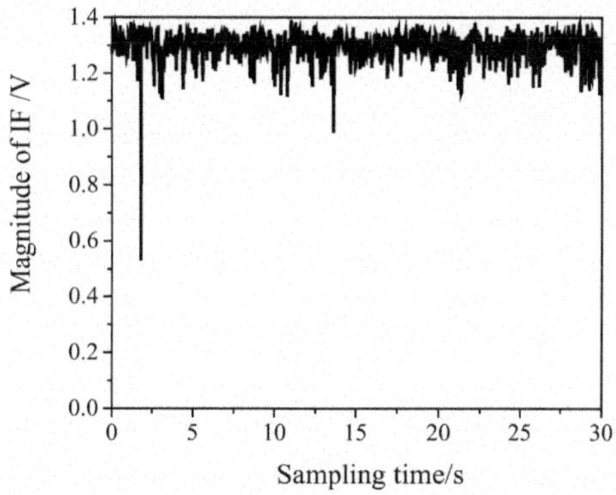

Fig. 4.46 Variation of IF signal amplitude with direct reset in practical system [34]

4.5.1 Experimental Setup

Figure 4.47 shows the schematic diagram of coherent detection with polarization control, in which the polarization control system mainly includes polarization controller, ARM and detector, in which the detector completes the extraction of intermediate frequency (IF) signal amplitude information, ARM is used for AD acquisition of amplitude information and generation of control signal by algorithm, and the polarization controller changes the polarization state of the signal light according to the control signal.

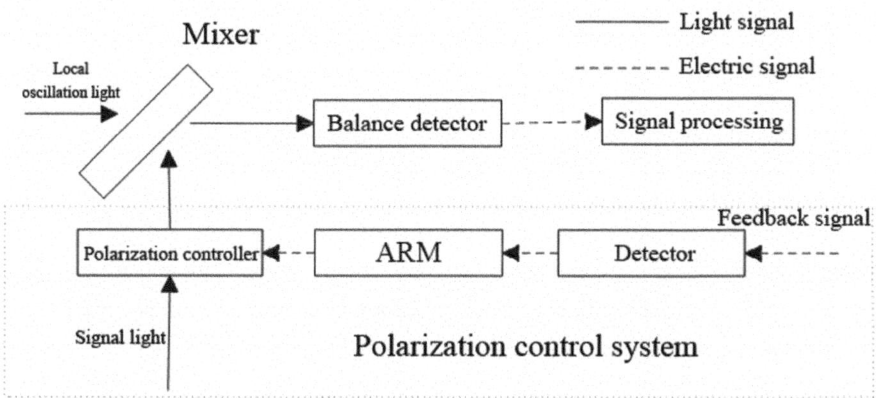

Fig. 4.47 Coherent detection schematic

4.5.2 Polarization-Controlled External Field Experiments

Coherent optical communication links of 600 m, 1.3 km and 10.2 km were built to study the specific changes of the signal optical polarization state at different distances, and the effects on the performance of the communication system when the polarization control system is open-loop and closed-loop were investigated.

(1) 600 m polarization control experiment

As shown in Fig. 4.48, we measured the variation of the polarization state of the signal light at 600 m in one hour at different time, weather and temperature, respectively. The average rate of change of the signal light polarization state on the bunga ball in (a) is 0.137 rad/s, and the maximum rate of change is 1.02 rad/s; the average rate of change of the signal light polarization state in (b) is 0.0742 rad/s, and the maximum rate of change is 0.484 rad/s; the average rate of change of the signal light polarization state in (c) is 0.081 rad/s and the maximum rate of change is 1.009 rad/s.

From Fig. 4.48, it can be seen that in a coherent optical communication link of 600 m, the rate of change of the signal light polarization state on the bunga ball is relatively slow, and in the Ref. [2] at a communication distance of 600 m, $C_n^2 = 2 \times 10^{-14} m^{-2/3}$, The mean value of the change in polarization angle is about 0.02 rad. The change in polarization is much faster compared to the theory, and the main reasons for this phenomenon are:

(1) The polarization state data is measured by the spectrometer, which comes with a measurement error of $\pm$ 0.087 rad, so the measurement instrument itself comes with a large measurement error.
(2) The signal light output from EDFA at the transmitting end and the transmitting antenna are connected by single-mode fiber, and the coupling from the fiber to the mixer at the receiving end is also connected by single-mode fiber, which will inevitably cause slight shaking of the fiber because it is exposed outside, thus causing irregular changes in the polarization state and introducing certain measurement errors, and we try to fix the exposed fiber to reduce the measurement errors when doing the experiment.

The polarization state of the signal light measured in Fig. 4.48 is different each time, mainly because the optical fiber used in the whole system is single-mode fiber, the bending and jitter of the fiber will introduce random changes in the polarization state of the signal light, it can be seen that although the measured polarization state is different each time, the rate of change of the polarization state of the signal light is relatively slow, and it has less effect on the IF signal.

The variation of IF signal amplitude in the open-loop and closed-loop states of the polarization control system within 20 s of the 600 m coherent optical communication experiment is measured as shown in Fig. 4.49. The polarization control system is in the open-loop state for the first 10 s and the polarization controller is in the closed-loop state for the last 10 s in Fig. 4.49a and b. Although the variation range of the polarization state of the signal light in Fig. 4.49 is not large and changes slowly,

Fig. 4.48 Changes in the polarization state of signal light in 600 m field experiment [34]

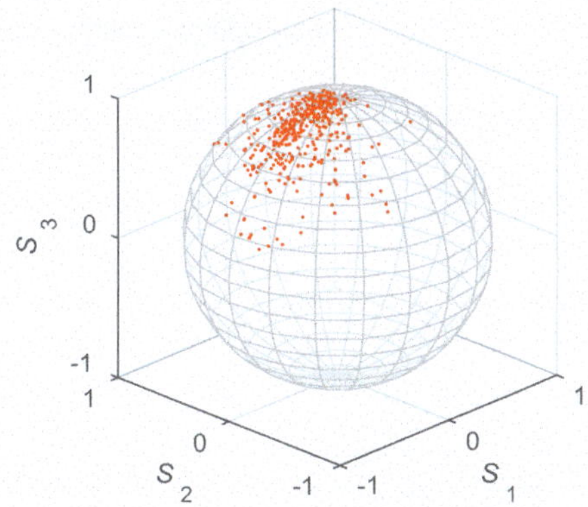

(a)2018-04-09,00:00-01:00, sunny, 15℃.

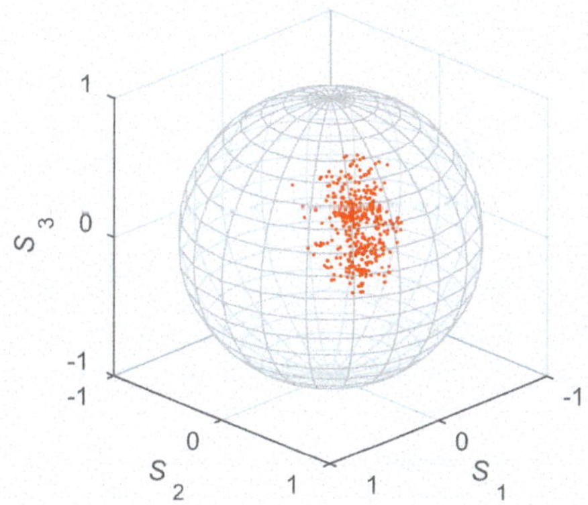

(b) 2018-04-10,15:30-16:30, cloudy, 18℃.

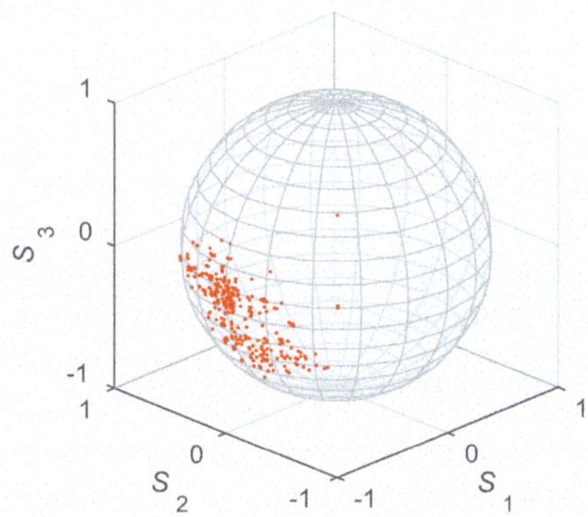

(c) 2018-04-13,10:00-11:00, light rain, 10℃.

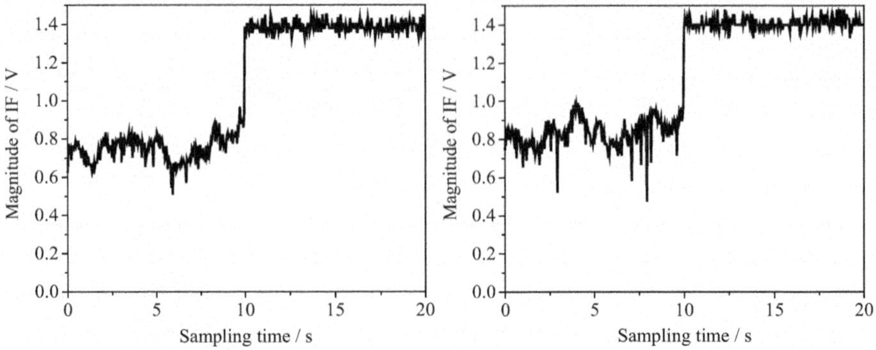

Fig. 4.49 The influence curve of polarization control system on the amplitude of IF signal in 600 m field experiment [34]

it can be seen from Fig. 4.49 that the uncertainty of the polarization state when the polarization controller is in closed loop has a greater impact on the IF signal, indicating that the impact of the change of polarization on the IF signal amplitude at 600 m can be neglected, and only the uncertainty of the polarization state of the signal light arriving at the front end of the mixer needs to be considered, and the polarization state of the signal light is The polarization of the signal light can be controlled as a line polarized light along the x-axis.

In Fig. 4.49a, the amplitude of the IF signal fluctuates between 0.51 and 0.97 V with a fluctuation variance of 4.5×10^{-3} when the polarization controller is open-loop, and the amplitude of the IF signal fluctuates between 1.33 and 1.45 V with a fluctuation variance of 7.4×10^{-4} after closed-loop, which improves the mixing efficiency by 70.6%. In Fig. 4.49b, the amplitude of the IF signal fluctuates between 0.47 V and 0.99 V with a fluctuation variance of 5.7×10^{-3} when the polarization controller is open-loop, and the amplitude of the IF signal fluctuates between 1.33 V and 1.48 V with a fluctuation variance of 6.7×10^{-4} after closed-loop, and the mixing efficiency is improved by 65%, as shown in Fig. 4.49a and b. 600 m communication distance, the amplitude of the IF signal can increase rapidly and the mixing efficiency is improved after the polarization control system is closed loop.

(2) 1.3 km polarization control experiment

Figure 4.50 shows the changes of the signal light polarization state in one hour measured at 1.3 km for different time periods and weather conditions, respectively: (a) the average rate of change of the signal light polarization state on the bunga ball in (a) is 0.04 rad/s, and the maximum rate of change is 0.3 rad/s; (b) the average rate of change of the signal light polarization state in (b) is 0.31 rad/s, and the maximum The average rate of change of the signal light polarization state in (b) is 0.31 rad/s and the maximum rate of change is 0.98 rad/s; the average rate of change of the

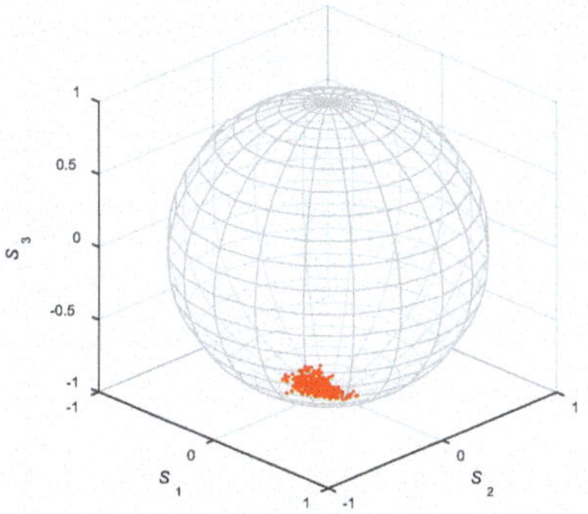

(a) 2018-04-23,21:30-22:30, sunny, 16℃;

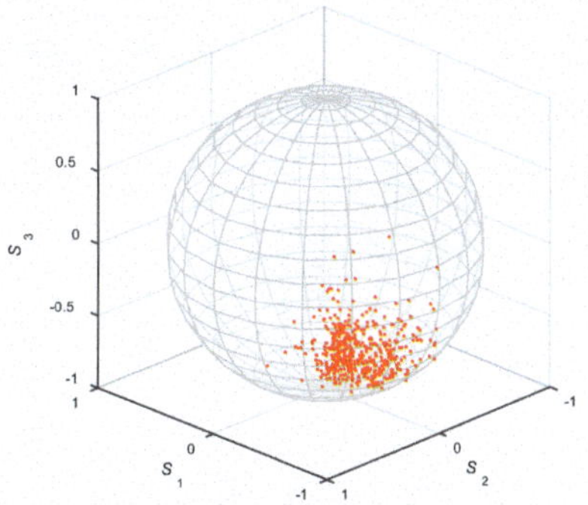

(b) 2018-04-25,17:00-18:00, cloudy, 17℃;

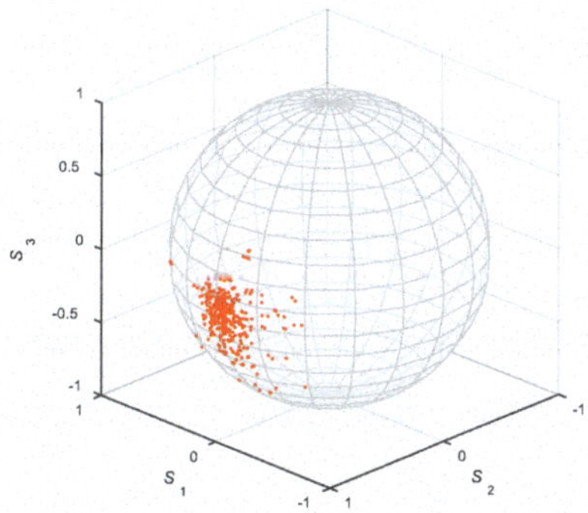

(c) 2018-05-06,22:00-23:00, light rain, 14℃.

Fig. 4.50. Changes in the polarization state of signal light in 1.3 km field experiment [34]

signal light polarization state in (c) is 0.087 rad/s and the maximum rate of change is 0.82 rad/s.

From Fig. 4.50, we can see that at the communication distance of 1.3 km, the average change rate of signal light polarization state is almost the same as the average change rate of signal light polarization state at 600 m, and the received signal light polarization state is slowly changing in a small area on the bunga ball, and the measured signal light polarization state is not the same every time, so in the process of control, the polarization controller only needs to Therefore, the polarization controller only needs to control the polarization state of the signal light as horizontal linearly polarized light.

Figure 4.51 shows the changes of the IF signal amplitude in open-loop and closed-loop conditions of the two polarization control systems measured at a communication distance of 1.3 km. In Fig. 4.51b, the amplitude of the IF signal fluctuates between 0.4 ~ 0.91 V with a fluctuation variance of 1.2×10^{-2} when the polarization control system is open-loop, and after closed-loop, the amplitude of the IF signal fluctuates between 0.96 ~ 1.15 V with a fluctuation variance of 1×10^{-3}. The mixing efficiency can be improved by about 64%. As can be seen from the figure, after the closed-loop

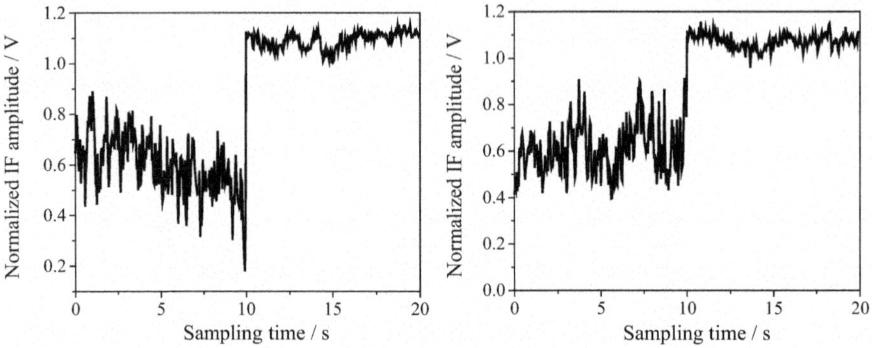

Fig. 4.51 The influence curve of the polarization control system on the amplitude of the IF signal in 1.3 km field experiment [34], **a** 2018–04-23, **b** 2018–04-24

polarization control system, the polarization control system basically eliminates the amplitude jitter caused by polarization, and the signal quality is improved, completing the experiment of polarization control in 1.3 km coherent optical communication, and the jitter of IF signal amplitude after closed-loop can be regarded as caused by the light intensity flicker caused by atmospheric turbulence.

(3) 10.2 km polarization control experiment

Figure 4.52 shows the change of the polarization state of the signal light measured at the communication distance of 10.2 km, the average change rate of the polarization state of the signal light on the bunga ball in figure (a) is 1.06 rad/s and the maximum change rate is 2.89 rad/s, the average change rate of the polarization state of the signal light in (b) is 1.1 rad/s and the maximum change rate can reach 3.14 rad/s, (c) the average change rate of the signal light is 0.72 rad/s, and the maximum change rate can reach 2.97 rad/s. From the figure, it can be seen that the change rate of the polarization state of the signal light is significantly greater than 600 m and 1.3 km, and at the communication distance of 10.2 km, the optical power of the signal light received at the receiving end generally fluctuates between 0μw ~ 3μw, and when the signal When the signal light power state is low, it will be lower than the response threshold of the spectrometer, which will introduce a certain measurement error, the actual signal light polarization state change rate is less than the measured, but from Fig. 4.52 can be seen that the signal light polarization state change range is significantly greater than 600 and 1.3 km. at this time, the polarization has a greater impact on the IF signal.

Figure 4.53 shows the changes of the IF signal amplitude in the open-loop and closed-loop states of the polarization controller at 10.2 km for two groups. (a) in the open-loop, the IF signal amplitude fluctuates between 0.39 ~ 0.95 V with a fluctuation variance of 0.0132, after the closed-loop, the IF signal amplitude fluctuates between 1.04 ~ 1.43 V with a fluctuation variance reduced to 0.0065, and the mixing efficiency The amplitude of the IF signal fluctuates between 0.34 ~ 1.09 V with a fluctuation variance of 0.0105 and a mean value of 0.8708 in the open loop, and the amplitude of

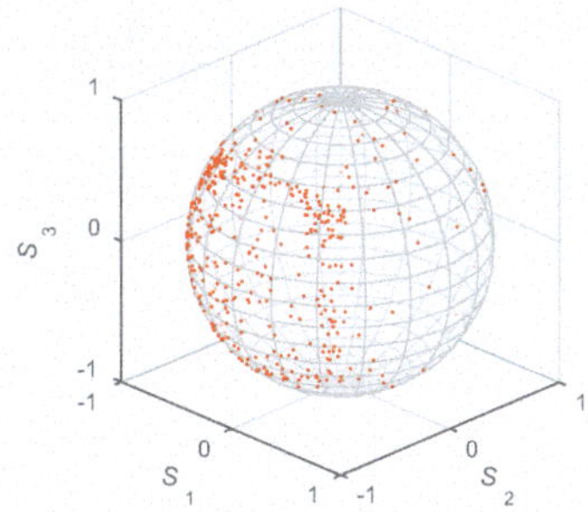

(a)2018-07-25,03:00-04:00, sunny, 30°;

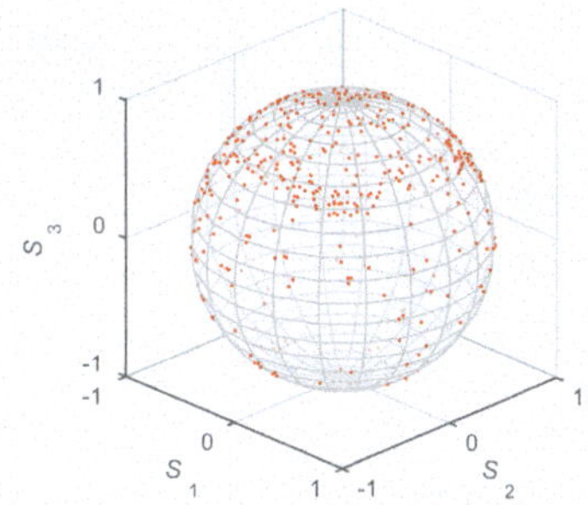

(b) 2018-07-25,04:00-05:00, sunny, 30°;

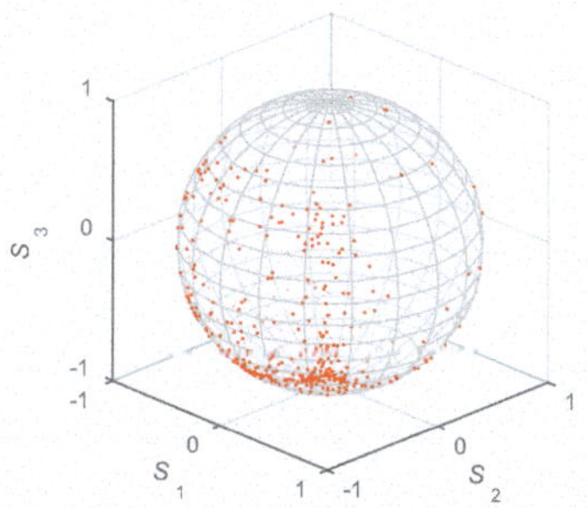

(c) 2018-08-20,00:00-01:00, sunny, 29°.

Fig. 4.52 Changes in the polarization state of signal light in 10.2 km field experiment [34]

the IF signal fluctuates between 1.06~ 1.43 V with a fluctuation variance of 0.0065 and a mean value of 1.2906 after the closed loop, which improves the mixing efficiency by 54.48%. However, compared with 600 m and 1.3 km, the fluctuation variance of the IF signal amplitude is larger after closed loop, and the impact of light spot flicker and wavefront distortion on the signal is enhanced at 10.2 km distance. fluctuations.

When the communication distance is 600 m and 1.3 km, the polarization state of the signal light changes slowly and has a small impact on the coherent optical communication system. When the communication distance increases to 10.2 km, the change rate of the polarization state of the signal light increases significantly and has a serious impact on the performance of the coherent optical communication system, but the polarization control system significantly improves the performance of the coherent optical communication system after the polarization control system closes the loop.

The performance of the polarization control system was tested with the communication distances of 600 m, 1.3 km and 10.2 km. The experimental results show that the amplitude of the IF signal increases rapidly, the fluctuation variance decreases and the mixing efficiency increases after the polarization control system closes the loop, completing the polarization control experiments at different distances and improving

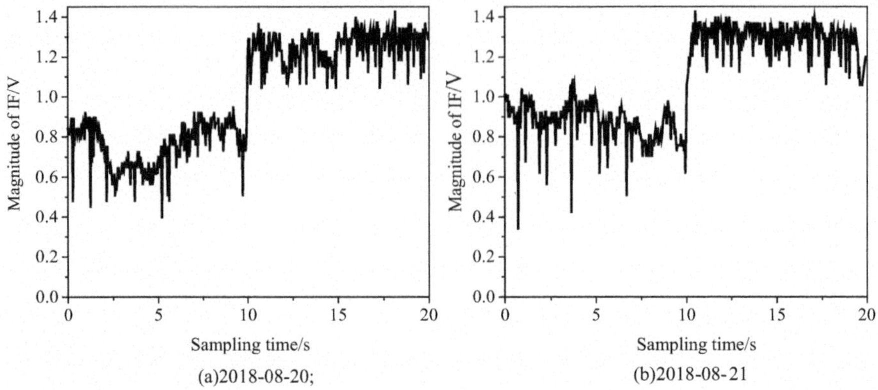

Fig. 4.53 The influence curve of the polarization control system on the amplitude of the IF signal in 10.2 km field experiment [34]

the performance of the communication system. The changes of signal light polarization state under different distances and weather conditions are measured. The data show that the average change rate of signal light polarization state gradually increases with the increasing communication distance, and the influence of polarization on the communication system is gradually enhanced.

References

1. James DFV (1994) Change of polarization of light beams on propagation in free space. J Opt Soc Am A 11(5):1641–1643
2. Zhang JK, Ding SL, Zhai HL, Dang AH (2014) Theoretical and experimental studies of polarization fluctuations over atmospheric turbulent channels for wireless optical communication systems. Opt Expr 22(26):32482
3. Wang MJ, Jia WG, Zhang SY, Qiao HL, Nei-Mu-Le M (2015) Effect of Raman gain on the state of polarization evolution in a low-birefringence fiber. Acta Physica Sinica 64(3):034212
4. Ulrich R (1979) Polarization stabilization on single-mode fiber. Appl Phys Lett 35(11):840
5. Granestrand P, Thylen L (1984) Active stabilisation of polarisation on a single-mode fibre. Electron Lett 20(9):365–366
6. Lefevre HC (1980) Single-mode fibre fractional wave devices and polarisation controllers. Electron Lett 16(20):778–780
7. Kidoh Y, Suematsu Y, Furuya K (1981) Polarization control on output of single-mode optical fibers. IEEE J Quantum Electron 17(6):991–994
8. Imai T, Kiyoshi N, Yamaguchi H (1985) Optical polarisation control utilising an optical heterodyne detection scheme. Electron Lett 21(2):52–53
9. Barnes WL (1988) 2×2 optical fibre polarisation switch and polarisation controller. Electron Lett 24(23):1427–1429
10. Wang T, Cao MC, Luo FG (2001) All-optical polarized beam controller. Semiconductor Optoelectron 22(06):394–400
11. Shen B, Wang P, Polson R, Menon R (2014) Ultra-high-efficiency metamaterial polarizer. Optica 1(5):356–360

12. Chen S, Wen PJ, Zhang HM, Yan M, Wang H, Yao MY (2004) Improved adaptive fiber squeezer polarization controller. Chinese High Technol Lett 14(8):1–4
13. Li WW, Zhang XM, Chen KS, Zou YY (2005) Application of simulated annealing algorithm on endless polarization controller. Acta Photonica Sinica 34(06):820–824
14. Zheng Y, Shen Y, Zhang JZ (2004) Particle swarm optimization used as a control algorithm for adaptive PMD compensation. IEEE Photonics Technol Lett 17(1):85–87
15. Liu Y, Chi H, Zhang XM, Jin XF, Zheng SL (2009) A novel control scheme for-four late retardation polarization controller. Microw Opt Technol Lett 51(1):124–128
16. Liu WY, Cao L, Chen XW, Zhang L (2016) Polarization compensation algorithm for quantum key distribution. J Infrared Millimeter Waves 35(02):210–213
17. Li T, Ke XZ, Chen J, Ning WG (2012) Polarization control in the coherent optical detection system. Infrared and Laser Eng 41(11):3069–3074
18. Chen K, Zheng JY, Zhou JH, Zhang XM (2017) Design of real-time fast polarization control algorithm. Laser Technol 41(05):738–742
19. Yin M, Li RT, Wang XL, Ma YX, Ma PF (2018) Theoretical and experimental research on polarization control of fiber laser based on SPGD algorithm. Laser and Optoelectron Progress 55(09):96–103
20. Rysdale LJ (1986) Method of overcoming finite-range limitation of certain state of polarisation control devices in automatic polansation control schemes. Electron Lett 22(2):100–101
21. Noé R (1986) Endless polarisation control in coherent optical communications. Electron Lett 45(22):772–773
22. Walker NG, Walker GR, Davidson J (1988) Endless polansation control using an integrated optic lithium niobate device. Electron Lett 24(5):266–268
23. Martinelli M, Chipman RA (2003) Endless polarization control algorithm using adjustable linear retarders with fixed axes. J Lightwave Technol 21(9):2089–2096
24. Li WW, Zhang XM, Chen KS, Zou YY (2005) A study for phase-shift characteristics of polarization controller based on simulated annealing algorithm. Acta Optica Sinica 25(04):449–453
25. Rong WF, Dupont L, Wang YX, Yeo YK, Chen J, Kuang W, Paulose V (2011) An endless reset-free polarization control method using two wave plates. IEEE Photonics J 2(6):865–872
26. Zhang XG, Fang GQ, Zhao XY, Zhang WB, Xi LX, Xiong QJ, Li XX, Zhang GY (2010) A polarization stabilizer up to 12.6 krad/s with an additional function of stable state of polarization transformation. Chin Phys B 19(04):311–317
27. Koch B, Noé R, Mirvda V, Sandel D (2011) 100 krad/s endless polarisation tracking with miniaturised module card. Electron Lett 47(14):813–814
28. Yu DY, Tan HY (2000) In: Engineering optics. Machinery Industry Press, pp 322–323
29. Xin GLY, Fan AY, Kang CH (1994) In: Polarized light. Atomic Energy Press, pp 49–50
30. Bohen M, Wolff E, translated by Yang JS (2009) In: Principles of optics, 7th edn, Electronic Industry Press, pp 524–525
31. Smith AM (1980) Single mode fibre pressure sensitivity. Electron Lett 16(20):773
32. Ma TT, Tong SF, Nan H, Jia XY (2007) Effect of polarization property of signal light on mixing efficiency of space coherent detection. Laser & Optoelectronics Progress 54(2):020604
33. Zhang QY, Zhu Y, Su Y, Zhou H, Jing JS (2013) Research on the polar blind zone of the fiber squeezer polarization controller. Acta Optica Sinica 33(05):23–28
34. Ma BB (2019) In: Experimental study of polarization control in coherent optical communication systems, Xi'an University of Technology
35. Wang L (2001) In: Intelligent optimization algorithms and their applications. Tsinghua University Press, pp 17–29
36. Kennedy J, Eberhart R (1995) Particle swarm optimization. In: Proceedings of ICNN'95—international conference on neural networks, vol 27. Pp 1943–1948
37. Shao Q (2017) Research on particle swarm algorithm and its engineering application cases, Jilin University
38. Wang DF, Meng L (2016) Performance analysis and parameter selection of PSO algorithms. Acta Automatica Sinica 42(10):1552–1561

39. Hu W, Li ZS (2007) A simpler and more effective particle swarm optimization algorithm. J Softw 04:861–868
40. Min SQ, Huang DX (1996) Experimental investigation on endless polarization controller. Chinese J Scientif Instrum 04:357–375
41. Li WW (2005) Design and research of polarization controller and its algorithm based on transparent electro-optic ceramic, Zhejiang University

Chapter 5
Double Balanced Detection.-Wavefont Correction System

Contents

The intensity modulation/direct detection (IM/DD) of the composition of communication systems is simple and easy in engineering applications. However, the background light and thermal and scattered particle noise have a significant impact on the performance of the communication system, and the IM/DD can not reach the theoretical limit of detection sensitivity. Under the identical link distance, bit error rate (BER), and code rate conditions, the sensitivity of coherent detection is 20 dB higher when compared with simple IM/DD, and the improved detection sensitivity effectively reduces the size, quality, and power consumption of the coherent optical communication system.

X. Ke and J. Wu, *Coherent Optical Wireless Communication Principle and Application*, Optical Wireless Communication Theory and Technology,
https://doi.org/10.1007/978-981-19-4823-7_5

5.1 Domestic and International Development: History and Current Situation

The primary, more mature detection methods that are presently being used are as follows [1]:

(a) Single-tube detection. This detection method can meet the basic requirements of coherent detection directly by irradiated the mixed optical on the photodetector (PD). However, a higher signal-to-noise ratio (SNR) can only be achieved by increasing the local oscillation optical power, but the local oscillation optical power is too high and introduces a strong residual RIN.

(b) Balanced detection. Balanced detection technology is to receive two channels of light with opposite phases output by the optical mixer through two PDs with similar performance, and the common mode noise can be suppressed by the differential amplifier to obtain the intermediate frequency (IF) signal. And then the sensitivity and SNR of coherent optical communications can be improved effectively. (c) Double-balanced detection. Double-balanced detection technology is developed based on balanced detection technology. After the signal and local oscillator (LO) light beams pass through the 90° optical mixer coupling beam splitter, a four-way signal light with a four-way phase difference of 90° is obtained. Thereafter, beams of opposite phases are obtained through two-by-two combinations of the two balanced detectors for reception. Finally, two orthogonal IF signals are obtained, which contain the signal light intensity, frequency, phase, polarization, and so on. The same double-balanced outlier detection also demonstrates excellent SNR performance and stable environmental adaptability [2]. This technology significantly improves the sensitivity and bandwidth of coherent optical communication. In addition to applications in coherent optical communication, it has a very wide application prospect and research value in the fields of coherent LIDAR and external aberration interferometry.

5.1.1 Foreign Developments: History and Current Situation

In 1985, Gregory proposed the concept of balanced detection, in which two PDs with nearly identical parameters are used simultaneously in a photodetection unit, and the signal and local oscillation light beams are incident on the two PDs separately. And, the high-frequency signal is filtered out to obtain the desired IF signal [3]. Balanced outlier detection overcomes the shortcomings of single-tube detection of the low utilization of the local oscillation light, so that the coherent optical communication receiver can produce a good SNR at low local positive power. This technology also effectively suppresses the interference in common-mode noise, which can improve the SNR of the coherent optical communication system to a level close to the quantum noise limit.

Based on the theory of balanced detection, the double-balanced detection technique was developed in 1987 by Kuwaharra [4] using a 90° optical mixer to couple the local oscillation and signal light beams for beam splitting processing to obtain four output light beams within phase differences of 90° using four photoelectric detectors with extremely similar parameters for photoelectric conversion. The final two channels are obtained as the in-phase and orthogonal components. The phase difference between these two components is compared and analyzed, and a stable optical signal reception is obtained. In addition to the advantages of balanced detection technology, the double-balanced detection technology also includes a variety of modulation and demodulation methods. It also demonstrates good environmental adaptability and better system stability.

In 1990, Caldera [5] developed the first successful 3-dB optical coupler based on InGaAlAs/InP. This coupler can split a 1.47–1.53 μm laser beam into two identical beams, which addresses one of the key problems of balanced detection technology.

In 1998, Arvizu [6] developed and fabricated a prototype of the balanced detection system based on the principle of balanced outlier detection technology and used it to conduct experiments using the technology. The results indicated that the common-mode rejection ratio of the prototype could reach 26 dB, verifying the superiority of the technology.

Around 1995, owing to the development of the wavelength-division multiplexing and optical direct amplification technologies, particularly the maturity and application of the erbium-doped fiber amplifier (EDFA) technology, the bandwidth and relay-free optical transmission distance of optical communication systems were significantly improved. The detection technology of the IM/DD system was sufficient to meet the demand of laser communication performance at that time. Therefore, the research community gradually lost interest in coherent detection and its application technology [7]. In wireless optical communication, it is impossible to use EDFA for link relaying and online amplification, and, thus, coherent detection technology gained attention again after more than a decade of oblivion.

In 2009, Karafolas [8] conducted a coherent binary phase-shift keying (BPSK) communication experiment on the German TerraSAR-X satellite using the balanced detection technique for a bidirectional satellite link with a data rate of 5.6 Gbit/s. In 2009, Painchaud [9] analyzed the double-balanced detection technique theoretically and experimentally and studied the effect of the optical characteristics of double-balanced detector devices on detection performance.

In 2014, Caplan [10] realized a space optical communication receiver using the double-balanced detection technique, which can operate effectively in a free-space channel with a SNR close to the quantum limit, demonstrating the excellent performance of the double-balanced detection technique.

Between 1980 and 1986, the SILEX system of the European Space Agency (ESA) employed the IM/DD detection regime with a maximum communication rate of 150 Mbit/s [11]. In the early 1990s, ESA launched the short range interplanetary optical link (SROIL) research project dedicated to the study of free-space coherent optical communication. In 2008, ESA applied balanced detection technology between the German TerraSAR-X and US N-FIRE satellites at a distance of 5000 km. This experiment employed the BPSK modulation format with a BER of less than 10^{-9} and a data transmission rate of 5 Gbit/s [12].

In parallel with the SILEX program, the ESA also launched the SROIL program in Switzerland in 1989, which focused on the development of miniaturized and lightweight terminals that could be used in short-range communication links between small satellites operating in low orbit. In 1998, the SROIL terminal mock-up was successfully developed, and the terminal receiver used a simple phase-shift monitoring modulation method and zero-difference coherent detection, the transmitting antenna aperture of the communication system is 3.5 cm, and the BER is below 10^{-6} with a data rate of 1.5 Gbit/s. Based on this program, SROIL plans to develop a medium- and long-range intersatellite optical communication system in addition to short-range low-orbit satellite communication [13], which still adopts binary phase-shift monitoring modulation along with the zero-difference coherent detection system and can realize communication between two satellites at a distance of 6000 km; the communication between two satellites may be able to reach a date rate of 6.5 Gbit/s with a BER below 10^{-9}.

NASA's Jet Propulsion Laboratory (JPL) of Caltech and MIT's Lincoln Laboratory have also been committed to research on coherent detection technology. In as early as 1991, the JPL completed the intersatellite coherent optical communication demonstration system of PSK with a code speed of 100 Mbit/s. However, at that time, it was applied mainly to the IM/DD system [14]. With the development and maturity of narrow linewidth lasers, the advantages of coherent detection technology gradually appeared. After 1999, JPL also shifted its research focus to coherent detection technology. Simultaneously, Lincoln Laboratory studied the performance of various coherent communication schemes, such as the SNR and BER, on the intersatellite optical communication platform. Their experimental device can achieve a communication distance of 3000 km using coherent detection technology with a BER of 10^{-6} at a bit rate of 2 Gbit/s [15].

In March 2006, the ground optical station in Tokyo, Japan, completed a low earth orbit (LEO) satellite–ground laser communication experiment. In 2011, the National Institute of Information and Communication Technology (NICT) of Japan developed a dual-band digital zero-difference coherent optical communication receiver, which is suitable for laser communication systems in the 1060 and 1550 nm band and can achieve a rate of 6 Gbit/s or more. In 2011, the Japan Aerospace Exploration Agency (JAXA) realized a zero-difference coherent optical communication experiment on the ground with a data transmission rate of 2.5 Gbit/s and introduced an analog frequency shift of 7.5 GHz in the experiment [16], considering the Doppler shift effect owing to the relative movement between LEO and geostationary (GEO) satellites. The coherent optical communication has been proven to have a strong anti-interference capability, and experiments have indicated that its reception sensitivity is better than −52 dBm.

5.1.2 *Domestic Developments: History and Present Situation*

In 1990, Man [17] applied the balanced detection technology to the coherent optical-fiber communication system and verified the feasibility of the technology through experiments. In 1991, Chen [18] designed and tested the corresponding balanced receiver. In 1993, Pei [19] analyzed the characteristics of the balanced detector and the effect of detector asymmetry on the system. In 1995, Chen [20] analyzed the theoretical model of coherent optical heterodyne balanced reception technology and verified the feasibility of balanced mixing polarization diversity reception technology through experiments.

In 2010, Gao [21] analyzed the effect of polarization mixing error caused by optical element defects such as the polarizing beam splitter and polarization factor of incident light in the balanced detection system on signal demodulation. In 2011, Yang [22] analyzed the effects of the coupler beam splitting ratio, matching the degree of PD quantum efficiency and local oscillator optical power in balanced detection.

In 2012, Chen [23] conducted a theoretical analysis and experimental research on the factors of the atmospheric channel that impact the balanced heterodyne detection system and proposed key solutions to overcome this impact. In the same year, Pang [24] established the mathematical model of the in-phase and quadrature (IQ) demodulation of double-balanced heterodyne detection. The simulation results indicate that balanced detection demonstrates higher SNR when compared with ordinary detection. In 2013, Huai [25] conducted simulations on the double-balanced detection system, verified its principle, and analyzed the impact of the internal devices of the optical mixer on the system. In 2014, Liu realized a prototype of the balanced detector using the mixing magic T and conducted experiments.

In 2015, Dai [26] realized a prototype of the balanced detector using the differential amplifier and built a platform for testing coherent optical communication. Its sensitivity was tested to verify the theoretical analysis. In 2016, Kong [27] studied the impact of various detector characteristic parameters and LO optical power on the SNR in the double-balanced detection system and verified its theory through experiments.

5.2 Structure and Principle of Double-Balanced Detection System

The dual-balanced detection system consists of two parts: the optical mixing part at the front and the balanced detection part at the back, as illustrated in Fig. 5.1.

The core device of the optical mixing part is a 90° optical mixer, whose primary function is to coherently mix the signal light and local oscillation light and finally generate four output light beams within phase differences of 90°. Thereafter, the output beams with opposite phases are combined in pairs and input to the two balanced detectors. The balanced detector is the main device of the balanced detection

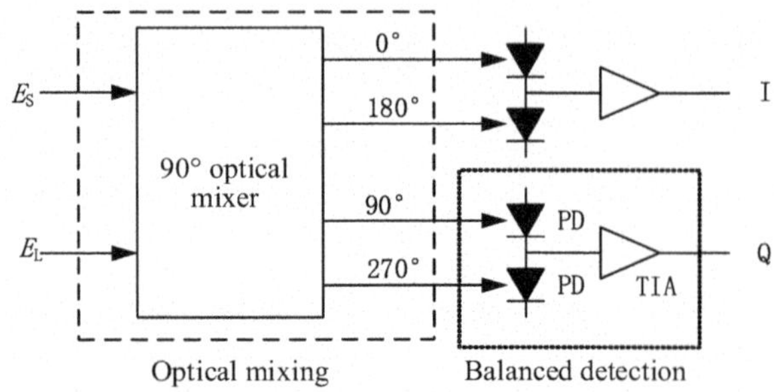

Fig. 5.1 Structure of double-balanced detection system

part; it is accompanied by two similar-performing PDs cascaded with each other from the middle tap following the transimpedance amplifier (TIA) to obtain the required IF signal. The main role of the balanced detector is to effectively suppress or even eliminate the effect of common-mode noise on the system. The double-balanced detection system finally generates two IF signals with 90° phase difference, referred to as the in-phase (I) and vertical (Q) components. The processing of these two signals yields the required information [28]. The optical mixer is used to accurately generate a phase difference of 90° in the output light beam. Further, for each branch, balanced optical signal photoelectric conversion and processing are the key aspects of the performance of the double-balanced detection system.

5.2.1 Classification of 90° Optical Mixers Used in Double-Balanced Detection Techniques

The primary component used in double-balanced detection technology is the 2 × 4 (two input, four output) 90° optical mixer. The mixer can be realized through various principles and structures, and several 2 × 4 90° optical mixers have been developed based on various applications and scenarios. Because coherent optical communication technology can be applied in fiber-optic and space laser communication systems, two types of 90° optical mixers can be introduced: fiber-optic transmission type and free-space transmission type.

(1) **Optical-fiber transmission type 90° optical mixer**

The optical-fiber transmission optical mixer is mainly used in optical-fiber communication systems, which are usually realized using optical fibers and waveguide devices. Based on the various implementation principles, these mixers can be classified as the following types.

a. **3-dB coupler type**

The 3-dB coupler type optical mixer [29] is one of the most widely used in fiber-optic communication systems. It mainly consists of a 3-dB coupler and a polarizing beam splitter (PBS). Its schematic is presented in Fig. 5.2: the signal light is linearly polarized and the local oscillation light is circularly polarized.

The role of the 3-dB coupler is to coherently couple the local oscillation and signal light beams and separate the two light paths such that they have a phase difference of 90°. These two light beams cross through the PBS for polarization separation; a phase difference of 180° is introduced in the process, and the final output is four light beams with phase-shift relationships of 0°, 90°, 180°, and 270°. The 3-dB coupler is the key component of this type of 90° optical mixer, which is realized mainly through optical-fiber or waveguide devices and easily affected by environment and temperature. Its phase output performance is unstable, producing a phase difference between the two arms. This is the main drawback of this type of optical mixer. As illustrated in Fig. 5.2, because the four beams split by the two PBSs are sent through the balanced detector, the splitting ratio of the 3-dB coupler is required to balance the consistency of each branch of the double-balanced detection system.

b. **Directional coupler type**

Figure 5.3 presents the schematic of a 2 × 4 90° optical mixer implemented using directional couplers and a phase-shifter. This type of 90° optical mixer mainly consists of four directional couplers (DCs) and two phase shifters (PSs) with a continuously adjustable phase.

The signal light is input from port 1 of DC1, which divides it, and the phase difference between the two output light beams is adjusted by phase-shifter PS1. Similarly, the local oscillation light is input from port 2 of DC2 and divided in it, and the phase difference of the two light outputs from DC2 is adjusted by phase-shifter PS2. The phase-shifters PS1 and PS2 are adjusted so that the phase of the input light

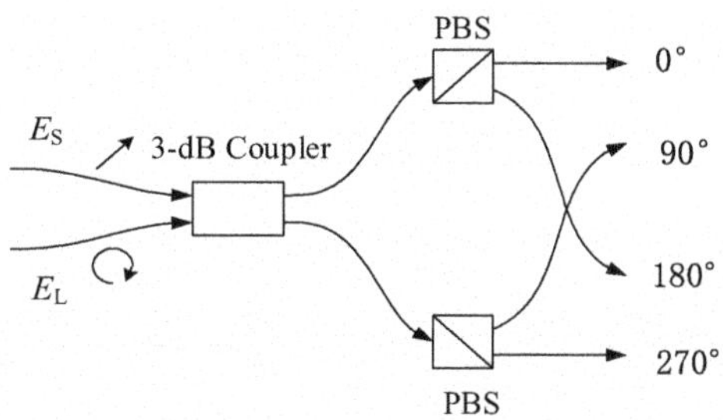

Fig. 5.2 Schematic of 3-dB coupling type 2 × 4 90° optical mixer

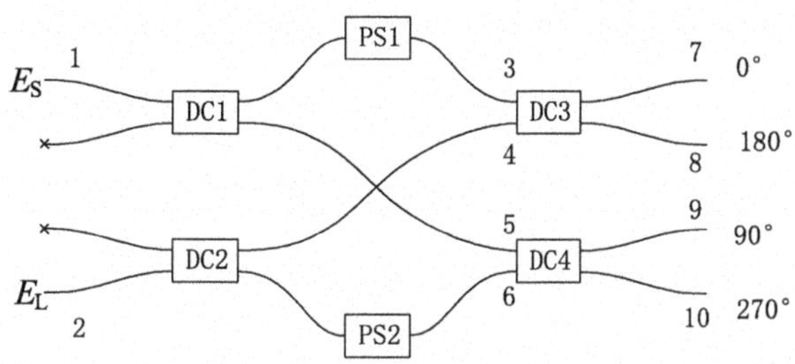

Fig. 5.3 Waveguide integrated 2 × 4 90° optical mixer [30]

obtained from ports 1 and 3 is the same and the phase of the input light obtained from ports 2 and 4 is 90° apart. Thereafter DC3 and DC4 are coupled again to split the light to get four output light beams; the phase of the light beam output from port 7 and those of the light beam output from ports 8–10 differ by 180°, 90°, and 270°, respectively. Thus, the function of the 90° optical mixer is realized. This type of optical mixer can ideally produce the desired phase accurately because the phase is continuously adjustable, but in practice it is affected by the environment and causes jitter and drift of the phase.

c. **Multimode interference coupler type**

Multimode interference (MMI) couplers work by exploiting the self-imaging effect of multimode waveguides. The self-imaging effect refers to the phenomenon that in a multimode waveguide, the input optical field periodically replicates one or more images along the propagation direction of the waveguide. The 90° optical mixer realized by MMI can be further classified as the 4 × 4, 2 × 2, and 2 × 4 MMI coupler combination schemes. The latter two schemes require additional phase-shifters to augment the MMI coupler to complete the 90° optical mixing function, increasing the complexity of the system. For the 2 × 2 MMI coupler combination, the phenomenon of waveguide crossover must also be considered, which causes crossover loss and signal interference, affecting the quality of signal transmission. The 4 × 4 MMI coupler is a more ideal and widely used scheme because there is no waveguide crossover phenomenon and it does not require any additional phase-shifters [31]. Figure 5.4 presents a schematic of a four-input, four-output multimode interference coupler structure.

As illustrated in Fig. 5.4, the input ports are denoted by $n = 1$–4 and the output ports are denoted by $m = 1$–4. $\phi_{n,m}$ denotes the phase shift from the input port n to the output port m. The phase-shift equation from the input port to the output port is expressed as follows [32, 33]:

$$\phi_{n,m} = -\beta_0 \frac{3L_\pi}{4} + \frac{15}{32}\pi - \frac{\pi}{2}(-1)^{n+m} + \frac{\pi}{16}\left[n + m - n^2 - m^2 + (-1)^{n+m}\left(2nm - n - m + \frac{1}{2}\right)\right] \tag{5.1}$$

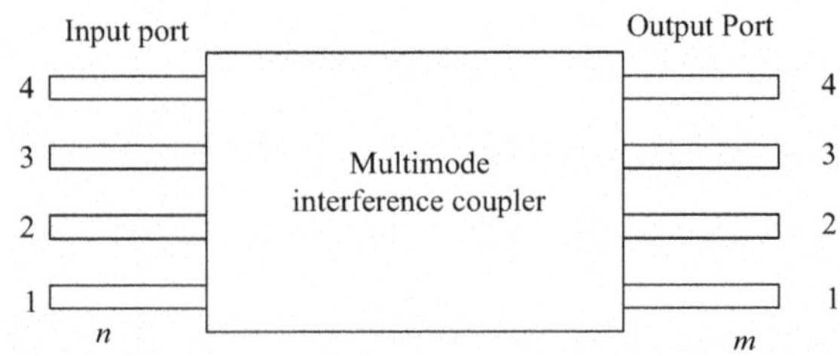

Fig. 5.4 Schematic of multimode interference coupler structure

where β_0 denotes the propagation constant of the fundamental mode and L_π represents the beat length. Assuming that the signal and local oscillation light beams are input from the input ports 1 and 3, respectively, the phases of the optical field at each output port are as follows:

$$
\begin{aligned}
(\phi_{1,1} - \phi_{3,1}) - (\phi_{1,1} - \phi_{3,1}) &= 0^\circ \\
(\phi_{1,2} - \phi_{3,2}) - (\phi_{1,1} - \phi_{3,1}) &= 90^\circ \\
(\phi_{1,3} - \phi_{3,3}) - (\phi_{1,1} - \phi_{3,1}) &= 180^\circ \\
(\phi_{1,4} - \phi_{3,4}) - (\phi_{1,1} - \phi_{3,1}) &= 270^\circ
\end{aligned}
\tag{5.2}
$$

It can be seen that the relative phase differences between the signal and local oscillation light beams are 0°, 90°, 180°, and 270° for the four light outputs following MMI, thus achieving 90° optical mixing.

d. **Hybrid 90° optical mixer**

The 90° optical mixer in fiber-optic communication systems is susceptible to environmental impact, leading to drawbacks such as unstable phase output. To overcome this problem, a scheme using a dielectric film beam splitter and calcite has been proposed [34], as illustrated in Fig. 5.5. This scheme is an improvement over the 3-dB coupled optical mixer, in which the 3-dB coupler is replaced by a dielectric film beam splitter. The signal and local oscillation light beams are coupled and split by the dielectric film beam splitter. Thereafter, polarization separation is performed using calcite, and finally four output light beams with 90° phase difference are obtained from ports 3 to 6.

(2) **Free-space transmission type 90° optical mixer**

a. **90° optical mixer based on optical beam splitting film**

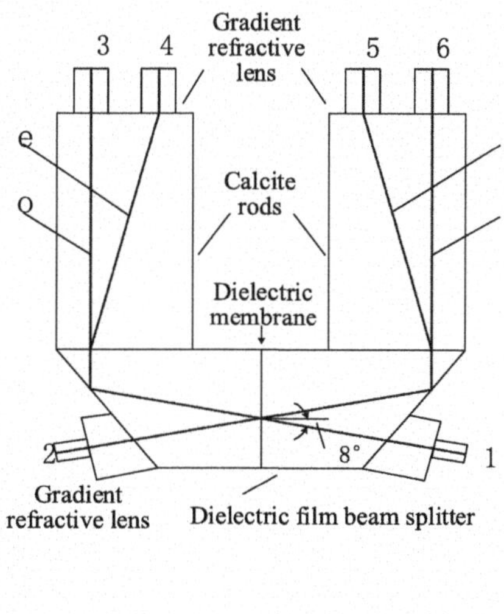

Fig. 5.5 Hybrid 2 × 4 90° optical mixer

This type of 90° optical mixer is composed mainly of a beam splitter and a waveplate. The beam splitter can be divided into two different implementations depending on the nature of the beam splitting film used: polarizing and nonpolarizing beam splitters.

(1) **Nonpolarizing beam splitter type**

As depicted in Fig. 5.6, the core device in this scheme is a nonpolarizing beam splitter (NPBS), in which the signal light is incident at a polarization angle of 45°, and circularly polarized localized light is passed through the NPBS for coupling and beam splitting. Thereafter, the outgoing light is polarized and separated by the two PBSs, respectively. The principle behind and function of this type of mixer are similar to those of the 3-dB coupler, which couples and splits the localized and signal light beams that are then made incident on the PBS separately to produce the four light beams of desired phases [35]. However, phase control in this scheme is entirely realized by the beam splitter; therefore, both the unpolarized beam splitter and the polarizer are more demanding, requiring a 1:1 ratio of reflected and transmitted light to make the phase difference of the final output light beam 90°.

(2) **Polarizing beam splitter type**

This type of 90° optical mixer is the most widely and maturely applied spatial optical mixer in the double-balanced detection system. Its schematic is presented in Fig. 5.7; the mixer is composed mainly of PBS and waveplate. The role of the latter is to convert the polarization of the local oscillation light from linear to circular. The signal light with a polarization angle of 45° is made to pass through the PBS for coherent mixing and beam splitting, that is, the P light and S light are separated using the PBS. Thereafter, the beam is polarization-state-rotated and phase-delayed by the waveplate at an appropriate deflection angle and passed through the two PBSs again for polarization separation to obtain four output light beams with 90° phase difference [36].

Fig. 5.6 Nonpolarizing beam splitter type of 90° optical mixer

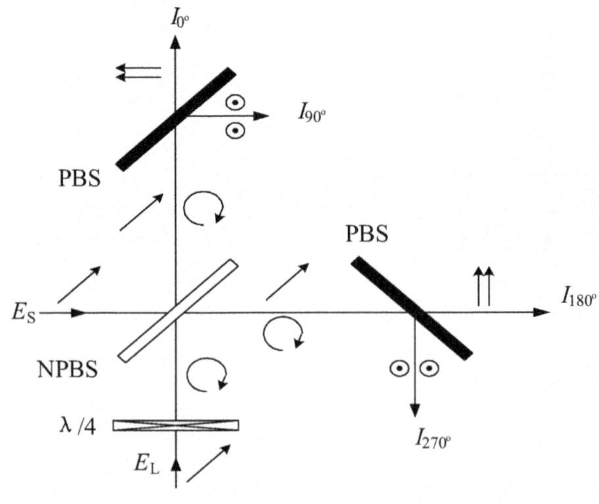

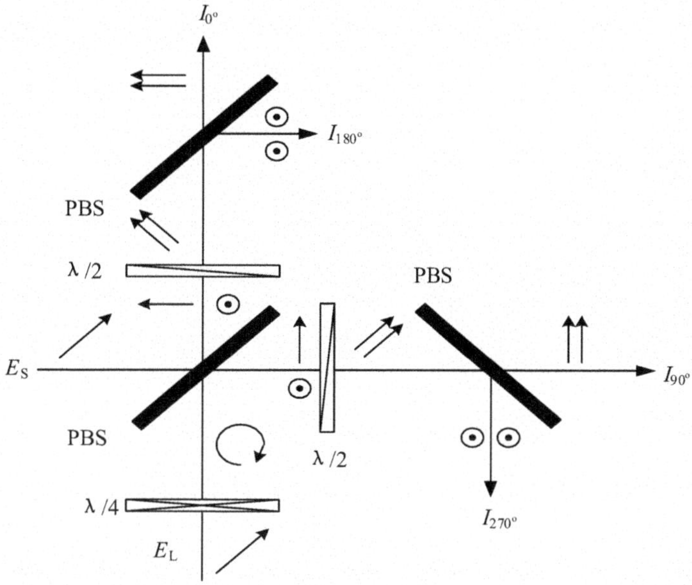

Fig. 5.7 Polarizing beam splitter type of 90° optical mixer

b. **90° optical mixer based on crystal birefringence effect**

Although the 90° optical mixer based on PBSs and waveplates can generate beams of the desired phase, it is not easy to miniaturize and integrate the device owing to its large number of components. Thus, the insertion loss and transmission conditions such as the optical range difference of the beam need to be strictly controlled.

To address this problem, some researchers have proposed to realize 90° optical mixers through coupling beam splitting using the birefringence effect of certain crystals [41], for example, the birefringent spatial light mixer realized using calcite [38] and the electronically controlled phase-shifted spatial light mixer realized using lithium niobate crystal [37].

(1) **Birefringent free-space optical mixer**

As illustrated in Fig. 5.8, this type of spatial light mixer is composed of four birefringent optical plates (A, B, C, D) with identical structures, a detecting birefringent plate (E), and a phase-delay plate (P). The main cross section of the detecting birefringent plate is placed at an angle of 45° with the main cross section of the birefringent optical plate. The polarization angles of the signal and localization light beams are 45° incident to the upper part of the optical plate A and the lower part of the optical plate B, respectively. Because of the birefringence effect of the crystal, the two beams of light are divided into LO o-light and e-light, signal o-light and e-light, after the four beams pass through the phase-delay plate, the polarization between the two beams of mutually orthogonal phase generates a delay of $\pi/4$, and then again incident to the birefringent optical plate C and D, which the main cross-sectional planes

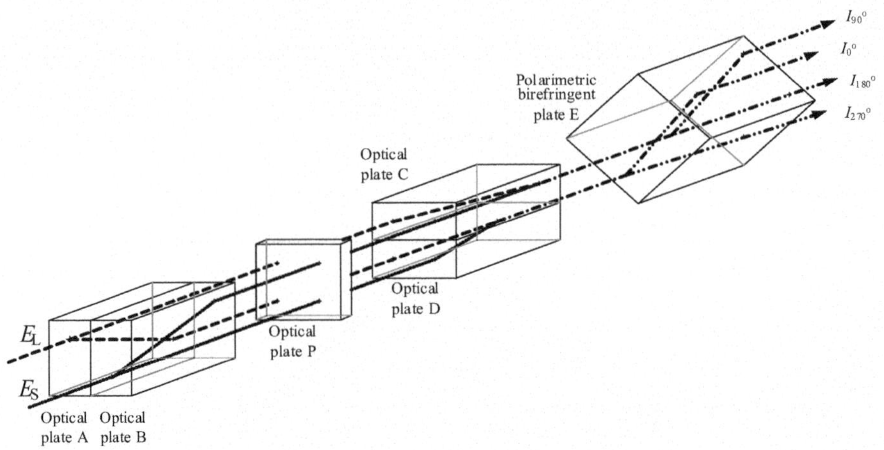

Fig. 5.8 Crystal birefringent optical mixer

are perpendicular to the main cross-sectional planes of A and B, so that the original o-light in A and B is converted into e-light in C and D, and the e-light is converted into o-light, so that the o-light and e-light from different optical plates are combined in C and D, respectively. The two synthesized beams are then made to pass through the birefringent element E of the detector to obtain four sequentially phase-shifted 90° output light beams, thus realizing a 2 × 4 90° optical mixer [38].

(2) **Electronically controlled phase-shifted spatial light mixer**

The phase control of the birefringent spatial light mixer is achieved through the placement angle between the phase-delay sheet and the detuned birefringent plate. To be able to control the phase of the output light more precisely, an electronically controlled phase-shift scheme has been proposed to add a control voltage to the lithium niobate crystal to generate the desired phase shift through electro-optical modulation, so that the relative phase difference of the finally obtained four-way light beam is 90° [39]. The schematic structure is presented in Fig. 5.9.

The birefringent optical plates A, B and C, D depicted in Fig. 5.9 are cut from the same wafer, with identical optical structure and properties, to achieve consistent phase delay. The main cross section of the check-deflection birefringent plate is placed at 45° to the main cross section of the birefringent optical plate. The directions of polarization of LO and signal light are 45° to the main cross section of the birefringent crystal plate. Owing to the birefringence effect of the crystal, the local vibration and signal light beams each form two beams of mutually perpendicular line polarization light, and the upper two of the four light beams obtained from the birefringent optical plates A and B are passed through the birefringent optical plate C to synthesize a beam of output light. Similarly, the lower two beams are passed through the birefringent optical plate D for combined beam output. The two beams from C and D are projected onto the birefringent plate to obtain four beams with different phases. The phase of the

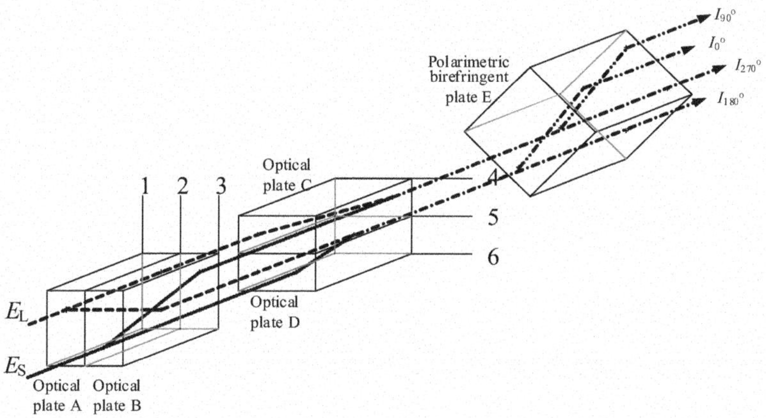

Fig. 5.9 Structure of electronically controlled phase-shifted spatial light mixer

beams is controlled by adding control electrodes 1–6 to the birefringent optical plates A–D, as illustrated in the figure. An appropriate voltage is added so that the output light can generate four beams, I_{90°, I_{0°, I_{180°, and I_{270°, with 90° phase difference in sequence, thus realizing a 2 × 4 90° optical mixer [40].

5.2.2 *Classification of Balanced Detectors*

The balanced detector is also one of the core devices in double-balanced detection technology. Compared with the traditional direct detection, the use of a balanced detector significantly improves the sensitivity of reception and can effectively suppress the impact of receiver and electronic line noise on the detection of weak optical signals [41]. Earlier coherent optical communication systems used a simple single-tube detector, in which the local oscillation and signal light beams in the optical field of the optical mixer interfere to produce IF signals. One of the two signals incident on the PD, and only half of the optical power input by the system is used. However, in the coherent optical communication system of the balanced detector, optical signals from the two output ports of the optical mixer are incident on the two PDs of the balanced detector. Therefore, the optical power of the input signal is fully utilized, thus improving the effective utilization of light and offsetting a large portion of the noise [42]. Considering its basic structure, the balanced detector can be divided into two types: parallel balanced detector and series balanced detector.

(1) **Parallel balanced detector**

In a parallel structure, each of the two PDs of the balanced detector independently perform the photoelectric conversion of the input signal light, thus generating the required IF current signal through the differential and amplifier circuits [43]. Based on the type of differential circuit used, there can be two types of devices: mixing magic T type and differential amplifier type. Figures 5.10 and 5.11 present the schematics of the two circuits.

The structures presented in the two figures are quite similar; the difference is that the optical current is processed differentially by the hybrid magic T type (Microwave hybrid) [44] circuit to produce the IF current signal, but because the hybrid magic T circuit is a passive device, the synthesized signal is attenuated. Therefore, it should be followed by a current amplifier. The mixing magic T circuit is a significant device commonly used in microwave RF circuits; its primary role is to carry out power distribution and synthesis. It has four ports: two inputs and two outputs. The input and output at each port are indicated in Fig. 5.12.

In the mixing magic T type balanced detector, the signal light output from the optical mixer is received by two separate photoelectric detectors, and the photoelectric stream after photoelectric conversion is fed into the mixing magic T circuit. The current containing the IF signal is output after differencing, and then the IF current signal required for the demodulation of the signal is obtained after passing through an amplifier.

When the balanced detection technique was first proposed, a balanced detector of the mixed frequency magic T type was used. The balanced detector used by Gregory et al. is depicted in Fig. 5.13.

To overcome the drawback that the hybrid magic T is a passive device that causes signal attenuation, an improved version was proposed by adding an amplifier after the PD to increase the amplitude of the signal, as illustrated in Fig. 5.14.

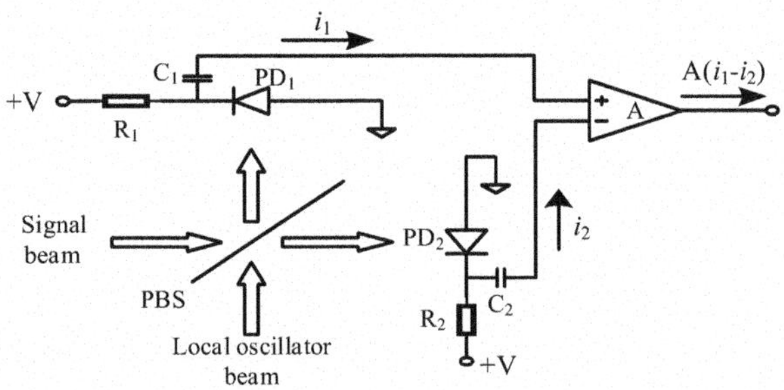

Fig. 5.10 Differential-amplifier-type balanced detector

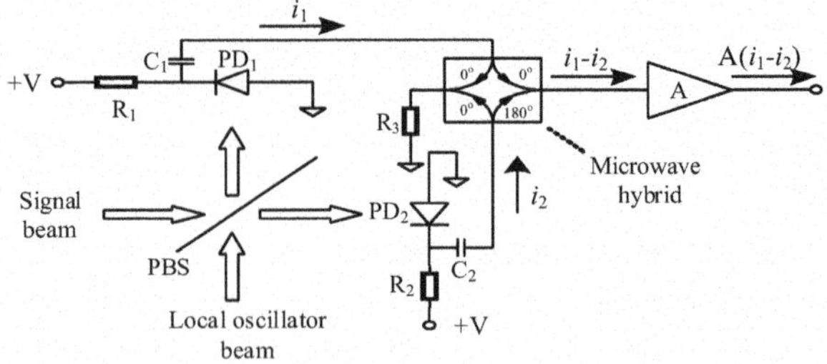

Fig. 5.11 Hybrid magic T type balanced detector

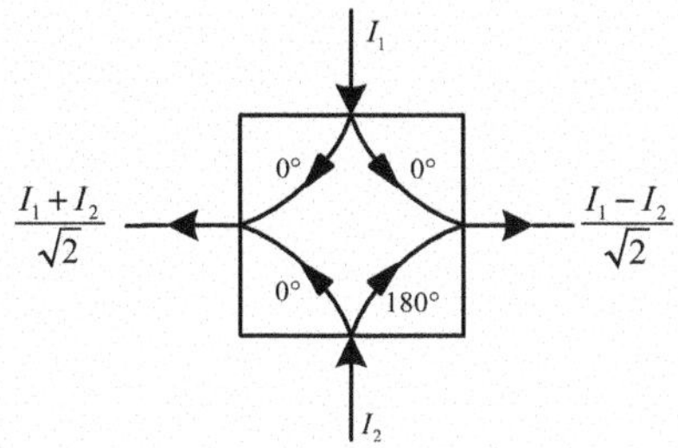

Fig. 5.12 0°–180° mixing magic T circuit

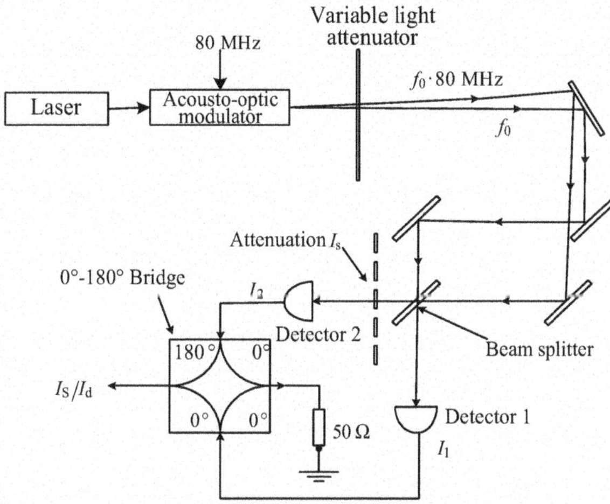

Fig. 5.13 Mixing magic T type balanced detection system [3]

Figure 5.15 depicts the use of two mixing magic T circuits to construct the differential balanced detector. By adding the same bias to two PDs with front-end amplifiers, the isotropic end PD1+ of the differential output of the front-end amplifier and the reverse end PD2− of the differential output of the other front-end amplifier are coupled by the mixing magic T circuit as the isotropic and reverse inputs of the limiting amplifier, respectively, in opposite directions. Thus, the IF signal is obtained.

As illustrated in Fig. 5.16, a differential amplifier is used to directly replace the mixing magic T circuit to overcome the signal attenuation. First, a front-end amplifier is connected directly after the PD and, thereafter, the IF signal is obtained from the differential amplifier.

However, these two structures strictly require the two arms of the devices to be consistent; the PDs in the two arms of the balanced detector and the supporting capacitance–resistance devices are required to be extremely similar to each other in terms of their performance parameters to meet the needs of balanced detection. In practice, it is difficult to find PDs with identical performance parameters.

(2) **Tandem balanced detector**

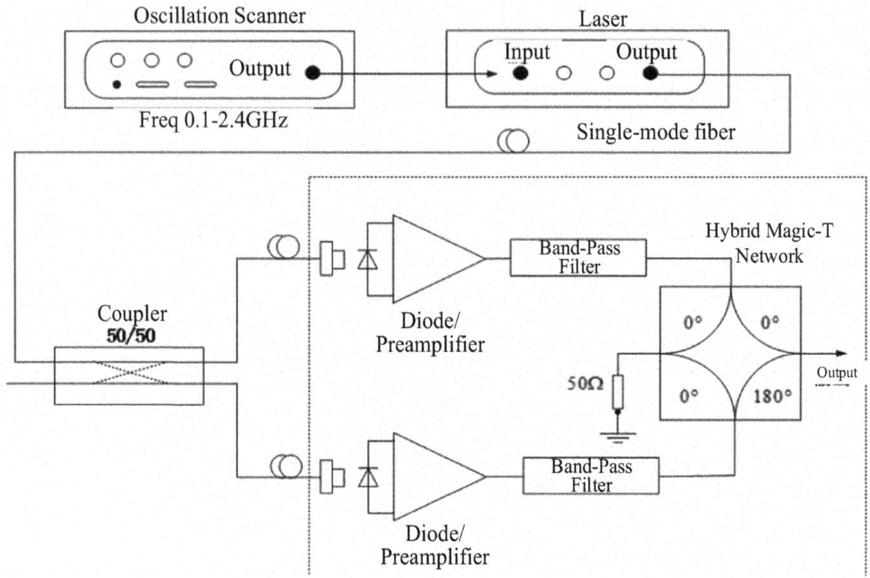

Fig. 5.14 Mixing magic T type balanced detector with front-end amplifier [6]

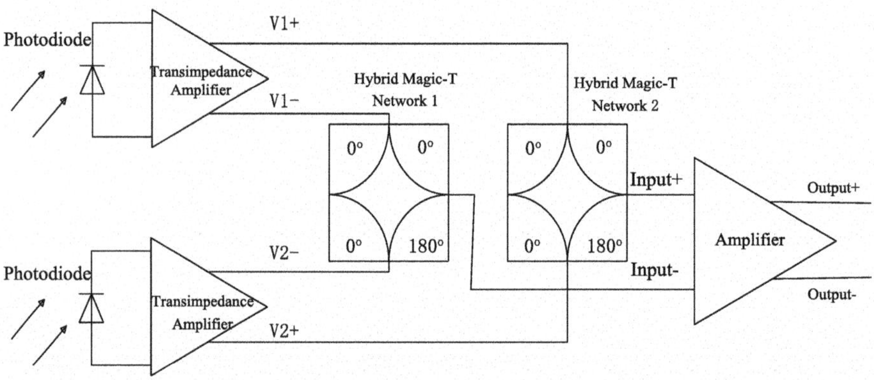

Fig. 5.15 Dual mixed frequency magic T type balanced detector [1]

Figure 5.17 depicts the structure of the tandem balanced detector. The two beams of light obtained after the coupling and beam splitting of the local oscillation and the signal light beams are converted by two photoelectric detectors cascaded with each other, and then the required IF current signal is obtained from the intermediate tap followed by the trans impedance amplifier.

A balanced detector with this structure has two PDs connected in series and a middle tap connected to a transimpedance amplifier. This reduces the requirement of device consistency for both arms of the balanced detector.

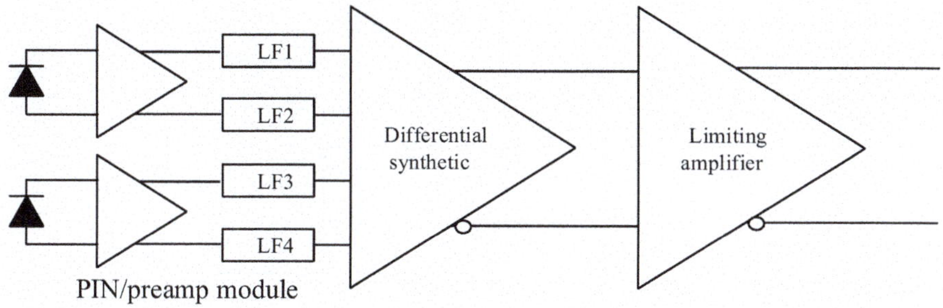

Fig. 5.16 Differential amplifier type balanced detector [1]

Fig. 5.17 Balanced detector in series

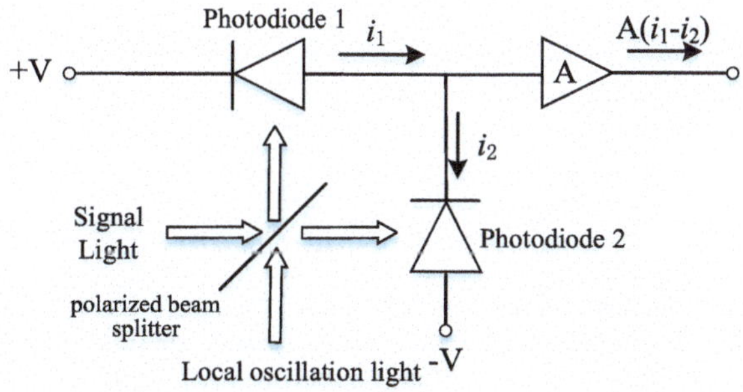

The series-type balancing amplifier structure is relatively simple when compared with the parallel type. Therefore, it is more widely used in practice; the schematic of its circuit is illustrated in Fig. 5.18. As can be seen in the figure, the structure of the photodetection part of the series-type balanced detector is relatively fixed, whereas the back-end amplifier is designed to improve the common-mode rejection ratio of the balanced detector. It can be used in high-speed coherent optical communication systems (above 10 Gbit/s).

Figure 5.19 presents the circuit schematic of the BPD-002 balanced detector manufactured by U2T (Germany). It can be seen that it uses a series-type balanced

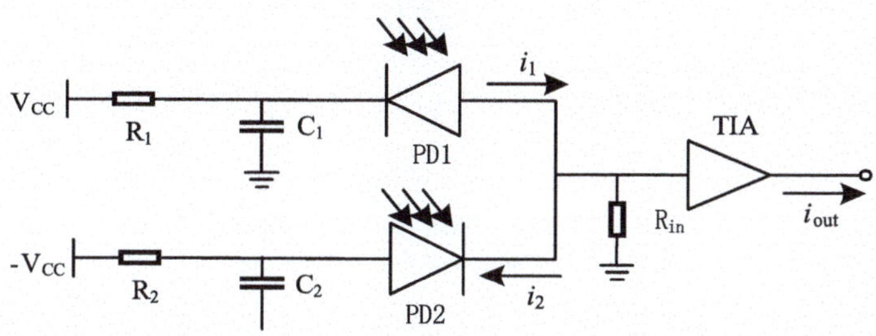

Fig. 5.18 Circuit schematic of series balanced detector

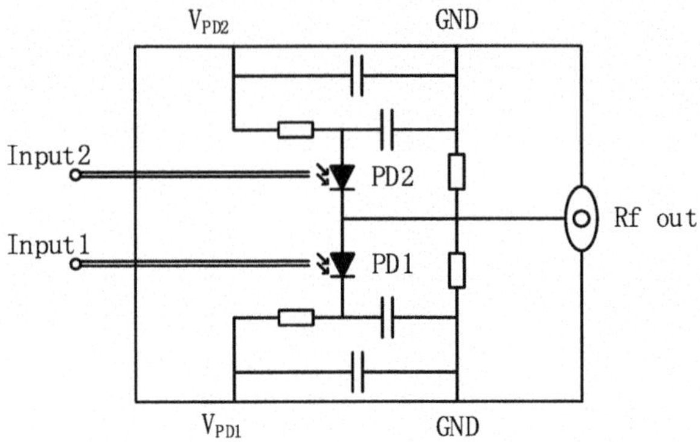

Fig. 5.19 Schematic of BPD-002-type balanced detector circuit

detector structure. This type of detector can be applied to the 40-Gbps optical communication system.

5.2.3 Principle of Double-Balanced Detection

The double-balanced detection system is developed based on the principle of balanced detection. An analysis of the principle of the former requires the analysis of the principle of balanced detection technology.

(1) **Balanced detection**

As mentioned, balanced detection technology is the theoretical basis of double-balanced detection technology. The core idea is to phase-subtract the input signal to the balanced detector; the IF signal is detected as a differential mode component, whereas the RIN and common-mode components are suppressed. We first analyze the principle of the balanced detection technique.

Figure 5.20 presents a schematic of the balanced detection technology. The reflection coefficient R and transmission coefficient T of the PBS are set as follows:

$$\begin{cases} R = \left(1/\sqrt{2}\right)e^{j\pi/4} \\ T = \left(1/\sqrt{2}\right)e^{-j\pi/4} \end{cases} \tag{5.3}$$

For the convenience of analysis, the local oscillator light E_L and signal light E_S are set as follows:

$$\begin{cases} E_L(t) = E_L \exp[j(\omega_L t + \phi_L)] \\ E_S(t) = E_S \exp[j(\omega_S t + \phi_S)] \end{cases} \tag{5.4}$$

After passing through the PBS, the two signals are represented as follows:

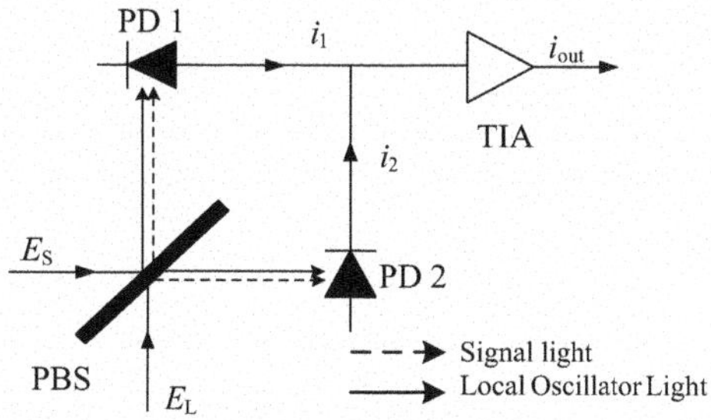

Fig. 5.20 Schematic depicting the principle of balanced detection technology

$$\begin{bmatrix} L_1 \\ L_2 \end{bmatrix} = E_L(t)\begin{bmatrix} T \\ R \end{bmatrix} = \frac{1}{\sqrt{2}} E_L e^{j(\omega_L t)} \begin{bmatrix} 1 \\ -j \end{bmatrix} \tag{5.5}$$

$$\begin{bmatrix} S_1 \\ S_2 \end{bmatrix} = E_S(t)\begin{bmatrix} T \\ R \end{bmatrix} = \frac{1}{\sqrt{2}} E_S e^{j(\omega_S t)} \begin{bmatrix} 1 \\ -j \end{bmatrix} \tag{5.6}$$

In the optical domain, mixing the signals L_1 and S_1 and L_2 and S_2 generates photocurrent in the balanced detection arms, which can be expressed as follows:

$$i_1 = \alpha\left[|E_L|^2 + |E_S|^2 + \frac{|E_L E_S|}{2}\cos(\omega_{if} t + \varphi)\right] + n_1(t) \tag{5.7}$$

$$i_2 = \alpha\left[|E_L|^2 + |E_S|^2 - \frac{|E_L E_S|}{2}\cos(\omega_{if} t + \varphi)\right] + n_2(t) \tag{5.8}$$

where α represents the responsiveness of the PD; w_{if} and φ are the frequency and phase differences between the local oscillation and signal light beams, respectively ($w_{if} = w_S - w_L, \varphi = \phi_S - \phi_L$); the first two terms in the brackets indicate the DC components of the power generation of the local oscillation and signal light beams; the third term indicates the medium-frequency component of the photodetection; and $n_1(t)$ and $n_2(t)$ represent the RIN current terms introduced by the local oscillation light beam. The waveforms of i_1 and i_2 are displayed in Fig. 5.21.

After determining the difference, the expression for the IF current at the output of the balanced detector is as follows:

$$i_{out} = i_1 - i_2 = \alpha|E_L E_S|\cos(\omega_{if} t + \varphi) + \Delta n(t) \tag{5.9}$$

where $\Delta n(t)$ represents the residual RIN in the output signal, $\Delta n(t) = n_1(t) - n_2(t)$. The waveform of i_{out} is displayed in Fig. 5.22.

A comparison of Figs. 5.21 and 5.22 indicates that, in the ideal case, the DC component, that is, the common-mode signal and the RIN, can be eliminated after the photocurrent generated by the two PDs is differenced and the amplitude of the IF signal containing information is doubled, thus effectively improving the SNR.

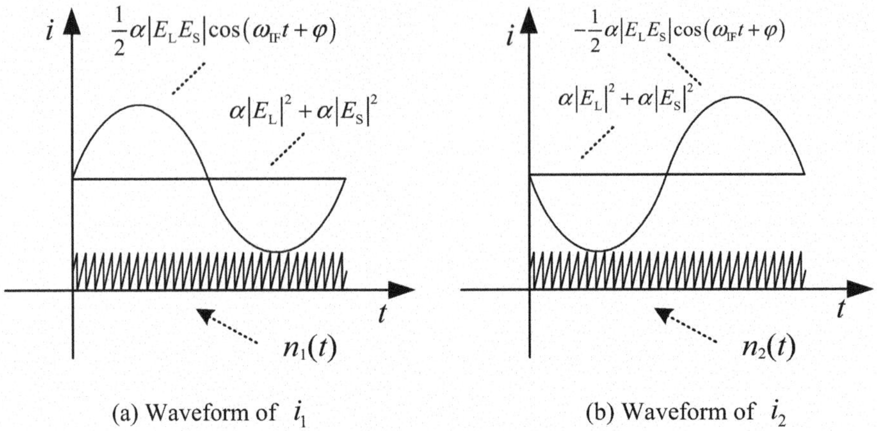

Fig. 5.21 Waveform of photocurrent detected in the PD

Fig. 5.22 Waveform of IF current signal

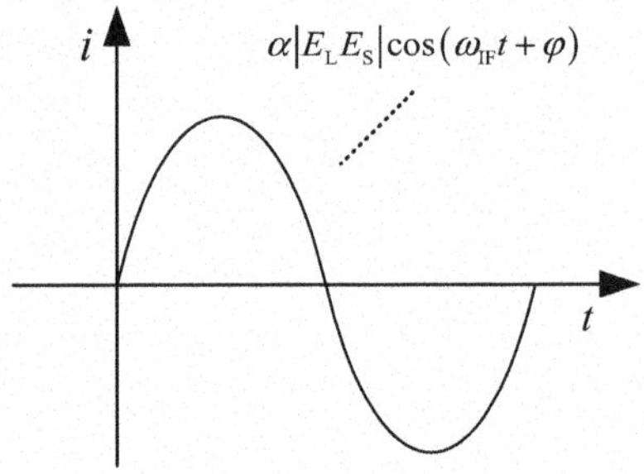

(2) **Double-balanced detection technology**

Figure 5.23 presents the schematic of the double-balanced detection technique. The local oscillation and the signal light beams are coherently mixed in the 90° optical mixer to generate four signal light beams with relative phase differences of 0°, 90°, 180°, and 270°. The two beams with phase differences of 0° and 180° are input to balanced detector 1 and the output branch becomes in-phase (I); similarly, the two beams with phase differences of 90° and 270° are input to balanced detector 2 and the output branch becomes quadrature (Q). A balanced detector with two input light beams is used in the PD to generate photocurrent following photoelectric conversion. The DC components of the two signals are equal, following differential processing to generate two coherent signals, and the phase difference between these two signals is 90°.

The signal light beam E_S and local oscillation light beam E_L enter from the two input ports of the 90° optical mixer. Based on the principle of the mixer, the local oscillation light is converted into circularly polarized light by the line polarization light following the quarter waveplate, and the angle between the quarter waveplate and the direction of polarization of the local oscillation light is 45°. The function of

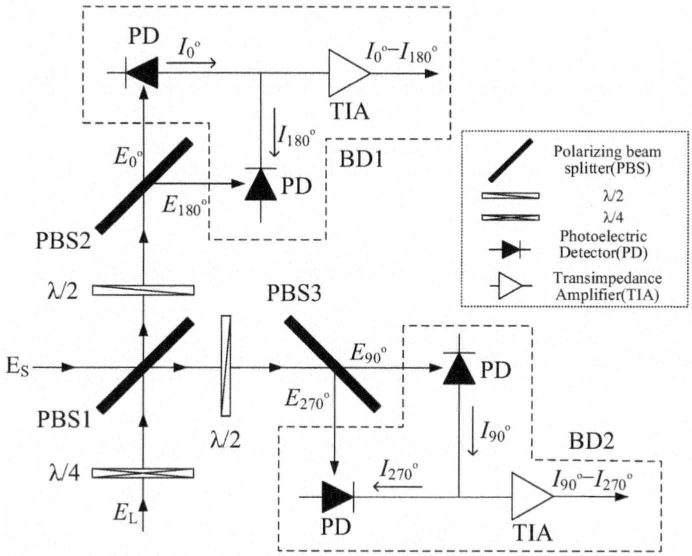

Fig. 5.23 Schematic of double-balanced detection system

the PBS is to project the input light (*p*-light), reflection of *s*-light, local oscillation light, and signal light generated by the PBS for mixing and splitting the beams and then input them to the half waveplate, the angle between half waveplate and *p*-(*s*-)light is 22.5°, after *p*-light pass through the half waveplate, the angle between Linearly polarized light and *p*-light angle is 45°, four beams with relative phase difference of 0°, 90°, 180° and 270° can be obtained by PBS. The two beams with opposite phases are combined and fed into two balanced detectors to obtain two orthogonal IF current signals.

The signal light beam can be expressed as follows [44]:

$$E_S = E_S(x, y, z)\begin{bmatrix} k_1 \\ k_2 \end{bmatrix}\exp(i\phi(t)) \tag{5.10}$$

where $E_S(x, y, z)$ is the complex amplitude of the signal light, k_1 and k_2 are the components of the two optical paths after the signal light beam passes through the beam splitter ($k_1^2 + k_2^2 = 1$), and $\phi(t)$ is the phase modulation of the signal light. The local oscillation light beam can be expressed as follows:

$$E_L = E_L(x, y, z)\begin{bmatrix} k_3 \\ k_4 \end{bmatrix}\exp(i\varphi) \tag{5.11}$$

where $E_L(x, y, z)$ is the complex amplitude of the local oscillation light, k_3 and k_4 are the components of the signal light passing through the two optical paths of the beam splitter ($k_3^2 + k_4^2 = 1$), and φ denotes the initial phase of the local oscillation light.

The matrix expression for the quarter wave slice, when parallel to the *p*-light, can be expressed as follows:

$$\Lambda_{1/4} = \begin{bmatrix} \exp(i\pi/4) & 0 \\ 0 & \exp(-i\pi/4) \end{bmatrix} \tag{5.12}$$

Following the passage of the local oscillation light through the quarter waveplate, the matrix can be expressed as follows:

$$E_L' = \Lambda_{1/4} \cdot E_L = E_L \begin{bmatrix} k_3 \exp(i\pi/2) \\ k_4 \end{bmatrix} \exp i(\varphi - \pi/4) \tag{5.13}$$

The Jones matrix for the PBS can be expressed as follows:

$$\begin{cases} R = \begin{bmatrix} 0 & 0 \\ 0 & r_\perp \exp(i\rho_\perp) \end{bmatrix} \\ T = \begin{bmatrix} t_\parallel \exp(i\tau_\parallel) & 0 \\ 0 & 0 \end{bmatrix} \end{cases} \tag{5.14}$$

where R and T represent the reflection and projection matrices for the PBS, and $r_\perp$ ($t_\parallel$) and $\rho_\perp$ ($\tau_\parallel$) represent the amplitude (phase) in the reflection and projection directions, respectively; ideally, $t_\parallel + \tau_\parallel = 1$.

Following passage through the PBS, the optical path of the local oscillation light beam can be expressed as follows:

$$E_1 = RE_S + TE_L' = \begin{bmatrix} t_\parallel \exp(i\tau_\parallel) k_3 E_L \exp[i(\varphi + \pi/4)] \\ r_\perp \exp(ip_\perp) k_2 E_S \exp[i\phi(t)] \end{bmatrix} \tag{5.15}$$

$$E_2 = TE_S + RE_L' = \begin{bmatrix} t_\parallel \exp(i\tau_\parallel) k_1 E_S \exp[i\phi(t)] \\ r_\perp \exp(ip_\perp) k_4 E_L \exp[i(\varphi - \pi/4)] \end{bmatrix} \tag{5.16}$$

Let the angle between the fast axis of the half waveplate and the direction of polarization of light be α. Therefore, the transmission matrix of the half waveplate can be expressed as follows:

$$\Lambda_{1/2} = \begin{bmatrix} \cos 2\alpha & \sin 2\alpha \\ \sin 2\alpha & -\cos 2\alpha \end{bmatrix} \tag{5.17}$$

In the 90° optical mixer, the angle between the half waveplate and the direction of polarization of the input light is 22.5°. Therefore, the transmission matrix of the half waveplate can be expressed as follows:

$$\Lambda_{1/2} = \frac{\sqrt{2}}{2}\begin{bmatrix} 1 & 1 \\ 1 & -1 \end{bmatrix} \tag{5.18}$$

Following passage through the half waveplate, the optical path can be expressed as follows:

$$E_1' = \Lambda_{1/2} \cdot E_1 = \frac{\sqrt{2}}{2}\begin{bmatrix} t_{\parallel}\exp(i\tau_{\parallel})k_3 E_L \exp[i(\varphi + \pi/4)] + r_{\perp}\exp(ip_{\perp})k_2 E_S \exp[i\phi(t)] \\ t_{\parallel}\exp(i\tau_{\parallel})k_3 E_L \exp[i(\varphi + \pi/4)] - r_{\perp}\exp(ip_{\perp})k_2 E_S \exp[i\phi(t)] \end{bmatrix} \tag{5.19}$$

$$E_2' = \Lambda_{1/2} \cdot E_2 = \frac{\sqrt{2}}{2}\begin{bmatrix} t_{\parallel}\exp(i\tau_{\parallel})k_1 E_S \exp[i\phi(t)] + r_{\perp}\exp(ip_{\perp})k_4 E_L \exp[i(\varphi - \pi/4)] \\ t_{\parallel}\exp(i\tau_{\parallel})k_1 E_S \exp[i\phi(t)] - r_{\perp}\exp(ip_{\perp})k_4 E_L \exp[i(\varphi - \pi/4)] \end{bmatrix} \tag{5.20}$$

Thereafter, passage through the PBS generates the following four outputs of the optical mixer:

$$\begin{cases} I_{0^\circ} = 1/2\{k_2^2|E_S|^2 + k_3^2|E_L|^2 + 2k_2k_3|E_S E_L|\cos(\phi(t) + \rho_{\perp} - \tau_{\parallel} - \pi/2) - (\varphi - \pi/4)\} \\ I_{180^\circ} = 1/2\{k_2^2|E_S|^2 + k_3^2|E_L|^2 - 2k_2k_3|E_S E_L|\cos(\phi(t) + \rho_{\perp} - \tau_{\parallel} - \pi/2) - (\varphi - \pi/4)\} \\ I_{90^\circ} = 1/2\{k_1^2|E_S|^2 + k_4^2|E_L|^2 + 2k_1k_4|E_S E_L|\cos(\phi(t) - \rho_{\perp} + \tau_{\parallel}) - (\varphi - \pi/4)\} \\ I_{270^\circ} = 1/2\{k_1^2|E_S|^2 + k_4^2|E_L|^2 - 2k_1k_4|E_S E_L|\cos(\phi(t) - \rho_{\perp} + \tau_{\parallel}) - (\varphi - \pi/4)\} \end{cases} \tag{5.21}$$

and because $\varphi = \rho_{\perp} - \tau_{\parallel} - \pi/4$, the final IF signal can be obtained as

$$\begin{cases} I_{0^\circ} - I_{180^\circ} = 2k_2k_3|E_S E_L|\cos(\phi(t)) \\ I_{90^\circ} - I_{270^\circ} = 2k_1k_4|E_S E_L|\cos(\phi(t) + \pi/2) \end{cases} \tag{5.22}$$

These two signals represent the in-phase (I) and vertical (Q) components required to demodulate the information. By combining these two components, the effective information carried by the signal light beam can be obtained.

5.3 Balance Mismatch Analysis of Double-Balanced Detection Technology

Ideally, the signal power and transmission time in the balanced connection of the dual-balanced detection system are equal. Balance mismatch reduces the suppression of common-mode signals and amplification of differential signals in coherent optical communication by the double-balanced detection system. The balance mismatch mainly includes the power mismatch caused by the uneven light splitting on each branch in the 90° optical mixer and the inconsistent response of the two photodetectors on the balance detector, and the time mismatch caused by the different optical range length of the 90° optical mixer in each branch and the PD response time. In this section, we analyze the primary sources of the balance mismatch from two aspects: the mixer and the balanced detector.

5.3.1 *Effect of Mixer*

A variety of implementation schemes based on different principles can be used for the 90° optical mixer, among which the most widely used in space optical communication is the scheme based on PBS. The beam splitting ratio of the PBS [45] and the polarization mixing error caused by internal defects owing to the manufacturing process [46] are significant factors that affect double-balanced detection.

(1) **Beam splitting ratio of PBS**

The double-balanced detection system is composed of two balanced detectors; a simplified schematic is presented in Fig. 5.24.

Following passage of the signal and local oscillation light beams through the beam splitter, the signal light is split into two beams, e_{S1} and e_{S2}. The two beams obtained from the beam splitter are e_{L1} and e_{L2}. The expressions for the beam splitting ratios of these four beams are as follows:

$$\begin{cases} e_{S1} = \sqrt{(1-\varepsilon)}E_S\cos(\omega_S t + \varphi_S) \\ e_{S2} = \sqrt{\varepsilon}E_S\cos(\omega_S t + \varphi_S) \\ e_{L2} = \sqrt{\varepsilon}E_L\cos(\omega_L t + \varphi_L) \\ e_{L1} = \sqrt{(1-\varepsilon)}E_L\cos(\omega_L t + \varphi_L) \end{cases} \tag{5.23}$$

where ε is the beam splitting ratio of the beam splitter. Therefore, we obtain two PDs, that is, the intermediate frequency currents generated on the two arms of the balanced detection system are as follows:

$$i_1 = i_2 = \alpha\sqrt{\varepsilon(1-\varepsilon)}E_S E_L\cos(\omega_{IF}t + \varphi_S - \varphi_L) \tag{5.24}$$

The power of the IF signal of the corresponding two arms is expressed as follows:

$$P_{IF1} = P_{IF2} = 2M^2\alpha^2\varepsilon(1-\varepsilon)P_S P_L R_L \tag{5.25}$$

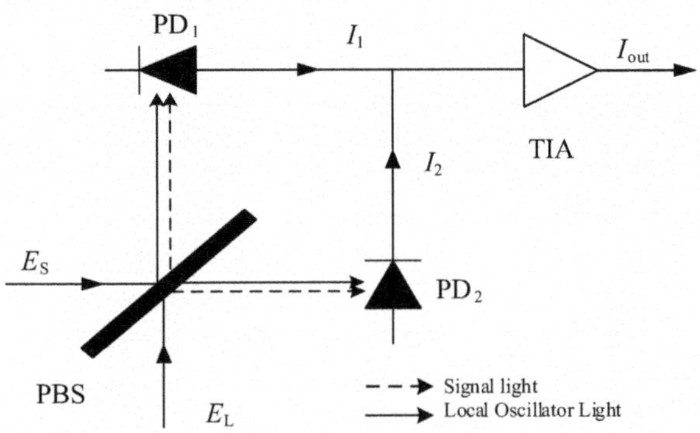

Fig. 5.24 Schematic of double-balanced detection single circuit

where P_S and P_L represent the powers of the signal and local oscillation light beams, respectively, R_L represents the load resistance, and $\alpha = \eta e/hv$ represents the photoelectric conversion efficiency. The noise powers in the two arms of the balanced detection system are expressed as follows:

$$\begin{cases} P_{D1} = 2M^2e\{\alpha[\varepsilon P_L + (1-\varepsilon)P_S + P_b] + i_d\}\Delta f_{IF} R_L \\ P_{D2} = 2M^2e\{\alpha[(1-\varepsilon)P_L + \varepsilon P_S + P_b] + i_d\}\Delta f_{IF} R_L \end{cases} \tag{5.26}$$

where M represents the photoelectric conversion efficiency coefficient [54]; $M = 1$ (2) when the detector is a photodiode (photoconductive detector). In the two expressions of Eq. (5.26), the first halves of P_{D1} and P_{D2} represent shot noise. The latter halves represent thermal noise. Under the limit conditions, the shot noise introduced by the local oscillator is much greater than the thermal noise. Therefore, Eq. (5.26) can be simplified as follows:

$$\begin{cases} P'_{D1} = 2M^2e\alpha\varepsilon P_L \Delta f_{IF} R_L \\ P'_{D2} = 2M^2e\alpha(1-\varepsilon)P_L \Delta f_{IF} R_L \end{cases} \tag{5.27}$$

Therefore, the SNRs in the two arms in the balanced detection system are derived as follows:

$$\begin{cases} f_{SNR1} = \frac{\eta(1-\varepsilon)P_S}{hv\Delta f_{IF}} \\ f_{SNR2} = \frac{\eta\varepsilon P_S}{hv\Delta f_{IF}} \end{cases} \tag{5.28}$$

The two IF signals enter the adder following the phase shift, and the output IF current and power are expressed as follows:

$$\begin{cases} I_{out} = i_1 + i_2 = 2\alpha\sqrt{\varepsilon(1-\varepsilon)}E_S E_L \cos(\omega_{if}t + \varphi_S - \varphi_L) \\ P_{out} = 4M^2\alpha^2\varepsilon(1-\varepsilon)P_S P_L R_L \end{cases} \tag{5.29}$$

Because the scattered noise has the same phase, the total noise is obtained as follows:

$$P_{noise} = P'_{D1} - P'_{D2} = 2M^2e\alpha\left[1 - 2\sqrt{\varepsilon(1-\varepsilon)}\right]P_L \Delta f_{IF} R_L \tag{5.30}$$

Therefore, the SNR of the balanced detection system is expressed as follows:

$$f_{SNR\_B} = \frac{P_{out}}{P_{noise}} = \frac{2\varepsilon(1-\varepsilon)\eta P_S}{1 - 2\sqrt{\varepsilon(1-\varepsilon)}hv\Delta f_{IF}} \tag{5.31}$$

and the SNR equation for ordinary outlier detection is as follows:

$$f_{SNR\_S} = \frac{\eta P_S}{hv\Delta f_{IF}} \tag{5.32}$$

As depicted in Fig. 5.25, Point A in the figure indicates the thresholds of the SNR of the balanced detection system and the SNR of the single arm. Point C indicates that the SNRs of the two arms of the balanced detection system are the same when the beam splitting ratio is 0.5. The horizontal line in the figure indicates the relative SNR of normal aberration detection under identical input conditions, and the intersection point B of the curve with the balanced detection curve corresponds to the smallest beam splitting ratio when the SNR of balanced detection is larger than that of normal aberration detection. When the beam splitting ratio of polarization beam splitter is between 0.292 and 0.5, the SNR ratio of balanced detection system is larger than that of the normal heterodyne detection. The maximum value is achieved at a beam splitting ratio of 0.5. Therefore, in practical applications, to maximize the SNR of coherent optical communication, the PBS should be set at a suitable angle so that the beam splitting ratio is as close to 0.5 as possible.

(2) **Polarization mixing error**

A PBS uses a crystal or optical film to separate two beams of light, with their directions of polarization perpendicular to each other. The output light beam generated by an ideal PBS should be linearly polarized. As depicted in Fig. 5.26, ideally, the light field distribution of the output light should be linearly polarized, which is now assumed to be polarized at an angle of 45°. When the internal defects of the PBS are considered, its polarization changes to elliptical, as depicted in Fig. 5.27.

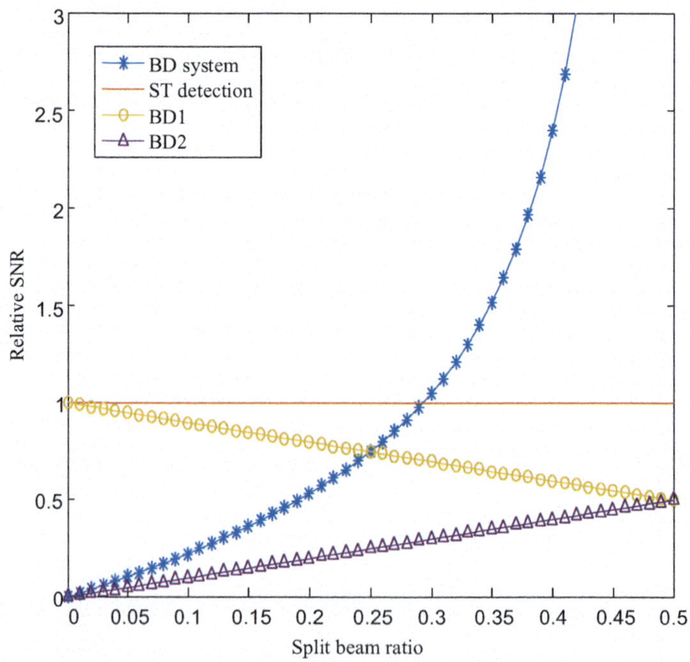

Fig. 5.25 Balanced detection system and ordinary outlier detection SNR comparison chart

To facilitate the analysis, we can replace the expressions Eq. (5.4) for the signal and principal vibration light beams (Sect. 5.2) as follows:

$$E_S = \begin{pmatrix} k_1 \exp[i(\omega_S t + a_x)] \\ k_2 \exp[i(\omega_S t + a_y)] \end{pmatrix} \quad (5.33)$$

$$E_L = \begin{pmatrix} k_3 \exp[i(\omega_L t + a'_x)] \\ k_4 \exp[i(\omega_L t + a'_y)] \end{pmatrix} \quad (5.34)$$

As depicted in Fig. 5.27, the coordinate system is converted into the $x' - y'$ coordinate system. At this point the electric field of the beam can be expressed as follows:

$$E = \begin{pmatrix} 1 \\ b \exp[i(\frac{\pi}{2} - \delta)] \end{pmatrix} \exp[i(\omega t + \phi)] \quad (5.35)$$

where b is the ellipticity, that is, the ratio of the short and long semi-axes of the ellipse and δ is the nonorthogonal factor coefficient, which indicates the degree of orthogonal deviation of the two incident beams. Converting the coordinate system back into the x–y coordinate system, the electric field of the beam is transformed as

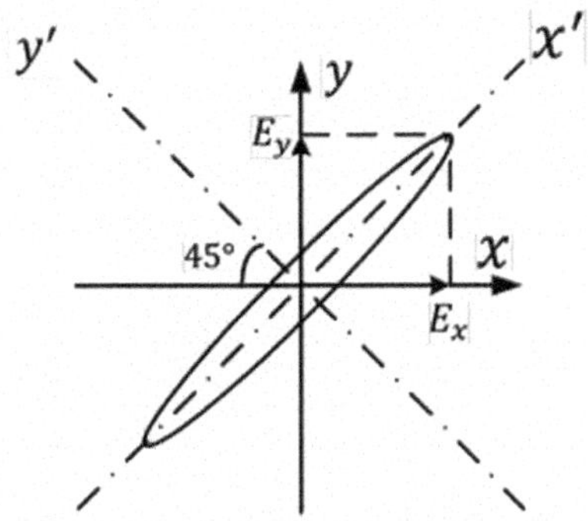

Fig. 5.26 Ideal case of the light field distribution of polarized light

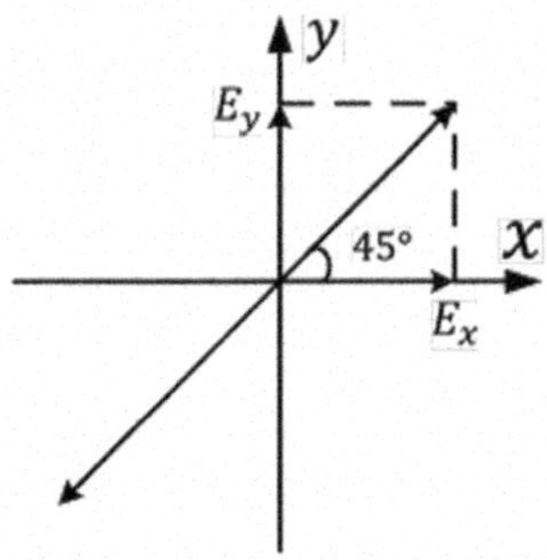

Fig. 5.27 Practical light field distribution of polarized light

follows:

$$E = \begin{pmatrix} E_x \\ E_y \end{pmatrix} = \frac{\sqrt{2}}{2}\begin{pmatrix} 1 - bi\cos\delta - b\sin\delta \\ 1 + bi\cos\delta + b\sin\delta \end{pmatrix}\exp[i(\omega t + \phi)] \tag{5.36}$$

which can be simplified as follows:

$$E = \begin{pmatrix} E_x \\ E_y \end{pmatrix} = \frac{\sqrt{2}}{2}\begin{pmatrix} 1 \\ \rho\exp(i\beta) \end{pmatrix}\exp[i(\omega t + \phi)] \tag{5.37}$$

where

$$\rho\exp(i\beta) = \frac{1 - bi\cos\delta - b\sin\delta}{1 + bi\cos\delta + b\sin\delta} \tag{5.38}$$

When considering the internal defects of the PBS, the matrix expression for the PBS is converted from Eq. (5.14) into the following expression:

$$\begin{cases} R = \begin{bmatrix} \gamma & 0 \\ 0 & r_\perp\exp(i\rho_\perp) \end{bmatrix} \\ T = \begin{bmatrix} t_\parallel\exp(i\tau_\parallel) & 0 \\ 0 & \gamma \end{bmatrix} \end{cases} \tag{5.39}$$

where γ denotes the degree of impact of the internal defects. Substituting Eq. (5.39) in the mathematical model of the double-balanced detection, the expression for the four-way photogenerated current at the output of the mixer can be obtained as follows:

$$\begin{cases} I_{0^\circ} = \frac{1}{8}\big[k_1^2 + k_3^2 + 2k_1k_3\cos(\omega_{if}t + \alpha_x - \alpha_x' + \rho_\parallel - \tau_\parallel) \\ \quad + k_2^2\rho^2\gamma^2 + k_4^2\rho^2\gamma^2 + 2k_2k_4\rho^2\gamma^2\cos(\omega_{if}t + \alpha_y - \alpha_y' + \rho_\perp - \tau_\perp)\big] \\ I_{90^\circ} = \frac{1}{8}\big[k_1^2\gamma^2 + k_3^2\gamma^2 + 2k_1k_3\gamma^2\cos(\omega_{if}t + \alpha_x - \alpha_x' + \rho_\parallel - \tau_\parallel) \\ \quad + k_2^2\rho^2 + k_4^2\rho^2 + 2k_2k_4\rho^2\cos(\omega_{if}t + \alpha_y - \alpha_y' + \rho_\perp - \tau_\perp)\big] \\ I_{180^\circ} = \frac{1}{8}\big[k_1^2 + k_3^2 + 2k_1k_3\cos(\omega_{if}t + \alpha_x - \alpha_x' + \tau_\parallel - \rho_\parallel) \\ \quad + k_2^2\rho^2\gamma^2 + k_4^2\rho^2\gamma^2 + 2k_2k_4\rho^2\gamma^2\cos(\omega_{if}t + \alpha_y - \alpha_y' + \tau_\perp - \rho_\perp)\big] \\ I_{270^\circ} = \frac{1}{8}\big[k_1^2\gamma^2 + k_3^2\gamma^2 + 2k_1k_3\gamma^2\cos(\omega_{if}t + \alpha_x - \alpha_x' + \tau_\parallel - \rho_\parallel) \\ \quad + k_2^2\rho^2\gamma^2 + k_4^2\rho^2\gamma^2 + 2k_2k_4\rho^2\gamma^2\cos(\omega_{if}t + \alpha_y - \alpha_y' + \tau_\perp - \rho_\perp)\big] \end{cases} \tag{5.40}$$

The output IF current is expressed as follows:

$$\begin{cases} I_{0^\circ} - I_{180^\circ} = 2k_1k_3\cos(\omega_{if}t+\phi) + 2k_2k_4\rho^2\gamma^2\cos(\omega_{if}t+\phi) \\ I_{90^\circ} - I_{270^\circ} = 2k_1k_3\gamma^2\cos(\omega_{if}t+\phi) + 2k_2k_4\rho^2\cos(\omega_{if}t+\phi) \end{cases} \tag{5.41}$$

where γ is much smaller than 1 and ρ is close to 1. Therefore, Eq. (5.41) can be simplified as follows:

$$\begin{cases} I_0 - I_{180} \approx 2k_1k_3\cos(\omega_{if}t+\phi) \\ I_{90} - I_{270} \approx 2k_2k_4\rho^2\cos(\omega_{if}t+\phi) \end{cases} \tag{5.42}$$

It can be seen from Eq. (5.42) that when compared with the previous output IF current, the present formula has an additional influencing factor ρ, which is the polarization mixing error caused by the internal defect of the PBS.

Using Eq. (5.38), this factor can be expressed as follows:

$$\rho = \left(\frac{1 + b^4 + 2b^2\cos(2\delta)}{1 - 2b\sin\delta + b^2}\right)^{1/2} \tag{5.43}$$

Figure 5.28 is based on Eq. (5.43), from which the interrelationship among ρ, b, and δ can be obtained. ρ varies periodically with σ. The nonorthogonality factor represented by δ itself also varies periodically, with a period of 2π, indicating that the correction factor caused by defects inside the PBS is also periodic, but there is no periodic law between the correction factor and the ellipticity b.

In general, we require ρ^2 to be as close to 1 as possible, in which case the solution of Eq. (5.43) yields the following expression:

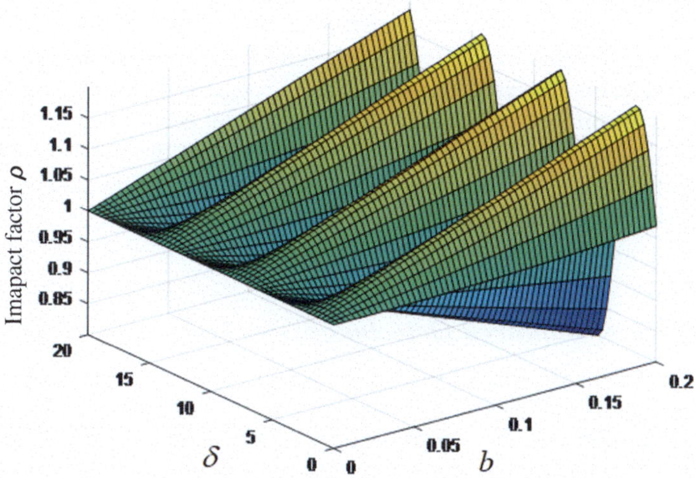

Fig. 5.28 Impact factor of polarization mixing error versus ellipticity and nonorthogonal factor

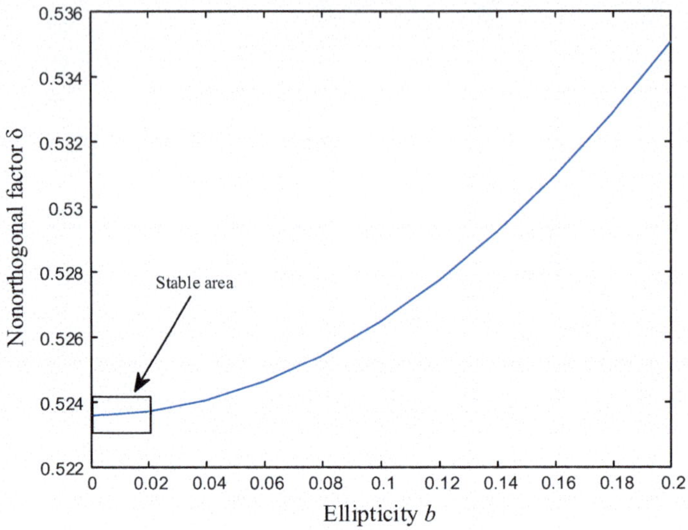

Fig. 5.29 Relationship between nonorthogonal factor δ and ellipticity b

$$b^3 - 3b + 4b\cos^2\delta + \sin\delta = 0 \tag{5.44}$$

b is generally very small, and Eq. (5.44) can be reduced to the following:

$$b^2 + 4\cos^2\delta - 3 = 0 \tag{5.45}$$

The relationship between b and δ can be obtained from Eq. (5.45), as depicted in Fig. 5.29. The figure indicates that when b is small, δ is relatively smooth, approximately 0.523. This part of the region is referred to as the stable region of the nonorthogonal factor. Larger the ellipticity factor caused by the polarization mixing error, larger is the orthogonality factor of the two output beams, that is, larger is the deviation of the phase difference between the two output beams of the PBS from 90°, which affects the balance consistency of the two arms of the double-balanced detection and limits the improvement of the system's SNR.

5.3.2 *Effect of Balanced Detectors*

The main source of the balance mismatch caused by the balanced detector is the asymmetry of the PD and other electronic devices, including factors such as different PD responsiveness and optical signal transmission delays.

(1) **Responsiveness of PD**

The difference between the balanced detector and the ordinary PD is that the requirement of consistency of the two PDs is a relatively harsh condition; the parameters of the two should be extremely similar to meet the requirements of balanced detection [47]. Several parameters of the PD are considered; the most significant ones are the responsiveness and the detection area. We now assume that the detection area is equal and consider mainly the effect of responsiveness. To separate the impact of the optical mixing part, we assume that the mixing part is the ideal case, that is, when the beam splitting ratio is 0.5, and set the conversion coefficients of the two detectors as α_1 and α_2, where $\alpha_1 = k\alpha_2$. Therefore, the expression for the IF signal based on the ratio of conversion coefficients can be derived from Eq. (5.24) as follows:

$$P_{out} = \frac{1}{4}\left(\frac{\eta e}{h\upsilon}\right)^2 E_S^2 E_L^2 (1+k)^2 R_L \tag{5.46}$$

From Eq. (5.30), the power expression for the noise can be obtained as follows:

$$P_{noise} = \frac{\eta e^2}{h\upsilon} E_L^2 \Delta f (k-1) R_L \tag{5.47}$$

Therefore, the equation of SNR for balanced detection can be transformed into the following expression:

$$f_{SNR} = \frac{\eta E_S^2 (1+k)^2}{4h\upsilon \Delta f (k-1)} \tag{5.48}$$

Equation (5.48) indicates the impact of the detector conversion coefficient ratio of the two arms of the balanced detector on the SNR. When $k = 1$, that is, the parameters of the two arms of the balanced detection device are identical, the system can achieve the best performance.

In practice, it is impossible to obtain two identical detectors, and we introduce the equilibrium consistency factor β, expressed as follows:

$$\beta = 1 - \left(\frac{2|\alpha_1 - \alpha_2|}{\alpha_1 + \alpha_2}\right) \times 100\% \tag{5.49}$$

Assuming $k < 5$, this equation translates into an expression related to the conversion factor as follows:

$$\beta = \left(\frac{3k-1}{k+1}\right) \times 100\% \tag{5.50}$$

The sources of noise in the balanced detector are primarily thermal, scattered particle, and RINs as well as the dark current in the PD. The thermal and scattered particle noises generated by the dark current are much smaller than the relative intensity and scattered particle noises generated by the local oscillation light beam.

As the scattered particle noise cannot be eliminated, the primary role of the balanced detector is to suppress the RIN, that is, the better the consistency of balance, the smaller the RIN. The expression for the rms current of the scattered particle noise of the PD is as follows:

$$\left\langle i_{shot}^{2}\right\rangle = 2e\alpha P_{L} B \tag{5.51}$$

The RIN can be expressed as follows:

$$\left\langle i_{\mathrm{RIN}}^{2}\right\rangle = \mathrm{RIN} \times B \times (\alpha P_{\mathrm{L}})^{2} \tag{5.52}$$

Based on the preceding analysis, it is clear that the communication requirements will be satisfied if the following inequality holds:

$$\mathrm{RIN} \times B \times [(1-\beta)\alpha P_{\mathrm{L}}]^{2} \leq 2ea P_{\mathrm{L}} B \tag{5.53}$$

where RIN denotes the RIN of the semiconductor laser; its range of values is generally -155 to -130 dB/Hz. Here, we consider RIN $= -135$ dB/Hz and local oscillation optical power $P_{L} = 5$ mW. By substituting these values in Eq. (5.53), we have $\beta \geq 95\%$. That is, the balance of the dual-balanced detection system must be at least greater than 95% with respect to the balance of the detector consistency to meet the requirements of communication.

(2) **Optical signal transmission delay**

The optical signal transmission delay refers to the time difference between the two PDs of the balanced detector receiving the optical signal inconsistently and the PD at the time of consistent photoelectric conversion. The transmission delay of the optical signal also impacts the suppression of the RIN.

Let the thermal, scattering, and RINs be represented by $n_S(t)$, $n_T(t)$, and $n_I(t)$, respectively. Therefore, the total noise of the balanced detector can be expressed as follows:

$$n(t) = n_S(t) + n_T(t) + n_I(t) \tag{5.54}$$

where $n_S(t)$, $n_T(t)$, and $n_I(t)$ are all calculated as differences. The specific expressions are as follows:

$$n_S(t) = n_{S1}(t) - n_{S2}(t) \tag{5.55}$$

$$n_T(t) = n_{T1}(t) - n_{T2}(t) \tag{5.56}$$

$$n_I(t) = n_{I1}(t) - n_{I2}(t) \tag{5.57}$$

The scattered particle and thermal noises are uncorrelated. The RIN can be completely eliminated in the ideal case, but in practice there would be some errors caused by the transmission delay of the optical signal. Therefore, the RIN of the balanced detector can be expressed as follows:

$$n_I'(t) = n_I(t) - n_I(t - \Delta t) \tag{5.58}$$

where Δt denotes the time difference between the two arms of the equilibrium detector.

If the RIN is assumed to be a smooth random process, then the autocorrelation function for $n_I(t)$ can be expressed as follows:

$$R(\tau) = \int_{-\infty}^{\infty} n_I(t + \tau)n_I(t)dt \tag{5.59}$$

The autocorrelation function for $n_I(t - \Delta t)$ can be expressed as follows:

$$R(\tau) = \int_{-\infty}^{\infty} n_I(t - \Delta t + \tau)n_I(t - \Delta t)dt \tag{5.60}$$

The difference between Eqs. (5.59) and (5.60) is as follows:

$$\begin{aligned} R'(\tau) &= \int_{-\infty}^{\infty} [n_I(t - \Delta t + \tau)n_I(t - \Delta t)][n_I(t + \tau)n_I(t)]dt \\ &= 2R(\tau) - R(\tau + \Delta t) - R(\tau - \Delta t) \end{aligned} \tag{5.61}$$

The autocorrelation function and its power spectral density are a pair of Fourier transforms, that is,

$$S(f) = \int_{-\infty}^{\infty} R(\tau)e^{-j2\pi f\tau}d\tau \tag{5.62}$$

Therefore, the power spectral densities of the last two terms of Eq. (5.61) can be expressed as follows:

$$S_1(f) = \int_{-\infty}^{\infty} R(\tau + \Delta t)e^{-j2\pi f\tau}d\tau = e^{j2\pi f\Delta t}S(f) \tag{5.63}$$

$$S_2(f) = \int_{-\infty}^{\infty} R(\tau - \Delta t)e^{-j2\pi f\tau}d\tau = e^{-j2\pi f\Delta t}S(f) \tag{5.64}$$

The power spectral density of the RIN difference can be obtained from Eqs. (5.63) and (5.64) as follows:

$$S'(f) = \int_{-\infty}^{\infty} R'(\tau)e^{-j2\pi f\tau}d\tau = 2S(f) - e^{j2\pi f\Delta t}S(f) - e^{-j2\pi f\Delta t}S(f)$$
$$= 2S(f)(1 - \cos(2\pi f\Delta t)) \quad (5.65)$$

Equation (5.65) indicates that when the time difference between the two arms of the equilibrium detector is zero, the power spectral density of the RIN difference is zero, which means that there is no time difference between the two arms of the equilibrium detector, and the RIN is completely eliminated. Although the cos function in Eq. (5.65) is an integral multiple of 2π, the power spectral density is also zero, but because the noise is random and aperiodic, the larger the time difference, the greater is the noise. The advantage of the double-balanced detection technology is that it can effectively suppress the RIN introduced by the local oscillation light beam when the time difference between the two arms of the balanced detector is large, that is, the transmission time delay of the optical signal is large and affects the sensitivity of the double-balanced detection system.

5.4 Common-Mode Rejection Ratio in Double-Balanced Detection System

5.4.1 Common-Mode Rejection Ratio

The capability of a double-balanced detection system to reject RIN can be expressed in terms of the common-mode rejection ratio (CMRR) [51]. However, the limitations of the production process, for example, uneven beam splitting through the PBS in the optical mixer, unequal optical range of each branch, and inconsistent parameters of the PD in the balanced detector, can cause a mismatch of balance in the double-balanced detection system [48].

In Fig. 5.30, α_p and α_n indicate the received optical power of the two PDs in the balanced detector, τ_p and τ_n represent the time required for the photoelectric conversion of each beam from the output of the 90° optical mixer to the two arms of the balanced detector, and I_p and I_n represent the output currents of the two PDs in the balanced detector. The CMRR characterizes the capability of the balanced detector to reject common-mode signals and is expressed as

$$\mathrm{CMRR} = -10\log\left(\frac{|I_P - I_N|}{I_P + I_N}\right)^2 \quad (5.66)$$

Balance mismatch can be classified into two main types: power mismatch and time mismatch. Power mismatch ($\alpha_p \neq \alpha_n$) is mostly caused by uneven splitting through the optical mixer and the different response of the balanced detector. Time mismatch ($\tau_p \neq \tau_n$) is mostly caused by the difference in optical range between the optical mixer and the detector and the response times of the two PDs in the balanced

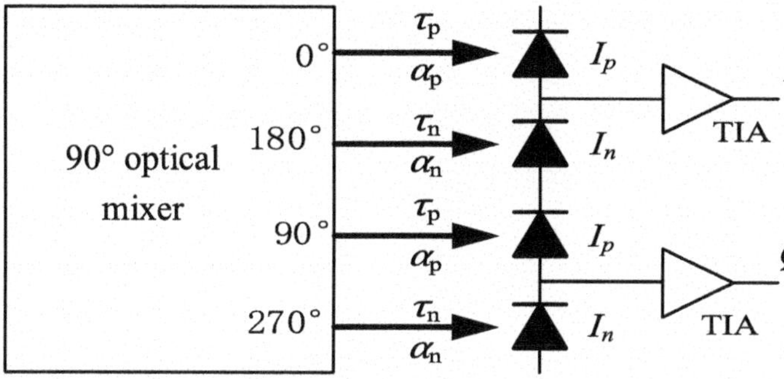

Fig. 5.30 Double-balanced detection of balance mismatch sources

detector. The expression for the common-mode rejection ratio can be converted as follows:

$$\mathrm{CMRR} = -10\log\left|\frac{\alpha^2 - e^{j2\pi f\tau}}{\alpha^2 + 1}\right|^2 \tag{5.67}$$

where α represents the degree of power mismatch, and its value is the product of the beam splitting ratio of the incident beam after mixing and the ratio of the responses of the two PDs in the balanced detector; τ characterizes the time mismatch, and its value is the sum of the time delay caused by the unequal optical ranges of the optical paths inside the optical mixer and the time delay caused by the varying response times of the PDs. Figure 5.31 depicts Eq. (5.67) for $\alpha = 1$ and $\tau = 0.1$–5 ps, that is, when the power balance, the CMRR is affected by the time mismatch curve with respect to frequency. In Fig. 5.32, $\tau = 0.01$ ps and $\alpha = 0.2$–1, that is, when the time mismatch is very small, the CMRR is affected by the power mismatch curve with respect to frequency.

Figures 5.31 and 5.32 indicate that the value of the CMRR increases as the degree of mismatch increases regardless of power or time mismatch. Figure 5.32 indicates that the slopes of the curves corresponding to various degrees of power mismatch vary significantly at low-frequency bands, whereas the curves tend to be close to each other at high-frequency bands, indicating that the CMRR at low-frequency bands is mostly affected by power mismatch. The value of the CMRR characterizes the ability of the balanced detection system to suppress the common-mode noise caused by the local oscillation light beam. Therefore, an increase in the CMRR causes the common-mode noise in balanced detection to increase, which causes the SNR of the communication system to decrease and, thus, affects the performance of the system.

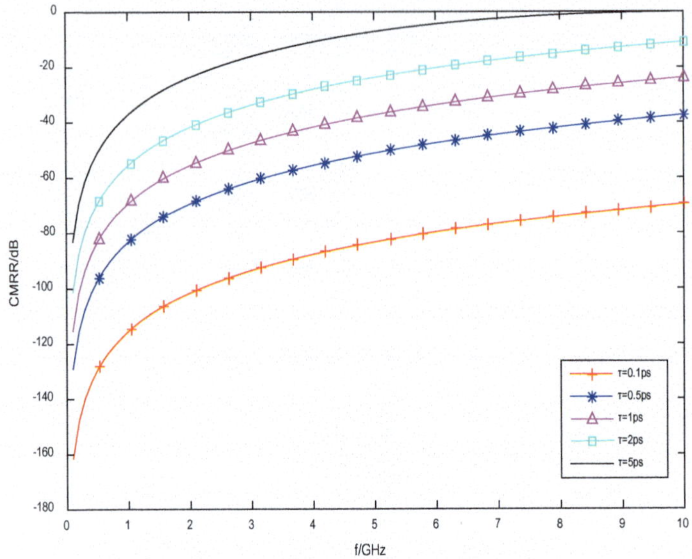

Fig. 5.31 Effect of time mismatch on the CMRR [54]

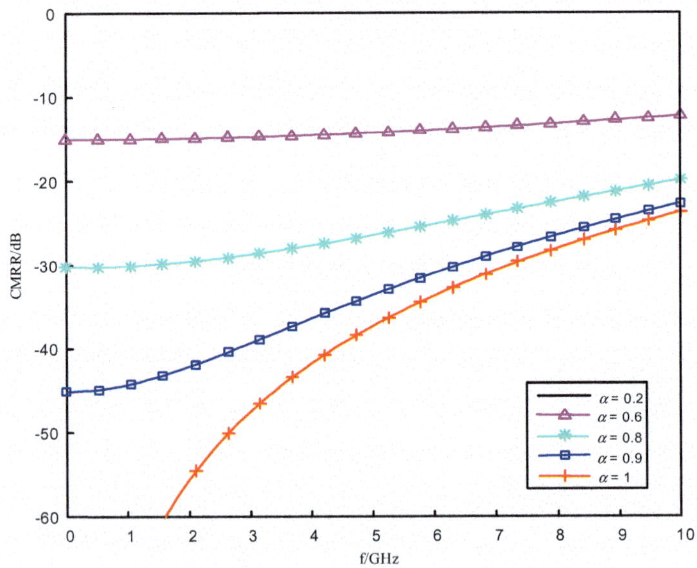

Fig. 5.32 Effect of power mismatch on the CMRR [54]

5.4.2 Signal-to-Noise Ratio

The primary cause of noise in the double-balanced detection system is the scattered particle noise generated by the PD, the thermal noise generated by the transimpedance amplifier (TIA), and the RIN introduced by the local oscillation light beam.

The scattered particle noise is generated mostly by the PD and is caused by uneven electron emission in the photoelectric conversion process. The single-sideband noise power spectral density representing the scattered particle noise generated by a PD can be expressed as follows [45]:

$$\sigma_{shot}^2 = 2 \cdot q \cdot (\eta \cdot P + I_d) \cdot \Delta f \tag{5.68}$$

where q is the electronic charge, η is the corresponding coefficient of the PD, I_d is the dark current, Δf is the single-sideband effective noise bandwidth, and P is the total power received. Because the fundamental optical power is high and the signal optical power is small, the total power can be approximated as the fundamental optical power.

As mentioned, the main source of thermal noise is the TIA, which is mainly caused by the random thermal motion of electrons in the resistance. The thermal noise can be simply calculated using the TIA-equivalent input reference noise current density. It can be expressed as follows [45]:

$$\sigma_{thermal}^2 = (i_{TIA})^2 \cdot \Delta f \tag{5.69}$$

where i_{TIA} is the one-sided differential input noise current density.

The RIN introduced by the local oscillation light beam is an unavoidable intensity fluctuation caused by relaxation oscillations in the output of the semiconductor laser. It can be expressed as follows [47]:

$$\sigma_{RIN}^2 = 2(\eta P_L)^2 \cdot RIN \cdot \Delta f \tag{5.70}$$

where P_L is the principal oscillation power.

The advantage of the double-balanced detection technique is that it can effectively suppress the RIN introduced by the local oscillation light beam, thus improving the SNR of the system. The CMRR characterizes the ability of double-balanced detection to suppress the RIN. It is introduced into the SNR equation to analyze its effect on the performance of the system. Based on its definition, the SNR at the output of the balanced detector can be expressed as follows:

$$SNR = \frac{\frac{1}{2} \cdot 8\eta^2 P_L P_S}{\sigma_{TIA}^2 + 8\eta^2 P_L P_{RIN} CMRR + \sigma_{shot}^2} \tag{5.71}$$

where P_S, P_L, and P_{RIN} are the powers corresponding to the signal light beam, local oscillation light beam, and RIN, respectively. The numerator represents the output signal power of the balanced detector, where the factor of 1/2 is introduced by the asynchronous phase noise caused by the signal and local oscillation light beams. The three noise variances included in the denominator are the thermal, scattered particle, and residual RINs generated by the TIA, PD, and local oscillation light beams, respectively.

After considering the insertion loss in the 90° optical mixer and substituting each noise variance formula in Eq. (5.71), the SNR formula is obtained as follows:

$$\mathrm{SNR} = \frac{P_{\mathrm{S}}}{\frac{8i_{\mathrm{TIA}}^2 e_{\mathrm{l}}^2 \Delta f/\eta^2}{P_{\mathrm{L}}} + 2P_{\mathrm{L}}(2\mathrm{RIN}\cdot\Delta f)\cdot\mathrm{CMRR} + 4qe_{\mathrm{l}}^2\Delta f/\eta} \tag{5.72}$$

where e_{l} is the insertion loss in the optical mixer, Δf is the effective noise bandwidth, η is the photoelectric conversion efficiency of the PD, and W is the electronic charge.

Equation (5.72) represents the interrelationship between the local oscillation optical power, CMRR, and SNR. The lower the CMRR, higher the suppression of the RIN and higher the SNR.

Let $N_{thermal} = 8i_{TIA}^2 e_l^2 \Delta f/\eta^2$, $N_{RIN} = 4P_L\cdot RIN\cdot\Delta f$, and $N_{shot} = 4qe_l^2\Delta f/\eta$. Therefore, Eq. (5.72) can be simplified as follows:

$$SNR = \frac{P_S}{\frac{N_{thermal}}{P_L} + P_L\cdot N_{RIN}\cdot CMRR + N_{shot}} \tag{5.73}$$

When the signal is phase-modulated, BER can be expressed in terms of SNR as follows [50]:

$$BER = \frac{1}{2}\left[erfc\left(Q/\sqrt{2}\right)\right] = \frac{1}{2}\left[erfc\left(\sqrt{SNR}\right)\right] \tag{5.74}$$

where Q is the quality factor (Q factor), used to characterize the performance of the communication system together with BER.

Under certain BERs, after determining the SNR, the sensitivity of detection can be determined from the noise power. The sensitivity of the balanced detector is the optical power of the input signal detectable at a given BER, which is one of the key technical indicators of the balanced detector [51].

As can be seen from Eq. (5.74), by introducing the concept of input reference noise variance, the scattered particle noise term is no longer correlated with the principal oscillation power but exists as a constant value. In addition, if the thermal noise introduced by the TIA and the RIN value from the local oscillation laser are known, we can derive an average CMRR value from the local oscillation power. This also provides us with a simple means to measure the quality of performance of the double-balanced detection system.

5.4.3 *Numerical Simulation*

Simulation is performed based on the schematic of the coherent optical communication system depicted in Fig. 2.1. In the simulation, the modulated signal is generated based on a pseudorandom bit sequence, the modulation method used is quadrature phase-shift keying, and the local oscillation light beam is set to a bandwidth of 100 kHz. The output branches of the default module are ideally balanced, and the insertion loss is set to 1 dB. The Q factor is used to characterize the performance of the system; a higher Q value indicates a lower BER. The Q factor is defined as the SNR of the receiver judgment circuit, which is employed because of its strong nonlinear effect in high-speed optical communication systems. Such an effect can have a significant impact on the final BER of the system, and the Q factor can effectively circumvent this effect.

(1) **Effect of relative intensity noise**

Based on Eqs. (5.72) and (5.74), the curve representing the effect of RIN on the performance of the system can be obtained as presented in Fig. 5.34. The initial conditions are as follows: $P_S = -33$ dBm, $i_{TIA} = 18.2\,\text{pA}/\sqrt{\text{Hz}}$, $\Delta f = 20$ GHz, $e_l = 1$ dB, $\eta = 0.75$ A/W, $\tau = 2$ ps.

As can be seen in Fig. 5.33, the larger the RIN, the larger the RIN of the system. When the principal oscillation power is small, the difference between the Q-factor surrogate values corresponding to different coefficients of RIN is small, indicating that the system is less affected by RIN when the principal oscillation power is small; the lowest point on each curve corresponds to the best system performance, and the horizontal coordinate of the lowest point corresponds to the maximum principal oscillation power value that can be introduced into the system. Any improvement in the system's performance affected by an increase in the principal oscillation power beyond the maximum value is offset by the RIN introduced by the principal oscillation light beam, and the system's performance deteriorates instead.

Based on Eq. (5.74), the curve corresponding to the decreased value of the system sensitivity with respect to the RIN at various values of the principal oscillation power can be obtained as depicted in Fig. 5.35. The initial conditions are as follows: $i_{TIA} = 18.2\text{pA}/\sqrt{\text{Hz}}$, $\Delta f = 20$ GHz, $e_l = 1$dB, $\eta = 0.75$ A/W, $\tau = 2$ ps.

Figure 5.34 indicates that as the RIN increases, the sensitivity drop also increases. The higher the power of the local oscillation light, the steeper the sensitivity drop curve. Further, as the RIN increases, the sensitivity drop curve gradually exceeds the curve of lower local oscillation power. This indicates that the larger the RIN value, the higher the RIN introduced by the local oscillation and the higher the effect of the RIN on the sensitivity of the system as the local oscillation power increases.

(2) **Effect of thermal noise**

Equations (5.72) and (5.74) indicate that when different input reference noise current densities are considered, data on the Q-factor with respect to the variation in the principal oscillation power can be obtained. Considering the maximum Q-factor as the

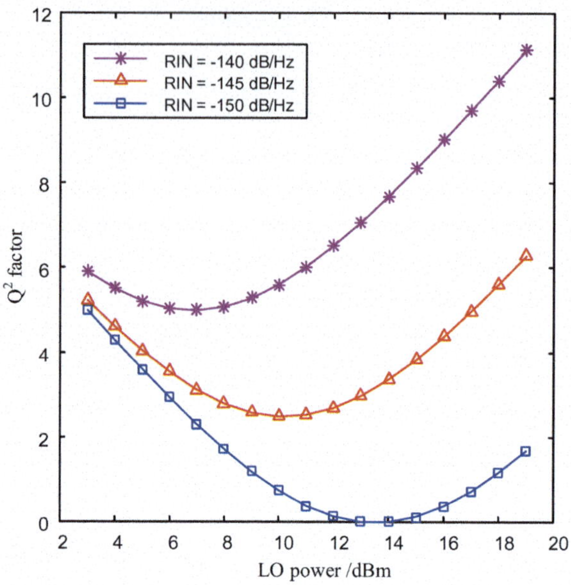

Fig. 5.33 Variation in Q-factor cost with respect to the LO power for various values of RIN [54]

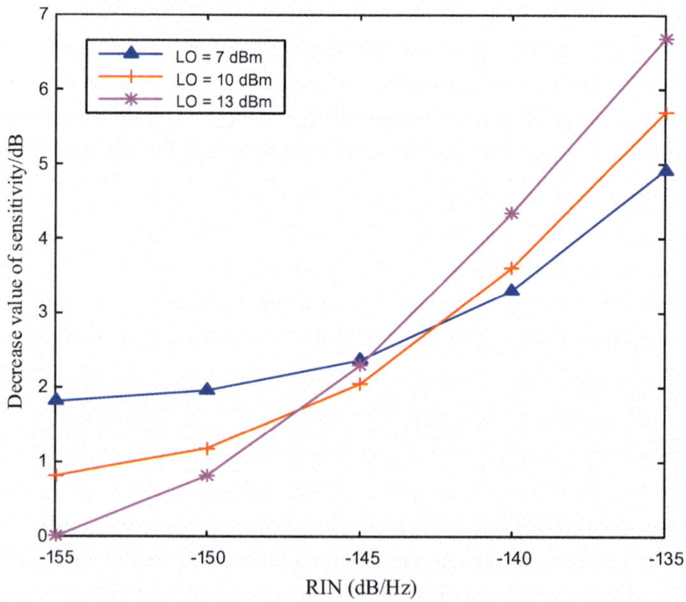

Fig. 5.34 Variation in system sensitivity with respect to the RIN value at various values of the principal oscillation power [54]

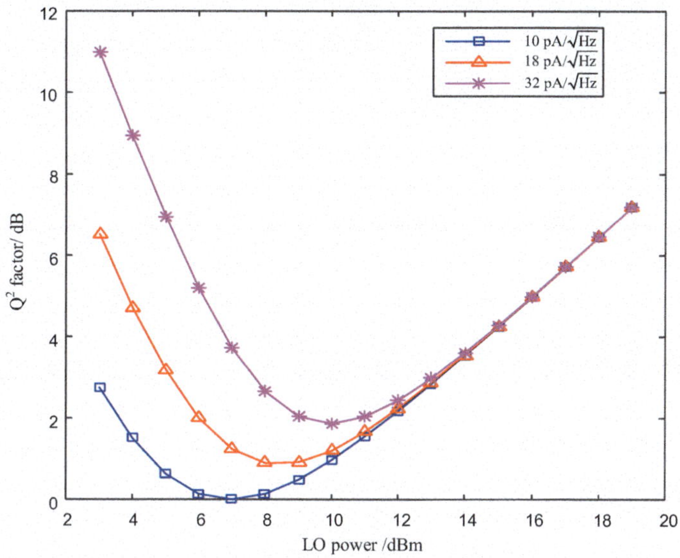

Fig. 5.35 Variation in system performance with respect to the principal vibration power at various values of thermal noise [54]

reference value, the difference between the Q-factor and the reference value as the vertical coordinate, and the principal oscillation power as the horizontal coordinate, the curves depicting the impacts of different degrees of thermal noise on the performance of the system can be obtained, as presented in Fig. 5.36. The initial conditions are as follows: $P_S = -33$ dBm, $\Delta f = 20$ GHz, $e_l = 1$dB, $\eta = 0.75$ A/W, RIN $= -145$ dB/Hz, $\tau = 2$ ps.

Figure 5.35 indicates that with an increase in the local oscillation power, the values of the curves corresponding to different noise current densities tend to be the same. This implies that when the local oscillation light beam becomes gradually larger, the extent to which the system performance is affected by the thermal noise will become smaller; the larger the thermal noise, higher would be the corresponding local oscillation power required for the system to achieve the best performance. This is because the larger the thermal noise, the higher the total noise power at the output of the system. It is necessary to improve the power of the output signal to suppress the impact of thermal noise on the SNR, that is, a higher local oscillation power would enhance the output signal power.

Using Eq. (5.72), we can obtain the system sensitivity drop values corresponding to the various values of the principal oscillation power with respect to the thermal noise input–reference noise current density change curve presented in Fig. 5.37. The initial conditions are $\Delta f = 20$ GHz, $e_l = 1$dB, $\eta = 0.75$ A/W, RIN $= -145$ dB/Hz, and $\tau = 2$ ps.

Figure 5.36 indicates that as the input reference noise current density increases, the degree of sensitivity decreases. The change curve is very steep when the local

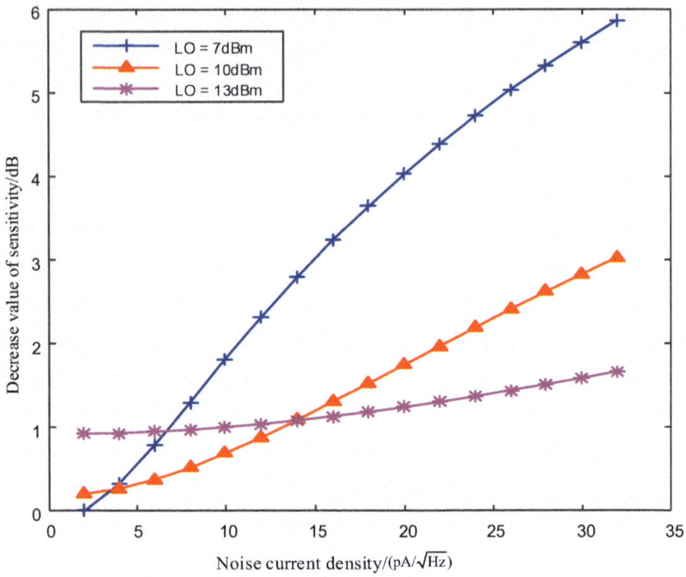

Fig. 5.36 Variation in system sensitivity with respect to thermal noise at various values of principal oscillation power [54]

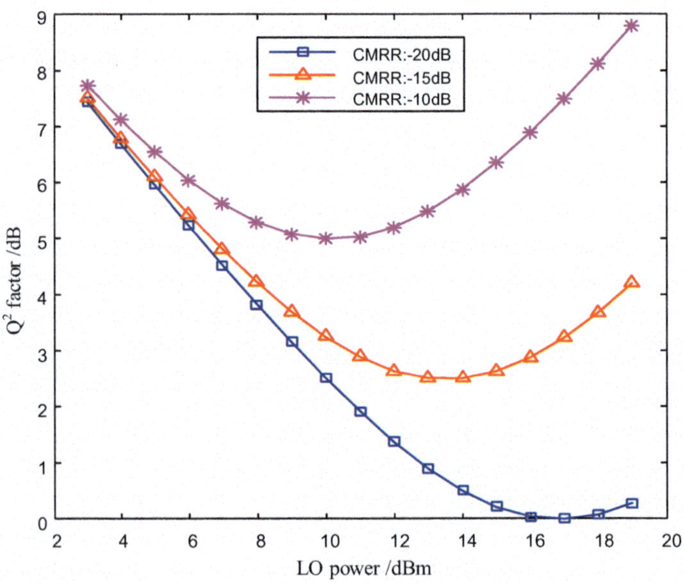

Fig. 5.37 Variation in system performance with respect to LO power at various CMRRs [54]

oscillation power is low, and as the input–reference noise current density increases, the sensitivity drop curve gradually exceeds the curve of lower local oscillation power.

(3) **Effect of CMRR**

The CMRR can effectively suppress the RIN. To obtain a balance between the thermal noise and the remaining RIN with the aim of achieving the best system performance, the factor of CMRR cannot be ignored.

Based on Eqs. (5.72) and (5.74), the curves depicting the effect of various CMRRs on system performance can be obtained as presented in Fig. 5.37. The initial conditions are as follows: $i_{\text{TIA}} = 18.2\,\text{pA}/\sqrt{\text{Hz}}$, $\Delta f = 20$ GHz, $e_{\text{l}} = 1$ dB, $\eta = 0.75$ A/W, RIN $= -145$ dB/Hz, and $\tau = 2$ ps.

Figure 5.37 indicates that the difference among the Q-factor generation values at various CMRRs is smaller when the local oscillation power is low. This is because when the local oscillation power is small, the system is mainly affected by thermal noise, and the effect of the CMRR on the suppression of system RIN performance is not evident. Therefore, when the CMRR of the double-balanced detection system is higher, the amplitude of the output signal can be increased using the higher local oscillation power, whereas the RIN does not increase. However, the excessive remaining RIN increases the total noise power and affects the improvement of the SNR of the system.

Based on Eq. (5.72), the curves corresponding to the decrease in the sensitivity of the system with respect to the CMRR at various values of the principal oscillation power can be obtained, as depicted in Fig. 5.40. The initial conditions are as follows: $i_{\text{TIA}} = 18.2\,\text{pA}/\sqrt{\text{Hz}}$, $\Delta f = 20$ GHz, $e_{\text{l}} = 1$ dB, $\eta = 0.75$ A/W, RIN $= -145$ dB/Hz, and $\tau = 2$ ps.

Figure 5.38 indicates that as the CMRR increases, the degree of decrease in the sensitivity also becomes larger. This happens because when the degree of balance mismatch in the balanced detection system is high, the CMRR also increases, that is, the CMRR characterizes the degree of balance mismatch in the system; the ratio is proportional to the degree of mismatch. Further, the figure also indicates that the slope of the curve increases with an increase in the principal oscillation power.

5.5 Optisystem Simulation of Double-Balanced Detection System

Optisystem is a powerful simulation software used for optical communication systems, including the design, test, and optimization of various types of optical networks, physical layers, virtual optical circuit construction, and so on. Optisystem has a very powerful simulation environment with real devices and hierarchical system definitions, and it can be extended through the additional user device library and complete interface.

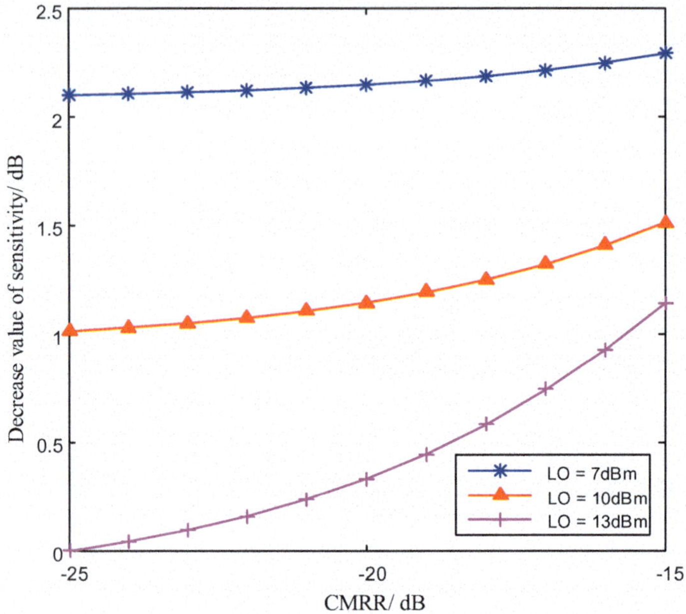

Fig. 5.38 Variation in sensitivity with respect to CMRR at various values of local oscillation power [54]

5.5.1 Simulation of Double-Balanced Detection System

The simulation system was built based on the block diagram of double-balanced detection depicted in Fig. 5.15. Figure 5.39 displays the time domain simulation waveforms of the In-phase and Quadrature output signals of the system, from which it can be seen that the phase difference between the In-phase and Quadrature channel is $\pi/2$. The SNR is 41 dB at this point.

(1) **Impact of quarter waveplate**

When the incident angle between the quarter waveplate and the local oscillator is not 45°, the local oscillator is elliptically polarized. Therefore, the phase difference between the two output signals is no longer $\pi/2$. When the rotation angle of the quarter waveplate is varied from 45° to 30°, the phase difference between the two output signals is again no longer $\pi/2$. The phase difference between the In-phase and Quadrature channel is also no longer $\pi/2$, which is consistent with the theoretical analysis. At this point, the SNR is 38 dB (Fig. 5.40).

(2) **Effect of half waveplate**

The *s*- and *p*-component of the signal and fundamental light beams, respectively, are mixed and coupled into one beam in any balanced detection branch of the double-balanced detection system, the direction of polarization is rotated by 45° following

Fig. 5.39 Time domain waveforms of in-phase and quadrature channel at the receiver [54]

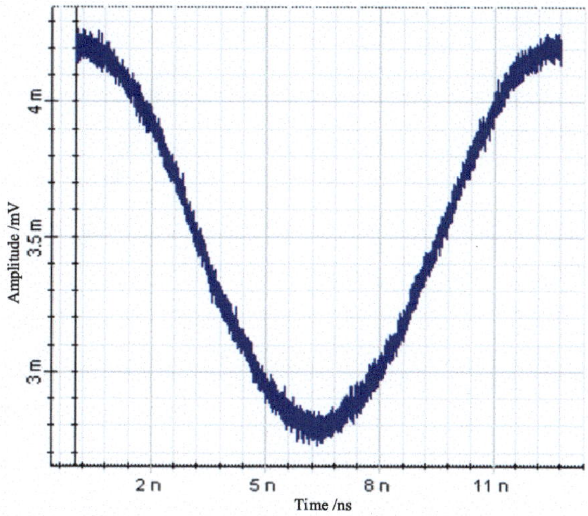

(a) In-phase channel

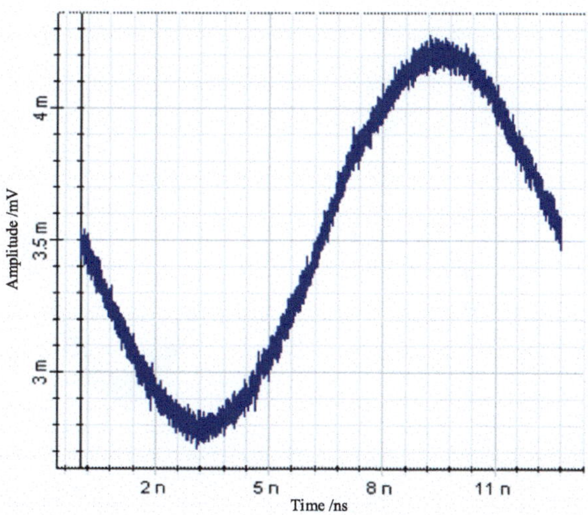

(b) Quadrature channel

the half waveplate with a rotation angle of 22.5°, and, thereafter, the beam is divided into four equal components by the PBS. Therefore, when the rotation angle of the half wavelength is not strictly 22.5°, it will cause the components after beam splitting to be different, and then the final current generated in the PD will not be equal.

When the rotation direction of the half waveplate is varied from 22.5° to 30°, other input conditions remaining constant, the corresponding amplitudes of the I and

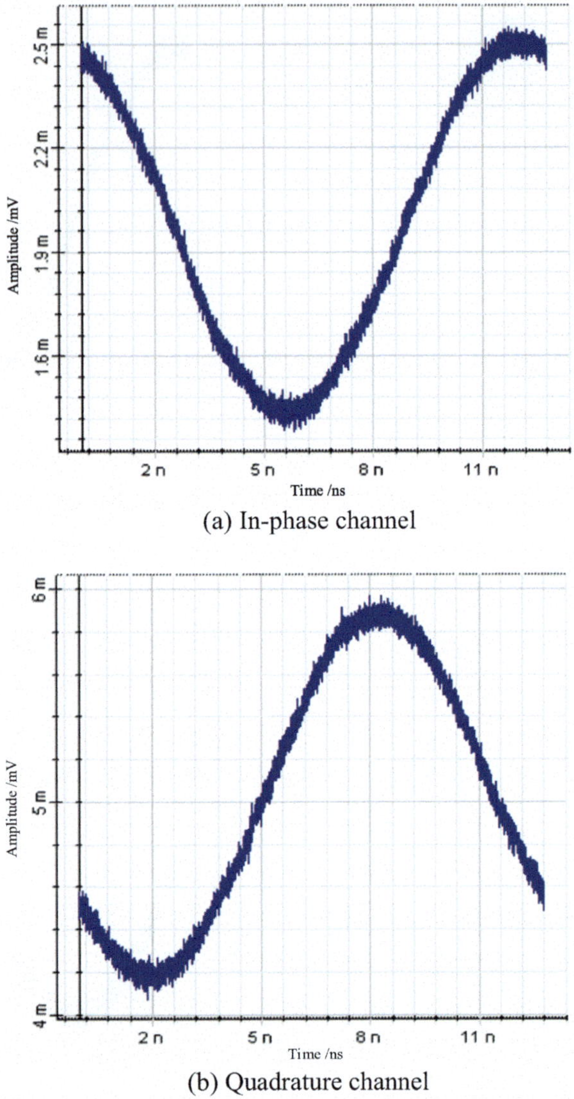

Fig. 5.40 Time domain waveforms of the in-phase and quadrature channel after rotating the quarter waveplate [54]

Q components will be reduced, as depicted in Fig. 5.41. At this point, the SNR is 37 dB.

(3) **Effect of difference frequency**

The difference frequencies of the signal and local oscillation light beams are changed to 150 MHz, other input parameters remaining constant; thus, the time domain waveforms of the In-phase and Quadrature channel change to those displayed in Fig. 5.42. The SNR is 44 dB at this point.

Fig. 5.41 Time domain waveforms of the in-phase and quadrature channel after rotating the half waveform [54]

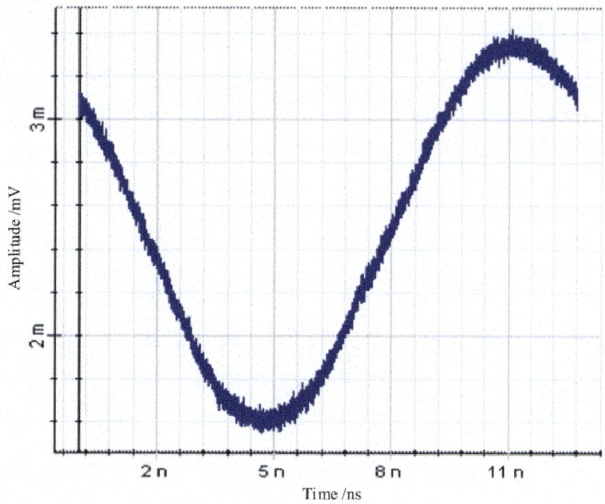

(a) In-phase channel

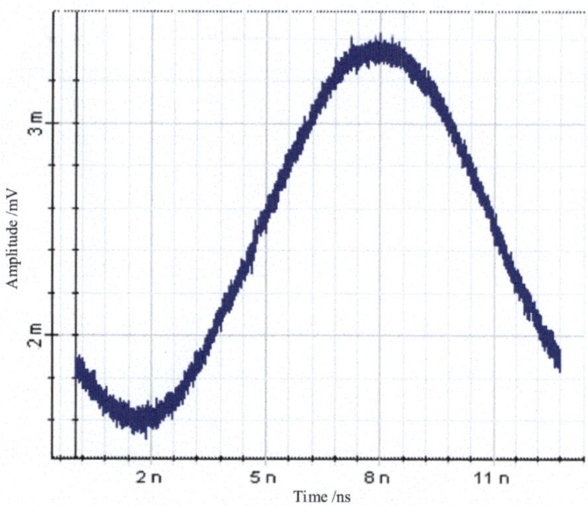

(b) Quadrature channel

Figure 5.42 indicates that when the difference frequency between the signal and local oscillation light beams becomes 150 MHz, the phase and amplitude of the time domain waveform of the output signal remain the same, and the SNR is improved by approximately 3 dB. Double-balanced detection is developed based on outlier detection, and zero-difference detection has a higher SNR than outlier detection. Therefore, the smaller the difference frequency, the closer the system is to zero-difference detection and the higher is the SNR. Thus, the simulation results are basically consistent with the theoretical analysis.

Fig. 5.42 Time domain waveforms of in-phase and quadrature channel after varying the differential frequency [54]

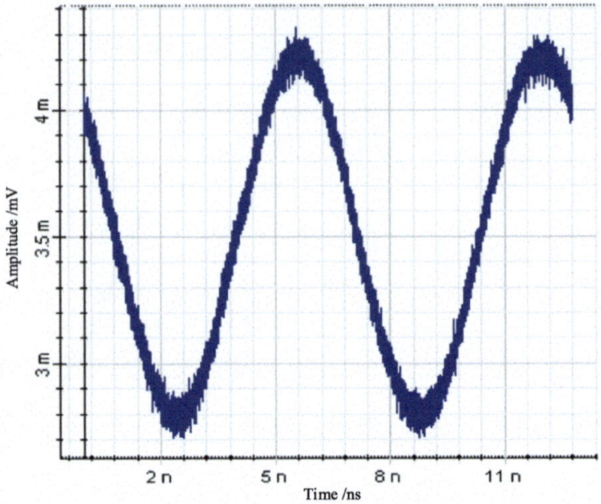

(a) In-phase channel

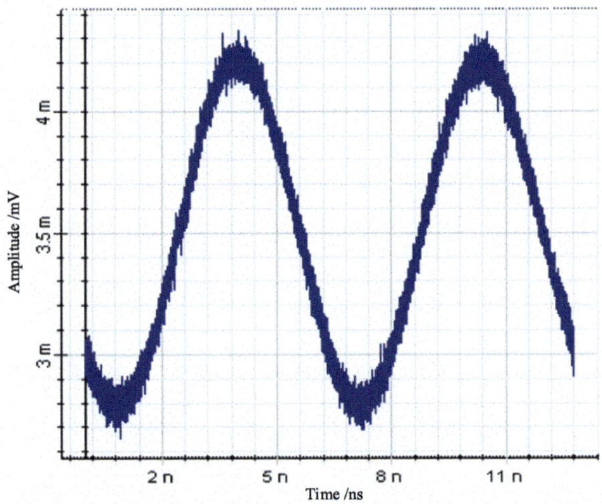

(b) Quadrature channel

5.5.2 *Effect of Power Mismatch on the SNR of Double-Balanced Detection*

In practice, the deviation of the angle between the waveplate and the optical axis, unequal beam splitting by the PBS, and different response of the detector causes unequal response photocurrent, which leads to a power mismatch between the two arms of the balanced detector and affects the SNR of the system.

(1) **Simulation experiment**

In the experiment, an optical attenuator each is added to the two branches of the balanced detector to simulate the unequal photocurrents caused by the various factors, as illustrated in Fig. 5.43.

By adjusting the attenuation coefficients α_1 and α_2 of the two attenuators, the simulated PD produces photocurrents of various magnitudes. When the attenuation coefficient is 1, it means that there is no attenuation in the optical path. Maintaining other device parameters constant, the attenuation coefficients α_1 and α_2 of the two optical attenuators are set to 1, that is, the initial state is that of no attenuation. Thereafter, as α_1 is decreased in steps of 0.2, the variation in the SNR with respect to α_2 from high to low is observed and recorded. That is, we obtain a relationship reflecting the effect of unequal response photocurrent on the SNR of the balanced detection system. Based on the simulation experiment results presented in Fig. 5.44, the following conclusions can be drawn:

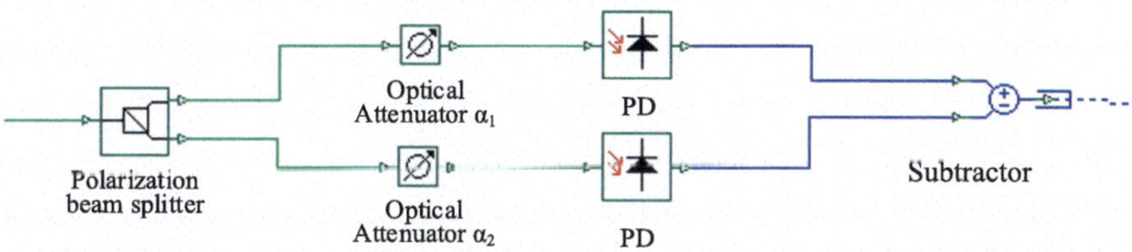

Fig. 5.43 Simulation of the branch of the balanced detector in the case of power mismatch [54]

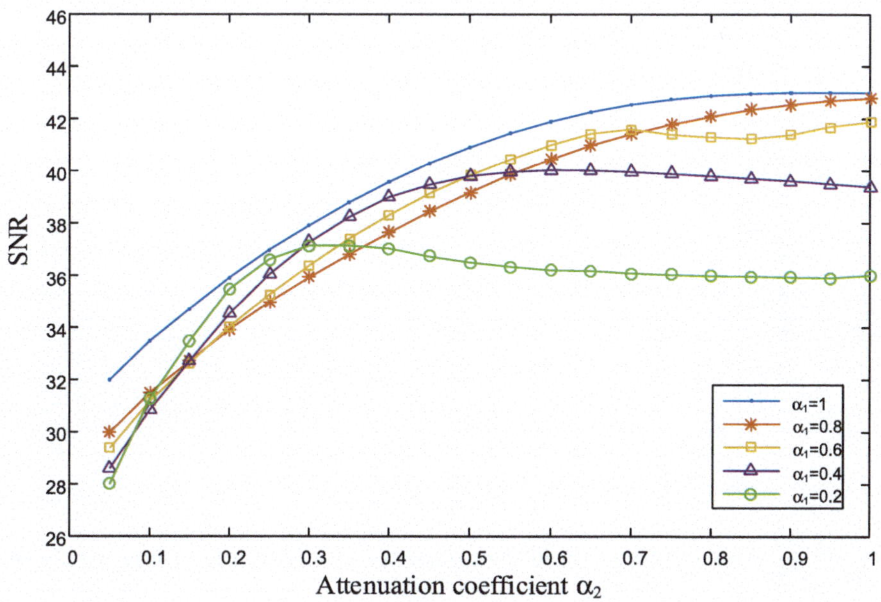

Fig. 5.44 Effect of the attenuation coefficients of the two arms of the balanced detector on the SNR [54]

(1) When the difference between the response currents of the two arms of the balanced detection is not large (for example, $\alpha_1 = 0.8$ in the figure), the SNR decreases continuously as the value of the attenuation coefficient of optical path 2 decreases.
(2) When there is a certain difference between the response currents of the two arms (for example, $\alpha_1 = 0.6$ in the figure), a peak is obtained in the SNR as the value of the attenuation coefficient of optical path 2 decreases.

(2) **Optimal parameter optimization**

As depicted in Fig. 5.45, let $k = I_1/I_2$, representing the consistency of balanced detection, where I_1 and I_2 are the response currents in the two arms of the balanced detector, from which the relationship curve between the coefficient of balanced consistency and the attenuation coefficient can be obtained at the optimum SNR.

Figure 5.45 indicates that, when $k > 0.7$, the optimal attenuation coefficient of optical path 2 is close to 1. The following conclusions can be drawn.

(1) When $k > 0.7$, the SNR of the balanced detection system optical path 2 without attenuation is higher than the SNR when using an attenuator. Therefore, in practice, there is no need to increase the number of attenuation devices in the optical path at this point.
(2) When $k < 0.7$, the SNR of the balanced detection system optical path 2 without attenuation is smaller than the SNR when using an appropriate size of attenuator. Therefore, an attenuation device of appropriate size should be added to optical path 2 to balance the two arms.

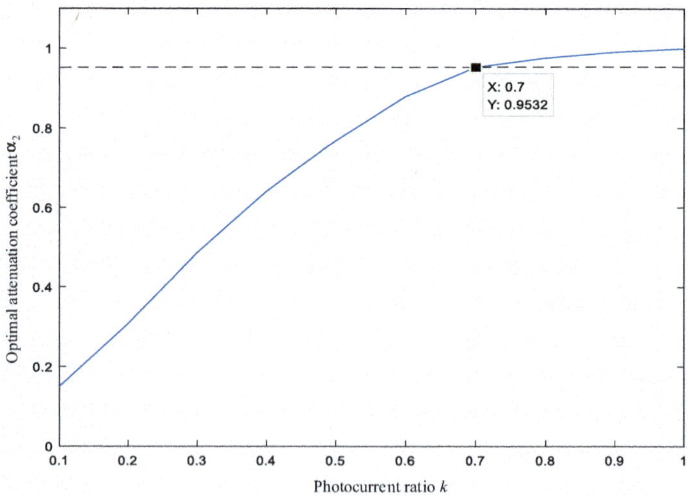

Fig. 5.45 Relationship between equilibrium consistency coefficient and optimal attenuation coefficient [54]

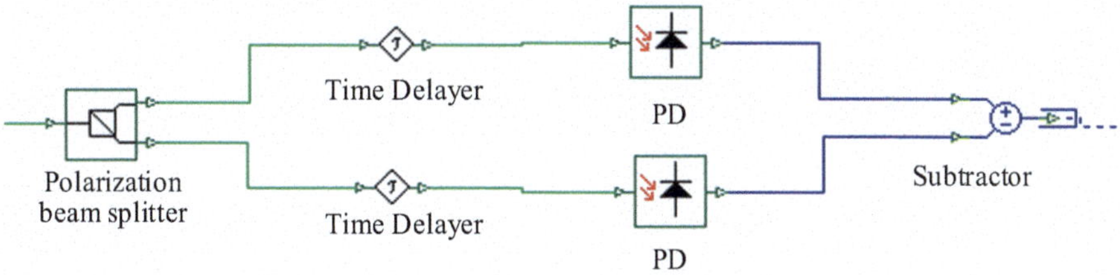

Fig. 5.46 Balanced probe branch with the addition of time delay [54]

5.5.3 Effect of Time Mismatch on SNR of Double-Balanced Detection

The time mismatch in the dual-balanced detection system is caused mainly by the difference in optical range between the PBS and the detector and the different response times of the detector. Assuming that the time mismatch leads to a time delay T, the phase difference between the two IF signals generated by the balanced detector is expressed as $\omega_{IF} \cdot T \cdot 360°/2\pi$, which leads to a phase shift of the RIN in the two detectors, without suppressing the RIN, which affects the SNR of the system.

Figure 5.46 depicts a time delayer in each arm of the balanced detector to simulate the time mismatch; by varying the time delay T, the corresponding phase difference can be obtained.

As depicted in Fig. 5.47 Maintaining other device parameters constant, the initial value of the two times delays is 0. By varying the time delay T all the way in the two arms to adjust the phase difference between the two arms and examining the variation in the SNR with respect to the phase difference from low to high, the relationship between the phase difference between the response currents of the two arms of the balanced detector and the SNR of the balanced detector can be obtained.

Figure 5.47 indicates that the SNR gradually decreases as the phase difference between the two arms of the balanced detection photocurrent increases. When the phase difference is small, the variation in SNR is relatively flat and negligible, whereas when the phase difference is large, the SNR decreases rapidly. Therefore, as long as the phase shift caused by the time mismatch is controlled within a certain range, the communication demand can be satisfied.

When the power mismatch between the two arms of the balanced detector exceeds a certain degree, the SNR of the system can be improved by adding the corresponding attenuation devices to the system to balance the two arms. The time mismatch is caused primarily by the difference in optical range between the PBS and the detector and the different response times of the detector, in practice, as long as the difference in optical range between the two arms can be controlled up to the centimeter level to ignore its impact.

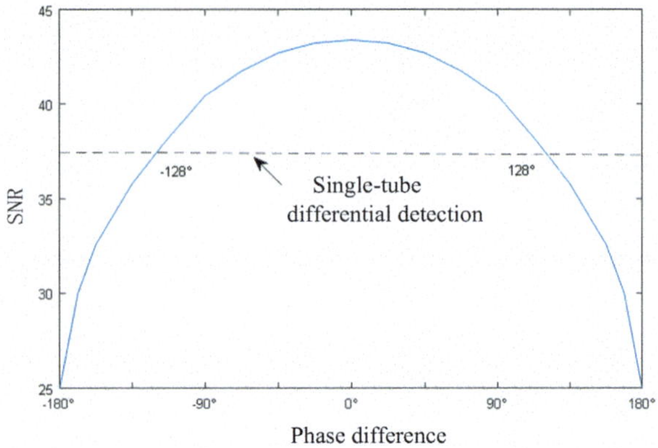

Fig. 5.47 Relationship between the phase difference and SNR between the two arms of the balanced detection photocurrent [54]

References

1. Liu HY, Zhang YG, Ai Y et al (2014) Design and implementation of balance detector used in coherent optical communication system [J]. Adv Lasers Optoelectron 51(7):23–29
2. Chan VWS (2006) Free-space optical communications [J]. J Lightwave Technol 24(12):4750–4762
3. Abbas GL, Chan VWS, Yee T (1985) A dual-detector optical heterodyne receiver for local oscillator noise suppression [J]. J Lightwave Technol 3(5):1110–1122
4. Kuwahara H, Chikama T, Kiyonaga T et al (2015) Research on dual-balanced heterodyne detection of 2 μm coherent lidar for wind measurement [D]. Beijing Institute of Technology, Beijing
5. Caldera C, Bernardi CD, Destefanis G, Meliga M, Morasca S, Rigo C, Stano A (1990) Balanced optical mixer integrated in InGaAlAs/InP for coherent receivers. Coherent Lightwave Commun 1372:82–87
6. Arvizu, Mendieta FJ, Chavez R (1998) Balanced photoreceiver for coherent optical communications [J]. Instrum Dev 3:3–14
7. Yao ZS, Yu H (2005) Progress in inter-satellite coherent optical communications [J]. Opt Commun Technol 29(8):44–46
8. Karafolas N, Sodnik Z, Annengol JMP et al (2009) Optical communications in space [C]. In: International conference on optical network design and modeling, vol 28, no 6. IEEE Press, pp 19–24
9. Painchaud Y, Poulin M, Morin M et al (2009) Performance of balanced detection in a coherent receiver [J]. Opt Express 17(5):59–72
10. Caplan DO, Carney JJ, Fitzgerald et al (2014) Multi-rate DPSK optical transceivers for free-space-application [C]. In: SPIE, San Francisco, CA, USA, 2014, 8971:8971K
11. Nielsen TT (2002) In orbit test result of an operational optical intersatellite link between ARTEMIS and SPOT4 SILEX. Proc SPIE 4638(68):1–15
12. Lange R (2008) In-orbit verification of optical inter-satellite communication links based on homodyne BPSK [J]. Proc SPIE Int Soc Opt Eng 6877:687702-687702-6
13. Pribil K (1998) Laser communication terminals: a key building block for the new broadband satellite networks. In: Broadband European networks and electronic image capture and publishing. International Society for Optics and Photonics, Switzerland, pp 172–177
14. Boroson DM (1993) LITE engineering model-I: operation and performance of the communications and beam-control subsystems [C]. In: Optics, electro-optics, & laser applications in science & engineering. International Society for Optics and Photonics, CA, USA, pp 73–82
15. Pillsbury AD (1993) LITE engineering model-II: the flight-qualified subsystems [J]. Proc SPIE 1866:58–72

16. Fujiwara Y, Mokuno M, Jono T et al (2007) Optical inter-orbit communications engineering test satellite (OICETS) [J]. Acta Astronaut 61(1):163–175
17. Yamakawa S, Hanada T, Kohata H (2011) R&D status of the next generation optical communication terminals in JAXA [C]. In: International conference on space optical systems and applications. IEEE, Santa Monica, CA, USA, pp 389–393
18. Man JL (1990) Research on digital broadband frequency stabilization system and balanced detection technique in coherent fiber optic communication [D]. Chengdu Institute of Telecommunications Engineering, University of Electronic Science and Technology
19. Chen Y (1991) An experimental study on a balanced receiver in coherent optical fiber communications [D]. Beijing University, Beijing
20. Pei AH, Zhang SR (1993) Research on a balanced detector of coherent optical fiber communication system [J]. J Changchun Inst Opt Precis Mach 16(4):12–16
21. Chen SQ, Guan KJ (1995) Study on polarization independent heterodyne balanced receiver [J]. J Beijing Univ Posts Telecommun 2:68–72
22. Gao L, Wang CH, Li YC et al (2010) Theory analysis of polarization mixing error in dual-balanced heterodyne detection [J]. Infrared Laser Eng 03:422–426
23. Yang YL, Li YC, Gao L et al (2011) Numerical simulation of balanced heterodyne detection for coherent lidar [J]. Infrared Laser Eng 40(10):1918–1922
24. Chen JN, Ke XZ (2012) Research on factors and key techniques of space optical heterodyne detection [J]. Semicond Optoelectron 33(4):548–552
25. Pang YJ, Wang CH (2012) Principle of dual-balanced coherent receive and its application in heterodyne interferometer [J]. J Photonics 41(9):1015–1018
26. Samuel Y (2013) Simulation and study for dual balanced heterodyne laser detection system based on the 0 pti system [J]. Software 34(4):132–134
27. Liu HY, Zhang YG, Ai Y et al (2015) Research on balanced detection technique for high-speed weak optical signals [J]. Laser Technol 39(2):182–184
28. Dai YH, Ai Y, Xiao W et al (2015) Impact of local oscillator power on SNR in space coherent optical communications [J]. J Photonics 44(1):173–179
29. Kong YS, Ke XZ, Yang Y (2016) Effect of local oscillation optical power on signal-to-noise ratio in spatially coherent optical communication [J]. Infrared Laser Eng 45(2):234–239
30. Zheng Y, Jiang HL, Tong SF et al (2013) Establishment of space light mixer mathematical model based on the space coherent laser communication [J]. J Opt 33(7):142–147
31. Seimetz M, Weinert CM (2006) Options, feasibility, and availability of 2 × 4 90° hybrids for coherent optical systems [J]. J Lightwave Technol 24(3):1317–1322
32. Han Y, Liang ZC, Zhai YM et al (2014) Polymer PLC-based hybrid integrated coherent receiver [J]. Optical Commun Res 40(5):55–58
33. Zhai YM, Wang J, Xu J et al (2015) Large-tolerance coherent optical mixer based on polymer multimode interferometer [J]. Photonics-Laser 14(2):233–238
34. Sun YL, Jiang XQ, Wang MH (2014) Analysis of general self-imaging positions in multimode interference couplers [J]. J Opt 24(6):759–762
35. Jeong SH, Morito K (2010) Compact optical 90° hybrid employing a tapered 2 × 4 MMI coupler serially connected by a 2 × 2 MMI coupler [J]. Opt Express 18(5):4275–4288
36. Hertz H (1989) Compact bulk optical 90° hybrid for balanced phase diversity receivers [J]. Electron Lett 25(22):1518–1519
37. Seery BD (1990) Experimental determination of power penalty contributions in an optical Costas-type phase-locked loop receiver [C]. International Society for Optics and Photonics, Los Angeles, CA, USA, pp 10–18
38. Cho PS, Harston G, Kerr CJ et al (2004) Improvement of coherent homodyne detection performance using time-gated amplification and LiNbO3 optical 90° hybrid [C]. In: SPIE, Orlando, FA, USA, vol 5403, pp 762–773
39. Su SD (2010) Electro-optical effects of lithium niobate crystals under non-near-axis propagation and its application in spatial optical bridges [D]. Guangxi University
40. Liu LR, Yan AM, Luan Z et al (2006) Birefringent free-space optical bridge [P]. Chinese patent, CN200610026338. X

41. Hou PP, Zhou Y, Zhi YN et al (2010) Free space optical 2 × 4 90 hybrid based on crystal birefringence [J]. J Opt 30(12):3413–3418
42. Wan LY, Su SD, Liu LR et al (2009) Design of a new 90′2 × 4 hybrid based on the birefringence and the electro-optic effect of crystal [J]. China Laser 36(9):2358–2361
43. Wang CH, Gao L, Pang YJ et al (2011) Experimental investigation for relation between beam splitter coefficient and signal-to-noise ratio of 2 um balanced coherent system [J]. J Opt 12(11):53–58
44. Li Y, Zhou J, Nie XM et al (2016) Research of laser velocimeter balanced detection system [J]. Adv Laser Optoelectron 16(2):52–57
45. Alexander S (1987) Design of wide-band optical heterodyne balanced mixer receivers [J]. J Lightwave Technol 5(4):523–537
46. Shi QY, Ai Y, Liang HX et al (2016) Analysis and test of balanced detector in coherent optical communication [J]. Sci Technol Eng 16(16):207–211
47. Kikuchi K (2008) Coherent optical communication systems [M]. Academic Press, Manhattan, pp 95–129
48. Jin L, Zhu YG (2016) Beam-splitting ratio impact on the SNR for the balanced heterodyne [J]. Infrared Phys Technol 76:122–125
49. Gao L, Wang CH, Li YC et al (2010) Theory analysis of polarization mixing error in dual-balanced heterodyne detection [J]. Infrared Laser Eng 39(3):422–426
50. Zhang P, Qin KY, Jiang DG et al (2016) Coherent satellite reception performance under optical front-end balance mismatch [J]. J Univ Electron Sci Technol 45(2):306–311
51. Painchaud Y, Poulin M, Morin M et al (2009) Performance of balanced detection in a coherent receiver [J]. Opt Express 17(5):3659–3672
52. Takara H, Shake I (2002) Averaged Q-factor method using amplitude histogram evaluation for transparent monitoring of optical signal-to-noise ratio degradation in optical transmission system [J]. 20(8):1367–1373
53. Dai YH, Liu YF, Zhou HT et al (2015) Sensitivity testing and experiment of balanced photodetectors in the space coherent optical communication [J]. Infrared Laser Eng 44(10):3110–3116
54. Liu X (2017) Research on double balance detection technology in coherent optical communication [D]. Xi'an University of Technology, Xi'an

Chapter 6
Adaptive Optics Correction

Contents

The application of the adaptive optics (AO) technology to the atmospheric laser communication system is discussed in this chapter. The working principles of the classical AO structure, wavefront sensor, and deformable mirror are introduced. Finally, the AO system is analyzed based on the characteristics of the coherent optical communication system and validated through a field experiment.

X. Ke and J. Wu, *Coherent Optical Wireless Communication Principle and Application*, Optical Wireless Communication Theory and Technology,
https://doi.org/10.1007/978-981-19-4823-7_6

6.1 Research Status of Adaptive Optics System

In 1953, American astronomer Babcock first proposed the basic concept of AO [1]. He proposed the use of real-time measurement and compensation correction methods on ground-based astronomical telescopes to suppress dynamic wavefront distortion caused by atmospheric turbulence. The unique feature of AO is its ability to compensate for wavefront distortion caused by dynamic disturbance of light waves. AO revolutionizes the passive working mode of the traditional optical system, which is ineffective in the presence of atmospheric turbulence, and gives the optical system dynamic and variable characteristics [2]. Another feature of AO is its real-time applicability [3]. The AO system can measure the distortion caused by the dynamic disturbance of the light wavefront in real time and feed this information to the wavefront corrector so that the wavefront distortion can be corrected simultaneously and the beam quality close to the diffraction limit can be obtained at the receiving antenna.

Owing to the irregularity of atmospheric turbulence in time and space, severe wavefront distortion occurs when a laser signal is transmitted through the atmosphere to the optical receiving end. The degree of distortion is closely related to the intensity of atmospheric turbulence. In the United States, Tyson [3] studied the ability of AO to improve the bit error rate of space optical communication systems under various turbulence conditions. The results indicate that the AO technology can significantly improve the communication performance under moderate and weak turbulence and increase the bit error rate of the communication system by several orders of magnitude.

In 1972, researchers in the United States developed a real-time atmospheric compensation imaging experimental system [4] and used it to detect and compensate for atmospheric turbulence disturbances at a level of 300 m. The compensated image resolution was close to the diffraction limit. In 1982, a 1.6 m telescope of the U.S. Air Force at the Maui Optical Station in Hawaii realized the observation of space targets operating in the low-earth orbit. An interferometer was used as a wavefront sensor to perform atmospheric turbulence distortion compensation and correction experiments [5].

In 1999, the Communication Research Laboratory of the Ministry of Posts of Japan used an AO system to perform a laser communication demonstration experiment between a ground station and the International Space Station. The results indicate that only low-order aberrations can be corrected to a significant extent. The uplink improved the coupling efficiency of the downlink and suppressed the beam drift in the uplink [6]. In 2002, Lawrence Livermore National Laboratory used the AO technology for numerical simulations of optical communication with a slope of 28.2 km near the ground. Under moderate turbulence conditions, AO was corrected. The fiber coupling efficiency was increased from 20 to 55% [7]. In 2004, Weyrauch of the University of Maryland performed a 2.3 km atmospheric horizontal transmission communication experiment and performed tilt and atmospheric aberration corrections. Their results indicate that the maximum intensity of the signal obtained by adopting the AO technology is 2.3 times that when only the tilt aberration is corrected [8].

In 2005, the Jet Propulsion Laboratory (JPL) used AO to compensate for the wavefront distortion caused by the atmospheric turbulence in the received beam in a satellite-to-ground optical communication link. The receiving aperture used in the experiment was 0.3 m, and the communication wavelength was 633 nm. After the AO closed-loop correction, the wavefront distortion root mean square (RMS) was reduced from 1.4 to 0.05 λ, and the Strehl ratio (SR) increased from 0.08 to 89% [9].

In 2008, the JPL studied the impact of AO systems with different correction capabilities on the communication system using an AO experimental platform for atmospheric laser communication. Experimental results indicate that the signal received by the coherent receiver after the AO closed-loop correction under strong turbulence conditions has a gain of at least 56 dB, and the bit error rate of the system is reduced by at least three orders of magnitude [10].

In 2008, the Johns Hopkins University Applied Physics Laboratory (JHU/APL) and AOptix Technology Co., Ltd. jointly demonstrated a 147 km long-distance two-way RF and FSO communication experiment using OOK modulation. The transmission rate reached 10 Gbit/s [11]. In this experiment, the transmitting end pre-correction and receiving end correction were used to improve the SR and single-mode fiber coupling efficiency, respectively. The anamorphic mirror had 35 drivers, and the control bandwidth was >1 kHz.

In 2009, the European Space Agency designed an AO system for the satellite-to-ground laser communication ground station in Spain. The installation and commissioning were completed in May 2010. The system adopted multi-core CPU control and was equipped with optional Field Programmable Gate Array (FPGA) components. The operating frequency of this system can be as high as 2000 Hz [12].

In 2014, Hashmi of the National University of Sciences and Technology, Pakistan, performed a numerical simulation experiment of inter-satellite AO communication in the laboratory [13]. His experimental results indicate that the SR of the system subject to noisy and strong turbulence conditions can be increased from 0.30 to 0.75 with the AO closed-loop correction.

In 2015, the JPL completed the test of the AO system in laser communication system prototype. In this test, video data were sent from the International Space Station to a ground station in California. Each test lasted 2.5 min, and the total test period was up to 3 months. The test results indicate that the AO closed-loop correction can improve the fiber coupling efficiency and compensate for the impact of atmospheric scintillation on the optical signal [14].

In 1979, the Key Laboratory of Adaptive Optics of the Chinese Academy of Sciences performed research in this field [15]. In 1985, the Chengdu Institute of Optoelectronics Technology, Chinese Academy of Sciences, developed a laser wavefront correction system for "Shenguang I" laser fusion device using the "hill-climbing method" optimization principle. After correction, the focal spot energy level was increased by three times [16].

In 1990, the Chengdu Institute of Optoelectronics Technology, Chinese Academy of Sciences successfully researched 21-unit dynamic wavefront error correction system and applied it to a 1.2 m astronomical telescope at Yunnan Observatory. Real-time atmospheric wavefront correction was achieved for the entire range of stellar targets and obtained resolution, and obtaineda clear photo of the double star [17].

In 2009, the Changchun Institute of Optics and Mechanics made a breakthrough in the 21-element deformable mirror manufacturing technology. In 2010, 97-element deformable mirror was developed, and 137-element deformable mirror was designed in 2011. These mirrors were successfully applied to a 1.23 m telescope. After AO correction, the half-height width of the image can be of high resolution [18].

In 2009, Xia performed the wavefront correction experiment of atmospheric optical communication based on AO [19]. The experimental results indicate that after applying AO technology to atmospheric optical communication for wavefront correction, the higher communication quality can be achieved with a smaller emission power.

In 2010, Han used the AO technology to compensate for distortion in the free-space optical communication system, and analyzed the bit error rate of the communication system [20].

In 2011, the Institute of Optoelectronics Technology of the Chinese Academy of Sciences and Wuhan University used existing AO systems and optical communication equipment to jointly perform 2.5 Gbit/s high bit rate atmospheric optical communication AO correction simulation turbulence experiments. The experimental results indicate that the use of optical correction can help suppress deep fading significantly, improve the quality of the optical communication system, and reduce the light intensity fluctuation [21].

In 2014, Liu experimentally analyzed the impact of the AO technology on the mixing efficiency of a coherent optical communication system [22]. The experiment used a communication wavelength of 1550 nm. The results indicate that the RMS of the wavefront is less than $\lambda/4$, the coherent mixing can be greater than 40% after the AO closed up. It is proved that AO can effectively improve the performance of a coherent optical communication system and reduce its bit error rate.

6.2 Adaptive Optics System in Coherent Optical Communication

6.2.1 Principles of Adaptive Optics

As depicted in Fig. 6.1, an AO system mainly comprises three basic components: the wavefront detector, wavefront controller, and wavefront corrector. Owing to the effect of atmospheric turbulence, wavefront distortion occurs in light when it propagates in the atmosphere. This reduces the beam quality. The wavefront detector can detect the phase of the distorted wavefront in real time, and the wavefront controller can calculate the phase based on the distorted wavefront information. The wavefront

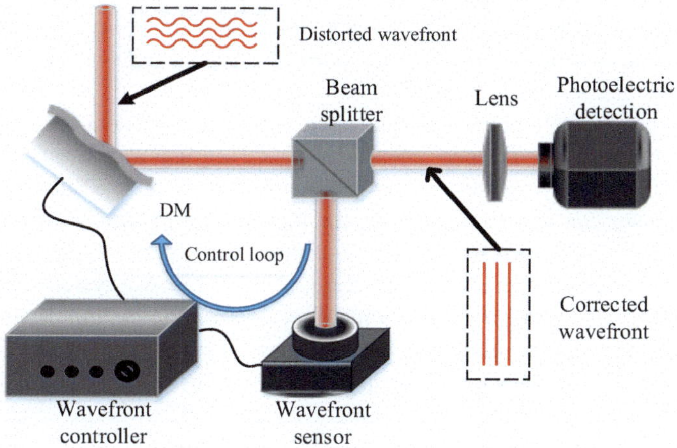

Fig. 6.1 Structure of adaptive optics system

corrector compensates the wavefront error in real time based on the applied voltage and improves the beam quality.

The basic principle of the AO technology is based on the phase conjugation technology [23]. For the light field with phase error

$$E_1 = E_0 e^{i\varphi} \tag{6.1}$$

where E_0 is the optical field amplitude, φ is the wavefront phase of the beam distortion caused by atmospheric turbulence. The function of the wavefront correction system is to produce a wavefront that is conjugated to its phase:

$$E_2 = E_0 e^{-i\varphi} \tag{6.2}$$

Phase error can be compensated by superimposing two light fields. The AO system is a compensation system, which can change the optical path of the distorted wavefront in real time through the wavefront corrector to correct the phase of the wavefront.

6.2.2 *Wavefront Sensor*

The wavefront sensor is the eye of the AO system. It is used to detect the wavefront distortion of the servo loop of the system. It measures the phase distortion of the optical wavefront on the entrance pupil of the system in real time and provides a real-time voltage control signal to the wavefront corrector. The system obtains an image close to the diffraction limit after closed-loop correction. To correct the wavefront distortion caused by atmospheric turbulence, the spatial and temporal resolutions of the wavefront sensor must match the temporal and spatial scales of the disturbance signal.

The commonly available wavefront sensors include the curvature [24], grating shear interference [25], reference [26], and Shaker-Hartman wavefront sensors [27]. A comparison of these wavefront sensors is listed in Table 6.1. The Shaker-Hartmann wavefront sensor can intuitively display the intensity and phase distribution information of wavefront distortion, and it has the characteristics of simple operation and high real-time detection accuracy. Therefore, it is used in AO systems, laser beam quality diagnosis, and optical components. It is also used in system detection, laser pulse wavefront shaping, atmospheric disturbance measurement, and other fields. Figure 6.2 depicts the detection principle diagram of the Shaker-Hartmann wavefront sensor, which is composed of a microlens array and a charge coupled device (CCD). The microlens array divides a spot into multiple tiny sub-spots, and each sub-spot is focused on the focal plane by a corresponding microlens and finally imaged onto the CCD detection target surface. The distortion in the wavefront is estimated by comparing the actual focus position of the sub-aperture with the ideal focus position. Based on the definition of the spot centroid, the coordinate position of the spot centroid in the case of discrete sampling is

$$x_c = \frac{\sum x_i E_i}{\sum E_i} \tag{6.3}$$

$$y_c = \frac{\sum y_i E_i}{\sum E_i} \tag{6.4}$$

where E_i is the light energy signal received by the first CCD pixel, (x_i, y_i) is the coordinate of the first pixel, and (x_c, y_c) is the centroid coordinate of the sub-spot.

Prior to measuring the deformed wavefront, the wavefront sensor must be calibrated using aberration-free parallel light and the ideal centroid of the sub-spots of each microlens must be recorded, as depicted in Fig. 6.2a, by comparing the centroid positions of the sub-spots. The standard centroid position can help calculate the average slope of the wavefront within the sub-aperture. The relationship between the wavefront aberration in the sub-aperture and the slope of the wavefront detected by the Shaker-Hartman detector is as follows:

$$\mathrm{S}_x = \frac{1}{A_s} \iint\limits_{A_s} \frac{\partial \phi(x, y)}{\partial x} \mathrm{d}x\mathrm{d}y = \frac{x_c - x_{c0}}{f} \tag{6.5}$$

$$\mathrm{S}_y = \frac{1}{A_s} \iint\limits_{A_s} \frac{\partial \phi(x, y)}{\partial y} \mathrm{d}x\mathrm{d}y = \frac{y_c - y_{c0}}{f} \tag{6.6}$$

Table 6.1 Comparison of four widely used wavefront sensors

Wavefront sensor	Output data type	Principle	Advantage	Disadvantage
Transverse shear interferometer	Slope	Wavefront shearing interference pattern produced by grating diffraction effect	High signal-to-noise ratio	Low utilization rate of light energy; 2π uncertainties exist
Shack-Hartmann	Slope	Microlens array focuses the incident wavefront on the photosensitive surface of the CCD to form a spot array image and calculate the centroid offset of each sub-aperture spot	High utilization rate of light energy; there is no uncertainty 2π; detection range is large;	The sub-aperture size limits the spatial resolution. When using the mode method for wavefront reconstruction, the optimal reconstruction order needs to be selected
Point diffraction interferometer	Phase	The measured beam is focused on a translucent mask with a pinhole in the center, and the phase information of the measured beam is contained in the interference pattern of the measured wave surface passing through the mask and the reference spherical surface produced by the pinhole diffraction	The requirement for coherence is not high; it is a common optical-path-type interferometer; the reference light comes from itself; the anti-interference performance is good	Low light source utilization
Wavefront curvature sensor	Curvature	Obtain the desired signal by comparing the light intensity distribution on the two symmetry planes behind the focal front	Can directly control the amount of deformation of the deformable mirror, with good real-time performance and low price	Low measurement accuracy

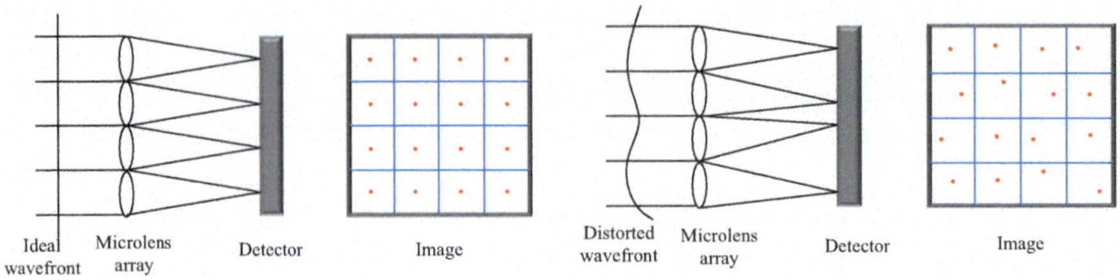

(a) Ideal wavefront (b) Distorted wavefront

Fig. 6.2 Principle of Shark-Hartman wavefront detector

where (x_{c0}, y_{c0}) is the ideal center of mass of the wavefront sensor, f is the focal length of the microlens, and A_s is the area of the microlens.

6.2.3 Working Principle of Wavefront Corrector

The wavefront corrector achieves the objective of correcting wavefront distortion by changing the optical path difference of the beam. Presently, wavefront correctors based on the movement of the mirror surface, known as deformable mirrors [28], are widely used. This type of the wavefront corrector has excellent characteristics of fast response speed, large deformation displacement, and working spectrum bandwidth. The wavefront corrector based on the refractive index change of the transmission medium, such as the liquid crystal spatial light modulator, has a high spatial resolution; however, its light energy utilization rate is low, response frequency is low, correction dynamic range is small, and applicable band is narrow [29]. The phase conjugation phenomenon of nonlinear optics can automatically correct the wavefront distortion of light waves, such as phase conjugate nonlinear optical crystals [30] and so on. The deformable mirror is composed of a deformable mirror and driver. The control signal changes the displacement of the mirror by controlling the movement of the actuator and corrects the wavefront distortion by changing the optical path of the incident light. The number of actuators of the deformable mirror directly determines the ability of the AO system to adjust the wavefront distortion. The higher the number of actuators, the stronger the ability of the system to compensate for the distorted wavefront. Figure 6.3 depicts the arrangement and sequence diagram of the actuators of a deformable mirror. Its theoretical model is as follows:

$$\phi(x, y) = \sum_{k=1}^{l} V_k \cdot f_k(x, y) \tag{6.7}$$

where l is the number of deformable mirror drivers; V_k is the control voltage of each driver; $f_k(x, y)$ is the shape response function of the deformable mirror, which is

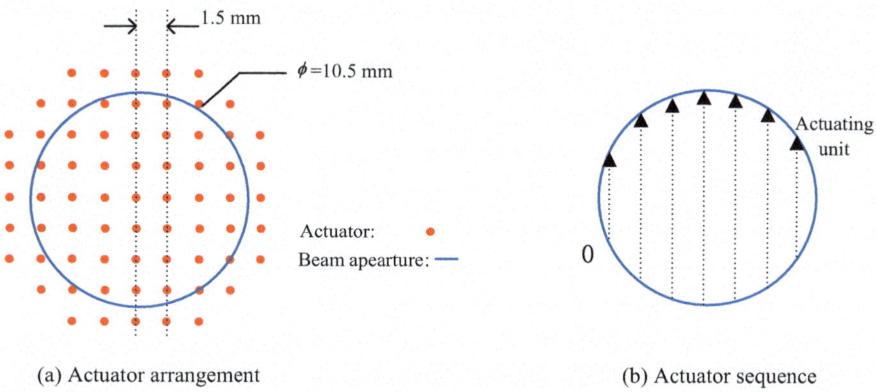

Fig. 6.3 Arrangement and sequencing of DM69 actuators

generally regarded as a Gaussian function [31]:

$$f(x, y) = \exp\left[\ln \omega(\sqrt{(x - x_i)^2 + (y - y_i)^2}/d)^{\alpha}\right] \tag{6.8}$$

where (x_i, y_i) is the center coordinate position of the first actuator of the deformable mirror; α is the Gaussian function index; $d = 1.5$ mm is the distance between the actuators of the deformable mirror; ω is the cross-linking value of the driver, generally takes a value in the range 0.05–0.12.

6.3 System Error Analysis

6.3.1 Error Analysis of Adaptive Optics System

a. **Principle of non-common optical path error formation**

Figure 6.4 depicts the AO optical path based on a coherent optical communication system. It is evident from the figure that the light output from the deformable mirror first divides the dichroic beam into two paths: one path enters the wavefront sensor for wavefront detection and the other enters the optical communication terminal through the convergent lens. Owing to the difference between the measured and communication optical paths, the different optical components contained therein lead to different aberrations. This, in turn, leads to the wavefront detected by the wavefront sensor and the wavefront entering the optical communication terminal being inconsistent. This difference is the non-common optical path error [32]. The AO system cannot measure and compensate for the optical path difference of the communication optical path in the actual working process, and the closed-loop control of the wavefront detector

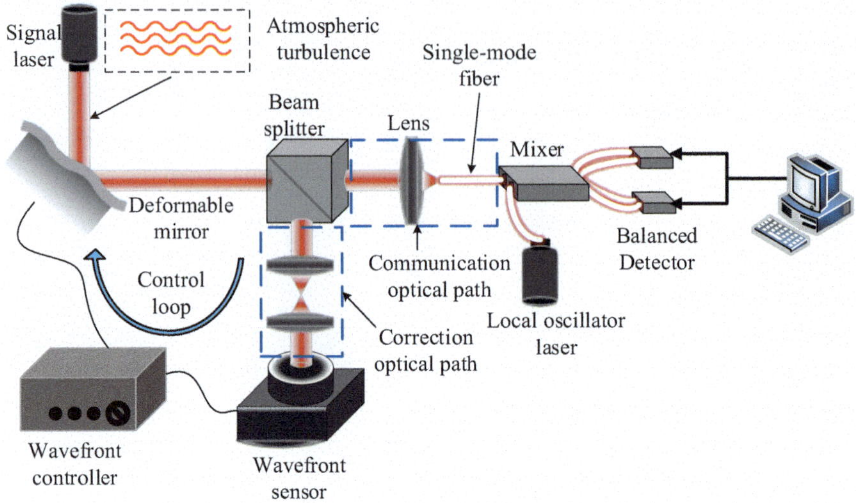

Fig. 6.4 Optical path of adaptive optical system based on coherent optical communication system

and corrector will add static aberration of its own optical path when correcting the wavefront distortion of the optical path, this part of the aberrations must be corrected.

b. **Effect of non-conjugate aberration on the system**

In AO systems, the commonly used 4*f* optical system makes the entrance pupil of the system and the wavefront detector and deformable mirror in a conjugate position, so as to ensure the consistency in detection, correction and communication wavefront. The two lenses in the ideal 4*f* system can transform the plane wave into a strictly spherical wave without introducing other additional aberrations. Therefore, using the optical imaging theory, it is easy to deduce that the conjugate position can completely reproduce the wavefront. However, owing to the complexity of the system, it is often difficult to ensure that the wavefront sensor and the deformable mirror are in a strict conjugate state during the installation and debugging process. Suppose the entrance pupil plane of the system is at position A, the wavefront to be corrected is ϕ_0, the strict conjugate position of A through the system is at B, and the wavefront is ϕ_1 at distance L from the conjugate position B' (see Fig. 6.5). At point ϕ_1' where the wavefront changes, the difference between wavefront ϕ_1' at point B' and wavefront ϕ_1 at point B is the phase error caused by non-strict conjugate.

The non-common optical path error and non-conjugate static aberration together constitute the static error of the system. The static aberration of the system is added to the communication optical path, resulting in the uncertainty of the wavefront of the communication optical path, which affects the communication quality of the communication system and must be corrected. Because of the static error of the system, when the system is closed, the optical power coupled into the single-mode fiber decreases. Figure 6.6 depicts the optical power variation curve of 1550 nm

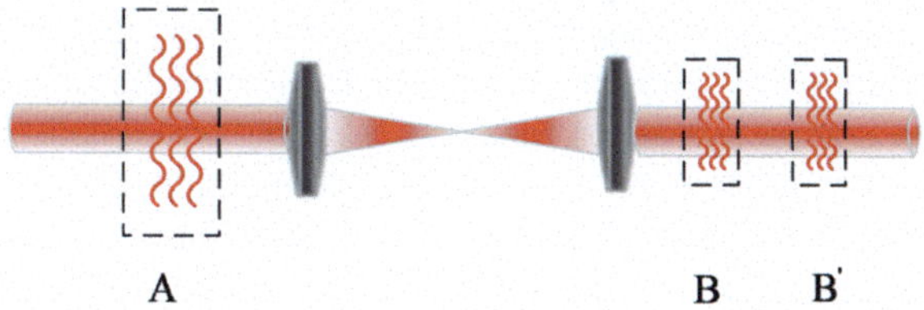

Fig. 6.5 Schematic of wavefront conjugation

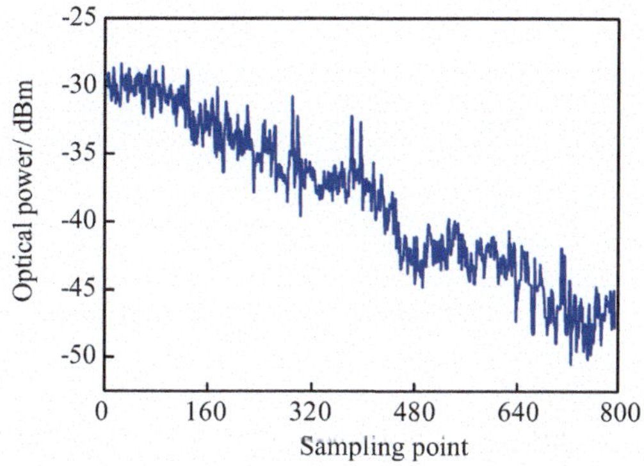

Fig. 6.6 Optical power change during the AO closed-loop process when non-common optical path error is not suppressed

signal light coupled into a single-mode fiber after the system is closed in a certain experiment. It is evident from the figure that the optical power drops significantly when the system is closed because the system static error is not suppressed.

6.3.2 *Methods to Suppress Systematic Errors*

The method uses optical power coupled into the single-mode fiber or the peak-to-peak value of the intermediate frequency voltage as the evaluation index of the system. First, the initial control voltage of the deformable mirror is set to zero, and a random disturbance voltage is applied to the actuator of the deformable mirror. The optimal algorithm determines the initial value of the control voltage of the deformable mirror and the reference wavefront of the system. The basis for the determination is as follows: the intermediate frequency voltage value (or the optical power coupled into the single-mode fiber) reaches the threshold or maximum value set by the system, and the iteration is terminated by the process. Before the end of the iteration, the Hartmann wavefront sensor is used to measure the wavefront. The measured wavefront is the reference wavefront of the subsequent wavefront correction system. At this time, the voltage value applied by the deformable mirror is the initial value of the system.

The error suppression module of the AO system based on random iteration mainly includes the following: deformable mirror, fiber coupling module, intermediate frequency voltage acquisition module (including mixer, double-balanced detector, intermediate frequency voltage peak-to-peak detector) or photoelectric detection module, and deformable mirror. Figure 6.7 depicts a schematic of the error suppression process with the peak-to-peak value of the intermediate frequency voltage as the system evaluation index. Its working principle is as follows: after the distorted wavefront is coupled into the single-mode fiber through the deformable mirror, it enters into coherent mixing with the local oscillator laser in the mixer. The double-balanced detector outputs I and Q intermediate frequency signals. The output intermediate frequency signal is detected by the envelope detector, A\D sampled, and transmitted to the initialization module of the control server through RS232. The balance detector, envelope detector, and A\D together constitute the performance evaluation index feedback module of the system. Figure 6.8 depicts a schematic of the error suppression optical path with the optical power coupled into the single-mode fiber as the system evaluation index. Its working principle is as follows: the distorted wavefront is coupled into the single-mode fiber through the deforming mirror and converted into an electrical signal by a photodetector. Next, it is passed to the initialization module of the control server via RS232. The control algorithms used in Figs. 6.7 and 6.8 use the stochastic parallel gradient descent (SPGD) algorithm.

Figure 6.9 depicts the system error suppression process using the intermediate frequency current as the evaluation index. It is evident that the use of the SPGD algorithm for static error correction increases the normalized intermediate frequency voltage value output by the coherent receiver. The normalized intermediate frequency voltage value increased from 0.75 before correction to 0.9 after correction, which is an increase of 0.15 (the intermediate frequency voltage value is approximately 300 mV). Figure 6.10 depicts the system error suppression process under the same atmospheric conditions as in Fig. 6.9. The optical power coupled into the single-mode fiber is used as the evaluation index. It is evident from the figure that the coupling into the single-mode fiber the optical power is increased from −28 dBm (IF voltage value is 1.0 V) before correction to −16 dBm (IF voltage value is 1.8 V) after correction, an increase of 12 dB (800 mV). Comparing Figs. 6.9 and 6.10, it clear

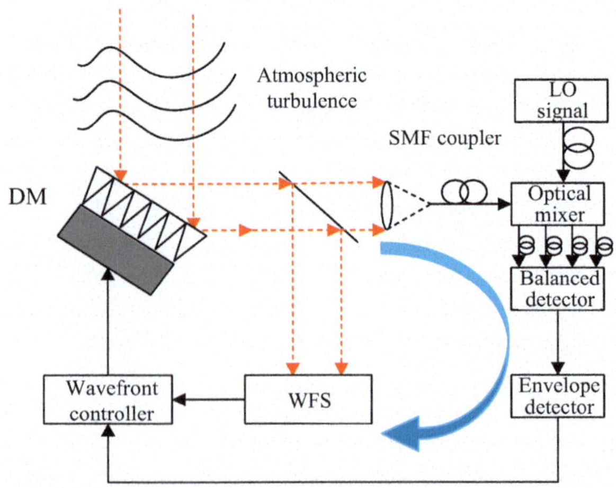

Fig. 6.7 Schematic of error suppression with the peak-to-peak value of the intermediate frequency voltage as the system evaluation index

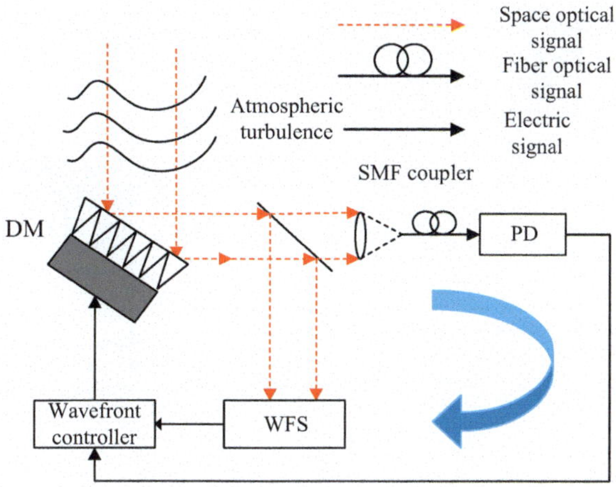

Fig. 6.8 Schematic of error suppression with coupling efficiency as the system evaluation index

that although the two schemes can suppress the system error to a certain extent, the error suppression method using optical power as the evaluation index is better than the method based on the normalized intermediate frequency. The error suppression method that uses the voltage peak-to-peak value as the evaluation index achieves a greater improvement. The main reason for this is that the output intermediate frequency voltage of the coherent receiver is not only affected by the optical power of the input single-mode fiber but also matched with the optical power of the local oscillator, frequency of the signal light and the local oscillator, polarization state, and degree of phase matching. These factors restrict the output of the intermediate frequency voltage of the coherent receiver, and when the optical power of the signal light reaches a certain level, the output intermediate frequency voltage of the system does not increase with the increase in the optical power of the single-mode fiber. Furthermore, for the coherent receiver of heterodyne detection, the frequency of the signal light and local oscillator light must be maintained within a certain range. The coherent receiver will have the output of the intermediate frequency information number, and this process can only be completed when the signal light power reaches a certain value. The system error suppression method that uses optical power as the evaluation index is not affected by the above factors. Therefore, system error suppression is relatively more thorough. This makes it more reasonable to use the signal optical power as the system evaluation index. In summary, the experiment used the coupling efficiency as the system evaluation index to suppress and thoroughly correct the system error.

Fig. 6.9 Error suppression process with coupled intermediate frequency voltage as evaluation index

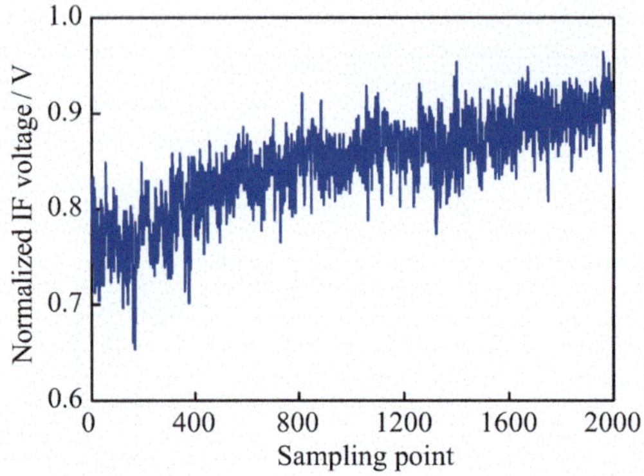

Fig. 6.10 Error suppression process with optical power as evaluation index

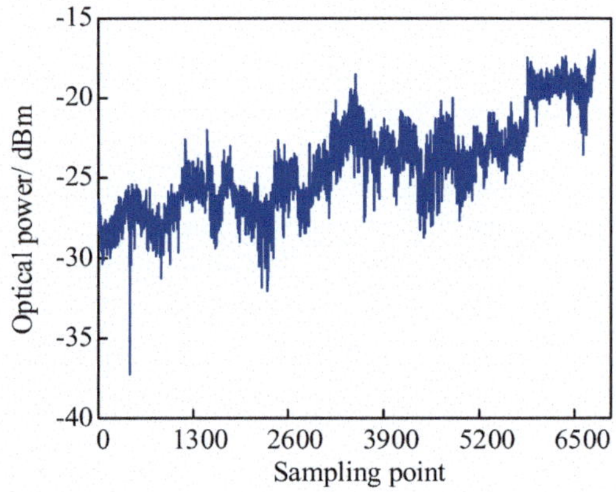

6.4 Implementation of Wavefront Controller

6.4.1 Wavefront Reconstruction Theory

In an AO system, wavefront reconstruction is the process of restoring the wavefront based on the discrete spot signal measured with a wavefront sensor. The purpose of wavefront reconstruction is to facilitate the calculation of the control voltage of the drive electrode of the wavefront corrector.

(1) **Reconstruction of wavefront using mode method**

The mode method expands the wavefront in the aperture into a weighted superposition of different aberration modes and then calculates the coefficients of each mode through the detected wavefront slope data to reconstruct the phase information of the wavefront. Zernike polynomials are a sequence of polynomials orthogonal on the unit circle, with symmetry and periodicity. Noll [33] reported that when the receiving aperture is a circular pupil, it can be described by Zernike polynomials in the wavefront analysis. The Zernike polynomial description of wavefront phase $\phi(x, y)$ is [33]

$$\phi(x, y)=\alpha_0+\sum_{k=1}^{n}\alpha_k Z_k(x, y)+\varepsilon \tag{6.9}$$

where α_k is the Zernike polynomial coefficient of the term; ε is the wavefront phase measurement error; and Z_k is the Zernike polynomial of the term, as follows:

$$\begin{cases} Z_{evenj} = \sqrt{n+1}R_n^m(r)\sqrt{2}\cos(m\theta) & m \neq 0 \\ Z_{oddj} = \sqrt{n+1}R_n^m(r)\sqrt{2}\sin(m\theta) & m \neq 0 \\ Z_j = \sqrt{n+1}R_n^0(r) & m = 0 \end{cases} \tag{6.10}$$

where $R_n^m = \sum_{s=0}^{(n-m)/2} \frac{(-1)^s (n-s)!}{s![(n+m)/2-s]![(n-m)/2-s]!}$, m, n are the axial and radial orders, respectively. Table 6.2 lists the first 28-order Zernike polynomials and their physical meanings.

Mode wavefront reconstruction is used to decompose the wavefront phase in the unit circle into different modes. Thereafter, the data measured by the wavefront sensor are used to solve each mode coefficient [34]. From Eq. (6.9), it is evident that the optical wavefront can usually be decomposed into multiple Zernike modes during optical wavefront analysis. Therefore, the Zernike mode wavefront reconstruction is actually the wavefront measured by the Zernike mode function series $Z_k(x, y)$

Table 6.2 First 28-order Zernike polynomials and their physical meanings

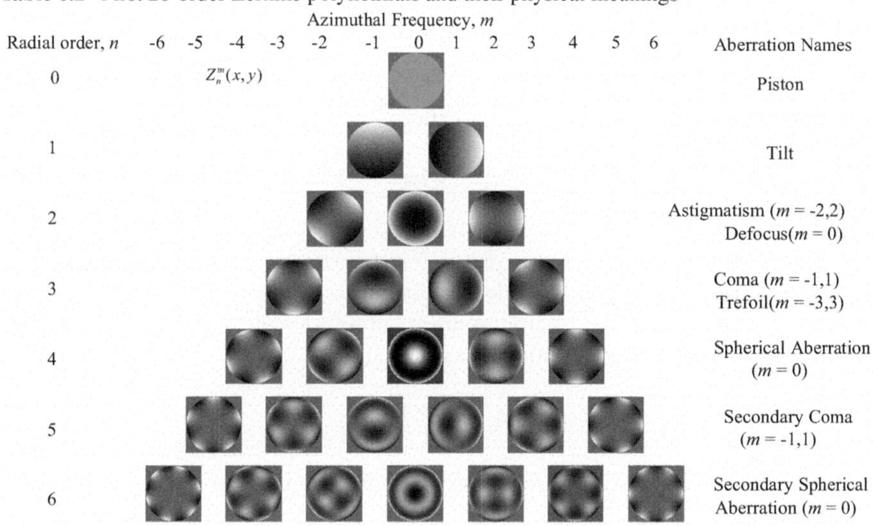

and the wavefront sensor. The linear relationship between the slopes is the process of solving the coefficients of each model term. Suppose the average slope of the wavefront phase of the incident beam measured by the wavefront sensor in the first sub-aperture is G*jx* and G*jy*, the relationship between the slope of the wavefront detected by the Hartmann sensor and the Zernike mode is [35]

$$\mathrm{G}_{jx} = \frac{1}{s_j}\iint\limits_{s_j}\left[\frac{\partial\phi(x,y)}{\partial x}\right]_j \mathrm{d}x\mathrm{d}y = \sum_{k=1}^{n}\frac{\alpha_k}{s_j}\iint\limits_{s_j}\left[\frac{\partial Z_k(x,y)}{\partial x}\right]_j \mathrm{d}x\mathrm{d}y = \sum_{k=1}^{n}\alpha_k Z_{jkx} \tag{6.11}$$

$$\mathrm{G}_{jy} = \frac{1}{s_j}\iint\limits_{s_j}\left[\frac{\partial\phi(x,y)}{\partial y}\right]_j \mathrm{d}x\mathrm{d}y = \sum_{k=1}^{n}\frac{\alpha_k}{s_j}\iint\limits_{s_j}\left[\frac{\partial Z_k(x,y)}{\partial y}\right]_j \mathrm{d}x\mathrm{d}y = \sum_{k=1}^{n}\alpha_k Z_{jky} \tag{6.12}$$

Zernike polynomials of *n* order are often used to facilitate computer processing. The pattern reconstruction can be expressed as a matrix:

$$\begin{bmatrix} \mathrm{G}_{1x} \\ \mathrm{G}_{1y} \\ \mathrm{G}_{2x} \\ \mathrm{G}_{2y} \\ \vdots \\ \vdots \\ \mathrm{G}_{mx} \\ \mathrm{G}_{my} \end{bmatrix} = \begin{bmatrix} Z_{11x} & Z_{12x} & \cdots\cdots & Z_{1nx} \\ Z_{11y} & Z_{12y} & \cdots\cdots & Z_{1ny} \\ Z_{21x} & Z_{22x} & \cdots\cdots & Z_{nnx} \\ Z_{21y} & Z_{22y} & \cdots\cdots & Z_{2ny} \\ \vdots & \vdots & \ddots & \vdots \\ \vdots & \vdots & \ddots & \vdots \\ Z_{m1x} & Z_{m2x} & \cdots\cdots & Z_{mnx} \\ Z_{m1y} & Z_{m2y} & \cdots\cdots & Z_{mny} \end{bmatrix} \bullet \begin{bmatrix} \alpha_1 \\ \alpha_2 \\ \vdots \\ \alpha_n \end{bmatrix} \tag{6.13}$$

where *m* is the number of sub-apertures of the S–H wavefront sensor; for order *n* of the Zernike mode for wavefront reconstruction, formula (6.13) can be simplified as

$$G = DA \tag{6.14}$$

where *G* is the wavefront phase slope vector, *D* is the wavefront reconstruction matrix, and *A* is the Zernike polynomial mode coefficients to be calculated.

Using formula (6.14), the minimum norm solution of the Zernike mode coefficients under the least square solution can be obtained [34]:

$$A = D^{-1}G \tag{6.15}$$

Next Eq. (6.15) is substituted into formula (6.9) to obtain the wavefront phase.

(2) **Reconstruction of wavefront using direct slope method**

The wavefront phase slope of the incident beam wavefront in the sub-aperture of the wavefront sensor approximately reflects the wavefront phase distortion information. Therefore, if the deformable mirror corrects the wavefront phase slope error in all sub-apertures, it can be effectively corrected. The wavefront phase is distorted. The direct wavefront slope method [36] entails solving the control voltage of the deformable mirror driver and establishing the relationship between the influence function of the deformable mirror driver and the wavefront phase slope measured by the Hartmann wavefront sensor. The wavefront sensor measures the average slope of the wavefront phase in the first sub-aperture as follows [37]:

$$\mathrm{G}_{jx} = \sum_{k=1}^{l} \frac{\mathrm{V}_k}{s_j} \iint_{s_j} \left[\frac{\partial f_k(x, y)}{\partial x} \right]_j \mathrm{d}x\mathrm{d}y = \sum_{k=1}^{l} V_k F_{jkx} \tag{6.16}$$

$$\mathrm{G}_{jy} = \sum_{k=1}^{l} \frac{\mathrm{V}_k}{s_j} \iint_{s_j} \left[\frac{\partial f_k(x, y)}{\partial y} \right]_j \mathrm{d}x\mathrm{d}y = \sum_{k=1}^{l} V_k F_{jky} \tag{6.17}$$

where s_j is the area of the first sub-aperture; V_k is the drive control voltage of the deformable mirror; l is the number of drives of the deformable mirror, the 69-unit deformable mirror used in this study, i.e., $l = 69$.

Assuming that the wavefront sensor has a total of m sub-apertures, all the slopes of the wavefront sensor can be expressed in the matrix form as [36]

$$\begin{bmatrix} \mathrm{G}_{1x} \\ \mathrm{G}_{1y} \\ \mathrm{G}_{2x} \\ \mathrm{G}_{2y} \\ \vdots \\ \vdots \\ \mathrm{G}_{mx} \\ \mathrm{G}_{my} \end{bmatrix} = \begin{bmatrix} F_{11x} & F_{12x} & \cdots\cdots & F_{1nx} \\ F_{11y} & F_{12y} & \cdots\cdots & F_{1ny} \\ F_{21x} & F_{22x} & \cdots\cdots & F_{nnx} \\ F_{21y} & F_{22y} & \cdots\cdots & F_{2ny} \\ \vdots & \vdots & \ddots & \vdots \\ \vdots & \vdots & \ddots & \vdots \\ F_{m1x} & F_{mnx} & \cdots\cdots & F_{mnx} \\ F_{m1y} & F_{m2y} & \cdots\cdots & F_{mny} \end{bmatrix} \bullet \begin{bmatrix} V_1 \\ V_2 \\ \vdots \\ V_n \end{bmatrix} \tag{6.18}$$

Formula (6.18) can be written simply as

$$G = FV \tag{6.19}$$

where G is the wavefront phase slope vector, which includes the wavefront phase slopes of all sub-aperture x direction and y direction measured by the wavefront sensor; F is the wavefront response matrix, which reflects the internal relationship between the wavefront sensor and deformable mirror, and can be obtained through field calibration; and V is the control voltage vector of the deformable mirror.

Using formula (6.19), the minimum norm solution of the control voltage vector under the least square solution can be obtained [34]:

$$V = F^{-1}G \tag{6.20}$$

(3) **Reconstruction of wavefront using zone method**

The phase between any two points on the wavefront obeys the following relationship [38]:

$$\phi(\rho) = \int_c \nabla\phi \mathrm{d}s + \phi(\rho_0) \tag{6.21}$$

where ∇ is the Hamiltonian, and c is the integration path (this integration has nothing to do with the path). Assuming that the wavefront gradient measured by the wavefront sensor g(x, y) is affected by the noise acting on the wavefront sensor, g(x, y) can be expressed as [34]

$$g(x, y) = \nabla\phi + n(x, y) \tag{6.22}$$

According to least squares Formula (6.22) can be expressed as:

$$\int_c (g(x, y) - \nabla\phi)^2 \mathrm{d}s = \min \tag{6.23}$$

Equation (6.23) satisfies Euler's equation [39]:

$$\nabla^2\hat{\phi}=\nabla g \tag{6.24}$$

where $\hat{\phi}$ is the estimate under ϕ least squares.

The wavefront is often discretized in computational processing. The entire wavefront is $(N - 1) \times (M - 1)$ divided into subintervals. Therefore, the area method for reconstructing the wavefront is transformed into using the measurement data (wavefront gradient or phase difference) around the sub-aperture to estimate the phase of the center point, thereby reconstructing the wavefront phase. Based on the position distribution of the phase measurement and reconstruction points of the wavefront sensor, the area method can be divided into a variety of models [40–42] such as the Hughton, Fred, and Sauchwell models. The red points in Fig. 6.11 represent the position of the reconstructed wavefront, and the crosses represent the position of the measured data.

In the Hughton model, the measured data are the phase difference between the grid points, and the points of the reconstructed phase are on the grid points. Because the subregion is small, the wavefront slope and phase for a subregion can be expressed

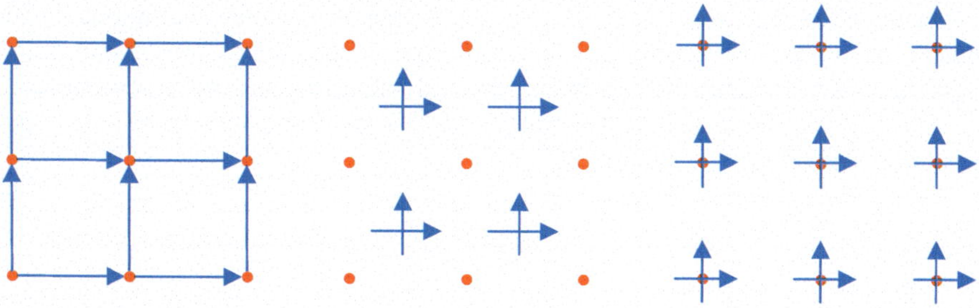

Fig. 6.11 Regional method wavefront reconstruction model

as

$$g_x = \Delta\varphi_x / h = \varphi_{i+1,j} - \varphi_{i,j} \tag{6.25}$$

$$g_y = \Delta\varphi_y / h = \varphi_{i,j+1} - \varphi_{i,j} \tag{6.26}$$

Equations (6.25) and (6.26) indicate that the slope is continuous in the region but not on the boundary, and the phase in the region changes regularly.

In the Fred model, the position of the measured wavefront slope is in the center of the area, and the point of the phase to be reconstructed is on the grid point. When calculating the slope at the center of the region, the wavefront slope and phase can be expressed by calculating the average value of the boundary phase as [43]

$$g_x = \frac{1}{h}\left[\frac{1}{2}\left(\varphi_{i+1,j+1} + \varphi_{i+1,j}\right) - \frac{1}{2}\left(\varphi_{i,j+1} + \varphi_{i,j}\right)\right] \tag{6.27}$$

$$g_y = \frac{1}{h}\left[\frac{1}{2}\left(\varphi_{i+1,j+1} + \varphi_{i,j+1}\right) - \frac{1}{2}\left(\varphi_{i+1,j} + \varphi_{i,j}\right)\right] \tag{6.28}$$

The wavefront slope of this model is also continuous within the region, the phase is still linear, and it is also discontinuous on the boundary.

In the Sauchiwell model, the measured wavefront slope and phase to be estimated coincide on the grid points, and the phase difference between adjacent grid points corresponds to the slope of the intermediate point. Then the wavefront slope and phase can be expressed as [44]

$$\frac{1}{2}\left(g^x_{i+1,j} + g^x_{i,j}\right) = \frac{1}{h}\left(\varphi_{i+1,j} - \varphi_{i,j}\right) \tag{6.29}$$

$$\frac{1}{2}\left(g^y_{i+1,j} + g^y_{i,j}\right) = \frac{1}{h}\left(\varphi_{i,j+1} - \varphi_{i,j}\right) \tag{6.30}$$

According to this model, the slope inside the region changes continuously and linearly. Therefore, the phase of the wavefront to be reconstructed changes according to the parabolic law. This helps to accurately estimate the phase of the wavefront.

For the above three models, the relationship between the wavefront slope and phase can be expressed as a matrix:

$$G = D\Phi \tag{6.31}$$

where G is a dimensional matrix composed of wavefront slopes $N(N-1) \times 1$ in the x or y direction, and ϕ is the wavefront phase matrix to be reconstructed, with $N^2 \times 1$. D is a matrix of coefficients with size $N(N-1) \times N^2$. Because the rank of matrix D is incomplete, it is a singular value matrix. Using the matrix theory, the minimum norm solution of formula (6.39) is [31]

$$\phi = \left(D^T D\right)^{-1} D^T G \tag{6.32}$$

(4) **Comparison of mode and area wavefront reconstruction algorithms**

The wavefront reconstruction algorithms of the mode and area methods were introduced in the preceding section. In this section, these two algorithms are compared. Figure 6.12a depicts the wavefront restored using the 69-order Zernike mode method; Fig. 6.12b depicts the restoration using the zone method. Figure 6.13 depicts the wavefront difference reconstructed by the Zernike mode and zone methods. Table 6.3 lists the PV and RMS statistics using different reconstruction algorithms. Figures 6.12, 6.13 and Table 6.3 indicate that the wavefront to be measured can be reconstructed correctly using both the Zernike mode and zone methods. The PV value error of the two wavefront reconstruction algorithms is 0.035, and the RMS value error is 0.048. In contrast, the PV and RMS values of the wavefront reconstructed using the Zernike wavefront reconstruction algorithm is slightly larger than that using the zonal method.

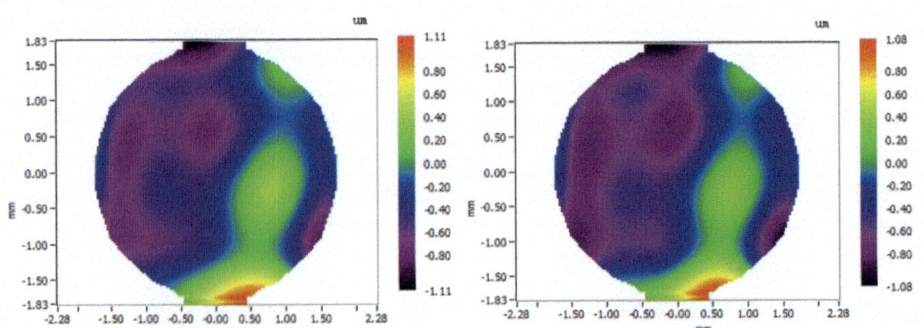

(a) Model method (b) zonal method

Fig. 6.12 Wavefront reconstructed using model and regional wavefront reconstruction algorithms

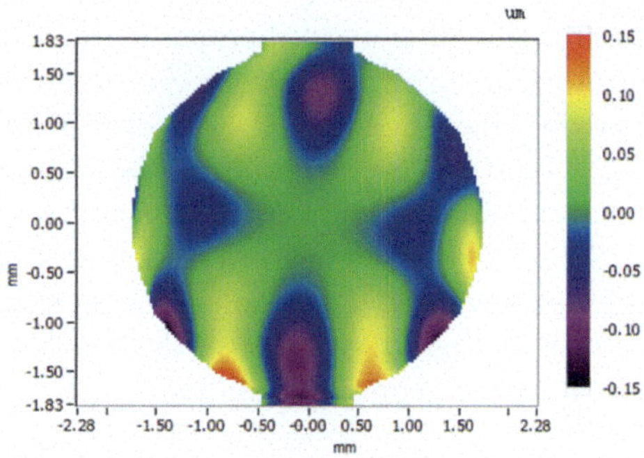

Fig. 6.13 Reconstruction of residual wavefront using zonal method

Table 6.3 PV and RMS values of reconstructed wavefront

Wavefront reconstruction method	PV (μm)	RMS (μm)
Zernike model	2.086	0.370
Regional law	2.051	0.355

6.4.2 Measurement of Influence Matrix of Deformable Mirror

The Shaker-Hartmann wavefront sensor in the AO system determines the ability of the system to detect aberrations in space. The wavefront corrector determines the ability of the system to correct aberrations in space. The matching relationship between these two devices determines the spatial compensation capability of the entire AO system. The influence function matrix of the deformable mirror directly reflects the relationship between the driving electrode voltage of the deformable mirror and the amount of deformation of the mirror surface. If each voltage of all driving electrodes and the combination of voltages between different electrodes are measured, it will be a significantly large number, which makes it is an unrealistic and impractical method. However, the deformation of the deformable mirror follows two important rules: it is the linear superposition of the deformation of the deformable mirror when each driving electrode acts alone [31]; under the action of a single electrode, the deformation of the deformable mirror is directly proportional to the voltage of the driving electrode [45]. Therefore, one only needs to measure the voltage of each drive electrode once to derive the influence function matrix of the deformable mirror. The push–pull method is the commonly used method to measure the influence matrix. The push–pull method states that, when measuring the influence matrix, the deformable mirror driver applies the control voltage twice: first positive and then negative or vice versa. Thereafter, it calculates the application of the positive and negative electrodes. However, the accuracy of the matrix in the measurement process is affected by the following: the detection noise of the wavefront sensor; the hysteresis, creep, nonlinearity and resonance characteristics of the deformable mirror; and the nonlinearity and saturation of the Hartmann wavefront sensor. Based on this, Kasper et al. [46] proposed the use of the multiple averaging effect of the deformable mirror to eliminate the effects of the measurement noise on the system. To reduce the measurement error, each drive electrode is pushed and pulled multiple times. Next, the average value of the multiple measurement results is obtained. The details of the measurement process of the influence matrix are as follows.

Suppose that the wavefront sensor includes a microlens, and the reflective deformable mirror (DM) includes m an actuator. Then, using the push–pull method to push and pull the reflective deformable mirror actuator for a second time, the average slope is given as follows:

$$s_m = \frac{1}{n} * \frac{1}{c} \sum_{i=1}^{n} \begin{pmatrix} s_{1m_x}(c) - s_{1m_x}(-c) \\ \vdots \\ s_{km_x}(c) - s_{km_x}(-c) \\ s_{km_y}(c) - s_{km_y}(-c) \\ \vdots \\ s_{km_y}(c) - s_{km_y}(-c) \end{pmatrix} \tag{6.33}$$

where c is the amount of control applied to the deformable mirror driver, and n is the number of pushs and pulls of the deformable mirror. Using formula (6.33), the influence matrix of the deformable mirror can be obtained as follows:

$$\mathrm{M}_{C2S} = [\, s_1 \cdots s_m \,] = \begin{bmatrix} s_{11x} & \cdots & \cdots & \cdots & s_{1mx} \\ s_{11y} & \cdots & \cdots & \cdots & s_{1my} \\ \vdots & \vdots & s_{ijx} & \vdots & \vdots \\ \vdots & \vdots & s_{ijy} & \vdots & \vdots \\ s_{k1x} & \cdots & \cdots & \cdots & s_{kmx} \\ s_{k1y} & \cdots & \cdots & \cdots & s_{kmy} \end{bmatrix} \tag{6.34}$$

Based on the relationship between the wavefront slope matrix and the response matrix:

$$S = \mathrm{M}_{C2S} V \tag{6.35}$$

Using the matrix operation, formula (6.35) can be written as

$$V = \mathrm{M}_{C2S}^{-1} S \tag{6.36}$$

where V is the voltage matrix, and V is the unit matrix in the calibration mode, S is the slope matrix or the wavefront phase matrix.

In the AO system, to convert the wavefront slope or phase value measured by the wavefront sensor into the control vector of the deformable mirror driver, it is necessary to calculate restoration matrix M_{S2C}, which converts the wavefront slope into deformation. The restoration matrix of the mirror control voltage can be obtained using formula (6.36):

$$M_{\mathrm{S2C}} = M_{\mathrm{C2S}}^{-1} \tag{6.37}$$

The wavefront reconstruction matrix can be obtained using singular value decomposition (SVD) to solve the pseudo-inverse of the response matrix. The SVD of response M_{C2S} is as follows [35]:

$$M_{\mathrm{C2S}} = U \Lambda V^{T} \tag{6.38}$$

where U is the dimensional matrix $2k \times m$, Λ is the dimensional diagonal matrix $m \times m$, and V is the dimensional unitary matrix $m \times m$. The elements on the main diagonal of matrix Λ are the singular values of M_{S2C}. The restoration matrix of M_{S2C} the deformable mirror control voltage can be expressed as [31]:

$$M_{S2C} = M_{C2S}^{-1} = V \Lambda^{-1} U^{T} \tag{6.39}$$

It is evident from formula (6.39) that the restoration matrix of the deformable mirror control voltage is obtained using the pseudo-inverse of the influence matrix. When the response matrix has a small singular value, the corresponding restoration matrix of the deformable mirror driver is a particularly large singular value. When the matrix is used to calculate the control voltage of the deformable mirror, the measurement error of the wavefront sensor is amplified, and the calculated driving voltage of the deformable mirror increases accordingly, making the closed-loop control unstable. To address this problem, Dubra proposed a method of filtering the singular values of the response matrix [47]. This method can filter out the characteristic modes with small singular values of the response matrix and improve the stability and accuracy of the system [48].

6.4.3 Realization of Wavefront Control Algorithm

(1) **Wavefront control model**

A neural network is combined with Proportional integral (PI) control, whose control parameters can be adjusted in real time through the self-learning and weighting coefficient self-adjustment characteristics of the neural network. This ensures that the system is always in the optimal control state. Figure 6.14 depicts the system model of the adaptive PI control algorithm. In this figure, G_r is the atmospheric turbulence disturbance, G_n is detection noise, S_n is the wavefront slope, $\Delta\phi$ is the real-time reconstructed wavefront, and $\Delta\phi$ is the residual voltage calculated using the sum response matrix E. The front response matrix is generally obtained through on-site calibration. To improve the stability of the system, singular value filtering is required for the calibrated response matrix; it is the actual control voltage of the deformable mirror output after PI operation, where Z^{-1} represents the system delay and generally takes a sampling period [49].

(2) **Real-time PI control algorithm implementation**

The PI control algorithm uses deviation to achieve control on application objects. Proportional control is mainly used to proportionally reflect the deviation signal of the control system. When deviation occurs, the controller immediately produces a control effect to reduce it. When the proportional control coefficient increases, the sensitivity of the closed-loop system increases, the adjustment speed increases, the steady-state error decreases, and oscillation increases. Therefore, when the proportional control coefficient exceeds a certain critical value, the closed-loop system becomes unstable. Integral control is generally used to eliminate the steady-state error in adaptive control systems and improve the degree of non-difference. The greater the integral control effect, the faster the system stability error is eliminated; however, when the value is significantly large, the system produces integral saturation, which causes system overshoot [49].

Digital PI algorithms are of two broad types: position-based and incremental algorithms. When using the incremental algorithm, only the amount of error between the current and previous times needs to be calculated. The position algorithm needs to calculate the error data at all previous times. Therefore, the calculation load of the incremental algorithm is significantly lower than that of the position algorithm. Furthermore, the incremental algorithm does not introduce integral saturation.

Control quantities $V_n(k)$ $V_n(k-1)$ at the first k and $k-1$ sampling moments are as follows:

$$V_n(k) = K_p[E(k) + \frac{T}{T_i}\sum_{j=0}^{k} E(j)] \tag{6.40}$$

$$V_n(k-1) = K_p[E(k-1) + \frac{T}{T_i}\sum_{j=0}^{k-1} E(j)] \tag{6.41}$$

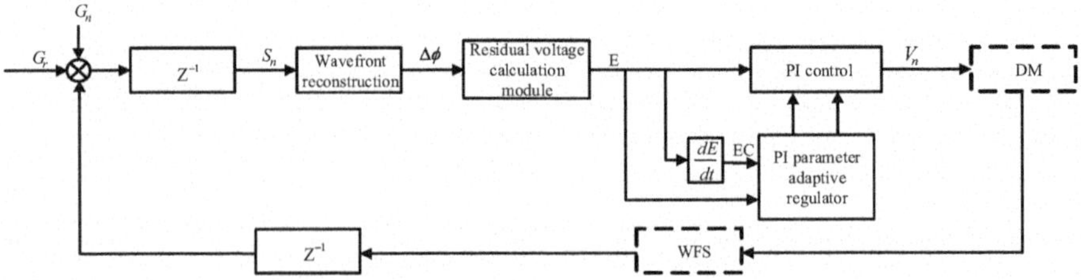

Fig. 6.14 Schematic adaptive optics control system

Subtracting formula (6.40) from (6.41) gives the increment of the control quantity at the first moment:

$$\Delta V(k) = K_p[E(k) - E(k-1)] + \frac{K_p T}{T_i} E(k) \tag{6.42}$$

To facilitate programming, formula (6.42) can be simplified as

$$\Delta V(k) = k_p E(k) - k_i E(k-1) \tag{6.43}$$

where $k_p = K_p(\frac{1}{T} + \frac{1}{T_i})$, $k_i = \frac{K_p}{T}$.

Suppose the maximum error vector of the system is $E_{\max}$. Then, the output vector of the controller is $V_{\max}$ and the minimum error vector of the system is $E_{\min}$. When $|E| \geq E_{\max}$, owing to the limitation of the performance parameters of the deformable mirror, the system error exceeds the correction capability of the system. Therefore, the output threshold is set up, and the system output is $V_n = sign(E) * V_{\max}$. Frequent control actions of the system may cause system oscillations. To prevent too frequent control actions and eliminate system oscillations, set $|E| \leq E_{\min}$. Therefore, the system output is $V_n = V_{n-1}$ when $E_{\min} \leq |E| \leq E_{\max}$. In this case, the system error is within the effective correction range, and the output control vector of the deformable mirror is

$$V_n = V_{n-1} + \Delta V \tag{6.44}$$

where ΔV is the actual output control vector through the wavefront controller:

$$\Delta V(k) = k_p E(k) - k_i E(k-1) \tag{6.45}$$

(3) **Closed-loop control parameter adjustment**

In the AO system, to achieve better control results, the PI controller must adjust the proportional and integral control functions based on the dynamic atmospheric disturbance [50], forming a relationship of mutual cooperation and mutual restriction in the control variables. However, this relationship is not necessarily a simple "linear combination", and it is necessary to obtain a set of optimal solutions from the ever-changing nonlinear combination. The neural network has the ability of arbitrary non-linear expression, and the two control parameters of the PI controller can be tuned to the best parameters in real time by learning the system performance to realize the optimal control. Figure 6.15 depicts the PI parameter adaptive control structure based on the back propagation (BP) neural network. The neural network can adjust the PI controller parameters based on the operating state of the system to achieve optimization of a certain performance index and make the output state of neurons in the output layer. For the two adjustable parameters of the PI controller, through the self-learning of the neural network and the adjustment of the weighting coefficients, the output of the neural network corresponds to the PI control parameters under a certain optimal control law.

The calculation formula for the hidden layer of the network is as follows:

$$\text{net}_i(k) = \sum_{j=1}^{2} \omega_{ij} x_j \tag{6.46}$$

where ω_{ij} is the weighting coefficient from the input layer to the hidden layer.

The calculation formula for the network output layer is as follows:

$$\text{net}_l(k) = \sum_{i=1}^{2} \omega_{li} net_i(k) \tag{6.47}$$

where ω_{li} is the weighting coefficient from the input layer to the hidden layer. Consider the performance index function of the following form:

$$J(k) = \frac{1}{2}(E(k))^2 \tag{6.48}$$

Adjust the weighting coefficient of the network based on the gradient descent and search and adjust based on the negative direction of the weighting coefficient. When the weight of the PI neural network uses only the gradient learning algorithm, the weighting coefficient of the neuron network is modified slowly and falls into the local optimum; therefore, it can be passed. The method of adding items improves the efficiency of network learning. The learning formula for the weighting coefficient of the PI neural network is as follows.

(1) The connection weight update formula from the input layer to the hidden layer is

$$\omega_{ij}(k) = -\eta \frac{\partial J(k)}{\partial \omega_{ij}} + \omega_{ij}(k-1) + \eta_1 \left[\omega_{ij}(k-1) - \omega_{ij}(k-2)\right] \tag{6.49}$$

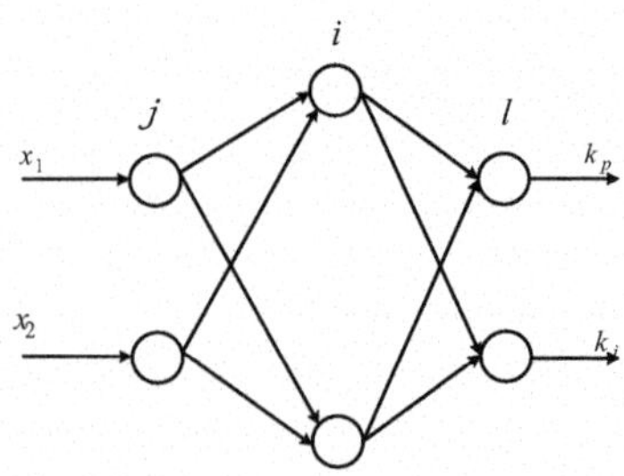

Fig. 6.15 Schematic of PI structure of BP neural network

(2) The update point formula of the connection weight from the hidden layer to the output layer is

$$\omega_{li}(k) = -\eta \frac{\partial J(k)}{\partial \omega_{li}} + \omega_{li}(k-1) + \eta_1 [\omega_{li}(k-1) - \omega_{li}(k-2)] \tag{6.50}$$

where η and η_1 are the learning rate.

6.5 Correction of Wavefront Distortion

6.5.1 Analysis of Closed-Loop Control Parameter Adjustment Process

Figures 6.16, 6.17, and 6.18 depict the control parameter adjustment, performance evaluation index change curve, and control tracking curve of the controller in the closed-loop control process, respectively. It is evident from Fig. 6.16 that, in the absence of any disturbance in the system, the system basically reaches a stable state after the initial 0.34 s of control parameters and adjustments; during the entire closed-loop control process, the integral coefficient is always maintained at approximately 0.075, while the proportional coefficient drops from the initial 0.08 to 0.07. With the deepening of the closed-loop control, theoretically the control error should continue to approach 0; however, it is evident from Figs. 6.17 and 6.18 that a certain residual error remains in the performance after the system is closed.

Table 6.4 lists the results of PI parameter tuning of the controller when the disturbance variances are 0.05, 0.10, 0.15 and 0.20, corresponding to small, medium, large, and large disturbances, respectively. It is evident that the proportional coefficient increases and the integral coefficient decreases when the atmospheric disturbance variance increases. This indicates that the proportional control has an enhanced

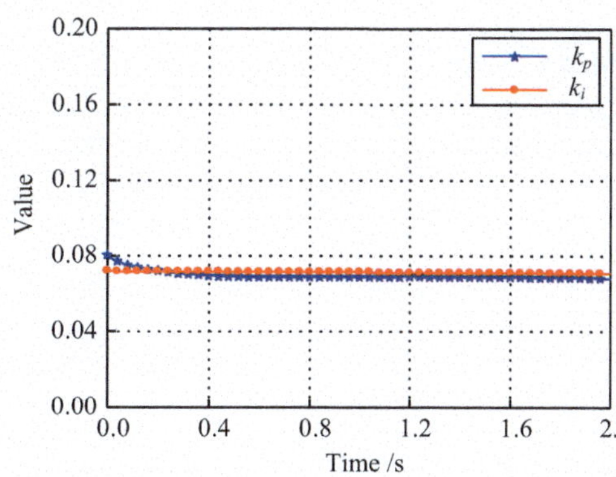

Fig. 6.16 Closed-loop control parameter adjustment process [50]

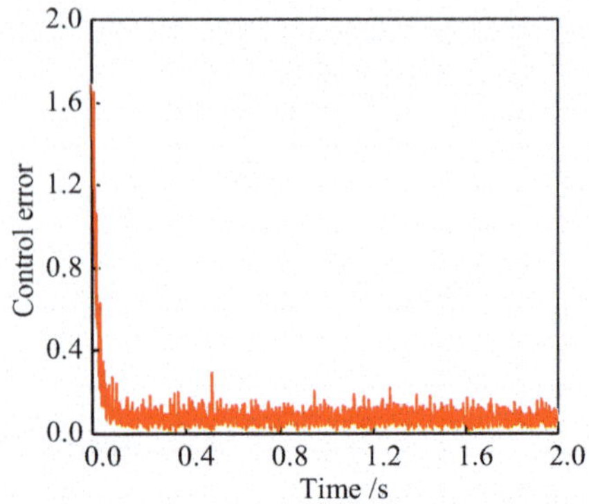

Fig. 6.17 Control error after closed-loop correction [50]

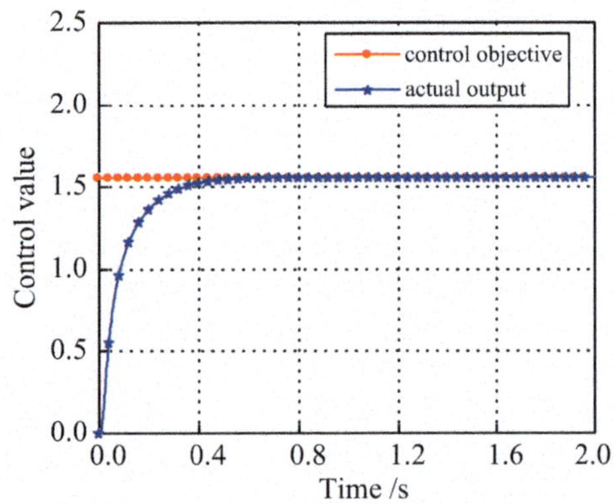

Fig. 6.18 Closed-loop control tracking curve [50]

control effect whereas the integral control has a weak effect when the atmospheric turbulence increases. This is because the wavefront fluctuation closed-loop adjustment time and the amount of closed-loop error correction increase when the wavefront disturbance variance increases. The residual error also increases slightly.

Table 6.4 Comparison of PI parameter tuning performance under various atmospheric disturbances

Atmospheric disturbance σ^2	k_p	k_i	Closed-loop time/s	Error before correction	Error after correction
0.05	0.145	0.085	0.15	0.07	0.003
0.10	0.079	0.072	0.25	0.326	0.0127
0.15	0.096	0.069	0.32	0.72	0.041
0.20	0.10	0.062	0.33	1.6	0.07

6.5.2 Impact of Wavefront Phase Distortion on Mixing Efficiency

The main factors affecting the magnitude of the intermediate frequency current are the average optical power P_S of the signal light, average optical power P_L of the local oscillation light, and frequency and phase differences between the signal light and the local oscillation light. In heterodyne detection, only background noise in the difference frequency band can enter the detection system; other noises are filtered out by the band-pass filter. The background noise does not produce additional coherent terms on the "difference frequency term" between the original signal light and the local oscillator light. The major noises in a coherent receiver are shot noise and thermal noise. The noise variance in the system can be expressed as

$$\sigma^2 = \sigma_s^2 + \sigma_T^2 \tag{6.51}$$

Because the optical power of the local oscillator is relatively large, only the impact of shot noise needs to be considered when analyzing the signal-to-noise ratio (SNR) of the coherent optical system. Therefore, the noise power of the detector can be expressed as

$$\sigma^2 = 2eI_{LO}\Delta f = 2e\Delta f\eta K\int_U A_{LO}^2 \mathrm{d}U \tag{6.52}$$

where U is the surface area of the photodetector, K is a constant, Δf is the frequency difference between the signal light and local oscillator light, and I_{LO} is the response current of the local oscillator light.

SNR is defined as the ratio of the power of the optical signal to the power of the noise [51]:

$$\mathrm{SNR} = \frac{\langle i^2(t)\rangle}{\sigma^2} \tag{6.53}$$

Therefore, for coherent detection, SNR can be expressed as [22]

$$SNR_{RF} = \frac{4\alpha\int_U A_S^2 \mathrm{d}U}{h\nu\Delta f}\cdot\frac{\left[\int_U \vec{A}_S\vec{A}_{LO}\cos(\Delta\varphi)\mathrm{d}U\right]^2+\left[\int_U \vec{A}_S\vec{A}_{LO}\sin(\Delta\varphi)\mathrm{d}U\right]^2}{\int_U A_S^2\mathrm{d}U\int_U A_{LO}^2\mathrm{d}U} \tag{6.54}$$

$$SNR_{RF0} = \frac{4\alpha\int_U A_S^2\mathrm{d}U}{h\nu\Delta f} \tag{6.55}$$

Then the coherent mixing efficiency of coherent detection can be expressed as [52]

$$\gamma = \frac{\mathrm{SNR_{RF}}}{\mathrm{SNR_{RF0}}} \tag{6.56}$$

Substituting formulas (6.54) and (6.55) into formula (6.56),

$$\gamma = \frac{\left[\int_U \underset{S}{\vec{A}}\,\underset{LO}{\vec{A}} \cos(\Delta\varphi)\mathrm{d}U\right]^2 + \left[\int_U \underset{S}{\vec{A}}\,\underset{LO}{\vec{A}} \sin(\Delta\varphi)\mathrm{d}U\right]^2}{\int_U A_S^2 \mathrm{d}U \int_U A_{LO}^2 \mathrm{d}U} \tag{6.57}$$

It is evident from Eq. (6.57) that mixing efficiency can directly reflect the phase matching degree of the signal light and local oscillator light, and it is a key parameter to describe the phase matching in the coherent detection system.

Next, we analyze the relationship between wavefront phase distortion and mixing efficiency to illustrate that the aberration caused by atmospheric turbulence causes the system SNR to decrease, thereby reducing the coherent mixing efficiency of the system. In the simulation, impacts of the fluctuation in light intensity and phase distortion caused by atmospheric turbulence on the mixing efficiency are considered. Antenna aberration and mismatch angle are not considered. Figures 6.19 and 6.20 depict the wavefront RMS and Strehl ratio (SR) and coherent mixing efficiency curve, respectively. Table 6.5 lists the corresponding statistics for Figs. 6.19 and 6.20. It is evident that the mixing efficiency of the coherent receiver decreases when the RMS of the wavefront increases, i.e., with the noise power constant, the system SNR drops sharply when the turbulence intensity increases. Furthermore, it also leads to an increase in system error codes. It is evident from Fig. 6.20 that the mixing efficiency of the coherent receiver also increases when SR increases, which indicates that the energy of the signal light beam is more concentrated when the local oscillator light remains unchanged. The greater the coupling efficiency, the higher mixing efficiency. It is evident from the data in Table 6.5 that when the RMS is greater than 0.8 and SR is less than 0.45, the coherent mixing efficiency is close to 0. This is because the intermediate frequency current output by the coherent receiver is almost pure noise, and the system has serious bit errors or even interruptions. communication; when the RMS value of the wavefront is 0.5 mμ the SR of the system is 0.75, and the coherent mixing efficiency is 0.3; when the RMS value of the wavefront is 0.6 mμ, the SR of the system is 0.68, and the coherent mixing efficiency is less than 0.2. When the coherent mixing efficiency is low, the probability of system error increases. Therefore, to reduce the system error rate, the corrected wavefront RMS must be less than 0.5 mμ.

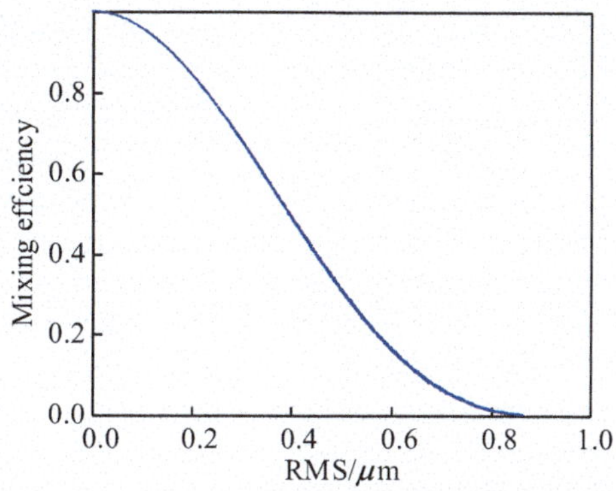

Fig. 6.19 Relationship between the root mean square of the wavefront and mixing efficiency [50]

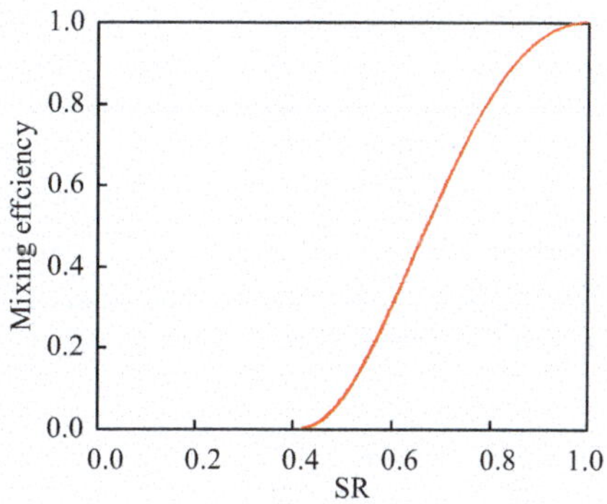

Fig. 6.20 Relationship between SR and mixing efficiency [50]

Table 6.5 Correspondence statistics of wavefront RMS, SR, and mixing efficiency

RMS (μm)	SR	Mixing efficiency
0.2	0.95	0.82
0.4	0.84	0.48
0.5	0.75	0.30
0.6	0.68	0.18

6.5.3 Impact of Mixing Efficiency on Coherent Optical Communication Systems

The mixing efficiency of the signal light and local oscillator light plays a vital role in coherent optical communication systems. The higher the mixing efficiency, the better the coherence between signal light and local oscillator light and the better the performance quality of coherent optical communication. Figure 6.21 depicts the curve of mixing efficiency versus turbulence intensity in various correction modes. It is evident that mixing efficiency gradually decreases when the intensity of atmospheric turbulence (D/r_0) increases. When the AO correction mode is gradually increased, mixing efficiency also increases. Without the AO correction, mixing efficiency is below 0.5. After AO corrects the first 30-order Zernike coefficient, the mixing efficiency is improved to above 0.5. Under weak turbulence conditions, such as $D/r_0 = 1$ time, only correcting the tilt aberration can effectively improve the mixing efficiency. After the tilt aberration is corrected, the mixing efficiency is close to 0.9. When the turbulence is strong, for example $D/r_0 = 10$, only correcting the tilt aberration cannot improve the mixing efficiency. In this case, more high-level corrections are needed to effectively improve the mixing efficiency.

Figure 6.22 depicts that the coherent mixing efficiencies before and after the AO wavefront correction with different D/r_0, in which the Zernike correction mode takes 30 orders. Table 6.6 lists the statistics of 100 consecutive frames of coherent mixing efficiencies before and after correction. The values of 1, 7, and 17 for D/r_0 correspond to weak, medium, and strong turbulence conditions. Comparing the open and closed coherent mixing efficiencies, it is evident that, under weak turbulence conditions, the coherent mixing efficiency increases from 0.59 to 0.72 after AO closed-loop correction; under moderate turbulence conditions, the coherent mixing efficiency increases from 0.59 to 0.72. When the loop is open, it increases from 0.10 to 0.70; and under strong turbulence conditions, the coherent mixing efficiency increases from 0.08 to 0.33 when the loop is closed. It is evident that AO has a significant correction effect under weak and medium turbulence conditions; under strong turbulence, the system must be further optimized to achieve better results.

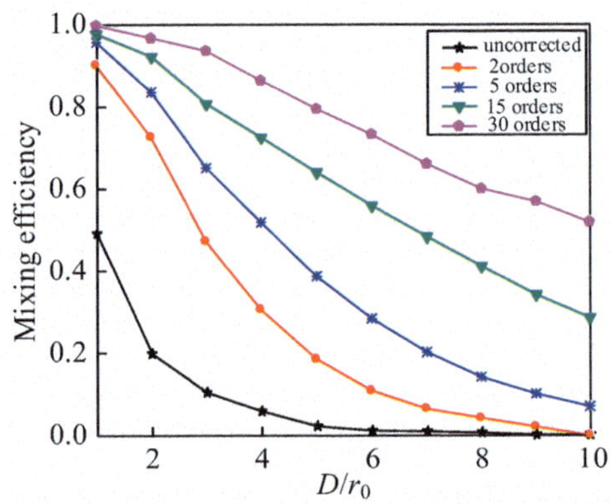

Fig. 6.21 Variation curve of coherent mixing efficiency with turbulence intensity [50]

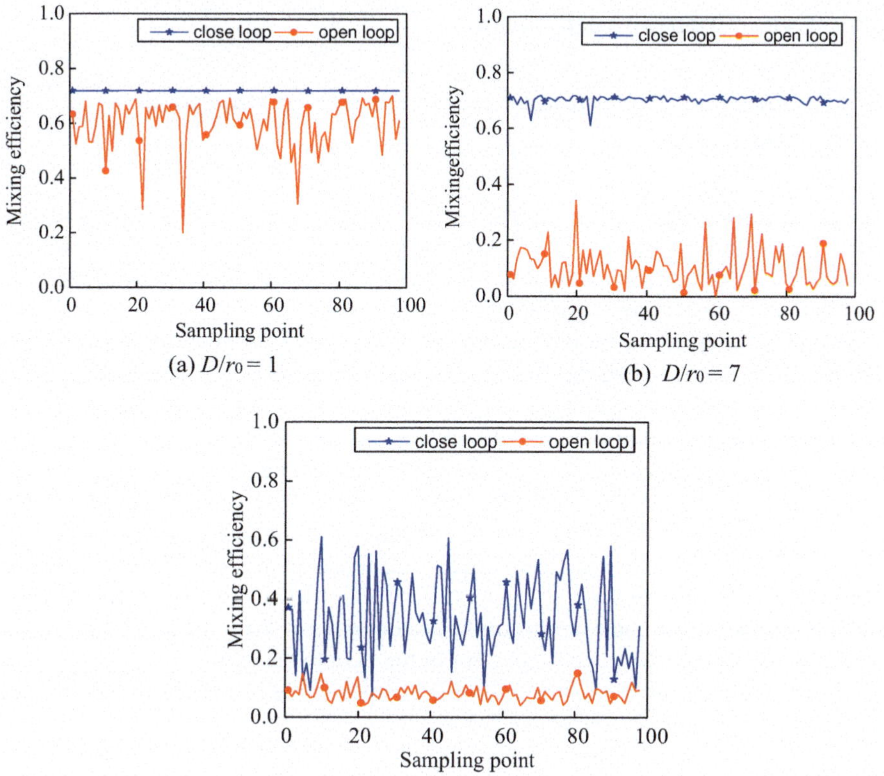

(a) $D/r_0 = 1$

(b) $D/r_0 = 7$

(c) $D/r_0 = 17$

Fig. 6.22 Coherent mixing efficiency for various timeframes [50]

Table 6.6 Average values of coherent mixing efficiency before and after calibration under various D/r_0 conditions

D/r_0	Before correction RMS	Coherent mixing efficiency in open loop	After correction RMS	Coherent mixing efficiency in closed loop
1	0.2	0.65	0.005	0.81
7	1.16	0.06	0.13	0.79
17	3.01	0.015	0.30	0.26

6.6 Experimental Verification

6.6.1 Analysis of Dynamic Characteristics of Wavefront Controller

The closed-loop control effect of wavefront control involves the closed-loop adjustment time, system control error, and effective correction bandwidth of the controller. The effective correction bandwidth of the controller can be obtained by comparing the open- and closed-loop control voltage power spectra.

Figure 6.23 depicts the time-domain control curve when the open-loop disturbance variance is 0, that is, when the atmospheric disturbance is static. The control target curve in Fig. 6.23a represents the control voltage when the loop is open, and the actual output curve represents the closed-loop control. It is evident from the real-time control curve that the output voltage of the controller coincides with the control curve after the initial 0.60 s adjustment. In other words, the system reaches a stable state. Therefore, when the static error is corrected, the closed-loop adjustment time is 0.6 s. Figure 6.23b is the closed-loop control error curve under static error. It reflects the residual error value of the system after closed-loop correction. Under static error, the initial control error of the system is 1.56. With the deepening of the control process, the residual control error of the system decreases. When the system reaches a stable closed loop, the closed-loop control error is reduced to 0.0001, and the system finally tends to be stable.

Figure 6.24 depicts the time-domain control curve when the open-loop disturbance variance is 0.015, that is, when the atmospheric disturbance is small (Fig. 6.24a). When the controller is adjusted for the first 0.3 s, the output value of the controller basically remains unchanged; the closed-loop adjustment time of the system is 0.3 s. Figure 6.24b depicts the control error curve when the atmospheric disturbance is small. The initial control error at this time is 0.075. When the system reaches a stable closed loop, the closed-loop control error is reduced to within 0.0045, which is 0.0705 lower than the initial time.

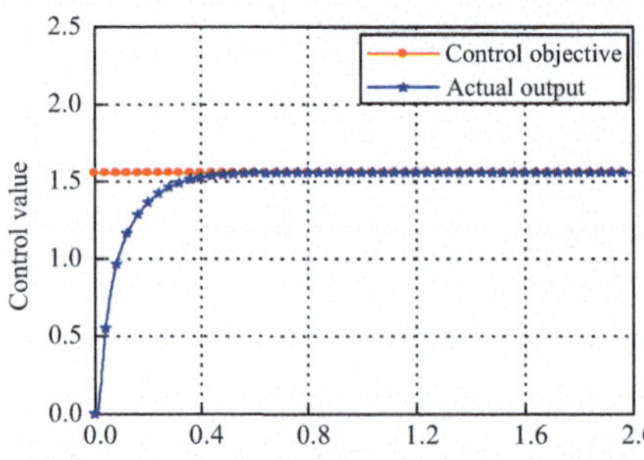

(a) Relationship between control objective and actual output

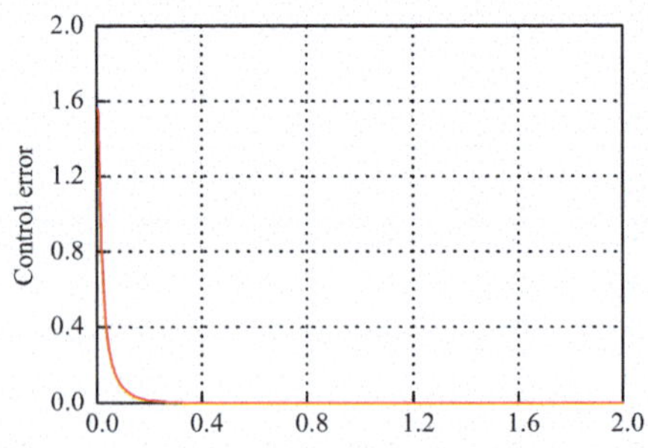

(b) Control error

Fig. 6.23 Time-domain control curve with disturbance variance $\sigma^2 = 0.015$ [50]

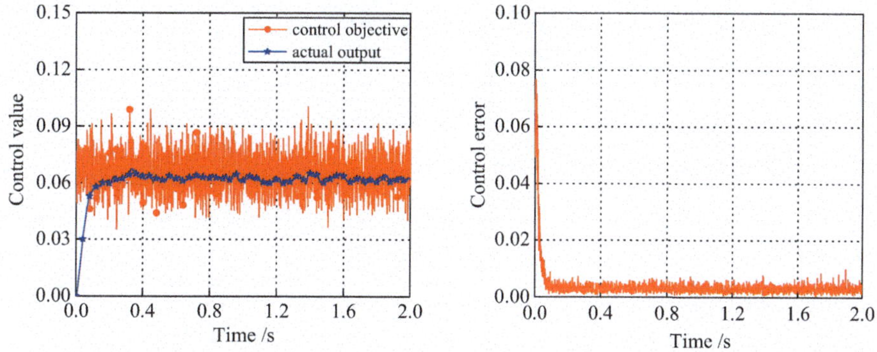

(a) Relationship between control objective and actual output

(b) Control error

Fig. 6.24 Time domain control curve with disturbance variance $\sigma^2 = 0.015$ [50]

Figure 6.25 depicts the open-loop disturbance variance of 0.06, that is, the time-domain control curve of the intermediate atmospheric disturbance. It is evident from Fig. 6.25a that the controller output value remains basically unchanged when it is adjusted after the initial 0.4 s, which means that the closed-loop adjustment time of the system is 0.4 s. Figure 6.25b depicts the control error curve when the atmospheric disturbance is small. The initial control error in this case is 0.32. When the system reaches a stable closed loop, the closed-loop control error is reduced to within 0.02, which is 0.3 less than the initial time.

Figure 6.26 depicts the open-loop disturbance variance of 0.128. It is evident from Fig. 6.26a that the stable closed-loop adjustment time of the system is 0.5 s. Figure 6.26b depicts the control error curve when the atmospheric disturbance is small. The initial control error at this time is 0.95. When the system reaches a stable

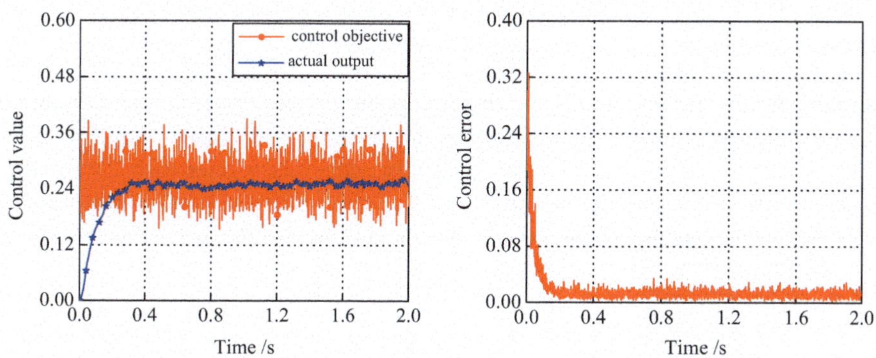

(a) Control objective versus actual output

(b) Control error

Fig. 6.25 Time-domain control curve with disturbance variance $\sigma^2 = 0.06$ [50]

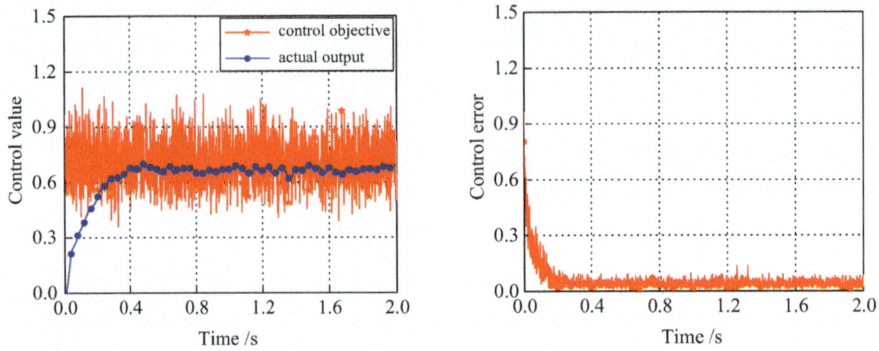

(a) Control objective versus actual output (b) Control error

Fig. 6.26 Time-domain control curve with disturbance variance $\sigma^2 = 0.128$ [50]

closed loop, the closed-loop control error is reduced to within 0.075, which is 0.875 lower than the initial time.

Figures 6.23, 6.24, 6.25 and 6.26 depict the time-domain control effects of the adaptive PI control algorithm under various atmospheric turbulence disturbances. By comparing the closed-loop adjustment time and control error, we can find that the closed-loop adjustment time of the static error is only related to the magnitude of the controlled quantity; the impact of the dynamic error time is related to not only the mean value of the controlled quantity but also the wavefront disturbance variance; for the dynamic error, although the closed-loop adjustment time increases with the disturbance variance, the residual control error value when the system is stable and closed loop is significantly smaller than the controlled variable indicates that the control system still has a good control effect against large disturbances. The algorithm has good convergence effect and closed-loop control effect for both static and dynamic errors.

Figure 6.27 depicts the power spectrum of the open-loop control signal and the corrected residual control signal when the adaptive PI control algorithm is used, with the open-loop disturbance variance being 0.128. As it shown, when the frequency is low, the difference between open-loop power spectral density and closed-loop power spectral density is large. Furthermore, as the frequency increases, the difference between the two gradually decreases; when frequency is greater than 25 Hz, the power spectrum between the two overlaps. The controller has good compensation ability for low frequency signal components, and the correction error bandwidth of the system is approximately 25 Hz.

6.6.2 Analysis of Wavefront Distortion Correction Effect

(1) **Impact of AO closed-loop correction on PV and RMS of wavefront**

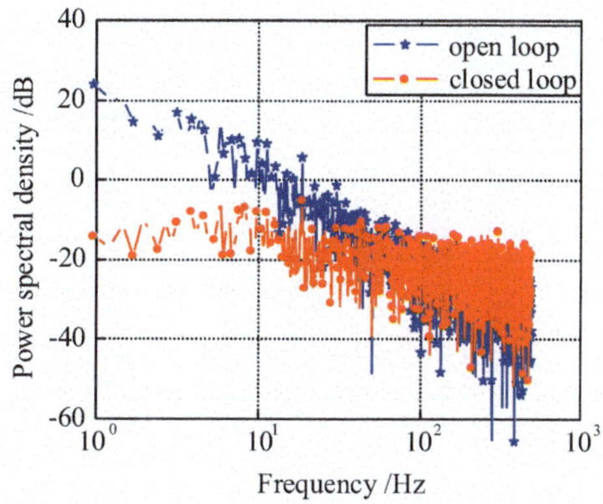

Fig. 6.27 Control voltage power spectrum when using adaptive PI control algorithm [50]

Figure 6.28 depicts the change in the RMS and SR of the wavefront over time in the open and closed loops of the AO system at various times of the day. Figure 6.28a depicts the average wavefront. The RMS can reflect the changes in atmospheric turbulence at various times of the day. It is evident from the figure that the average wavefront RMS in the open loop from 8:00 to 10:00 am and 18:00 to 21.00 pm is 2.97, 3.83 and 3.62 mm, respectively, which is large when compared with other times of the day, and indicate that the atmospheric turbulence in these two periods is stronger than that in other periods. By observing the RMS of the wavefront after the closed loop, it is evident that the mean square of the wavefront is generally reduced when the AO system is closed. This indicates that the AO closed-loop correction can adequately compensate the wavefront caused by atmospheric turbulence. The best correction effect of the system is between 12:00 noon and 14:00 pm. After correction, the average wavefront RMS of the system is 0.48. Figure 6.28b depicts the wavefront SR corresponding to Fig. 6.28a, which reflects the concentration of light intensity and energy. When the atmospheric turbulence is stronger, the wavefront distortion is more significant. The smaller the beam SR, the larger the SR. After comparing the SR values before and after the closed loop, it is evident that the correction ability of the AO is directly related to the concentration of the beam energy. When the RMS is small, the corresponding average open-loop SR is relatively large. After the AO closed-loop correction, the SR can be increased to 0.6 at most. It is evident from the above results that the AO system proposed in this study has long-term stable closed-loop working capability. After closed-loop correction, the wavefront becomes flatter and the beam energy is more concentrated.

Figure 6.29a, b depict the wavefront performance curves of AO in open and closed loops on a cloudy day, respectively. When the loop is open, the mean value of the wavefront PV is 12.99, and the variance is 0.50; the mean value of the RMS of the wavefront is 3.0, and the variance is 0.03. In the closed loop, the wavefront PV is 1.4, which is 11.55 lower than that in the open loop; the variance is 0.13, which is 0.37 lower than that in the open loop. The mean value of the wavefront RMS is 0.28, which is 2.72 lower than that in the open loop. The variance is 0.0065, which is 0.027 less when compared with that in the open loop. From the above data, it is evident

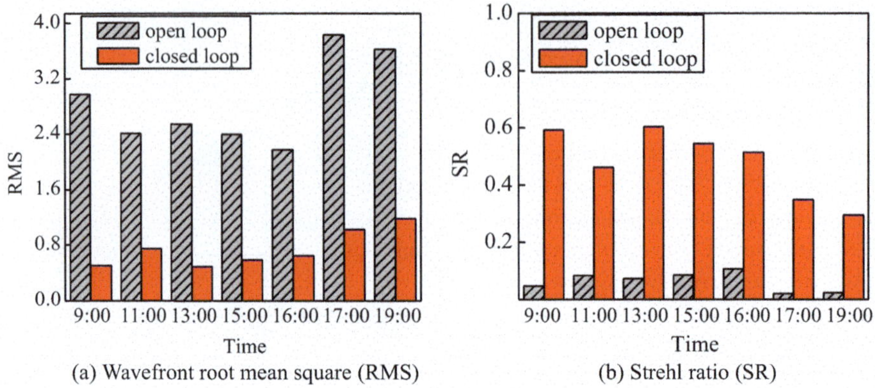

Fig. 6.28 RMS and SR at various times of the day of open and closed loops on April 23 [50]

that the mean value of the wavefront PV and RMS are large and the variance is small when the loop is open on a cloudy day. This indicates that the rate of change of the wavefront is slow, and the atmosphere at this time is relatively quiet; after the AO closed-loop correction, the wavefront PV and RMS become significantly small, and the wavefront becomes extremely flat, almost close to the ideal plane. This indicates that the AO system has a good correction effect on the wavefront on cloudy days.

Figure 6.30a, b depict the performance curves of the wavefront of AO in open and closed loops on a sunny day, respectively. When the loop is open, the mean value of the wavefront PV is 8.1, and the variance is 1.19; the mean value of the RMS of the wavefront is 2.27, and the variance is 0.09. In the closed loop, the wavefront PV is 2.3, which is reduced by 5.8 when compared with that in the open loop; the variance is 0.41, which is 0.78 less compared with that in the open loop. The mean value of the wavefront RMS is 0.50, which is 1.77 lower than that in the open loop. The variance

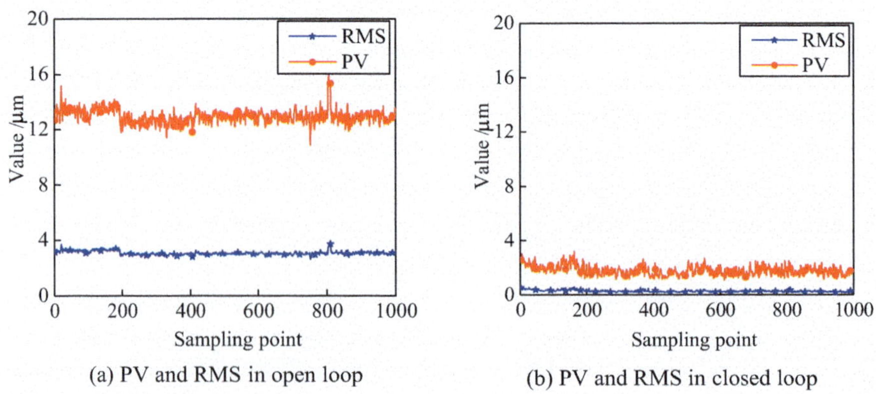

Fig. 6.29 PV and RMS changes on cloudy days [50]

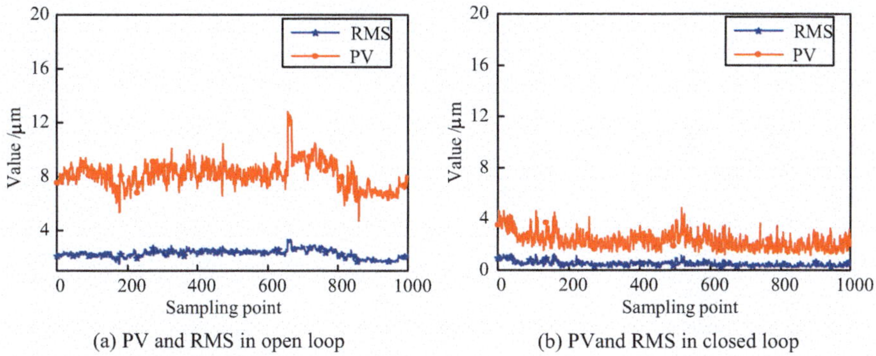

Fig. 6.30 PV and RMS changes on sunny days [50]

is 0.03, which is 0.05 less when compared with that in the open loop. Compared with the cloudy day, the open-loop PV and RMS mean values are smaller; however, the variance is large and the wavefront disturbance is significant. After the AO closed-loop correction, the mean and variance of the wavefront PV and RMS are slightly larger than those on sunny days. However, the open-loop ratio is significantly smaller, indicating that the AO system can significantly suppress wavefront fluctuations on sunny days.

Figure 6.31a, b depict the performance curves of the wavefront of AO in open and closed loops on rainy days. When the loop is open, the mean value of the wavefront PV is 9.31, and the variance is 1.73; the mean value of the wavefront RMS is 2.3, and the variance is 0.125. In the closed loop, the wavefront PV is 1.6, which is 7.72 less when compared with that in the open loop; the variance is 0.25, which is 1.47 less when compared with that in the open loop; the mean value of the wavefront RMS is 0.34, which is 2.0 less when compared with that in the open loop. The variance is 0.011, which is a decrease of 0.112 when compared with that in the open loop. The above data indicate that, after the AO closed-loop correction is used, the wavefront PV and RMS decrease, wavefront fluctuations are significantly improved, and the wavefront tends to be flat. In addition, it is evident from Fig. 6.31a that the deep fading phenomenon of the wavefront is substantial when the loop is open, which is one of the main factors affecting the stability of the coherent optical communication system. As is evident from Fig. 6.31b, after the AO closed-loop correction, the wavefront performance is basically maintained in a relatively good state, and the deep fading phenomenon of the fast wavefront performance is significantly improved. It is evident from the above results that the system can effectively suppress wavefront distortion caused by turbulence and can effectively suppress wavefront fluctuations and deep fading while reducing the wavefront peak, valley value PV, and wavefront RMS.

Figures 6.29, 6.30, and 6.31 depict the open- and closed-loop wavefront correction curves for cloudy, sunny, and rainy days, respectively, corresponding to weak, medium, and strong turbulence conditions. Comparing the open-loop wavefront, it is

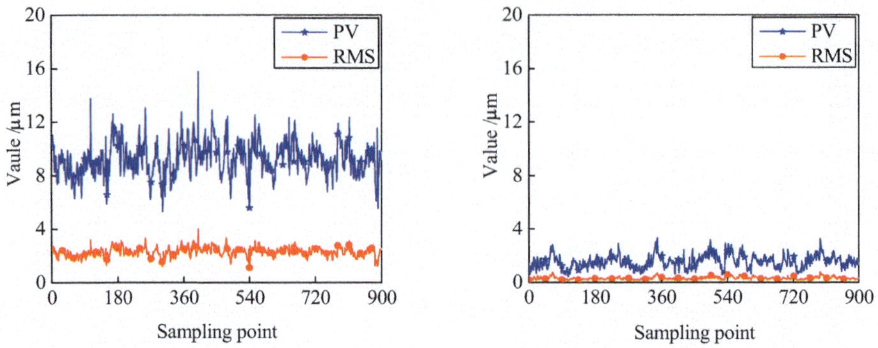

(a) PV and RMS in open loop

(b) PVand RMS in closed loop

Fig. 6.31 PV and RMS changes on rainy days [50]

evident that the wavefront fluctuation is substantial under strong turbulence, particularly the deep fading phenomenon, which makes the optical power coupled into the single-mode fiber severely jittered. Under the medium and weak turbulence conditions, the wavefront fluctuation is relatively small, and the wavefront average value is significant, which seriously affect the optical power coupled into the single-mode fiber. Through the closed-loop correction effects of weak, medium, and strong turbulence conditions, it is evident that the mean and variance of the wavefront PV and RMS are reduced after the closed-loop correction, indicating that the wavefront fluctuations are adequately suppressed. Distorted wavefronts caused by turbulence have acceptable correction effects.

(1) **Impact of AO closed-loop correction on coupling effect and intermediate frequency signal**

a. **Impact of closed-loop correction on optical power of single-mode fiber**

The closed-loop correction effect directly affects the optical power coupled into the single-mode fiber, and the size and fluctuation of the optical power directly affect the coherent efficiency of the coherent receiver.

Figures 6.32a–c depict the optical power curves of the single-mode fiber when the loop is open and closed in cloudy, sunny, and rainy days, respectively. It is evident from Fig. 6.32a that, during the closed-loop process, the optical power coupled into the single-mode fiber continues to increase, and finally reaches a stable state after the 70th frame. Based on statistical analysis, after AO stabilizes the closed loop, the optical power is increased from −41.54 dBm in the open loop to −30.03 dBm in the closed loop, which is an increase of 11.51 dB; the variance is reduced from 0.270 to 0.052, which is a decrease of 0.218. It is evident from Fig. 6.32b that, in the closed-loop process, the system basically reaches a stable state after the initial 76-frame closed-loop adjustment. Based on statistical analysis, after AO stabilizes the closed loop, the optical power increases from −44.20 dBm in the open loop to −33.41 dBm in the closed loop, which is an increase of 10.79 dB; the variance is reduced from 1.81 to 0.97, which is a decrease of 0.84. It is evident from Fig. 6.32c that, in the closed-loop process, the optical power coupled into the single mode reaches a stable state after the closed-loop correction after the first 127 frames. Based on statistical analysis, after AO stabilizes the closed loop, the optical power increases from −34.72 dBm in the open loop to −34.60 dBm in the closed loop, which is an increase of 9.12 dB; the variance is reduced from 2.82 to 1.35. Comparing the improvement effects of AO closed-loop correction on cloudy, sunny, and rainy days, the improvement effect on cloudy days is significantly better than that on sunny and rainy days, and it is better than rainy days on sunny days. This improvement effect is mainly reflected in the increase in the average optical power and reduction in wavefront jitter. It is evident from the above data that when the loop is open, owing to atmospheric turbulence, the wavefront phase distortion results in low optical power coupled into the single-mode fiber and is accompanied by severe fast fading. This fast fading phenomenon is atmospheric turbulence. It is caused by frequency jitter and other reasons introduced by the telescope antenna rack. Experimentally, it is found that the intermediate frequency signal output by the coherent receiver also drops sharply when the optical power coupled into the single-mode fiber drops sharply, thereby reducing the communication quality or even interrupting the communication altogether. The optical power increases after the AO closed-loop correction, whereas the fast fading phenomenon is suppressed.

b. **Impact of closed-loop correction on peak-to-peak value of intermediate frequency voltage.**

Figure 6.33 depicts the instantaneous value of the intermediate frequency voltage output by the coherent receiver on a cloudy day. Figure 6.33a depicts the instantaneous change curve of the intermediate frequency voltage in the open loop. The peak-to-peak instantaneous intermediate frequency voltage is 200 mV in this case. Figure 6.33b depicts the intermediate frequency voltage in the closed loop. The peak-to-peak intermediate frequency voltage can be increased to 640 mV, which is an increase of 440 mV when compared with that in the open loop. Figure 6.34 depicts the instantaneous value of the intermediate frequency voltage output by the coherent receiver on a sunny day. Figure 6.34a depicts the instantaneous change curve of the intermediate frequency voltage in the open loop. The peak-to-peak instantaneous intermediate frequency voltage is 160 mV in this case. Figure 6.34b depicts the intermediate frequency voltage in the closed loop. The peak-to-peak intermediate frequency voltage can be increased to 390 mV, an increase of 230 mV when compared with that in the open loop. Figure 6.35 depicts the instantaneous value of the intermediate frequency voltage output by the coherent receiver on a rainy day. Figure 6.35a depicts the instantaneous change curve of the intermediate frequency voltage in the open loop. The peak-to-peak instantaneous intermediate frequency voltage in this case is 50 mV. Figure 6.35b depicts the intermediate frequency voltage in the closed loop. The peak-to-peak intermediate frequency voltage can be increased to 260 mV, which is an increase of 210 mV when compared with that in the open loop. Comparing the effects of AO on the peak-to-peak value of the intermediate frequency voltage on cloudy, sunny, and rainy days, it is evident that the effect of raising the intermediate frequency voltage on cloudy days is significantly better than that on

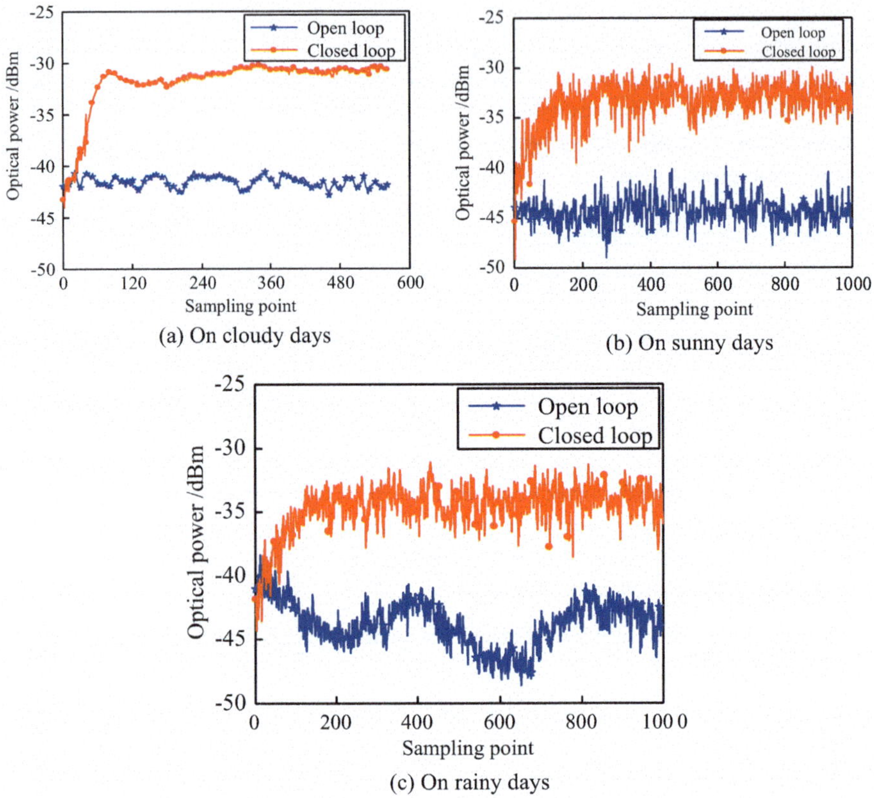

Fig. 6.32 Power coupled into single-mode fiber [50]

sunny and rainy days, and it is better on sunny days than on rainy days. Based on the experimental verification of the system, the peak-to-peak value of the intermediate frequency voltage is less than 220 mV, and the quality of the signal demodulated by the coherent demodulation system is poor and does not meet the communication requirements. In the open loop, the quality of the baseband signal recovered by the coherent demodulation system is poor. After the closed-loop correction, as the single-mode fiber coupling efficiency increases, the coherent mixing efficiency also increases so that the coherent demodulation system recovers a better quality baseband signal.

c. **Impact of closed-loop correction on power spectrum of intermediate frequency voltage**

Figure 6.36 depicts the power spectrum of the intermediate frequency voltage on a cloudy day. Figure 6.36a depicts the intermediate frequency voltage power spectrum when the loop is open. The main harmonic peak of the power spectrum is 3.6 dB/MHz at 117.1 MHz, and the multiple harmonic peak is not obvious enough. Figure 6.36b

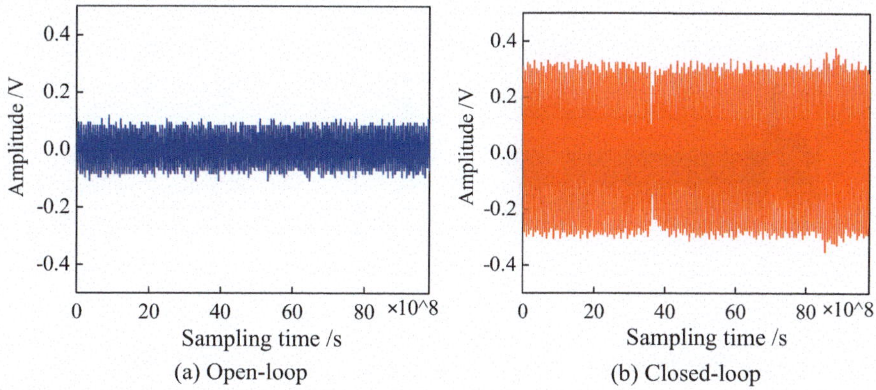

Fig. 6.33 Time-domain value of instantaneous IF voltage on cloudy days [50]

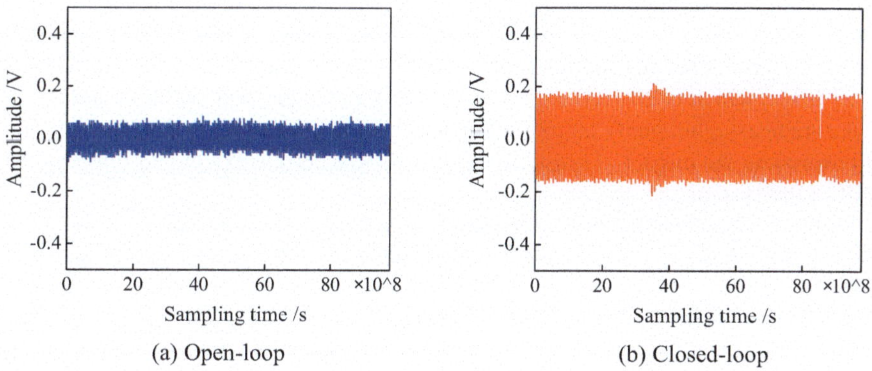

Fig. 6.34 Time-domain value of instantaneous IF voltage on a sunny day [50]

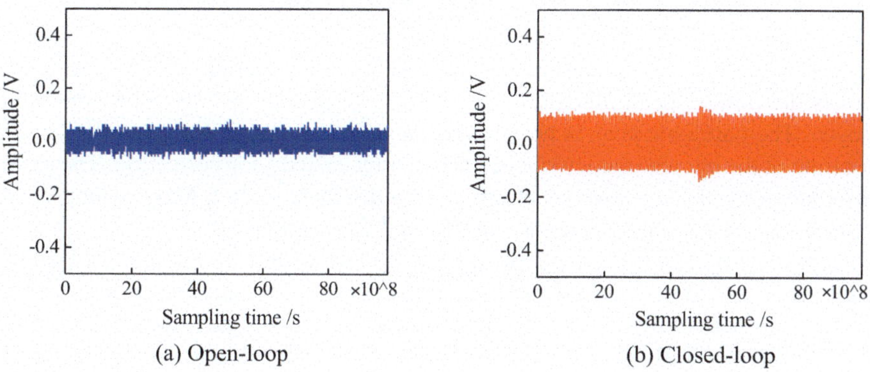

Fig. 6.35 Time-domain value of instantaneous intermediate frequency voltage on rainy day [50]

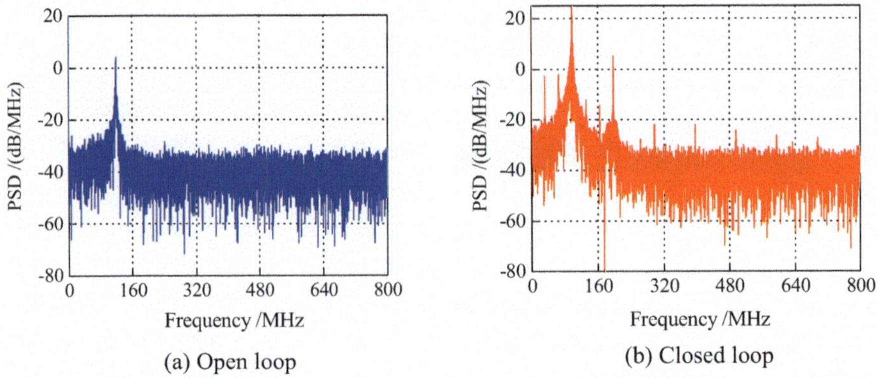

Fig. 6.36 Power spectrum of intermediate frequency signal on cloudy day [50]

depicts the power spectrum of the intermediate frequency signal after closed-loop correction. The main harmonic peak of the power spectrum is 23.8 dB/MHz at 102.1 MHz, which is 20.2 dB higher than that in the open loop. Multiple harmonic peaks are also significantly larger, indicating that the closed-loop control correction can overcome the impact of noise on the system to a certain extent and improve the SNR of the system.

Figure 6.37 depicts the power spectrum of the intermediate frequency signal on a sunny day. Figure 6.37a depicts the power spectrum of the intermediate frequency voltage when the loop is open. The peak value of the main harmonic of the power spectrum is 8.2 dB/MHz at 76.8 MHz. The peak value of the wave is not obvious enough. Figure 6.37b depicts the power spectrum of the intermediate frequency signal after the closed-loop correction. The main harmonic peak of the power spectrum is 19.6 dB/MHz at 74.1 MHz, which is an increase of 13.4 dB when compared with that in the open loop. For high-order harmonics, although the peak value is improved to a certain extent, the effect of the improvement is not evident. Figure 6.38 depicts the power spectrum of the intermediate frequency signal on a rainy day. Figure 6.38a depicts the power spectrum of the intermediate frequency voltage when the loop is open. The harmonic peak is not evident enough. Figure 6.38b depicts the power spectrum of the intermediate frequency signal after closed-loop correction. The main harmonic peak of the power spectrum is 15.5 dB/MHz at 76.6 MHz, which is an increase of 10.05 when compared with that in the open-loop dB.

Figures 6.36, 6.37, and 6.38 depict the IF signal optical powers on cloudy, sunny, and rainy days, respectively. Comparing the open-loop and closed-loop power spectra, it is evident that the gain of the coherent receiver is increased by 20.3 dB, 13.4 dB and 10.05 dB after closed-loop correction under cloudy, sunny and rainy conditions, respectively, and the closed-loop correction effect in cloudy days is better than that in sunny and rainy days. This is because the atmosphere on cloudy days is relatively quiet and the wavefront fluctuations are small. On sunny and rainy days, the spot drift caused by the temperature, many air masses with different refractive

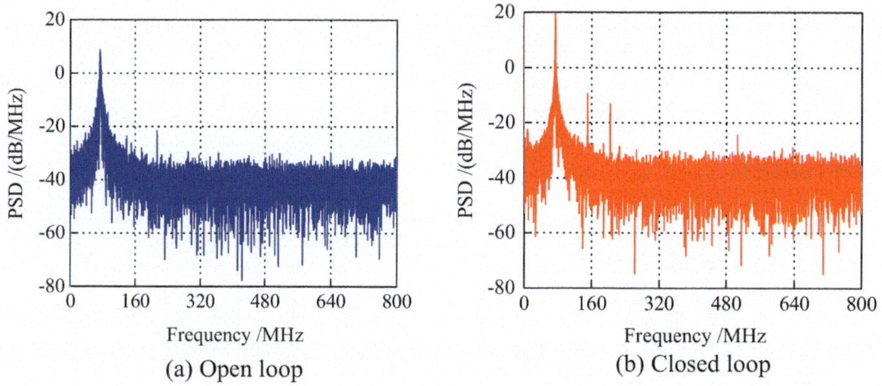

Fig. 6.37 Power spectrum of intermediate frequency signal on sunny day [50]

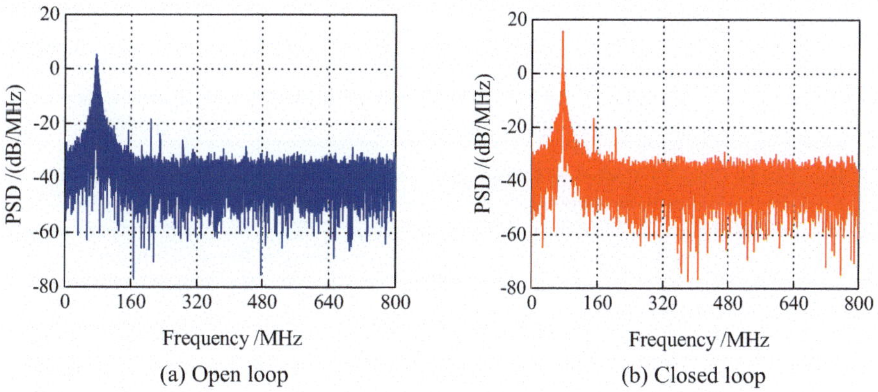

Fig. 6.38 IF signal power spectrum on a rainy day [50]

index formed by atmospheric humidity and the optical power attenuation caused by rain will reduce the coupling efficiency and increase the fluctuation of single-mode optical fiber, which substantially restrict the receiving performance of the coherent receiver.

A 1.3 km field experiment confirmed the effect of AO closed-loop correction on the performance of coherent optical communication under various weather conditions. The experimental results indicate that the system can effectively correct the wavefront distortion in a beam. By comparing the results of the AO closed-loop correction on cloudy, sunny and rainy days, the improvement effect of coherent optical communication performance is confirmed. In the closed loop, the optical power coupled into the single-mode fiber increased from −41.54 dBm, −44.20 dBm, and −43.72 dBm to −30.03 dBm, −33.41 dBm, and 34.60 dBm respectively; the peak-to-peak values of the intermediate frequency voltage increased, respectively, from 200 mV, 170 mV, and 50 mV to 640 mV, 380 mV, and 260 mV. The coherent receiving gain increased by 20.3 dB, 13.4 dB, and 10.05 dB respectively.

References

1. Huang LH, Rao C (2011) Wavefront sensorless adaptive optics: a general model-based approach. Opt Exp 19(1):371–379
2. Jiang WH (2006) Adaptive optics technology. Chin J Nat 28(1):7–13
3. Tyson RK (1991) Wizinowich P L. Principles of adaptive optics. Academic Press, Manhattan
4. Xie XC, Feng ZC (1991) Characteristics of IF signal current probability distribution for optical heterodyne receivers in the turbulent atmosphere. Appl Laser 5:209–213
5. Liu C, Chen M, Chen S et al (2016) Adaptive optics for the free-space coherent optical communications. Opt Commun 361:21–24
6. Kudielka KH, Hayano Y, Klaus W et al (2000) Low-order adaptive optics system for free-space lasercom: design and performance analysis. In: Adaptive optics for industry and medicine, international workshop, p 4–369
7. Wilks SC, Morrris JR (2002) Modeling of adaptive optics-based free-space communications systems. Proc SPIE 4821:121–128
8. Weyrauch T, Vorontsov MA (2005) Atmospheric compensation with a speckle beacon in strong scintillation conditions: directed energy and laser communication applications. Appl Opt 44(30):6388–6401
9. Hemmati H, Chen YJ, Crossfield I (2006) Telescope wavefront aberration compensation with a deformable mirror in an adaptive optics system. SPIE 105:123–127
10. Wright MW, Srinivasan M, Wilson K (2005) Improved optical communications performance using adaptive optics with an avalanche photodiode detector. Opt Eng 47(1):016003
11. Li Y (2014) Phase recovery method and experimental research based on optical intensity sensor. Xidian University, Xi'an
12. Berkefeld T, Soltau D, Sodnik Z (2010) Adaptive optics for satellite-to-ground laser communication at the 1 m Telescope of the ESA Optical Ground Station, Tenerife, Spain. In: Proceedings of SPIE—the international society for optical engineering, vol 7736(10), p 146
13. Hashmi AJ, Eftekhar AA, Adibi A et al (2014) Analysis of adaptive optics-based telescope arrays in a deep-space inter-planetary optical communications link between Earth and Mars. Opt Commun 333(4):120–128
14. Wright MW, Morris JF, Kovalik HM, et al (2014) Adaptive optics correction into single mode fiber for a low Earth orbiting space to ground optical communication link using the OPALS downlink. Opt Exp 23(26):33705
15. Jiang WH (1991) Adaptive optics technology progress. Chin J Quantum Electron 19(4):7–13
16. Jiang WH (1995) Adaptive optics technology of the Institute of Optoelectronics Technology. Optoelectron Eng 22(1):1–13
17. Jiang W, Li M, Tang G et al (1995) Adaptive optical image compensation experiments on stellar objects. Opt Eng 34(1):1–20
18. Lin XD, Liu XY, Wang JL, et al (2013) Performance test and experiment of correction capability 137-element deformable mirror. Opt Precis Eng 21(2):267–273
19. Xia LJ, Li XF (2010) Transmission Wave-front correction of atmospheric optical communication based on adaptive optics technology. J Terahertz Sci Electron Inform Technol 8(3):331–335
20. Han LQ, Wang Q, Xin TKG (2010) Free space optical communication based on adaptive optics compensation. J Appl Opt 31(02):301–304
21. Wu YY, Chen EH, Zhang Y et al (2012) Experiment on adaptive optics to improve the performance of FSO. Opt Commun Technol 04:15–18

22. Liu C, Chen S, Li XY et al (2014) Performance evaluation of adaptive optics for atmospheric coherent laser communications. Opt Exp 22(13):15554–15563
23. Hu SJ (209) The study of technology of the adaptive optics for all path aberration correction and double deformable mirrors adaptive optics. University of Electronic Science and Technology of China, Chengdu
24. Jiang ZF, Xi FJ, Xu XJ (2010) Theory and applications of defocus grating wavefront curvature sensing. Chin J Lasers 37(1):205–210
25. Xu XJ, Lu QS (2000) Comparison of wavefront restoration between shearing interferometer and Hartmann wavefront sensor. In: High power laser and particle beam, vol 12(3), pp 269–272
26. Bai FZ, Rao CH (2010) Effect of pinhole diameter on correction accuracy of closed-loop adaptive optics system using self-referencing interferometer wavefront sensor. Acta Phys Sin 59(11):8280–8286
27. Olivier SS (2002) Advanced adaptive optics technology development. In: International symposium on optical science and technology. Int Soc Opt Photonics 4494:1–10
28. Yang HF, Rao CH, Zhang YD et al (2010) Analysis of conjugate position requirements of deformable mirror and wavefront sensor in adaptive optics system. Opto-Electron Eng 36(4):27–34
29. Zhang HX, Zhang J, Wu LY (2008) Wavefront correction using liquid crystal spatial light modulator. Infrared Laser Eng 37(6):1062–1065
30. Zhang TS, Lei GY, Du XW (2000) Nonlinear optical phase conjugation compensation for laser atmospheric transmission beam distortion. In: Science and technology annual report of China academy of engineering physics, pp 269–270
31. Guo YM, Rao CH, Bao H (2014) Direct computation of the interaction matrix of adaptive optical system. Acta Physica Sinica 63(14):455–461
32. Wang L, Chen T, Liu XY et al (2015) Compensation of the non-common path aberrations in an adaptive optics system with a wavefront processor. Acta Photonica Sinica 44(5):116–120
33. Noll RJ (1976) Zernike polynomials and atmospheric turbulence. J Opt Soc Am (1917–1983) 66(3):207–211
34. Zhou RZ (1996) Theory of adaptive optics. Beijing Institute of Technology Press, Beijing, pp 62–64
35. Xia AL (2011) Research on adaptive optics system based on Shack-Hartmann wavefront sensor. Graduate School of Chinese Academy of Sciences (Xi'an Institute of Optics and Fine Mechanics), Xi'an, pp 23–24
36. Li XY, Jiang WH (2003) Analysis of the optimal-gradient reconstruction algorithm used in adaptive optical systems. Acta Optica Sinica 23(6):756–760
37. Chen SQ, Xu B, Hu SJ et al (2007) Error analysis of reconstruction matrix with direct gradient method. Opto-Electron Eng 34(2):5–10
38. Herrmann J (1980) Least-squares wave front errors of minimum norm. J Opt Soc Am 70(1):28–35
39. Hu MF (2003) Study on adaptive optical wavefront reconstruction algorithm. National University of Defense Technology, Changsha, pp 29–35
40. Hudgin RH (1977) Wave-front reconstruction for compensated imaging. J Opt Soc Am (1917–1983) 67(3):375–378
41. Zhao YF (2013) FPGA-based adaptive optics wavefront processing algorithm research and system optimization. Graduate School of Chinese Academy of Sciences (Changchun Institute of Optics, Fine Mechanics and Physics), Changchun
42. Hudgin RH (1977) Optimal wave-front estimation. J Opt Soc Am (1917–1983) 67(3):378–382
43. Fried DL (1977) Least-square fitting a wave-front distortion estimate to an array of phase-difference measurements. Opt Soc Am 67(3):370–375
44. Guo YM (2014) Optimization of wavefront reconstruction and wavefront control in adaptive optics systems. Graduate University of Chinese Academy of Sciences (Institute of Optoelectronic Technology), Chengdu
45. Kasper M, Fedrigo E, Looze DP et al (2004) Fast calibration of high-order adaptive optics systems. J Opt Soc Am A Opt Image Sci Vis 21(6):1004–1008

46. Dubra A (2007) Wavefront sensor and wavefront corrector matching in adaptive optics. Opt Exp 15(6):2762
47. Wild WJ (1998) Lyapunov stability criteria for zonal adaptive-optics systems. Opt Lett 23(8):570–572
48. Wang Z (2014) Research of real-time wavefront processing algorithm based on multi-core DSP. University of Chinese Academy of Sciences, pp 58–60
49. Liu ZW, Li ZD, Zhoug ZQ et al (2016) Adaptive optics correction technique based on fuzzy control. Acta Physica Sinica 65(1):131–138
50. Xu X (2018) Design and experimental research of wavefront correction system in wireless optical communication. Xi'an University of Technology, Xi'an
51. Chen M (2016) Research of phase matching in free-space coherent optical communication. Graduate School of Chinese Academy of Sciences, Chengdu, pp 29–30
52. Wu YY (2013) Application research of adaptive optics technique in atmospheric optical communications. University of Chinese Academy of Sciences, Beijing

Chapter 7
Wavefont Sensorless Adaptive Optics Correction

Contents

In wireless coherent optical communication, the range of optical transmission of the signal light beam changes randomly when transmitted through the atmospheric channel, leading to a distortion of the wavefront. Spatial angular collimation cannot be guaranteed at the receiver end when the aberrated signal light is mixed with the local oscillation light, which reduces the mixing efficiency and signal-to-noise ratio of the system. Adaptive optics (AO) technology without wavefront detection costs less, has a simple system structure, and does not require wavefront detectors to measure the aberration in the signal light beam. In this chapter, the principle of the abovementioned AO technology is investigated based on the stochastic parallel gradient descent (SPGD) algorithm, and the impact of this principle on the correction of aberrations in the optical wavefronts of signals in coherent optical communication systems is verified through numerical analysis and experiments.

X. Ke and J. Wu, *Coherent Optical Wireless Communication Principle and Application*, Optical Wireless Communication Theory and Technology,
https://doi.org/10.1007/978-981-19-4823-7_7

7.1 Fundamentals of Adaptive Optics

Figure 7.1 presents a schematic of the AO-based coherent optical communication system. As illustrated in the figure, the signal light beam is transmitted under atmospheric turbulence to produce aberrated light; the signal light beam is compensated by the conjugate phase at the receiver using AO technology. Thereafter, the compensated signal light beam is mixed with the local oscillation light beam and connected to the detector [1]. The wavefront sensor detects the aberration in the wavefront information of the signal light beam in real time, and this information is processed by the wavefront controller to determine the control quantity of the wavefront corrector. The wavefront corrector generates the conjugate quantity of the detected wavefront aberration as the correction quantity based on the size of the control quantity; further, the wavefront corrector performs real-time closed-loop correction of the aberrated signal light beam, thus improving the quality of the optical image obtained at the receiver. Several experiments have verified that AO technology can effectively correct a distorted signal light beam. Therefore, the study of channel compensation technology for AO-based coherent optical communication systems is significant.

Based on the principle used for the correction of the wavefront aberration, AO can be divided into two categories: AO with wavefront detection and AO without wavefront detection, the former also referred to as the conventional AO. Conventional AO mostly uses a wavefront detector for the real-time detection of the optical aberration information contained in the signal. In general, the greater the number of wavefront detector sub-apertures, the higher the accuracy of the detected wavefront information. However, in an actual system, the conventional AO structure is complex and expensive, and the wavefront detector sub-aperture diverts a part of the system light intensity, leading to the system failure when the light intensity is zero. AO without wavefront detection does not require to measure the aberration in the signal light beam using the wavefront detector, although the accuracy of the correction of the wavefront aberration is not as high when compared with conventional AO. However, considering the low cost of the system, the use of AO without wavefront detection for the correction of the wavefront aberration has the prospects of wide ranging applications. Figure 7.2 presents a block diagram to illustrate the principle of AO without wavefront detection.

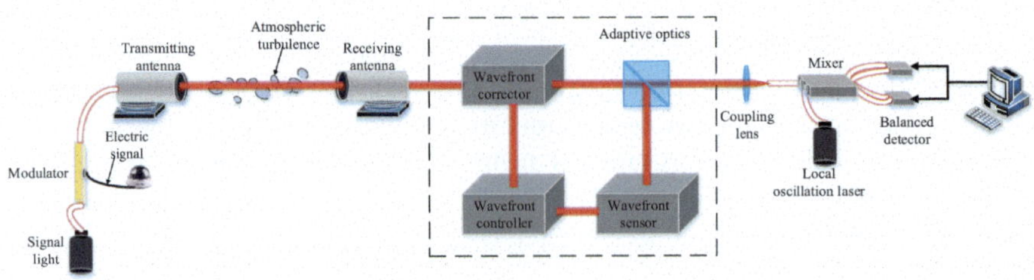

Fig. 7.1 Schematic of AO-based coherent optical communication system technology

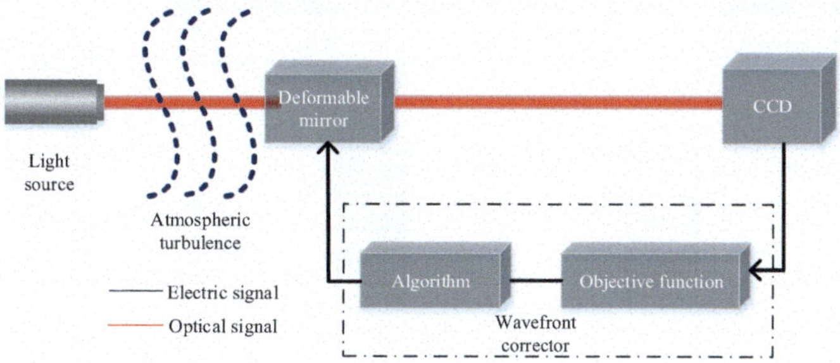

Fig. 7.2 Schematic of AO without wavefront detection

As illustrated in the figure, the system for AO without wavefront detection mainly consists of a wavefront controller, wavefront corrector (distorting mirror), and imaging detector (charge-coupled device, CCD) [2]. The parallel light from the light source is transmitted under atmospheric turbulence to produce an aberrated beam. The aberrated beam is incident on the deformation mirror, which corrects the aberrated beam and reflects the residual aberrated beam to the CCD. The wavefront controller drives an intelligent algorithm to regenerate the control signal for the deformation mirror, based on the system performance index values collected by the CCD, and achieve multiple closed-loop corrections of the aberrated beam.

7.1.1 Wavefront Corrector

The purpose of AO technology is to correct the wavefront aberration and obtain an ideal wavefront, whose performance directly determines the overall AO performance. At present, two types of wavefront correctors are commonly used: deformation mirrors and spatial light modulators.

(1) **Deformation mirror**

Deformation mirrors can be further divided into two types based on the distribution of the shape of the surface: continuous and discrete. Continuous-surface anamorphic mirrors are widely used in AO systems owing to their advantages of a continuous surface shape and high wavefront aberration compensation. The structure of the continuous-surface deformation mirror of a discrete actuator is depicted in Fig. 7.3.

Figure 7.3 indicates that the continuous-surface deformation mirror consists of three main parts: the base, actuator, and mirror surface. When a control signal is applied to the actuator, being based on a Gaussian function, it can deform the mirror surface owing to the up and down movements, and the total deformation is expressed as follows [3]:

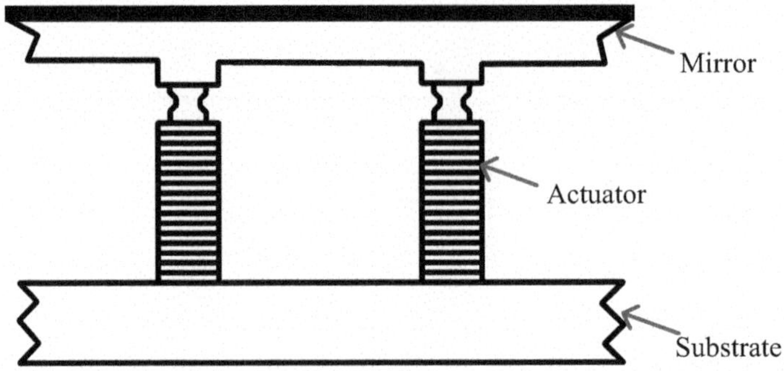

Fig. 7.3 Structure of the continuous-surface deformation mirror of a separation actuator

$$\psi(r) = \sum_{i=1}^{N} V_j S_j(r) \tag{7.1}$$

where N is the number of deformation mirror actuators, V_j is the control signal applied to the jth actuator, and $S_j(r)$ is the influence function for the jth actuator, which is expressed as follows:

$$S_j(r) = S_j(x, y) = \exp\left\{\ln w\left[\frac{1}{d}\sqrt{(x - x_j)^2 + (y - y_j)^2}\right]^{\alpha}\right\} \tag{7.2}$$

where w is the cross-connected value of the actuator, α is the Gaussian index, d is the normalized spacing between the actuators, and (x_j, y_j) are the coordinates of the jth actuator. The Gaussian index and cross-connected values have a more significant effect on low-order aberrations than on high-order aberrations; therefore, the appropriate values of w and α must be determined. Considering $w = 0.08$, $\alpha = 2$ and 69 actuator unit deformation mirrors in the simulation model, the total surface shape distribution of the deformation mirrors when the maximum control signal is applied to each actuator is depicted in Fig. 7.4, from which indicates that the surface shape of each actuator obeys a Gaussian distribution.

(2) **Liquid crystal–spatial light modulator**

A liquid crystal–spatial light modulator (LC-SLM) is a device that modulates the specific parameters of light waves by varying certain characteristics of liquid crystal molecules. Each molecule can modulate the light wave incident on its surface using electrical or optical signals.

The deformation mirror corrects the wavefront aberration phase by varying the difference in the optical range of the incident light wave, whereas the LC-SLM controls the phase of the incident light wavefront by modifying certain characteristics of the liquid crystal molecule. Under the birefringence effect of liquid crystal molecules, the incident light is divided into plain light and light obtained from the

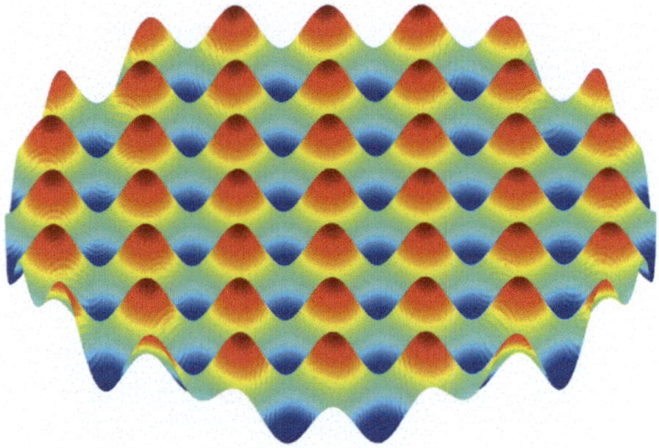

Fig. 7.4 Total surface shape distribution of deformation mirrors in the actuator unit

LC-SLM; the refractive indices of these two lights are denoted by θ, respectively. When different magnitudes of voltage are applied at the two ends of the LC-SLM, the liquid crystal molecules become angularly displaced in a specific direction under the action of an electric field. That is,

$$\theta = f(V) \tag{7.3}$$

where V is the applied voltage. The liquid crystal molecule gets deflected when the voltage loaded on the LC-SLM exceeds the voltage at which the liquid crystal molecule deflects itself. At this instant of time, the refractive index of the e-light changes and the refractive index of the o-light remains unaffected. The effective refractive index $n_e(V)$ of the e-light at this instant of time is determined based on physical optics and is expressed as follows [4]:

$$n_e(V) = \frac{n_e \cdot n_o}{\sqrt{n_e^2 \sin^2\theta + n_o^2 \cos^2\theta}} \tag{7.4}$$

To implement phase modulation, the LC-SLM must be in a state of pure phase modulation. The modulator would be in such a state when the beam incident on it is linearly polarized and the direction of polarization is parallel to the optical axis of the liquid crystal. The LC-SLM is designed to change the refractive index of the e-light by controlling the magnitude of the applied voltage, so that the refractive indices of the e- and o-light are changed. Furthermore, the phase of the incident light changes and the amount of phase change is expressed as follows [5]:

$$\delta_1 = \frac{2\pi d(n_e(V) - n_o)}{\lambda} \tag{7.5}$$

where δ_1 is the extent of phase change, d is the thickness of the liquid crystal layer, and λ is the wavelength of the incident light.

Because the liquid crystal layer has a certain thickness, the phase change over the entire surface of the LC-SLM can be determined by integrating Eq. (7.5) over the entire range of liquid crystal molecule thicknesses, that is,

$$\delta = \int_0^d \frac{2\pi d(n_e(V) - n_o)}{\lambda} \tag{7.6}$$

When the phase modulation is performed, the LC-SLM must be guaranteed to be in a state of pure phase modulation, which requires a highly optical environment. The LC-SLM is able to implement a limited amount of phase correction for the beam and is generally not suitable as a correction device for signal light wavefront distortion under strong turbulence.

7.1.2 Wavefront Controller

The wavefront controller converts the measured beam characteristic information into the control signal of the deformation mirror and controls the conjugate phase of the deformation mirror to produce a distorted signal light. In conventional AO, the wavefront sensor measures the slope of the beam wavefront. Therefore, the wavefront controller must reconstruct the wavefront to convert the wavefront slope information into wavefront phase information. Thereafter, the wavefront controller must generate a control signal for the deformation mirror to drive it to change the optical range of the beam. In AO without wavefront detection, the wavefront controller receives the far-field spot light intensity information measured by the CCD. The wavefront controller establishes a relationship between the light intensity information and the control signal of the deformation mirror through a specific intelligent optimization algorithm, and finally, the control signal size of the deformation mirror is determined based on the convergence and extreme value requirements of the optimization algorithm. The processing speed of the wavefront controller directly determines the operating speed of the AO system. To shorten the computing time as much as possible, a corresponding hardware processing module, such as the FPGA or microcontroller module, is generally added to the controller unit.

7.1.3 Stochastic Parallel Gradient Descent Algorithm

For AO without wavefront detection, intelligent optimization algorithms are divided into two main categories: mode-based and mode-free optimization algorithms. Mode-based optimization involves choosing a suitable base mode to describe the wavefront error and correcting the wavefront aberration by optimizing the base-mode coefficients, which are characterized by the requirement to obtain the influence function for each actuator face shape of the deformation mirror. There are two general methods based on mode optimization: one is based on the iterative mode optimization method, in which the wavefront aberration is represented by a suitable Zernike polynomial mode order, the relationship matrix between the influence function of the deformation mirror and the mode coefficients is established, and multiple iterations of the optimization algorithm are carried out to determine the size of the control signal for the deformation mirror actuators. The other method is based on a non-iterative mode, that is, a specific mode as a base function, such that the aberrated wavefront can be linearly expanded by the mode. Therefore, based on the mathematical relationship between the mode coefficients and the evaluation function, the mode coefficients and conjugate wavefront are determined to correct the wavefront aberration. The mode-free optimization algorithm directly controls the distortion mirror to correct the wavefront aberration and causes the system performance evaluation index to reach an optimal value. Commonly used mode-free optimization algorithms include the genetic algorithm (GA), simulated annealing algorithm (SAA), and stochastic parallel gradient descent (SPGD) algorithm.

A mode-based optimization algorithm requires the face shape influence function for each actuator of the deformation mirror in advance and determines the value of the optimal control signal of the deformation mirror by establishing the correspondence between the base-mode coefficients and the control signal of the deformation mirror. Therefore, a simpler and more convenient mode-free optimization algorithm is generally used, while the GA algorithm converges slowly and the SA algorithm gets trapped easily at the local extremes. The SPGD algorithm is discussed below.

Figure 7.5 presents the flowchart of the SPGD algorithm. First, the initial voltage signal of the deformed mirror and the random disturbance voltage signal conforming to the Bernoulli distribution are generated, and based on the initial voltage signal, positive and negative disturbance voltages are added to calculate the variation in the system objective function caused by the voltage. Finally, the voltage signal to be loaded in this iteration is calculated based on the iterative formula, and so on, until the objective function meets the conditions initially set by the system.

The steps for implementing the SPGD algorithm are as follows [6]:

(1) Initialization: A set of initial control voltage signals $U_0 = [u_1, u_2, \cdots, u_N]$ is generated randomly for the deformation mirror actuator, where $u_i = 0$, $i = 1, 2, \cdots, N$.
(2) Correction iteration module: The results of the jth iteration are known. In the ($j + 1$)th iteration, the random disturbance voltages $\Delta u_j = [\delta u_1, \delta u_2, \cdots, \delta u_N]$,

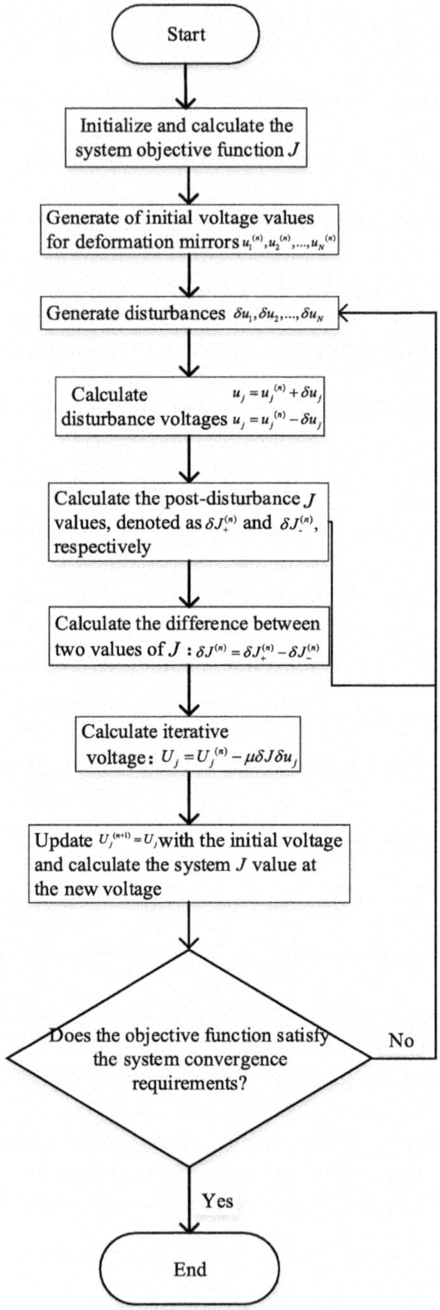

Fig. 7.5 Flowchart describing stochastic parallel gradient descent algorithm

which are independent of each other and conform to the Bernoulli distribution, are generated. The voltages $U_j + \Delta u_j$ and $U_j - \Delta u_j$ are applied to the deformable mirror, and the system performance evaluation functions $\delta J_+^{(j)}$ and $\delta J_-^{(j)}$, respectively, are calculated under the action of these voltages. The variation $\delta J^{(j)} = \delta J_+^{(j)} - \delta J_-^{(j)}$ in the system performance evaluation function is calculated using $\delta J_+^{(j)}$ and $\delta J_-^{(j)}$.

(3) Calibration update module: The new voltage quantity $U_{j+1} = U_j - \mu\delta J^{(j)}\Delta u_j$ and the corresponding system performance evaluation function $J^{(j+1)}$ are calculated using the voltage iteration formula U_{j+1}. If the current $J^{(j+1)}$ does not meet the system requirements, Step 2 is repeated; if the requirements are met, the loop is exited.

7.2 Correction of Wavefront of Aberrated Gaussian Beams Using SPGD Algorithm

7.2.1 *Optical Transmission Equation and Multiphase Screen Method*

The numerical simulation of light propagation in a random medium is based on the propagation equation of the light field. When the medium non-uniform scale is much larger than the wavelength λ of light, the propagation problem along the z direction can be approximated by the paraxial. The standard parabolic equation is expressed as follows [7]:

$$2ik\frac{\partial u}{\partial z} + \frac{\partial^2 u}{\partial x^2} + \frac{\partial^2 u}{\partial y^2} + 2k^2 n_1 u = 0 \tag{7.7}$$

where n_1 is the fluctuating value of the refractive index, $k = 2\pi / \lambda$, u is the complex amplitude of the optical field, and (x, y, z) denote the spatial position of the point source.

When the propagation of light only through vacuum is considered, the term related to the refractive index in Eq. (7.7) becomes zero and only the derivative term of the field survives. The solution of the point source located in space (x', y', z') in space (x, y, z) is expressed using the Green function, that is,

$$u(x, y, z) = \frac{u(x', y', z')}{z - z'} \exp\left[-ik\frac{(x - x')^2 + (y - y')^2}{2|z - z'|}\right] \tag{7.8}$$

When only the effect of the refractive index of the medium on the propagation of light is considered, so that the term in Eq. (7.8) independent of the refractive index becomes zero, the extent of the phase modulation of the light along the propagation direction is expressed as follows:

$$u(r,z) = u(r,z')\exp\left[ik\int_{z'}^{z} n_1(r,\zeta)d\zeta\right] = u(r,z')e^{iS} \tag{7.9}$$

Considering Eq. (7.9), the effect of vacuum and the refractive index of the medium on the propagation of light can be discussed separately when the phase S changes by a sufficiently small amount. That is, the transmission path z of the beam is considered to consist of free space, and the i phase screens z_i uniformly scattered in between, as depicted in Fig. 7.6. Free-space transmission is performed by Fresnel diffraction, and when the laser passes through the phase screen, the amplitude remains constant and only the phase changes. The optical field of the laser transmitted from the z_i plane to the $z_{i+1}= z_i+\Delta z$ plane is expressed as follows:

$$u(r,z_{i+1}) = f^{-1}\left\{ f\left[u(r,z_i)\exp[i\phi(r,z_i)]\right]\exp(ik\Delta z_{i+1} - i\frac{K_x^2 + K_y^2}{2k}\Delta z_{i+1})\right\} \tag{7.10}$$

where f and f^{-1} denote the Fourier and inverse Fourier transforms, respectively; Δz is the spacing between adjacent phase screens; K_x^2 and K_y^2 are the spatial wave numbers in the x- and y-direction of the null field, respectively; and the wave number $k = 2\pi/\lambda$ (where λ is the wavelength). When the optical field $u(r,z_i)$ and the phase screen $\phi(r,z_i)$ are known, the optical field $u(r,z_{i+1})$ after transmission can be determined using Eq. (7.10).

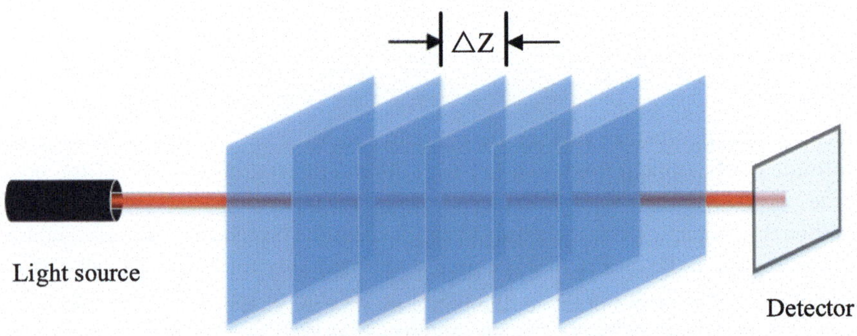

Fig. 7.6 Schematic of multiphase screen transmission

7.2.2 Simulation of Gaussian Beam Transmitted in Atmospheric Turbulence

The fundamental-mode Gaussian beam transmitted along the z-axis can be expressed as follows [8]:

$$u(x, y, z) = \frac{c}{w(z)} \exp\left\{\left[-\frac{r^2}{w^2(z)}\right] - i\left[kz + \frac{kr^2}{2R} - arctg\frac{z}{f}\right]\right\} \tag{7.11}$$

where c is the constant factor, $w(z)$ is the spot radius of Gaussian light at the point $w(z) = w_0[1 + (z/f)^2]^{1/2}$, w_0 is the waist radius of the beam, f is the confocal parameter ($f = \pi w_0^2/\lambda$), λ is the wavelength, and $R(z)$ is the radius of curvature of the Gaussian light at point z ($R = R(z) = z[1 + (f/z)^2]$).

Based on this theoretical analysis, the far-field spot light intensity map is simulated after the collimated Gaussian light is transmitted under atmospheric turbulence using the Zernike polynomial method, and the phase screen is simulated using a method based on the inversion of the Hill spectrum. The simulation parameters are the waist radius w_0 = 10 mm of the beam, transmission distance z = 6 km, wavelength λ=632.8 nm, phase screen size L = 0.4 m, phase screen grid number 256 × 256, phase screen interval 1.5 km, and far-field spot light intensity distribution after transmission at various turbulence intensities, as depicted in Figs. 7.7, 7.8, 7.9 and 7.10.

Figure 7.7 indicates that under the turbulence-free condition, the Gaussian beam displays a Gaussian-shaped distribution of the far-field spot light intensity after transmission through the atmospheric channel. Figures 7.8, 7.9 and 7.10 indicate that the far-field spot light will be distorted after transmission under atmospheric turbulence of various intensities, and the distortion becomes more significant as the intensity of turbulence increases. In addition to the described phenomenon, Figs. 7.10 indicate that the far-field spot mass drift and split occur after the light

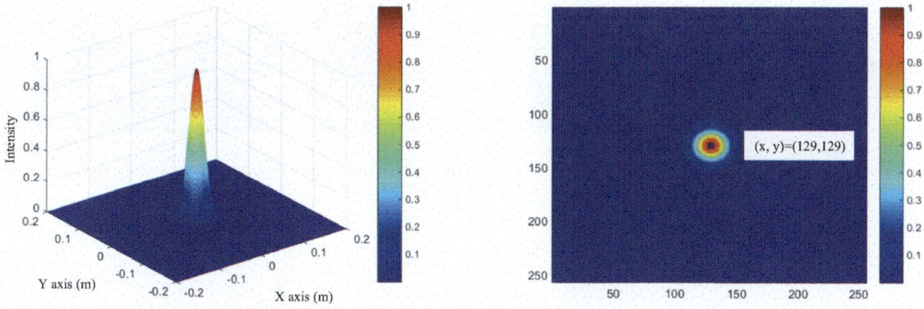

(a) Three-dimensional drawings (b) Two-dimensional top view

Fig. 7.7 Far-field spot light intensity distribution in the absence of turbulence

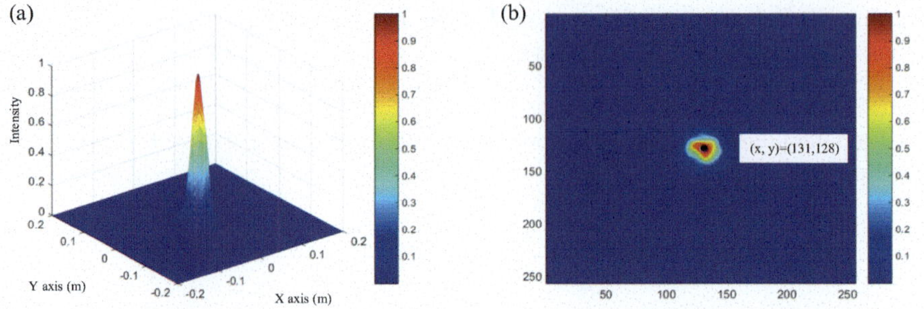

(a) Three-dimensional drawings (b) Two-dimensional top view

Fig. 7.8 Far-field spot light intensity distribution during weak turbulence

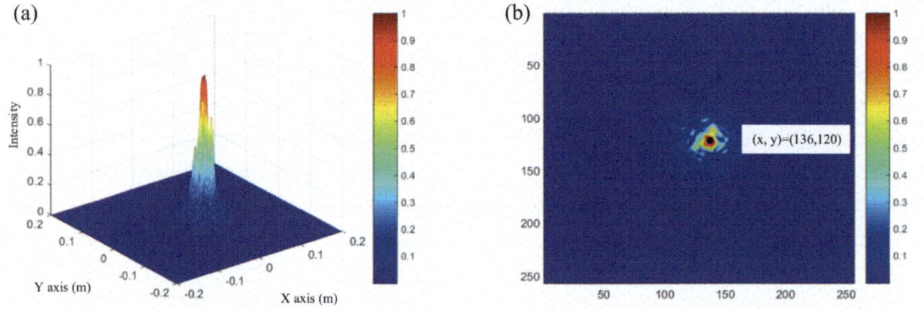

(a) Three-dimensional drawings (b) Two-dimensional top view

Fig. 7.9 Far-field spot light intensity distribution during medium-intensity turbulence

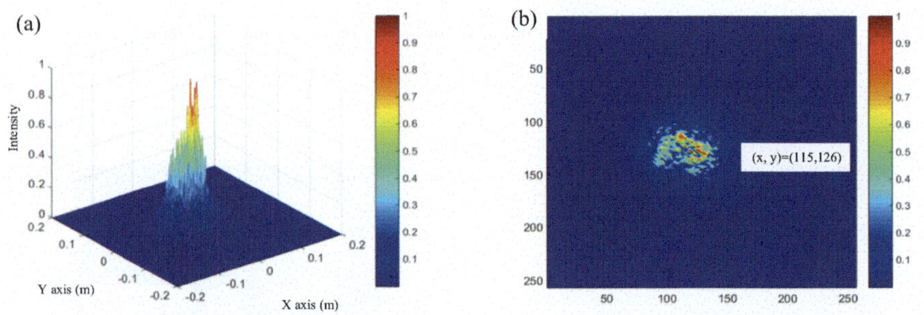

(a) Three-dimensional drawings (b) Two-dimensional top view

Fig. 7.10 Far-field spot light intensity distribution during strong turbulence

is transmitted through the atmospheric turbulence. In part (b) of each of the four figures, (x, y) represents the position of the center of mass, whose coordinates are $(x, y) = (129, 129)$,$(x, y) = (131, 128)$, and $(x, y) = (115, 126)$ under no turbulence, weak turbulence, and strong turbulence, respectively. These coordinates indicate that the effect of turbulence on the far-field spot is strengthened as the intensity of turbulence increases.

7.2.3 Signal Optical Wavefront Correction at Various Turbulence Intensities

The theoretical results indicate that the SPGD algorithm based on the variable gain coefficient can effectively improve the correction capability of the SPGD algorithm and make it converge quickly. Figures 7.11, 7.12, and 7.13 display the simulation results of the SPGD algorithm for the correction of the distortion in the optical wavefront under weak, moderate–strong, and strong turbulence, respectively. The simulation conditions are as follows: disturbance amplitude δu=0.006, gain coefficient $\gamma = 1.6/(1 + J)$, and 500 iterations.

Figures 7.11, 7.12, and 7.13 indicate that when the signal light is transmitted under weak, moderately strong, and strong turbulence, the value of the far-field spot normalized optical intensity increases from 0.7 to 0.9, 0.45 to 0.8, and 0.25 to 0.7, respectively, after AO correction. For various turbulence intensities, the far-field spot energy becomes more convergent, and stray light is suppressed well after AO correction without wavefront detection. The results of the numerical analysis indicate that AO without wavefront detection, based on the variable gain coefficient SPGD algorithm, effectively corrects the wavefront aberrations under various turbulence intensities.

(a) Uncorrected (b) Corrected

Fig. 7.11 Far-field spot map under weak turbulence

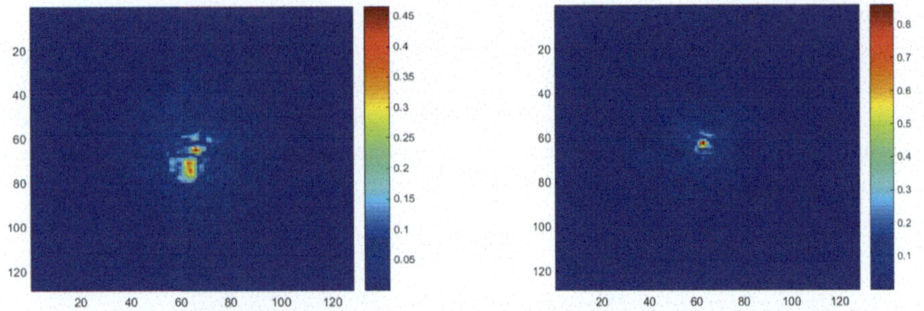

(a) Uncorrected (b) Corrected

Fig. 7.12 Far-field spot map under medium-intensity turbulence

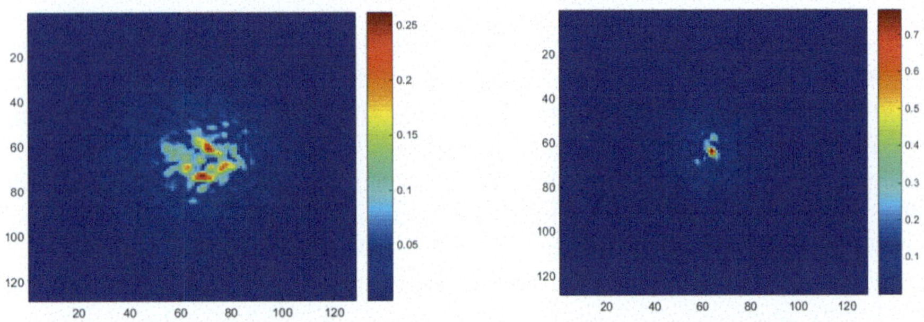

(a) Uncorrected (b) Corrected

Fig. 7.13 Far-field spot map under strong turbulence

7.2.4 AO Technology for Improvement of Performance of Coherent Optical Communication System

The higher the mixing efficiency, the better the coherence of the signal and local oscillation light beams. The impact of the AO technology on the mixing efficiency of the outlier detection coherent optical communication system based on the SPGD algorithm is analyzed below.

As depicted in Fig. 7.14, the mixing efficiency increases from 0.39 to 0.80 before correction as the number of algorithm iterations increases and stabilizes at 200 iterations. When the signal and local oscillation light beams are represented in the Airy + Gauss distribution model, the theoretical maximum value of the mixing efficiency of the system is 0.8145. A comparison of the results presented in Fig. 7.14 indicates that the AO technology can effectively improve the mixing efficiency of coherent optical communication systems.

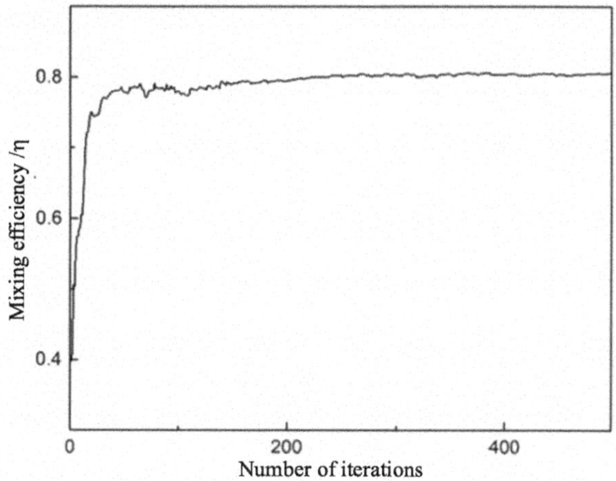

Fig. 7.14 Graph depicting variation in the mixing efficiency of the system

In coherent optical communication systems, the scattering and absorption of particles under atmospheric turbulence leads to very weak signal optical power at the receiver, and the local oscillation optical power has an amplifying effect on the signal optical power. The higher the local oscillation optical power, the better is the transmission gain of the system. Therefore, in a spatially coherent optical communication system, the local oscillation light power is much larger than the signal light power. Therefore, the internal noise of the receiver consists of mostly the scattered particle noise generated by the local oscillation light beam. The average noise current inside the receiver is expressed as follows [9]:

$$\langle i_N \rangle^2 = 2e\Delta f I_{LO} = 2e\Delta f R P_{LO} = 2e\Delta f RK \int_U A_{LO}^2 dU \tag{7.12}$$

where e is the charge, I_{LO} is the local oscillation photocurrent, Δf is the detector bandwidth ($\Delta f = 2/B$), B is the bit rate, R is the detector sensitivity ($R = e\eta/h\upsilon$), η is the quantum efficiency, h is the Planck constant, υ is the carrier frequency, P_{LO} is the local oscillation optical power, and K is the scaling factor of the detector.

The signal optical power $P_S = K \int_U A_S^2 dU$, local oscillation optical power $P_{LO} = K \int_U A_{LO}^2 dU$, and signal-to-noise ratio (SNR) of the coherent optical communication system is expressed as follows:

$$\begin{aligned} SNR &= \frac{\langle i_{RF} \rangle^2}{\langle i_N \rangle^2} = \frac{2\eta P_S}{h\upsilon B} \cdot \frac{\left[\int_U A_S A_{LO} \cos(\Delta\varphi) dU\right]^2 + \left[\int_U A_S A_{LO} \sin(\Delta\varphi) dU\right]^2}{\int_U A_S^2 dU \int_U A_{LO}^2 dU} \\ &= \frac{2\eta P_S}{h\upsilon B} \cdot \gamma_{RF} \end{aligned} \tag{7.13}$$

where $SNR_0 = 2\eta P_S/h\upsilon B$ is the SNR of the system at the limit of the scattered particle noise. Equation (7.13) can be simplified as follows:

$$SNR = SNR_0 \cdot \gamma_{RF} \tag{7.14}$$

The signal optical power P_S can be expressed as $P_S = N_P h\upsilon B$, where N_P is the average number of photons received per unit of bit time. Therefore, SNR_0 can be expressed as follows:

$$SNR_0 = \frac{2\eta P_S}{h\upsilon B} = \frac{2\eta}{h\upsilon B} N_P h\upsilon B = 2\eta N_P \tag{7.15}$$

Substituting Eq. (7.15) in Eq. (7.14), we obtain the following expression:

$$SNR = 2\eta N_P \cdot \gamma_{RF} \tag{7.16}$$

In digital communication systems, the bit error rate (BER) is an important indicator for measuring the accuracy of data transmission per unit time, and the lower the BER, the higher the accuracy of data transmission. Differential phase-shift keying (DPSK), also known as two-phase relative phase modulation, uses the difference between the initial phases of the front and rear code elements to transmit digital information. DPSK signal demodulation can recover the baseband signal correctly only by distinguishing the phase relationship between the front and rear code elements, which avoids the phenomenon of "inverted π" noted in binary phase-shift keying (BPSK). For an outlier detection coherent optical communication system based on DPSK modulation, the average BER of the system can be expressed as follows [10]:

$$BER = \frac{1}{2} e^{-SNR} \tag{7.17}$$

Substituting Eq. (7.14) in Eq. (7.17) yields the system BER in the outlier detection system as follows:

$$BER_{RF} = \frac{1}{2} e^{-SNR_0 \cdot \gamma_{RF}} \tag{7.18}$$

In the numerical simulation, the quantum efficiency η=1 corresponds to the average photon number N_P=18. Figure 7.15 presents the variation curve of the BER of the system after AO correction without wavefront detection under the condition of 500 iterations.

Figure 7.15 indicates that with an increase in the number of correction iterations, the system BER improves from the initial 3.4×10^{-8} to 1.1×10^{-13}, reaches the optimal value, and stabilizes at 200 iterations. This shows that the AO technology has a good correction effect on the BER of the coherent optical communication system with outlier detection, and can significantly reduce the probability of BER after the digital signal is transmitted under atmospheric turbulence.

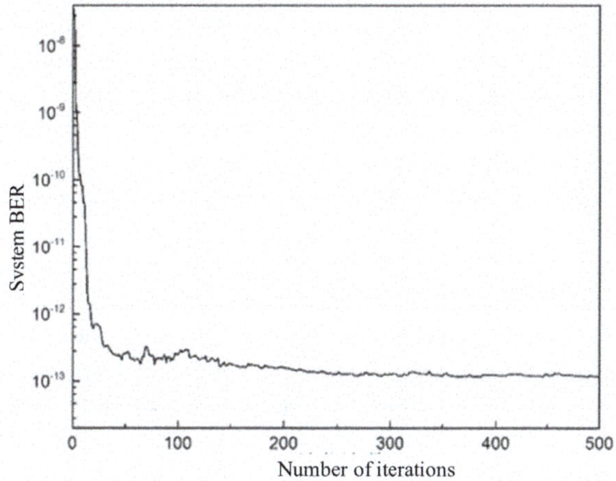

Fig. 7.15 Graph depicting variation in the system BER

7.3 Experimental Studies

7.3.1 Correction of Static Wavefront Distortion Using SPGD Algorithm

Figure 7.16 presents the experimental schematic of AO without wavefront detection. The experimental equipment includes an He–Ne laser, a spatial light modulator, a lens, a polarizer, a beam splitter, and a CCD camera.

As depicted in the figure, the parallel light from the laser is polarized by the polarizer P1 and is incident on a 4 focus (4F) system consisting of L1 and L2 for

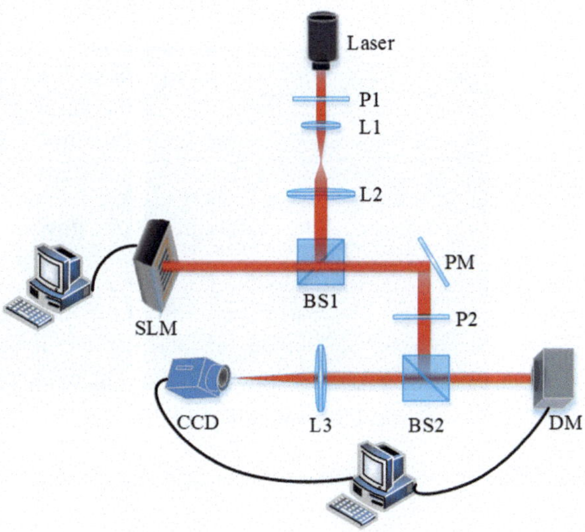

Fig. 7.16 Experimental schematic of AO without wavefront detection

beam expansion and collimation. The collimated beam is split by the beam-splitting prism BS1 and is incident on the spatial light modulator, which is loaded with a phase screen for the phase modulation of the parallel light. The modulated beam is reflected by the reflector PM and then split by BS2. The deformable mirror (DM) effects the initial correction of the aberrations in the modulated beam. The beam is then converged by lens L3 onto the CCD, which collects the intensity of the aberrations and transmits the information to the computer, which, in turn, generates the control signal for the deformation mirror to correct the wavefront aberrations several times using an optimization algorithm.

A semiconductor laser with an emission power of 35 mW and laser wavelength of 650 nm was used as the light source, and the wavefront distortion was corrected using the SPGD algorithm based on the variable gain coefficient.

The spot grayscale information collected using the CCD was used as feedback in the experiment. The average grayscale value of the pixel around the ring centered at the spot center of mass was used as the objective function, and the average grayscale value of the 10×10 pixel points centered at the spot center of mass was used. The following parameters were selected: phase screen size $L = 0.4$ m, perturbation amplitude δ=0.006 for the SPGD algorithm, and gain coefficient $\gamma = 1.6/(1 + J)$ (where J represents the objective function value of the system), and the distortion wavefront under various turbulence intensities was considered as the correction object for 300 closed-loop correction experiments. Figures 7.17, 7.18, and 7.19 depict the variations in the spot light intensity values under the weak, moderate–strong, and strong turbulence cases, respectively, before and after the correction of the wavefront aberrations using the AO technique.

The figures indicate that the AO technology effectively corrects the wavefront aberrations under various turbulence intensities. The grayscale value of the spot before correction is very small, and the spot distribution is not uniform; the stronger the turbulence intensity, the more significant the spot aberrations. The spot energy

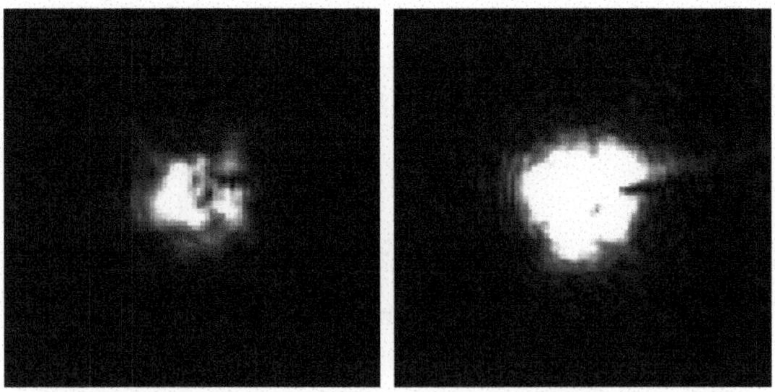

(a) Uncorrected (b) Corrected

Fig. 7.17 Intensity distribution of signal light under weak turbulence

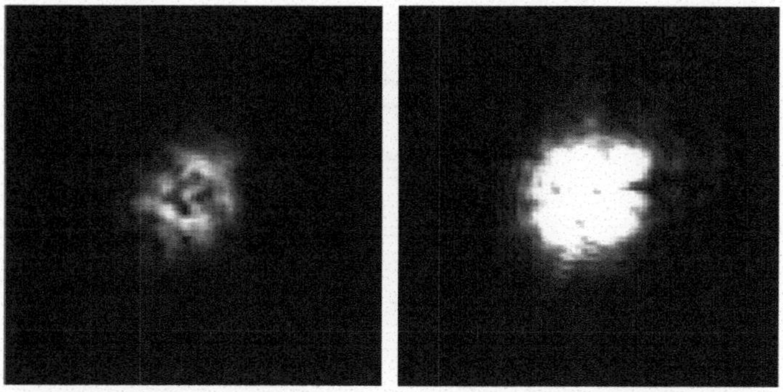

(a) Uncorrected (b) Corrected

Fig. 7.18 Intensity distribution of signal light under moderately strong turbulence

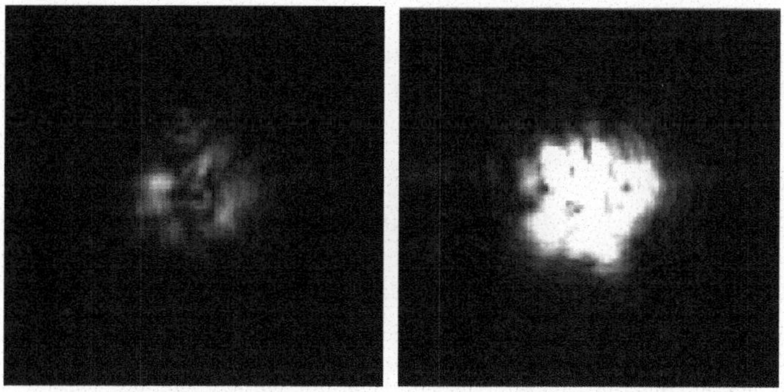

(a) Uncorrected (b) Corrected

Fig. 7.19 Intensity distribution of signal light under strong turbulence

becomes more uniform and convergent after the wavefront aberration is corrected using AO, and the central grayscale value is substantially improved.

As depicted in Fig. 7.20, the average grayscale value change curve of the 10 × 10 pixel points is centered at the center of mass of the spot following 300 closed-loop corrections of the system. The figure indicates that the average grayscale value of the system spot light intensity converges quickly and reaches the global optimum. Under weak, moderately strong, and strong turbulence, the value of the objective function (the average grayscale mean value of 100 pixel points at the center of the spot detected by the CCD camera) increases from 119.8, 84.6, and 35.7 after the AO correction of the optical wavefront distortion to 246.6, 249.1, and 245.7, which verifies the accuracy of the theoretical results.

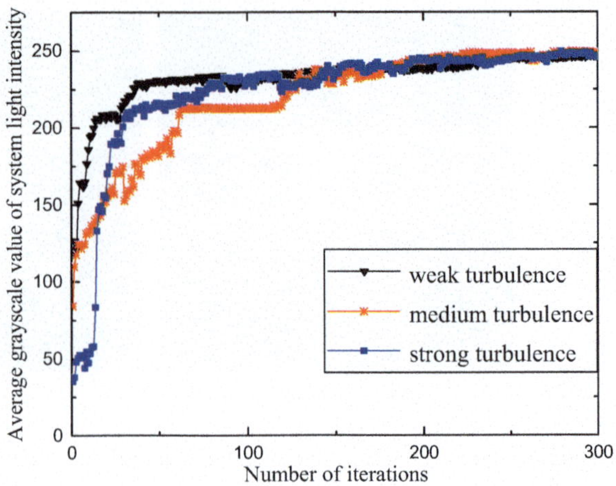

Fig. 7.20 Change curve of the average grayscale value of 10 × 10 pixel points of the system

7.3.2 SPGD Algorithm Wavefront Correction for Outlier Detection Coherent Optical Communication System

The schematic of a free-space outlier detection coherent optical communication system based on AO is presented in Fig. 7.21. The system uses a lithium niobate phase modulator with DPSK modulation on the transmitter side to modulate the video information onto the signal light emitted by the narrow-linewidth laser Adjustik E15 with a wavelength of 1550 nm and emitted through the Cassegrain antenna, which produces aberrated light when transmitted under atmospheric turbulence. The signal light reflected by the BS is coupled by a coupling lens into a single-mode optical fiber and then coherently mixed with the local oscillation light from a Basik E15 narrow-linewidth laser with a wavelength of 1550 nm in a mixer, which connects the output four-way phase-difference optical signals to a double-balanced detector. This transforms the optical signals into intermediate frequency (IF) current signals and passes them to the demodulation system. The high-speed digital-signal-processing unit in the demodulator enables the video to be viewed on a computer. A lower mixing efficiency leads to a particularly weak optical signal output from the mixer, making it impossible to recover the video signal from the reduced IF current signal of the system. Using the AO at the receiver side to correct the distorted signal light, the coupling efficiency between the signal and fundamental light beams can be improved by strengthening the IF signal and improving the quality of the communication system. The ability of AO to correct the distorted signals in coherent optical communications systems was verified in two separate out-of-field experimental scenarios.

(1) **Out-of-field 600-m experiment**

In these experiments, the SPGD perturbation amplitude δ=0.006 and gain coefficient $\gamma = 1.3/(90 + J)$ (where J represents the objective function value of the

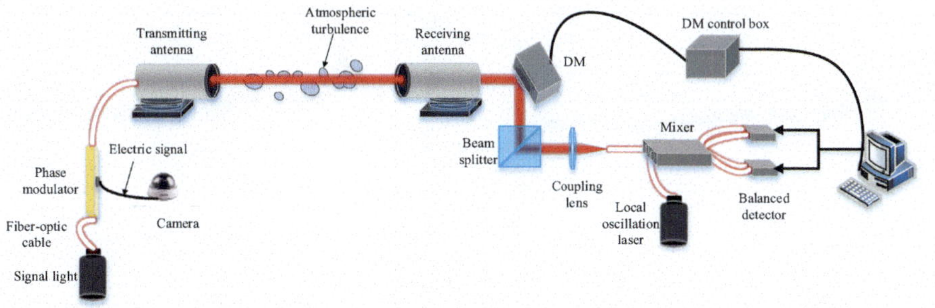

Fig. 7.21 Schematic of a coherent optical communication system based on AO without wavefront detection

system) were considered, and the optical power of the coupled-in single-mode fiber at the receiver was used as the objective function of the system for 500 closed-loop calibration experiments. Figure 7.22 presents the variation curve of the optical power of the coupled single-mode fiber during the calibration process.

Figure 7.22 indicates that the optical power increases from the initial −36.32 to −10 dBm as the number of iterations increases. At 310 iterations, the value of the optical power reaches a global optimum and stabilizes, indicating that AO can effectively improve the coupling efficiency of single-mode fibers. Table 7.1 presents the variation in the values of optical power variance over 100 adjacent iterations.

Fig. 7.22 Curve depicting the variation in optical power coupled into a single-mode fiber in 600 m field experiment

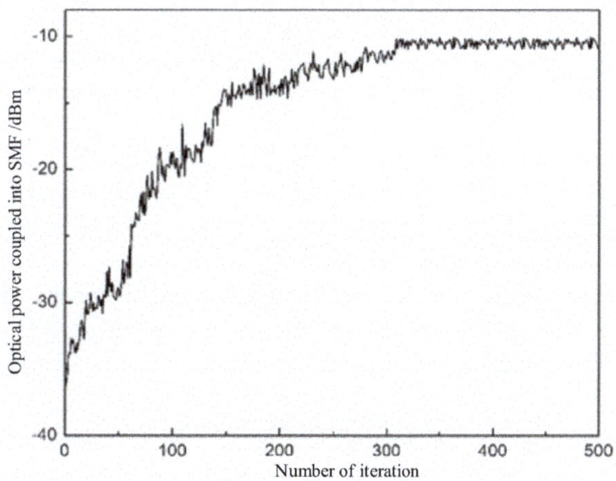

Table 7.1 Values of optical power variance per 100 cycles

Number of iterations	0–100	100–200	200–300	300–400	400–500
Variance	23.62	5.62	0.60	0.11	0.06

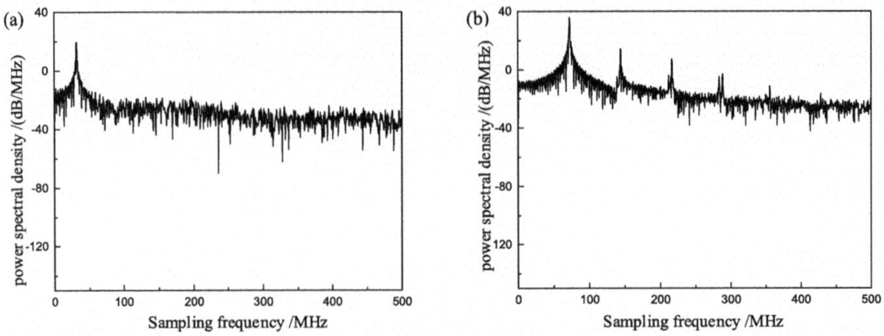

(a) Uncorrected (b) Corrected

Fig. 7.23 IF signal power spectrum before and after correction in 600 m field experiment

Table 7.1 indicates that in the first 100 iterations, the value of optical power variance is 23.62, illustrating that the optical power increases rapidly at this stage. As the number of iterations increases, the variance in optical power gradually decreases and tends to 0, indicating that the growth rate of the optical power gradually decreases at this stage and reaches a global optimum.

Figure 7.23 presents the power spectra of the IF signal before and after correction. The figure indicates that the main spectral peak becomes sharper after correction and the power spectrum density of the IF signal at this peak increases from 20 to 38 dB/Hz. The noise of the IF signal is filtered out and the SNR of the system is improved relative to that before correction.

Following correction using AO, the optical power coupled into the single-mode fiber at the receiver of the coherent optical communication system is significantly improved. This indicates that the signal optical wavefront distortion is corrected to improve the mixing efficiency of the system. Figure 7.24 presents graphs of the baseband signal generated by the demodulation system before and after correction. It can be seen that the IF signal of the system before correction is too small and the baseband signal output generated by the decoder is irregular in shape with significant burrs around it. The corrected IF signal meets the decoding requirements at the receiver. The shape of the baseband signal generated by the decoder is similar to that of a square wave so that the information can be decoded well.

(2) **Out-of-field 1300-m experiment**

Figure 7.25 depicts the optical power coupled into the fiber at a distance of 1300 m. As the number of iterations increases, the optical power increases from the initial −43.68 to −30.1 dBm, reaches the optimum value, and stabilizes at 120 iterations. This indicates that the optical power at the receiver of the coherent optical communication system coupled into the single-mode fiber is significantly improved after correction using the AO technique. Further, the mixing efficiency of the system is improved after the distortion in the optical wavefront of the signal is corrected.

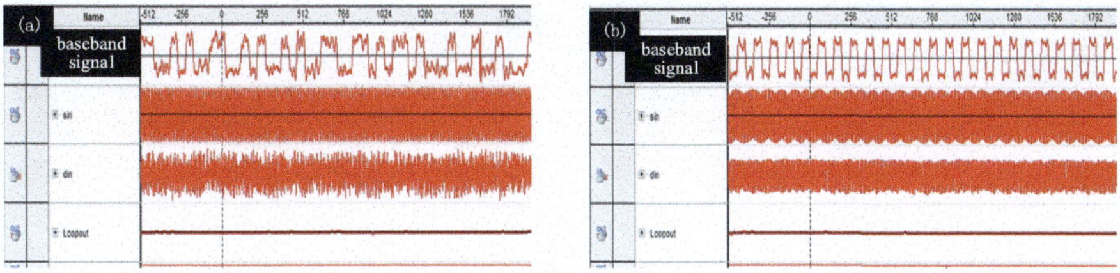

(a) Uncorrected (b) Corrected

Fig. 7.24 Baseband signal before and after AO correction in 600 m field experiment

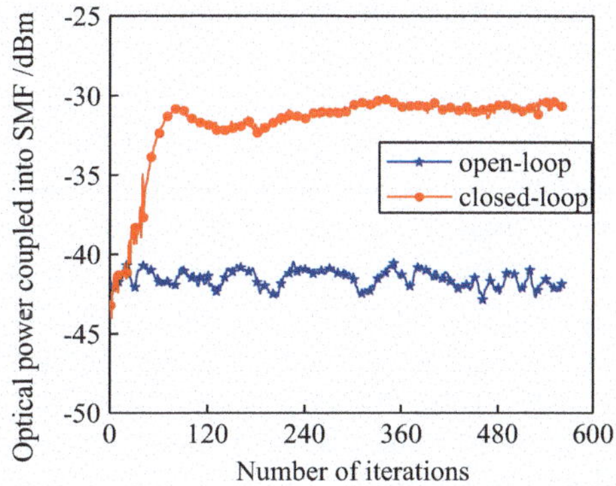

Fig. 7.25 Curve depicting the variation in optical power coupled into a single-mode fiber in 1.3 km field experiment

Figure 7.26 presents the plots of the power spectra of the IF signal before and after correction. The figure indicates that the main spectral peak becomes sharper after correction and the power spectrum density of the IF signal at the peak increases from 3 to 20 dB/Hz. The noise of the IF signal is filtered out, and the SNR of the system is improved when compared with that before correction.

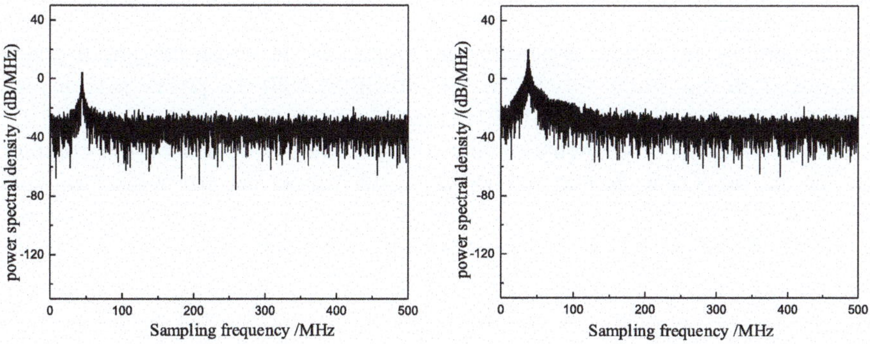

(a) Uncorrected (b) Corrected

Fig. 7.26 IF signal power spectrum before and after correction in 1.3 km field experiment

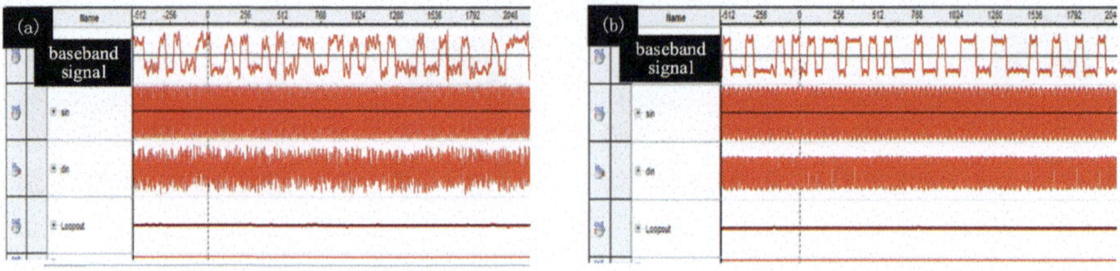

(a) Uncorrected (b) Corrected

Fig. 7.27 Baseband signal before and after AO correction in 1.3 km field experiment

The optical power coupled into the single-mode fiber at the receiver of the coherent optical communication system is significantly improved after correction using AO, indicating that the distortion of the optical wavefront of the signal may be corrected to improve the mixing efficiency of the system. Figure 7.27 presents graphs of the baseband signal generated by the demodulation system before and after correction. The figure indicates that the IF signal of the system before correction is too small to meet the decoding requirements and the baseband signal generated by the decoder is irregular in shape with significant burrs around it. The IF signal meets the decoding requirements at the receiver after the distortion of the optical wavefront of the signal is corrected using AO technology. Further, the shape of the baseband signal generated by the decoder is similar to that of a square wave, which can decode the video information well.

(3) **Out-of-field 10.2-km experiment**

The SPGD algorithm was used for 1250 closed-loop calibration experiments in a 10.2 km field with perturbation amplitude δ=0.006 and gain factor $\gamma = 1.3/(110 + J)$, where J represents the objective function value of the system, that is, the optical power coupled into the single-mode fiber at the receiver of the coherent optical communication system. Figure 7.28 depicts the calibration curve indicating the variation in the optical power of the coupled single-mode fiber during the calibration process.

The figure indicates that as the number of iterations increases, the optical power increases from the initial −41.66 to −16.80 dBm, with the power reaching an optimum value and stabilizing at 1050 iterations.

The results of the experiments conducted in the external field at various communication distances indicate that SPGD-based AO without wavefront detection can correct the distortion in the optical wavefront of the signal very well and that the experimental results are mostly consistent with the simulation results.

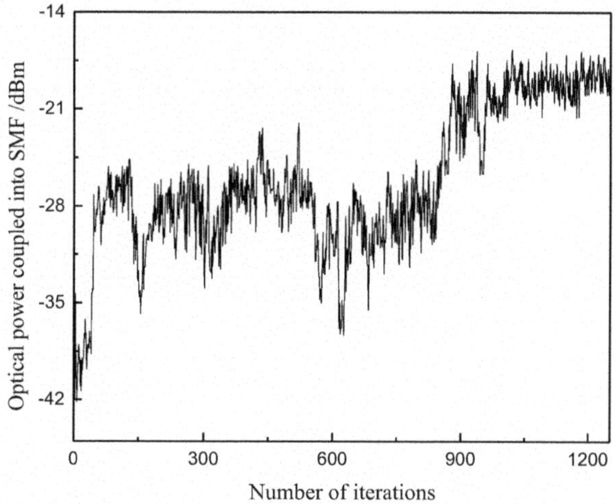

Fig. 7.28 Curve depicting the variation in optical power coupled into a single-mode fiber in 10.2 km field experiment

References

1. Ellerbroek BL, Rhoadarmer TA (2001) Adaptive wavefront control algorithms for closed loop adaptive optics. Math Comput Model 33(1):145–158
2. Booth MJ, Débarre D, Wilson T (2007) Image-based wavefront sensorless adaptive optics. In: Advanced wavefront control: methods, devices, and applications v. San Diego, CA, United states, pp 671102–671107
3. Rao XJ, Ling N, Jiang WH (1995) Experimental study on the measurement of deformation mirror influence function by digital interferometer. Acta Optica Sinica 15(10):1446–1451
4. Yariv A, Yeh P (1984) Optical waves in crystals: propagation and control of laser radiation. 39(4):657–661
5. Wu ST, Efron U, Hess LD (1984) Birefringence measurements of liquid crystals. Appl Opt 23(21):3911–3915
6. Spall JC (2002) Multivariate stochastic approximation using a simultaneous perturbation gradient approximation. IEEE Trans Autom Control 37(3):332–341
7. Rao RZ (2005) Light propagation in turbulent atmospheres. Anhui Science and Technology Publishing House, Hefei, pp 4–9
8. Han D, Liu YQ, Zhao X et al (2016) Effect of beam radius on the heterodyne efficiency for space coherent optical communication. J Changchun Univ Sci Technol 39(3):36–40
9. Liu C, Chen S, Li XY et al (2014) Performance evaluation of adaptive optics for atmospheric coherent laser communications. Opt Express 22(13):15554–15563
10. Fan CX, Cao LN (2012) Principles of communications. National Defense Industry Press, Beijing, pp 207–210

Chapter 8
Wavefont Correction Technique of Spatial Coherent Optical Communication with LC-SLM

Contents

In this chapter, the phase calibration of the transmissive liquid crystal spatial light modulator (LC-SLM-R) and reflective liquid crystal spatial light modulator (LC-SLM-R) is explained and experiments on the phase modulation characteristics of beams of various wavelengths incident on the LC-SLM are discussed. After calibration, the phase–grayscale relationship diagrams of the LC-SLM under various wavelengths are presented, and the feasibility of using the LC-SLM to correct the wavefront distortion is analyzed.

X. Ke and J. Wu, *Coherent Optical Wireless Communication Principle and Application*, Optical Wireless Communication Theory and Technology,
https://doi.org/10.1007/978-981-19-4823-7_8

8.1 Phase Calibration of LC-SLM

8.1.1 LC-SLM Phase Calibration

The LC-SLM is based on the principle of electronically controlling the birefringence of LCs to realize continuous modulation of the light wave phase by changing the electric drive signal of the LC [1]. Owing to the advantages of its low price, high resolution, fast response speed, low energy consumption, programmable control, the LC-SLM is used in adaptive optics [2], wavefront sensing [3], beam shaping [4, 5], and other optical fields [6, 7]. The birefringence effect of the LC-SLM is dependent on the incident light beams of the light waves. Therefore, the phase modulation characteristics of the light beams of various wavelengths vary. However, a manufacturer only tests and calibrates a specific wavelength of the LC-SLM during the factory test and calibration. In the optical experimental system, the LC-SLM must be used to measure the phase characteristics of the all the bands [8].

At present, interference and diffraction methods are used for the phase calibration of LC-SLMs to beams of various wavelengths. Some widely used interference methods are the double-slit [9, 10], Mach–Zehnder [11, 12], and Twyman-Green [13, 14] interference methods. The phase calibration of interferometry depends on the relative displacement of the fringes. The double-beam interference takes a long path before interference. Environmental factors and mechanical vibration affect the optical path difference, resulting in the instability of the collected fringes. The diffraction method [15, 16] is based on the transmission of light intensity and has high stability against environmental factors and mechanical vibration. However, the phase information must be characterized by the effective cosine value of the phase difference so that the calculation of the phase estimation is robust. Therefore, interferometry is the preferred method for calibrating the LC-SLM.

Two double slits are the main components of double-slit interferometry for phase calibration of the LC-SLM. One set of double slits, called the phase slit, is simulated by the LC-SLM as a calibration group. After getting divided into two halves, the phase of one half is fixed and that of the other half is changed based on the situation at hand. The other set of double slits is used as the reference group, and there is no phase difference anywhere on these double slits. When a beam passes through the two sets of double slits, two sets of interference fringes are generated. A certain relative movement occurs between the two sets of interference fringes by an appropriate change in the difference in the phase slits. The relative movement reflects the phase modulation amount of the LC-SLM. To bridge to the relationship between the added gray level [17], phase calibration using double-slit interference must be performed based on the reference group. Each measurement or calibration reflects the phase modulation on only the phase slit; the modulation characteristics do not reflect the phase modulation characteristics of the entire LC-SLM. The phase modulation characteristics of the LC-SLM are calibrated using the Twyman-Green interferometry. The idea is to divide a beam of light emitted by the laser into two beams using a beam splitting prism. The beam of light is incident on the beam combiner after being reflected by the LC-SLM. The two reflected beams are combined and interfered at the beam combiner, and the interference fringes are collected by a CCD. When the gray value loaded on the LC-SLM changes, the interference fringes collected by the CCD move by different degrees. The amount of the movement of these interference fringes is calculated through image processing. Subsequently, the amounts of phase modulation of the LC-SLM for various gray values are calculated based on the amounts of the movement of fringes [18, 19]. Theoretical analyses and experimental results have confirmed [20–23] that the Twyman-Green interferometry can perform good phase calibration for the reflective LC-SLM. For projected LC-SLM, Mach–Zehnder interferometry and shearing interferometry are used for phase measurement and calibration.

8.1.2 Structure of LC-SLM

Liquid crystal is an organic compound with a regular molecular arrangement between the solid and liquid components. Spatial light modulators used for wavefront correction in the field of adaptive optics usually use nematic LC-SLMs. The structure of a reflective LC-SLM is depicted in Fig. 8.1 [24–26]. The nematic LC layer is sandwiched between two polar plates of transparent and control electrodes. The upper and lower inner surfaces of the LC layer and electrode are closely attached to a transparent alignment film. The function of the electrode and alignment film is to make the LC molecules move parallel to the surface of the alignment film. Furthermore, the alignment of the LC molecules can be parallel or perpendicular to one another [26]. In the nematic LC-SLM, the alignment film helps to align the LC molecules on the upper and lower surfaces of the LC layer in parallel [26]. The top surface is covered with a layer of transparent glass, and the inner surface is coated with a transparent conductive layer electrode [26]. The bottom layer is a silicon substrate, etched with independently addressable electrodes, and a layer of high reflectivity aluminum film is plated on the electrodes [26]. The advantages of the reflective nematic LC-SLM are [26] as follows:

(1) The control electrode can be integrated onto a small silicon substrate. Therefore, the pixel size is significantly smaller than that of the transmissive type. Furthermore, the pixel resolution is high, and the fill factor is large.
(2) The reflecting surface uses an advanced metal reflecting film, which has high reflectivity.
(3) The thickness of the reflective nematic LC-SLM LC layer is smaller than that of the transmissive type, and the response speed is faster.

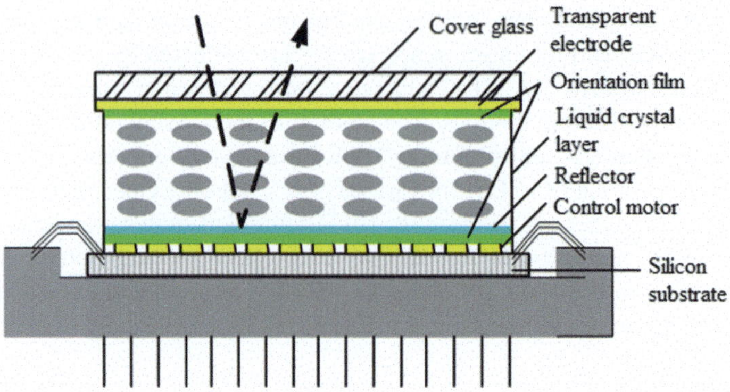

Fig. 8.1 Schematic of the structure of reflective liquid crystal spatial light modulator

8.1.3 Jones Matrix Analysis of LC-SLM Phase Modulation Principle

The optical path difference produced during the transmission of light can be calculated using the geometric optics theory:

$$l = z \times n \tag{8.1}$$

It is evident from Eq. (8.1) that optical length l is related to the geometric path z of the optical transmission and refractive index n of the transmission medium. Currently, the widely used phase modulators are deformable mirrors and spatial light modulators. Deformable mirrors achieve phase modulation by changing the z of light transmission, whereas spatial light modulators achieve phase modulation by changing the n of the transmission medium [27, 28]. The LC-SLM uses the birefringence effect of LCs to achieve phase modulation of the incident light; however, it must be operated in the pure phase modulation mode without being affected by amplitude modulation. When it is used to produce phase delay, the incident light and outgoing light must be sandwiched between two polarizers, as depicted in Fig. 8.2. The birefringence phenomenon occurs when light propagates in the LC and is divided into ordinary light, *o* light, and extraordinary light, *e* light. The corresponding refractive indices are n_o and n_e. The angles between the polarization directions of the polarizer and analyzer y are φ_1 and φ_2, respectively.

According to the wave optics theory [29], the LC-SLM produces certain phase retardation β after phase modulation of the incident light. Usually, β is called the birefringence coefficient, and its value is related to the thickness of the LC layer d, n_e, n_o, wavelength λ change. The expression for β is as follows [29]:

$$\beta = 2\pi d[n_e - n_o]/\lambda \tag{8.2}$$

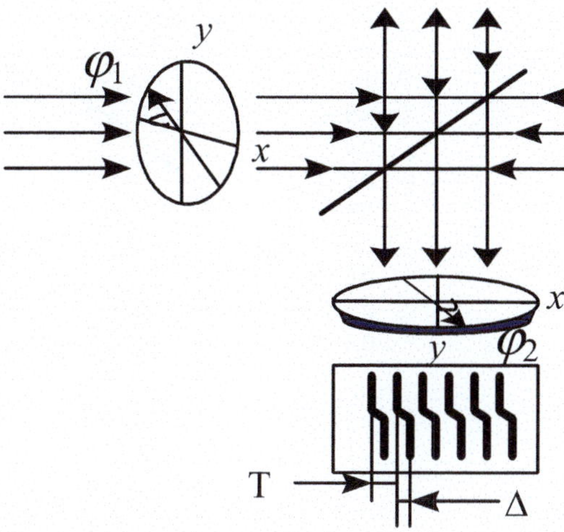

Fig. 8.2 Schematic of reflective spatial light modulator

For the normal crystal LC-SLM with quartz as the LC material, the refractive indices of extraordinary light and ordinary light are 1.55335 and 1.54424, respectively. When the thickness of the LC layer is from 0–300 μm, and the wavelength of the light incident on the LC-SLM is different, the relationship between the phase retardation β and the thickness of the LC layer is depicted in Fig. 8.3. It is evident from Fig. 8.3 that phase retardation increases with an increase in the thickness of the LC layer while the wavelength of the incident light beam is constant. In other words, the phase modulation ability of the LC-SLM increases with an increase in the thickness of the LC layer. When the thickness of the LC layer is constant, the phase retardation of the LC-SLM decreases as the wavelength increases.

The refractive index n_e of extraordinary light in formula (8.2) is called the equivalent refractive index of ordinary light, and its value changes based on θ, which is the angle at which the LC molecules deflect in the direction of the electric field when

Fig. 8.3 Relationship between phase retardation and thickness of liquid crystal layer

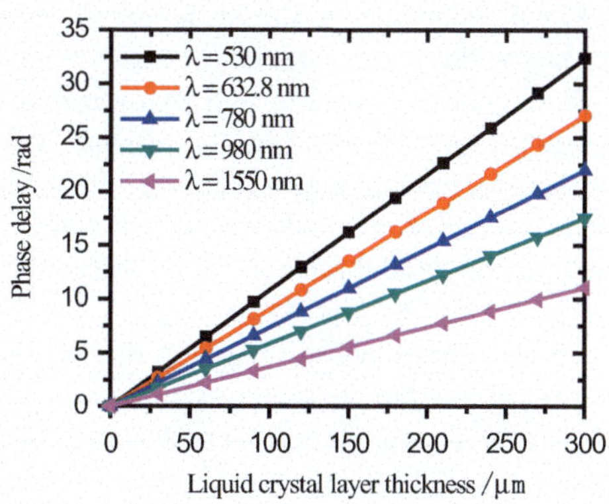

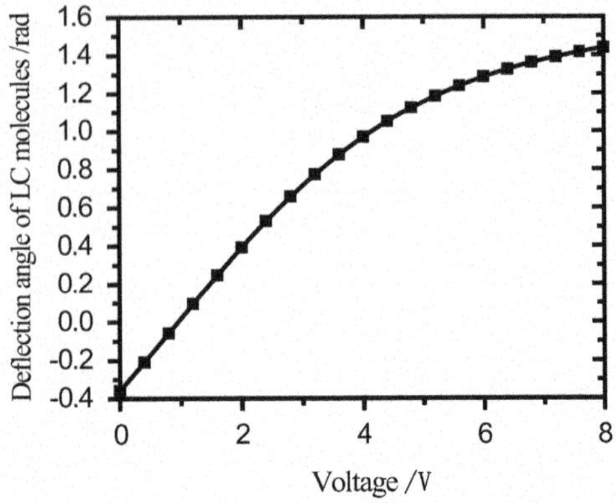

Fig. 8.4 Relationship between deflection angle of liquid crystal molecules and voltage

voltage V is applied to the LC layer. The relationship between θ and V is as follows:

$$\theta = \frac{\pi}{2} - 2\tan^{-1}\left\{\exp\left[-\left(\frac{V - V_{th}}{V_O}\right)\right]\right\} \tag{8.3}$$

where V_{th} is the threshold voltage, which is generally set to 0.95 V; V_O is the overload voltage at $\theta = 49.6°$, which is generally 2.6 V. The relationship between the deflection angle of the LC molecules and the voltage applied to the LC layer is depicted in Fig. 8.4. It is evident that θ of the LC molecules increases with an increase in V. When $V > V_{th}$, the value of n_e of the extraordinary light changes with the deflection of the LC molecules, which explains the ability of the LC to achieve phase retardation of the incident light wave. The equivalent n_e is a function of θ, and the relationship is [29]

$$n_e(\theta) = \frac{n_e \cdot n_o}{\sqrt{n_o^2 \sin^2(\theta) + n_e^2 \cos^2(\theta)}} \tag{8.4}$$

In formula (8.4), the deflection angle of the molecule is affected by the electric field voltage applied to the LC layer. Therefore, the value of n_e also changes with the applied voltage. Its relationship with the electric field voltage is depicted in Fig. 8.5.

It is evident from Fig. 8.5 that the equivalent refractive index of the LC layer is significantly affected by the applied and threshold voltages. When the applied voltage does not change, the greater the threshold voltage, the smaller the equivalent refractive index; when the threshold voltage does not change, the greater the applied voltage, the greater the equivalent refractive index.

From Eqs. (8.2)–(8.4) and Figs. 8.3, 8.4 and 8.5, it is evident that the change of the molecular deflection angle in the LC layer is controlled using the electric field voltage applied to the LC layer, and the voltage change causes equivalent refraction. The rate of change of light, which results in the change of the phase of the light wave incident on the LC layer, realizes electronically controlled modulation of the phase

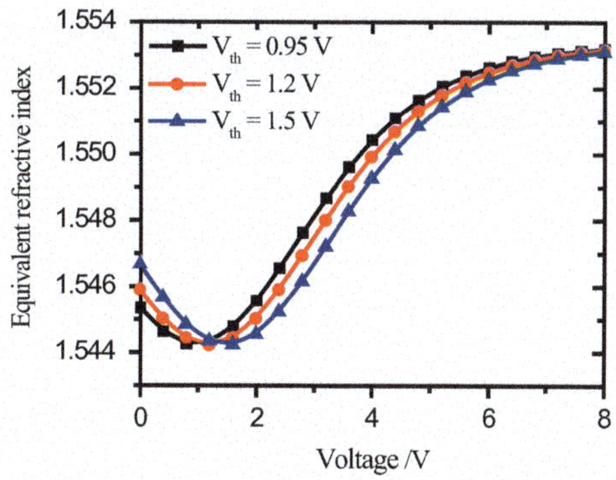

Fig. 8.5 Relationship between equivalent refractive index and voltage

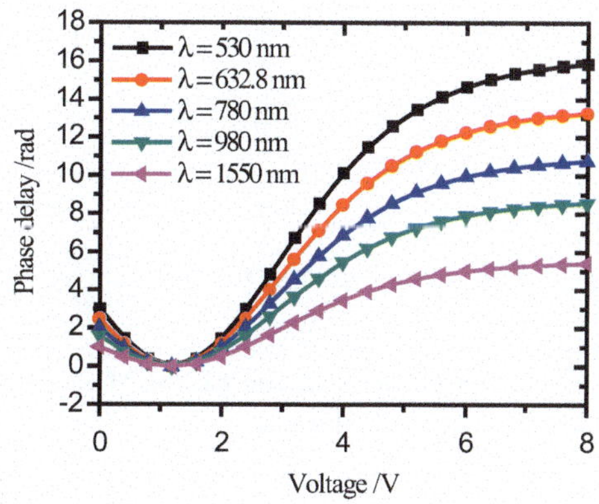

Fig. 8.6 Relationship between phase delay and voltage

retardation of the LC-SLM. Figure 8.6 depicts the relationship between the phase delay and voltage change.

It is evident from Fig. 8.6 that, when the applied voltage of the LC layer is constant, the phase retardation of the light beam passing through the LC layer decreases as the wavelength increases. Figure 8.6 depicts an ideal state without considering any additional device loss. In practice, the phase retardation of the LC layer to beams of various wavelengths cannot reach such a wide range. The actual phase retardation is significantly affected by the angle between the polarizer and optical axis. Therefore, the angle between the polarizer and optical axis must be considered when discussing the phase modulation capability of the LC-SLM.

The LC-SLM can be regarded as a polarization device, and the light intensity transmittance of the beam after passing through the polarizer, LC-SLM, and analyzer can be analyzed using the Jones matrix [30, 31]. Based on Fig. 8.2, the coordinate system of the direction of the propagation of light and the offset angle is established, as depicted in Fig. 8.7. The x–y plane is parallel to the incident plane of the LC layer, and the direction of the optical axis of the LC is parallel to the x-axis, which represents the direction of the propagation of light. In the Figure, the angle between the optical vibration direction of the polarizer and the x-axis is φ_1, and the angle between the optical vibration direction of the analyzer and the x-axis is φ_2. Then, the normalized Jones matrix of the incident light is.

$$E_{in} = \begin{bmatrix} 1 \\ 0 \end{bmatrix} \tag{8.5}$$

The change in the polarization direction of the beam after passing through the polarizer and analyzer can be represented as a rotation matrix. Set the rotation matrix as $R(\varphi)$, then

$$R(\varphi) = \begin{bmatrix} \cos\varphi & \sin\varphi \\ -\sin\varphi & \cos\varphi \end{bmatrix} \tag{8.6}$$

The Jones vector of the linearly polarized light after the incident light passes through the polarizer is

$$E_1 = R(\varphi_1)E_{in} = \begin{bmatrix} \cos\varphi_1 & \sin\varphi_1 \\ -\sin\varphi_1 & \cos\varphi_1 \end{bmatrix} \begin{bmatrix} 1 \\ 0 \end{bmatrix} = \begin{bmatrix} \cos\varphi_1 \\ -\sin\varphi_1 \end{bmatrix} \tag{8.7}$$

The optical axis of the LC-SLM is parallel to the x-axis. Therefore, the Jones matrix of the LC-SLM is

$$SLM(\beta) = \begin{bmatrix} 1 & 0 \\ 0 & e^{j\beta} \end{bmatrix} \tag{8.8}$$

The Jones matrix of the analyzer is

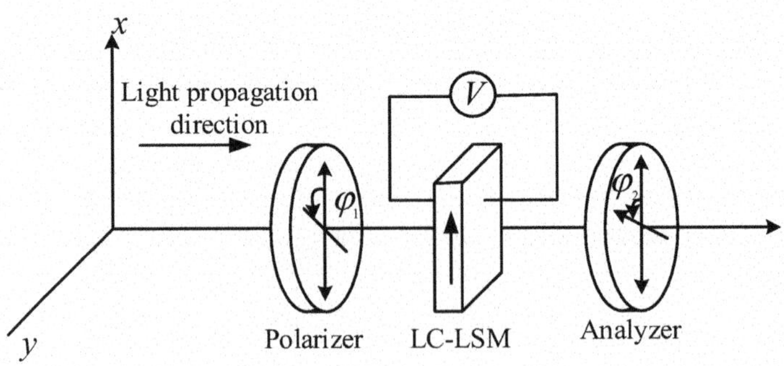

Fig. 8.7 Schematic of working principle of LC-SLM

$$P = \begin{bmatrix} 0 & 0 \\ 0 & 1 \end{bmatrix} \tag{8.9}$$

Therefore, the Jones matrix of the emitted light after passing through the analyzer is

$$E_2 = PR(-\varphi_2) = \begin{bmatrix} 1 & 0 \\ 0 & 0 \end{bmatrix}\begin{bmatrix} \cos\varphi_2 & -\sin\varphi_2 \\ \sin\varphi_2 & \cos\varphi_2 \end{bmatrix} = \begin{vmatrix} \cos\varphi_2 & -\sin\varphi_2 \\ 0 & 0 \end{vmatrix} \tag{8.10}$$

Based on the above analysis, it is established that the Jones vector of the incident light passing through the polarizer, LC-SLM, and analyzer is

$$\vec{E}_{\text{out}} = PR(-\varphi_2)SLM(\beta)R(\varphi_1)E_{\text{in}} \tag{8.11}$$

Incorporating Eqs. (8.7), (8.8), and (8.10) into Eq. (8.11), and after calculation, the complex amplitude of the emitted light is

$$\vec{E}_{\text{out}} = \cos(\phi_2)\cos(\phi_1) + e^{j\beta}\sin(\phi_2)\sin(\phi_1) \tag{8.12}$$

Next, Eq. (8.12) is simplified and the intensity of the emitted light and the phase delay caused by the LC-SLM are finally calculated as

$$\begin{aligned} I =& \frac{1}{4}\{\cos(\varphi_2+\varphi_1)+(\varphi_2-\varphi_1)+\cos\beta[\cos(\varphi_1+\varphi_2)-\cos(\varphi_2-\varphi_1)]\}^2 \\ &+\frac{1}{4}\sin^2\beta[\cos(\varphi_2+\varphi_1)-\cos(\varphi_2-\varphi_1)]^2 \end{aligned} \tag{8.13}$$

$$\delta = \beta + \tan^{-1}\frac{\sin\beta[\cos(\varphi_2+\varphi_1)-\cos(\varphi_2-\varphi_1)]}{\cos(\varphi_2+\varphi_1)+\cos(\varphi_2-\varphi_1)+\cos\beta[\cos(\varphi_2+\varphi_1)-\cos(\varphi_2-\varphi_1)]} \tag{8.14}$$

It is evident from Eq. (8.14) that the intensity and phase modulation characteristics of the LC-SLM depend on φ_1, φ_2 and birefringence coefficient β between the polarizer and analyzer and the optical axis of the LC. The first and second terms in the formula are the phase and intensity modulation components, respectively. To make the LC-SLM function in the best phase modulation state, the angle $= \varphi_1, \varphi_2$ and birefringence coefficient β must be adjusted. Using Eqs. (8.13) and (8.14), the normalized reflected light intensity and phase delay through the LC-SLM are simulated with the distribution curves of φ_1, φ_2, and β, and the results are depicted in Figs. 8.8 and 8.9.

It is evident from Fig. 8.9 that when φ_1 and φ_2 take a certain value, the normalized reflected light intensity and phase retardation of the LC-SLM are different based on the change in the birefringence coefficient. The normalized reflected light intensity presents a Gaussian distribution (except for $\varphi_1 = \varphi_2 = 0^{\circ}$), and the phase delay basically changes linearly with the change in the birefringence coefficient. However,

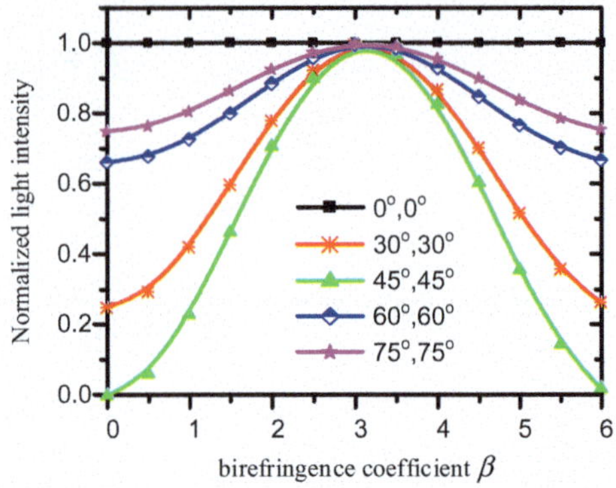

Fig. 8.8 Relationship curve between light intensity and polarization angle

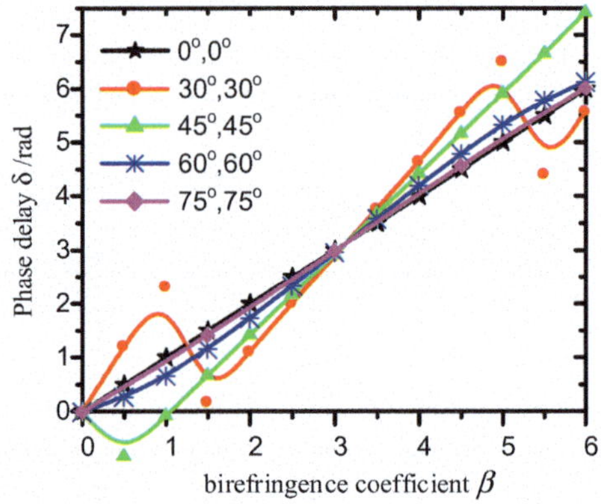

Fig. 8.9 Relationship between phase delay and polarization angle

when $\varphi_1 = \varphi_2 = 30^{\circ}$, the phase delay changes in oscillation and increases; when $\varphi_1 = \varphi_2 = 45^{\circ}$, the initial phase delay decreases. Thereafter, it increases with the increase in the birefringence coefficient. Therefore, when calibrating the LC-SLM or using it for phase measurement, the polarization angle of the polarizer and analyzer must be selected reasonably to achieve the best phase modulation state.

When $\varphi_1 = \varphi_2 = 0^{\circ}$ and the polarization directions of the polarizer and analyzer are parallel to the direction of the LC optical axis, the normalized reflected light intensity and phase retardation are, respectively, [26]

$$I = 1 \quad \delta = \beta = 2\pi d[n_e(\theta) - n_o]/\lambda \tag{8.15}$$

It is evident from Eq. (8.15) that the intensity modulation of the LC-SLM remains unchanged at the maximum value under this condition, that is, the intensity of the transmitted or reflected light is the strongest, the phase modulation reaches the best state, and it is in the pure phase modulation mode. This is also the best state for phase calibration or phase measurement of the LC-SLM. Figure 8.10 depicts the intensity and phase modulation characteristic curves in the best state. When the polarization

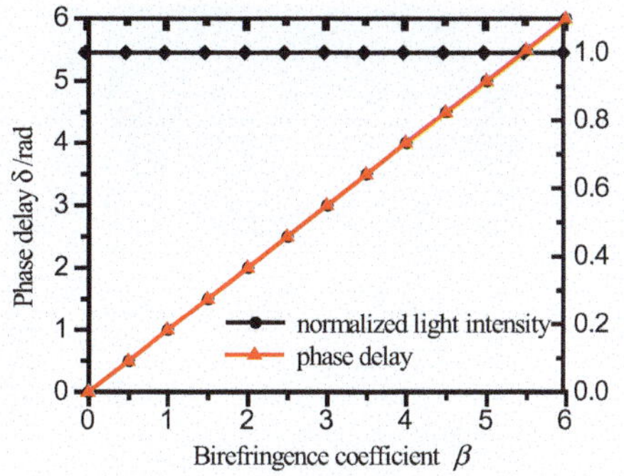

Fig. 8.10 Phase and intensity modulation curve when polarization angle of polarizer and analyzer is 0°

angles of the polarizer and analyzer take other values, although phase modulation can also be performed, the intensity of the reflected light is also subject to certain modulation, and the change in the phase delay is affected by the intensity, and the precise phase grayscale relationship cannot be given. Therefore, when using the phase characteristics of the LC-SLM for practical applications, the LC-SLM must be adjusted to operate in the best phase modulation mode.

8.2 Working Principle of Phase Calibration of LC-SLM

Based on the working principle and characteristics of the reflective LC-SLM, a calibration method based on the Twyman-Green interference principle is proposed. Based on the analysis and calculation of the Jones matrix in the preceding section, the LC-SLM is calibrated using the interference fringe movement method.

8.2.1 Interference Fringe Shift Method

The interference fringe movement method is widely used for phase calibration and phase measurement of the LC-SLM because the optical path of the method is simple and mature and the image processing is simple. The size of the phase modulation can be calculated using the amount of relative movement of the interference fringe. The direction of the movement of the interference fringes can also determine the direction of phase modulation.

From Eqs. (8.2)–(8.15), it is evident that the phase delay of the LC spatial light modulator is realized by changing the driving voltage applied to both ends of the modulator. The manufacturer maps the drive voltage of the modulator to the display gray value of the computer and loads the gray image that changes in the range of 0–255 into the LC-SLM driver to realize the control of the voltage, thereby realizing the control of the incident phase modulation of light. To accurately obtain the phase

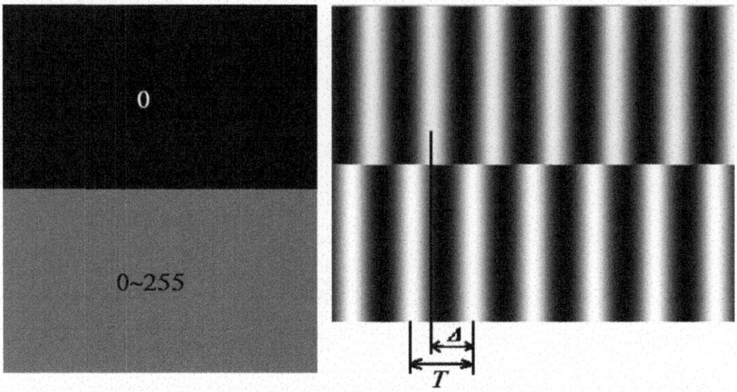

Fig. 8.11 Gradual grayscale image and interference fringes

shift amount, we loaded the step grayscale on the LC-SLM. The 0 Gy value added in the upper half remained unchanged, and the gray value in the lower half gradually changed from 0 to 255. The upper part of the interference fringe collected by the CCD remained unchanged, and the lower part had a certain absolute displacement owing to the difference in the gray values. The gray change and fringe movement are depicted in Fig. 8.11.

In Fig. 8.11, Δ is the absolute displacement and T is the period width of the interference fringe. Using the relationship between the phase and movement of the interference fringe [22], the phase delay can be calculated as follows:

$$\delta = 2\pi \Delta \big/ T \tag{8.16}$$

8.2.2 Working Principle of Interference Fringe Movement Method

In practice, the phase delay can be calibrated and calculated using the Twyman-Green interference principle [25], as depicted in Fig. 8.12. First, the LC-SLM is placed in the Twyman-Green interference light path. After passing through the pinhole filter, beam expander, polarizer, and collimator, a beam splitter splits the light of various wavelengths into two beams. The reflected reference light and phase modulated light reflected by the LC spatial light modulator interfere with each other after passing through the analyzer. The interference fringes of the LC-SLM-R at various step gray values are collected by the CCD. The amount of the movement decides the magnitude of the phase delay, and the calibration of the LC-SLM-R is completed.

Figure 8.13a depicts the step grayscale loaded on the LC-SLM in the experiment, and Fig. 8.13b depicts the interference fringes collected by the CCD. The absolute movement of the interference fringes is evident. When the gray value of 0 added

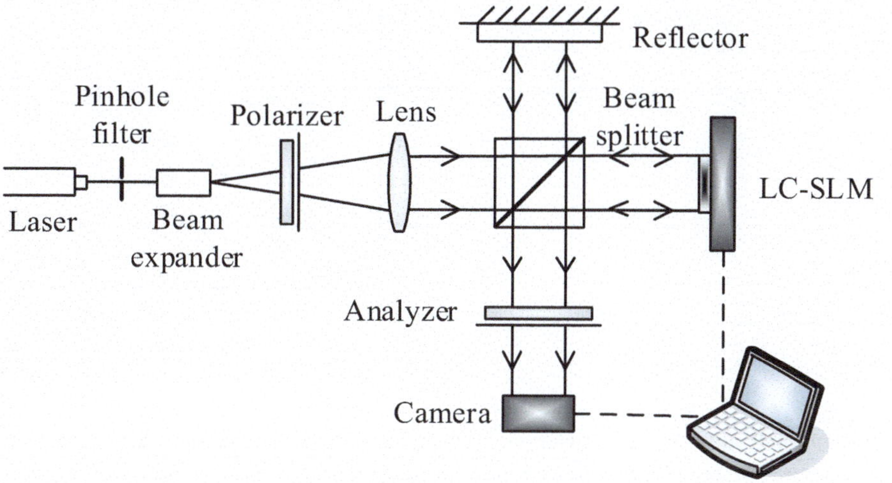

Fig. 8.12 Schematic of Twyman-Green interference

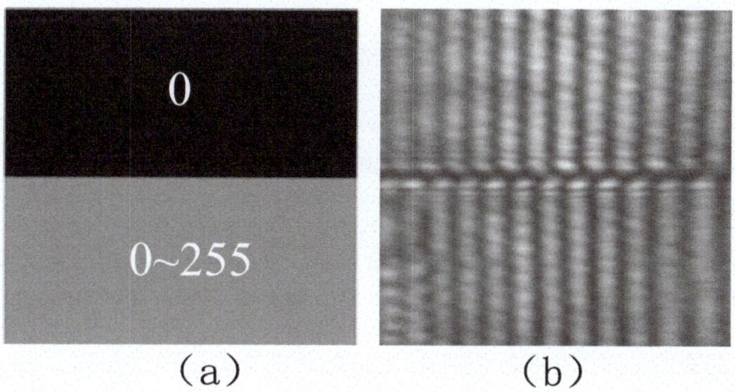

Fig. 8.13 Step grayscale and actual misaligned interference fringes

in the upper half remains unchanged, and the gray value of the lower half changes gradually from 0 to 255; the absolute amount of the movement between the upper and lower stripes is also different. Using the value of actual movement of the fringe and Eq. (8.16), the phase delay under various gray values can be calculated.

8.3 Phase Calibration Experiments

The phase modulation capability of the LC-SLM and the accompanying intensity modulation size in the phase modulation mode play a very important role in wavefront correction and simulation of the turbulent phase screen. Therefore, the LC-SLM must

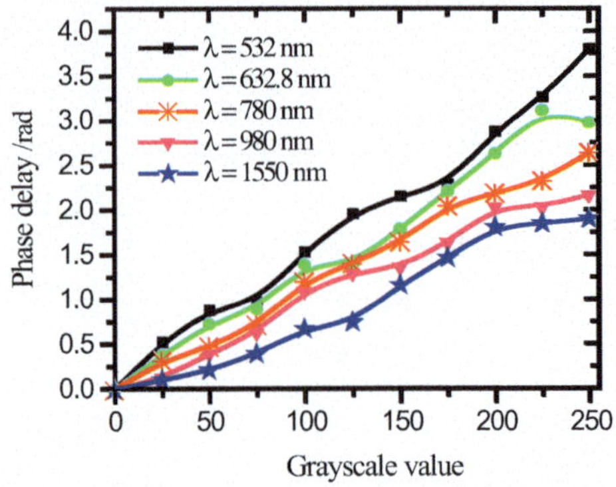

Fig. 8.14 Phase modulation characteristic curve of LC-SLM-R

be calibrated appropriately prior to its use. This section discusses the calibration experiments on the phase modulation of the reflective LC-SLM-R (LC-SLM-R) and transmissive LC-SLM (LC-SLM-T).

8.3.1 Phase Calibration Experiment of LC-SLM-R

Figure 8.14 depicts the phase modulation characteristic curves of the LC-SLM-R for various wavelengths of the incident light. It is evident that the phase modulation ability of the LC-SLM-R for various wavelengths of the incident light varies. The LC-SLM-R has strong phase modulation ability for light beams with small wavelengths but weak phase modulation ability for light beams with long wavelengths. When the wavelengths of the incident light wave are 532 and 1510 nm, the phase modulation ability of the LC-SLM-R reaches 3.798 and 1.906, respectively. The phase delay in the grayscale range of 0–255 is linearly distributed. Owing to certain environmental errors, measurement errors, human errors, and the phase difference of the device itself during the experiment, the measured phase modulation characteristic curve of the LC-SLM-R is not completely linear. Therefore, the least square method is used for curve fitting, with the objective being to find out when the LC-SLM-R needs to obtain the same phase modulation for two different beams of wavelengths 632.8 and 1550 nm. The LC-SLM-R adds the relationship between the gray levels to facilitate the wavefront reconstruction when the beacon light is used for wavefront correction later.

8.3.2 Least Squares Fitting

The objective of using least squares fitting is to interpolate and subdivide the phase between the measured phase data points and obtain the gray value corresponding to the interpolation point based on the fitted curve. The objective of the fitting method is to establish a polynomial fitting curve with the smallest fitting error to approximate the measurement data, thereby filtering or reducing the interference of the measurement error on the interpolation node and realizing the subdivision of the measurement phase.

Before fitting, the n th least squares fitting polynomial is constructed as follows:

$$\delta(g) = \sum_{j=0}^{n} a_j g^j \quad (j = 0, 1, 2 \cdots n) \tag{8.17}$$

The fitting mean square error of fitting value $\delta(g_j)$ and measured value $g_i(i = 0, 1, 2 \cdots m)$ of the corresponding node $\overline{f}_i$ in Eq. (8.17) is

$$\|r\|_2^2 = \sum_{i=0}^{m} w_i \left[\overline{f}_i - \varphi(g_j)\right]^2 \quad (i = 0, 1, 2 \cdots m) \tag{8.18}$$

Weighting coefficient $w_i = 1$ is used in formula (8.18), and the degree of the least square fitting polynomial is determined using the relatively stable minimum value of the mean square error. Using formula (8.17), the measurement data of the 632.8.nm and 1550 nm wavelength beams are obtained using the first to fifth degree polynomials fitting. The mean square error calculated during the fitting process is as follows:

$$\begin{gathered}\left\|r_{1(632.8)}\right\|_2^2 = 1.4851,\ \left\|r_{2(632.8)}\right\|_2^2 = 0.7092,\\ \left\|r_{3(632.8)}\right\|_2^2 = 0.0161,\ \left\|r_{4(632.8)}\right\|_2^2 = 0.0023,\ \left\|r_{5(632.8)}\right\|_2^2 = 0.0023\end{gathered}$$

$$\begin{gathered}\left\|r_{1(1550)}\right\|_2^2 = 1.1541,\ \left\|r_{2(1550)}\right\|_2^2 = 0.4845,\\ \left\|r_{3(1550)}\right\|_2^2 = 0.0263,\ \left\|r_{4(1550)}\right\|_2^2 = 0.005,\ \left\|r_{5(1550)}\right\|_2^2 = 0.005\end{gathered}$$

It is evident from the fitting error data that the mean square error no longer decreases significantly when $n = 4$, and it tends to be a stable value. Therefore, the best least squares approximation can be achieved using the fourth-order fitting polynomial. The fourth-order fitting polynomials for various wavelengths are

$$\begin{aligned}\delta(g)_{632.8} &= 6 \times 10^{-10} g^4 - 9 \times 10^{-8} g^3 - 2 \times 10^{-5} g^2 + 0.0169g - 0.0347\\ \delta(g)_{1550} &= -2 \times 10^{-10} g^4 + 7 \times 10^{-8} g^3 + 3 \times 10^{-5} g^2 + 0.0012g - 0.002\end{aligned} \tag{8.19}$$

Substituting gray value $g_i(i = 0, 5, 10 \cdots 255)$ into fitting polynomial (8.19), the phase node δ_i corresponding to the gray value can be obtained using the least square fitting. However, from Eq. (8.19), it is evident that $\delta(g_i)_{632.8} \neq \delta(g_i)_{1550}$ when the gray value added to the LC-SLM-R is the same. In actual experiments, to use the LC-SLM-R for wavefront correction, it is necessary to obtain the gray value added by the LC-SLM-R under the same phase distortion of beams of various wavelength, the distorted wavefront can be corrected by changing the LC-SLM-R gray value,. Therefore, when the phase of the distorted wavefront is known, the required gray value can be obtained by solving the inverse function of Eq. (8.19). Table 8.1 lists the gray levels added by the 632.8 and 1550 nm beams under the same phase retardation.

8.4 LC-SLM-R Spatial Coherent Optical Communication Wavefront Correction System

8.4.1 LC-SLM-R Wavefront Distortion Correction Principle

Phase conjugation is a widely used principle for wavefront correction in adaptive optical (AO) systems. First, the signal light wavefront is measured. Next, the wavefront controller is used to generate a phase conjugate wavefront to correct the signal light wavefront so that the signal light wavefront becomes the best possible plane wave, and the wavefront of the signal light and the local oscillator light achieves the best possible spatial phase matching at the surface of the optical mixer, thereby improving the mixing efficiency of the two beams and further improving the photocurrent of the difference frequency signal output by the heterodyne detection. The principle of phase conjugate wavefront correction is depicted in Fig. 8.15.

Coherent optical communication uses the AO system for wavefront correction. The objects to be measured and controlled are light wavefronts. The objective is to achieve appropriate spatial matching performance between the transmitted signal light wavefront and the local oscillator light wavefront. When the signal light is transmitted to the receiving end over a long distance, the attenuation due to atmospheric turbulence makes the received light energy significantly weak. To reliably detect the signal light, photon counting is often used for wavefront detection. These characteristics have brought a series of technical problems to the wavefront correction system and are also challenges in the practical application of the AO technology to the wavefront system of spatial coherent optical communication [32].

8.4.2 Structure of Wavefront Correction System

A schematic of the wavefront correction system for spatial coherent optical communication is depicted in Fig. 8.16. The beam splitter prism separates the signal light from the beacon light, which is used to measure the wavefront. The signal light is used to transmit information. The phase information provided by the front measurement

Table 8.1 Gray levels added by 632.8 and 1550 nm beams under the same phase retardation

Δ	Grayscale 632.8	Grayscale 1550	δ	Grayscale 632.8	Grayscale 1550	δ	Grayscale 632.8	Grayscale 1550	δ	Grayscale 632.8	Grayscale 1550
0	0	0	90	33	84	180	79	141	270	124	175
5	3	7	95	36	88	185	81	143	275	126	177
10	4	13	100	38	92	190	83	145	280	129	179
15	6	19	105	40	95	195	84	148	285	132	182
20	7	2 5	110	42	99	200	86	150	290	135	184
25	9	33	115	44	101	205	88	152	295	138	187
30	11	40	120	46	106	210	89	153	300	141	189
35	13	47	125	48	110	215	92	155	305	144	191
40	15	52	130	50	113	220	93	157	310	147	193
45	17	55	135	53	117	225	95	159	315	150	196
50	18	59	140	57	121	230	97	161	320	151	198
55	20	62	145	59	124	235	98	161	325	153	200
60	22	65	150	63	127	240	101	164	330	155	203
65	23	68	155	66	128	245	106	166	335	156	214
70	25	71	160	69	132	250	109	168	340	157	225
75	27	74	165	72	134	255	114	169	345	158	242
80	29	77	170	75	136	260	118	171	350	160	/
85	31	81	175	77	138	265	123	173	355	162	/

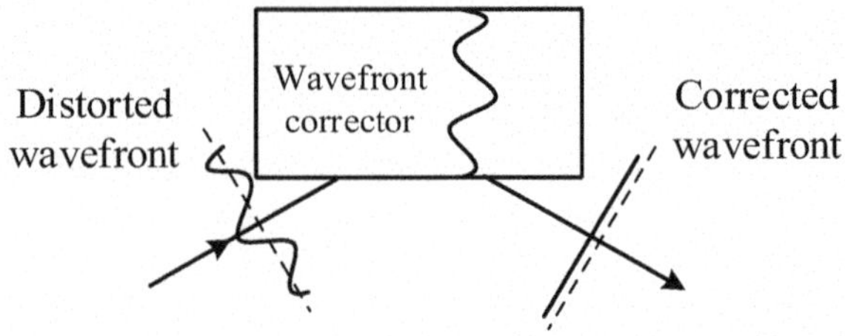

Fig. 8.15 Schematic of phase conjugate wavefront correction

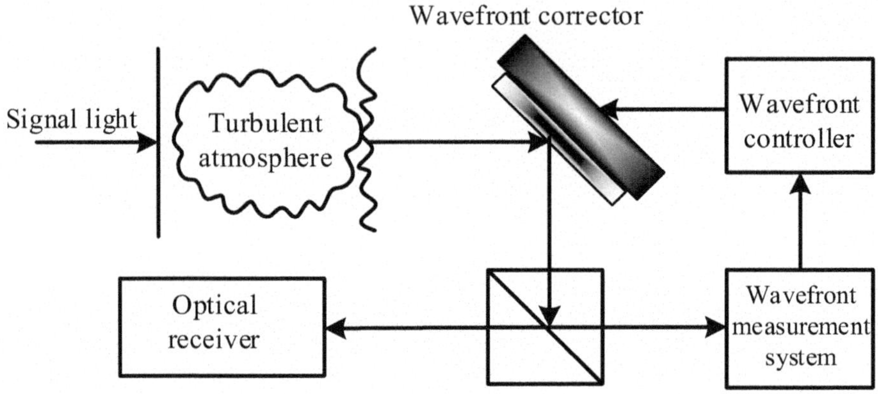

Fig. 8.16 Wavefront correction system for spatial coherent optical communication

system is used to reconstruct the distorted wavefront through the reconstruction algorithm and generate the conjugate wavefront required by the wavefront controller. The conjugate wavefront is sent to the wavefront corrector, and the wavefront is corrected. The wavefront of the beam behind the filter is close to the ideal wavefront [33].

(1) Wavefront detection system

The wavefront detection system is an important part of the spatial coherent optical communication system. Its main function is to detect reliable information on wavefront distortion and provide it to the wavefront controller [29]. The widely used wavefront detection systems mainly include the Shack-Hartmann (S-H), interference, curvature, and quadrangular pyramid wavefront sensors.

The SH wavefront sensor divides the detection wavefront into a series of sub-aperture areas, focuses the sub-aperture spot on the array surface of the photodetector, measures the spot centroid deviation of the sub-aperture area on each array surface, and calculates the corresponding sub aperture slope, the sub-aperture slope is used to reconstruct the distorted wavefront information and feed it to the wavefront controller. The S-H wavefront sensor has a simple structure, strong anti-interference ability, flexible optical path construction, and good real-time performance. It does not need the coherence of the detected light waves, and there is no 2π uncertainty problem.

However, the response wavelength is limited to the visible light, and the dynamic range and sampling frequency of the sensor are not adjustable.

The interferometric wavefront sensor uses an interferometer as a wavefront detector. The sensitivity of the detector can reach the wavelength level. At present, it is mostly used to detect the optical surface topography and evaluate the quality of beams. Self-referencing interferometers are widely used in AO systems. Shearing interferometer is a type of such self-reference interferometers. This interferometer shears the measured wavefront to make the wavefront of two beams form a certain dislocation and then make the dislocation wavefront interfere with the detected wavefront. A detection technology for the detected wavefrontcan be divided into lateral, radial, rotating and reverse shearing. The practical applications of the transverse and radial shear interference technology are particularly extensive and include the detection of wavefront distortion caused by atmospheric turbulence in real time.

The curvature wavefront sensor was proposed by Roddier, who used it to measure the curvature of the light wavefront [34]. The relationship between the wavefront curvature and phase distribution measured with the curvature wavefront sensor can be expressed by the Poisson equation. The wavefront curvature distribution information can be directly used for the compensation of wavefront distortion, and the signal output by the curvature wavefront sensor can be directly used as the control signal for the double piezoelectric deformable mirror, which has been successfully applied to the AO system of the astronomical telescope.

The basic objective of the quadrangular pyramid wavefront sensor [35] is to detect the relationship between the measured signal and wavefront using the light intensity difference of each sub-pupil on the detection surface. It is widely used in astronomical observations.

The detection of wavefront distortion is a challenge in wavefront correction, because there is no ideal wavefront related to the signal light wavefront as a reference. Therefore, in the experiment, the slope or curvature information of the distorted wavefront phase must be measured before reconstructing the distorted wavefront, and then the phase conjugation method must be used to correct the wavefront. In this study, the S–H wavefront sensor was used to measure the distorted wavefront of the signal light.

(2) Wavefront controller

The wavefront controller uses the reconstructed wavefront phase information as the control signal to realize a closed-loop control of the adaptive wavefront correction system of spatial coherent optical communication.

(3) Wavefront corrector

The wavefront corrector in the AO system is the most important core component. The wavefront corrector uses two methods to correct the distorted wavefront. The objective of one method is to achieve phase correction by changing the refractive index. The typical device used is LC-SLM. The birefringence of the LC is changed by changing the applied external electric field. Next, the phase modulation characteristics of the LC-SLM are changed to correct the wavefront. This method has the advantages of no mechanical inertia, low cost, small size, and low power consumption. By adjusting the polarizer and analyzer, the pure phase modulation characteristic of 2π can be realized. The phase change range is large in this case. In the other method, wavefront correction is performed by changing the optical path length. The typical device is a deformable mirror. It has the advantages of a wider correction dynamic range and high time–space bandwidth. It is widely used in AO systems. However, it has many driving parts and its manufacturing process is complex and costly. Therefore, this study used the LC-SLM as the wavefront correction device to correct the wavefront distortion of the signal light caused by atmospheric turbulence.

8.5 Principle of Wavefront Measurement

To accurately obtain the distortion information of the corrected wavefront, reconstructing the conjugate wavefront of the distorted wavefront is an important step toward the wavefront correction of the spatially coherent optical communication system. Therefore, it is necessary to study and discuss the use of the transverse shear interferometer and S-H sensor to measure the wavefront.

8.5.1 Static Wavefront Measurement of Transverse Shear Interferometer

(1) Principle of static wavefront measurement of transverse shear interferometer

Shearing interferometry does not require a standard reference wavefront, and the measured interferogram does not represent the phase distribution of the measured wavefront; rather, it represents the light intensity map. To obtain the phase information of the measured wavefront, it is necessary to effectively process the interference fringes measured using shearing interference. The processing process includes two steps. The first step is image processing, which extracts continuous and accurate three-dimensional coordinates from the interferogram and performs phase extraction. The extracted phase value is wrapped and limited within the range of $[-\pi, \pi]$. To obtain accurate wavefront phase information, the wrapped phase map must be unpacked to recover the three-dimensional map of the measured differential wavefront. The second step is to reconstruct the measured wavefront, calculate the phase information of the measured wavefront based on the measured three-dimensional map of the differential wavefront, and reconstruct the wavefront. The working principle for using the transverse shearing interferometer to measure the wavefront is depicted in Fig. 8.17.

A schematic of lateral shear interferometer is depicted in Fig. 8.18. The beam output by the laser is expanded and collimated into a parallel beam and then split

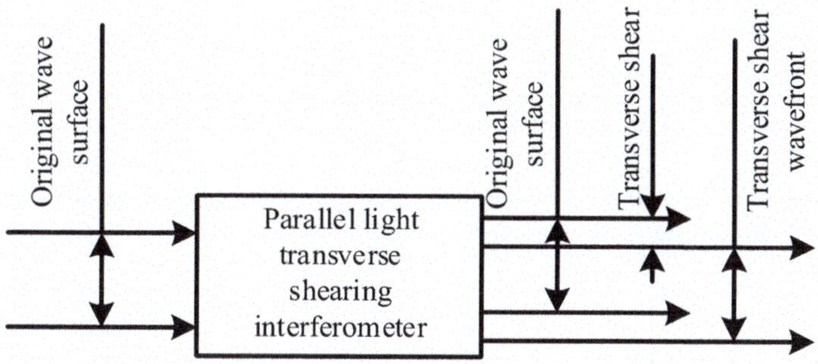

Fig. 8.17 Principle of lateral shear interferometer

by a beam splitting (BS) prism into two beams, namely, transmitted beam W_1 and reflected beam W_2. W_1 is reflected by the total reflection mirror M_1 and exits after the beam combiner (BC). W_2 is reflected by the total reflection mirror M_2 and exits after BC. To make the two beams emitted by the BC form a sheared beam, the position of M_2 must be adjusted in a manner that the distances between BS and M_1 and between BS and M_2 are no longer equal. In this way, the interference obtained after shearing interference can be detected in the overlapping area of the beam stripe. Assuming that the optical path difference between the wavefronts of the two beams is $\Delta W\ (x,y)$, then

$$\Delta W(x, y) = W(x + s, y) - W(x, y) \tag{8.20}$$

where s, $W(x, y)$, $W(x + s, y)$, and (x, y) represent the amount of shear, measured wavefront, shear wavefront, and rectangular coordinates, respectively. When $s = 0$, the optical path between the two wavefronts is equal, and no interference fringes are generated.

Because W_1 and W_2 are divided beams from the same laser, the amplitudes of W_1 and W_2 are equal. Therefore, the amplitude and light intensity expressions of the two beams can be obtained as

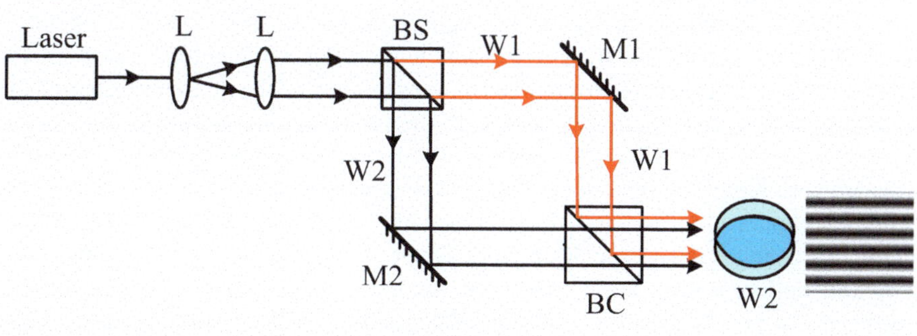

Fig. 8.18 Schematic of lateral shear interference

$$A(x, y) = \frac{I_{\max} - I_{\min}}{I_{\max} + I_{\min}} = 1 \tag{8.21}$$

$$I(x, y) = 2A(x, y)^2[1 + \cos \Delta\phi(x, y)] \tag{8.22}$$

Assuming that the wavefront generated after the transverse shear displacement is $W'(x, y)$, the optical path difference between the two wavefronts after shearing can be obtained as

$$\Delta W(x, y) = W(x, y) - W'(x, y) = \frac{\Delta\phi(x, y)}{k} = \frac{\lambda}{2\pi}\Delta\phi(x, y) \tag{8.23}$$

where $k = 2\pi/\lambda$ is the spatial angular frequency, λ is the wavelength of the measured light wave, $\Delta W(x,y)$ is the difference wavefront of the measured wavefront, and $\Delta\phi(x,y)$ is the difference of the measured wavefront phase function. The intensity distribution function of the transverse shear interferogram can be obtained using formulas (8.22) and (8.23).

(2) Extraction of wavefront phase by four step phase shift method

The interference fringe measured using the transverse shearing interferometry contains the phase information of the measured wavefront. Phase information must be extracted from the interference fringe to reconstruct the wavefront. First, it is necessary to effectively recover the phase information of the wavefront from the light intensity information of the interference fringes in a sinusoidal distribution in space. This objective can be achieved using two methods: intensity and phase methods.

The intensity method includes the fringe extreme value positioning method and the regularization method. The fringe extreme value positioning method obtains the wavefront phase information through an image processing algorithm. The regularization method inversely transforms the detection light intensity to obtain the phase information.

The phase method mainly includes two methods: the Fourier transform method (FFT), and time-domain phase shift method. The FFT method is also known as the space carrier frequency phase shift. In this method, a two-dimensional Fourier transform on the collected interferogram through image processing is used to obtain a positive first-order spectrum. Next, a two-dimensional Fourier transform is performed on the positive first-order spectrum. Inverse transformation, the phase information of the measured wavefront is finally obtained. The time-domain phase shift method is based on moving M1 in Fig. 8.18 regularly through a phase shifter thereby changing the beam path length on the optical path and obtaining sheared interference fringes with a different brightness. These interference fringes represent light intensity information that changes the cosine. Using a CCD to collect three or more shearing interferograms under various phase conditions, the phase distribution of the measured wavefront can be obtained by using the orthogonality of the cosine function. This study used a four-step phase shift method to extract the phase information of the measured wavefront.

The four-step phase shift method is a widely used phase extraction method in shearing interference technology. A piezoelectric ceramic stack is added to the total reflection mirror M_1 in Fig. 8.18 to control M_1 to sample at equal intervals less than the 1/4 period. Next, the shearing interference wavefront generated when M_1 is at various positions is collected. The phase of the wavefront changes sequentially by $\pi/2$ from 0 to 2π in a manner that four pairs of interference light intensity distribution diagrams in one period are obtained. The four pairs of light intensity distributions are

$$\begin{aligned} I_1(x, y) &= a(x, y)\{1 + A(x, y)[\Delta\phi(x, y)]\} \\ I_2(x, y) &= a(x, y)\left\{1 + A(x, y)\left[\Delta\phi(x, y) + \frac{\pi}{2}\right]\right\} \\ I_3(x, y) &= a(x, y)\{1 + A(x, y)[\Delta\phi(x, y) + \pi]\} \\ I_4(x, y) &= a(x, y)\left\{1 + A(x, y)\left[\Delta\phi(x, y) + \frac{3\pi}{2}\right]\right\} \end{aligned} \tag{8.24}$$

Using Eq. (8.24), the differential phase of the measured wavefront can be obtained as

$$\Delta\phi(x, y) = \arctan\left[\frac{I_4(x, y) - I_2(x, y)}{I_1(x, y) - I_3(x, y)}\right] \tag{8.25}$$

Equation (8.25) is the calculation formula for the four-step phase shift method to extract the phase. The formula uses subtraction and division operations. The advantage of these two operations is that the fixed noise in the interference field and the noise caused by the gain change of the photodetector are automatically eliminated. However, the phase extracted using the four-step phase shift method is obtained using the arctangent function. Therefore, the phase value of the measured wavefront is limited to the range of $[-\pi, \pi]$. To eliminate the impact of the phase jump, it is necessary to perform the unpacking operation on the obtained wavefront phase and then transform the phase into the range of $[0, 2\pi]$.

(3) Least square method phase unwrapping

The algorithm of phase unwrapping is divided into two types: time- and space-domain phase unwrapping algorithms. The objective of time-domain phase unwrapping is to perform phase unwrapping processing from a low frequency to a high frequency on the time axis and output the unwrapped phase of the measured wave surface picture. The spatial-domain unwrapping algorithm is the most widely used unwrapping method at present. It is also of two types: the path tracking and minimum norm methods [36]. The path tracking method is a partial phase unwrapping algorithm, which is essentially the process of selecting a suitable path to integrate the wrapped phase difference. The minimum norm method is a global algorithm that has nothing to do with the path. Its objective is to realize the optimization algorithm of the objective function. When norm $p = 2$, the minimum norm method solution package becomes the least square solution package operation, which is the most widely used phase unwrapping algorithm at present.

The least square method is used for the phase unpacking operation, which can not only suppress the impact of noise points but also ensure the continuity of the unpacked wave surface. This method starts from the whole image generated through the four-step phase shift method, uses the least square algorithm to obtain the phase surface with the smallest sum of square deviations from the package phase slope, and outputs the unpackaged wavefront phase surface. Its basic principle is as follows. For wrapped phase ϕ (x, y) extracted using the four-step phase shift, the corresponding unwrapped phase φ (x, y) is obtained in a manner that the value of the minimum error function J is the smallest [37]:

$$J = \iint \|\nabla\varphi - \nabla\phi\|^p dA = \min \tag{8.26}$$

Where $\Delta\varphi$ is the phase gradient value after unwrapping the measured wavefront, $\Delta\phi$ is the wrapped phase gradient value, dA is the integral surface element, and $\|\nabla\ \varphi - \nabla\ \phi\|^p$ is the norm of p. When $p = 2$, it is the minimum two-multiplication unwrapping algorithm [38].

Rewriting Eq. (8.26) as a discrete least squares expression, we get

$$J = \sum_{i=0}^{M-2}\sum_{j=0}^{N-1}\left(\varphi_{i+1,j} - \varphi_{i,j} - \Delta^x_{i,j}\right)^p + \sum_{i=0}^{M-2}\sum_{j=0}^{N-2}\left(\varphi_{i,j+1} - \varphi_{i,j} - \Delta^y_{i,j}\right)^p = \min \tag{8.27}$$

where M and N are the rows and columns of the two-dimensional phase data, respectively; Δ is the difference operator, $\Delta^x_{i,j}$ is the wrapped phase gradient in the x direction, $\Delta^y_{i,j}$ is the wrapped phase gradient in the y direction [39], $i = 0,1,2,\ldots,M - 1, j = 0,1,2,\ldots, N - 1$. To make J reach the minimum value in the phase unwrapping process, Eq. (8.27) must be derivated once. The first derivative of J is equal to zero. The following expression can be obtained:

$$dJ = p(d_1 + d_2) \tag{8.28}$$

In Eq. (8.28),

$$d_1 = \sum_{i=0}^{M-2}\sum_{j=0}^{N-1}\left(\varphi_{i+1,j} - \varphi_{i,j} - \Delta^x_{i,j}\right)\left|\varphi_{i+1,j} - \varphi_{i,j} - \Delta^x_{i,j}\right|^{p-2}\left(d\varphi_{i+1,j} - d\varphi_{i,j}\right) \tag{8.29}$$

$$d_2 = \sum_{i=0}^{M-2}\sum_{j=0}^{N-1}\left(\varphi_{i,j+1} - \varphi_{i,j} - \Delta_{i,j}^{y}\right)\left|\varphi_{i,j+1} - \varphi_{i,j} - \Delta_{i,j}^{y}\right|^{p-2}\left(d\varphi_{i,j+1} - d\varphi_{i,j}\right) \tag{8.30}$$

In Eqs. (8.29) and (8.30), for the convenience of calculation,

$$\begin{cases} a_{i,j} = \left(\varphi_{i+1,j} - \varphi_{i,j} - \Delta_{i,j}^{x}\right)\left|\varphi_{i+1,j} - \varphi_{i,j} - \Delta_{i,j}^{x}\right|^{p-2} & (0 \le i \le M-2; 0 \le j \le N-1) \\ 0 & (i = -1, M-1; 0 \le j \le N-1) \end{cases} \tag{8.31}$$

$$\begin{cases} b_{i,j} = \left(\varphi_{i,j+1} - \varphi_{i,j} - \Delta_{i,j}^{y}\right)\left|\varphi_{i,j+1} - \varphi_{i,j} - \Delta_{i,j}^{y}\right|^{p-2} & (0 \le i \le M-1; 0 \le j \le N-2) \\ 0 & (0 \le i \le M-1; j = 1, N-1) \end{cases} \tag{8.32}$$

Incorporating Eqs. (8.31) and (8.32) into Eqs. (8.29) and (8.30) and simplifying, we get

$$d_1 = \sum_{i=0}^{M-1}\sum_{j=0}^{N-1}\left(a_{i-1,j} - a_{i,j}\right)d\varphi_{i,j} \tag{8.33}$$

$$d_2 = \sum_{i=0}^{M-1}\sum_{j=0}^{N-1}\left(b_{i,j-1} - b_{i,j}\right)d\varphi_{i,j} \tag{8.34}$$

Incorporating Eqs. (8.33) and (8.34) into Eq. (8.27) and simplifying, we get

$$dJ = -p\sum_{i=0}^{M-1}\sum_{j=0}^{N-1}\left(a_{i,j} - a_{i-1,j} + b_{i,j} - b_{i,j-1}\right)d\varphi_{i,j} = 0 \tag{8.35}$$

where $d\varphi_{i,j}$ is an indeterminate number. When $dJ = 0$, we get

$$a_{i,j} - a_{i-1,j} + b_{i,j} - b_{i,j-1} = 0,\ \ 0 \le i \le M-1; 0 \le j \le N-1 \tag{8.36}$$

From the boundary conditions $a_{i,j}=0$, $i = -1, M-1$; $0 \le j \le N-1$ and $b_{i,j}=0$, $0 \le i \le M-1$; $j = -1, N-1$, the boundary conditions of the phase are

$$\begin{cases} \varphi_{-1,j} - \varphi_{0,j} = \phi_{-1,j} - \phi_{0,j} \\ \varphi_{M,j} - \varphi_{M-1,j} = \phi_{M,j} - \phi_{M-1,j} \end{cases} \quad 0 \le j \le N-1 \tag{8.37}$$

$$\begin{cases} \varphi_{i,-1} - \varphi_{i,0} = \phi_{i,-1} - \phi_{i,0} \\ \varphi_{i,N} - \varphi_{i,N-1} = \phi_{i,n} - \phi_{i,N-1} \end{cases} \quad 0 \le i \le M-1 \tag{8.38}$$

Substituting boundary conditions (8.37) and (8.38) into Eq. (8.36) and simplifying, the optimal solution of the equation can be obtained as

$$\left(\varphi_{i+1,j}-\varphi_{i,j}-\Delta_{i,j}^{x}\right)U_{i,j}+\left(\varphi_{i,j+1}-\varphi_{i,j}-\Delta_{i,j}^{y}\right)V_{i,j}$$
$$-\left(\varphi_{i,j}-\varphi_{i-1,j}-\Delta_{i-1,j}^{x}\right)U_{i-1,j}-\left(\varphi_{i,j}-\varphi_{i,j-1}-\Delta_{i,j-1}^{y}\right)V_{i,j-1}=0 \quad (8.39)$$

In formula (8.39),

$$\begin{cases} U_{i,j}=\left|\varphi_{i+1,j}-\varphi_{i,j}-\Delta_{i,j}^{x}\right|^{p-2}, & 0\le i\le M-2;\ 0\le j\le N-1 \\ 0, & \text{else} \end{cases} \quad (8.40)$$

$$\begin{cases} V_{i,j}=\left|\varphi_{i,j+1}-\varphi_{i,j}-\Delta_{i,j}^{y}\right|^{p-2}, & 0\le i\le M-1;\ 0\le j\le N-2 \\ 0, & \text{else} \end{cases} \quad (8.41)$$

When $p=2$, the above optimization algorithm is transformed into the least square method to solve the wrapping operation, and Eq. (8.39) is solved to obtain the discrete Poisson equation as follows:

$$\varphi_{i+1,j}+\varphi_{i,j+1}+\varphi_{i,j-1}+\varphi_{i-1,j}-4\varphi_{i,j}=\rho_{i,j},\quad 0\le i\le M-1; 0\le j\le N-1 \quad (8.42)$$

where $\rho_{i,j}=\left(\Delta_{i,j}^{x}-\Delta_{i-1,j}^{x}\right)+\left(\Delta_{i,j}^{y}-\Delta_{i,j-1}^{y}\right)$ is the value of the wrapped phase. To solve for the value of (8.42), the boundary conditions are

$$\begin{cases} \Delta_{-1,j}^{x}=\Delta_{M-1,j}^{x}=0,\ \ 0\le j\le N-1 \\ \Delta_{i,-1}^{y}-\Delta_{i,N-1}^{y}=0,\ \ 0\le i\le M-1 \end{cases} \quad (8.43)$$

Therefore, the main objective of the least square method to solve the wrapped phase under boundary conditions. The calculation process is as follows:

$$\varphi_{i,j}^{(k+1)}=\frac{1}{2}\left(\varphi_{i+1}^{(k)}+\varphi_{i-1}^{(k)}+\varphi_{i,j+1}^{(k)}+\varphi_{i,j}^{(k)}-\rho_{i,j}\right),\quad 0\le i\le M-2; 0\le j\le N-2 \quad (8.44)$$

The values of the four vertices are

$$\begin{cases} \varphi_{0,0}^{(k+1)}=\frac{1}{2}\left(\varphi_{1,0}^{(k)}+\varphi_{0,1}^{(k)}-\rho_{i,j}\right) \\ \varphi_{M-1,0}^{(k+1)}=\frac{1}{2}\left(\varphi_{M-1,1}^{(k)}+\varphi_{M-1,0}^{(k)}-\rho_{i,j}\right) \\ \varphi_{0,N-1}^{(k+1)}=\frac{1}{2}\left(\varphi_{1,N-1}^{(k)}+\varphi_{0,N-2}^{(k)}-\rho_{i,j}\right) \\ \varphi_{M-1,N-1}^{(k+1)}=\frac{1}{2}\left(\varphi_{M-1,N-2}^{(k)}+\varphi_{M-2,N-1}^{(k)}-\rho_{i,j}\right) \end{cases} \quad (8.45)$$

$$\begin{cases} \varphi_{0,j}^{(k+1)} = \frac{1}{3}\left(\varphi_{1,j}^{(k)} + \varphi_{0,j+1}^{(k)} + \varphi_{0,j-1}^{(k)} - \rho_{i,j}\right) \\ \varphi_{M-1,j}^{(k+1)} = \frac{1}{3}\left(\varphi_{M-2,j}^{(k)} + \varphi_{M-1,j+1}^{(k)} + \varphi_{M-1,j-1}^{(k)} - \rho_{i,j}\right) \end{cases} \quad 1 \le j \le N-2 \quad (8.46)$$

$$\begin{cases} \varphi_{i,0}^{(k+1)} = \frac{1}{3}\left(\varphi_{M-2,j}^{(k)} + \varphi_{M-2,j+1}^{(k)} + \varphi_{M-2,j-1}^{(k)} - \rho_{i,j}\right) \\ \varphi_{i,N-1}^{(k+1)} = \frac{1}{3}\left(\varphi_{i+1,N-1}^{(k)} + \varphi_{i-1,N-1}^{(k)} + \varphi_{i,N-2}^{(k)} - \rho_{i,j}\right) \end{cases} \quad i \le i \le M-2 \quad (8.47)$$

Next, the space domain $\varphi_{i,j}^{(0)} = 0$ is initialized and least squares phase unwrapping operation is performed. Finally, the phase information of the measured distortion wavefront is obtained.

8.5.2 Shack-Hartmann Real-Time Wavefront Measurement Principle

The main components of the S-H wavefront sensor are the microlens array and CCD. Its structure is simple and affected less by the environmental conditions. It can be used for real-time measurement and is widely used in optical inspection, adaptive optics, laser beam quality diagnosis, and so on. The working principle of the S-H wavefront sensor is depicted in Fig. 8.19. The CCD area detector corresponding to the microlens array is divided into several sub-areas. The transmission beam is decomposed by the microlens array into multiple basic beams. These secondary beams are focused on the detection surface of the CCD. When these secondary beams are ideal plane waves, the arrangement of the focal spot array obtained on the CCD focal plane is regular, and the center of the focal spot is on the optical axis of the sublens. The phase and intensity of the transmitted beam can be obtained by analyzing and processing the beam information on the CCD detection surface, and the first derivative of the phase can be measured. In fact, the measured beam has a certain wavefront aberration. As depicted in Fig. 8.19, this aberration causes the focus of the beam to shift and deviate from the optical axis of the sublens. The amount of the movement of each focal point is proportional to the local derivative of the wavefront of the incident beam, and this ratio is the focal length of the microlens. The beam intensity is directly proportional to the amplitude of each focus on the CCD detection surface.

In the experiment, each wavelet of the incident beam, sub-aperture of the microlens array, and corresponding subarea of the CCD constitute an intensity detection system with a limited receiving aperture. The fluctuations in the sum of the corresponding pixel gray values in each sub-aperture are counted. Next, long-term experimental measurements are performed to obtain fluctuations in the light intensity caused by the turbulent atmosphere. After the normalized variance statistics, the beam is transmitted by the turbulence in the turbulent atmosphere caused by the light intensity fluctuates are obtained.

The shearing interferometry measurement of the wavefront requires certain calculation and processing based on the measured light intensity information to obtain

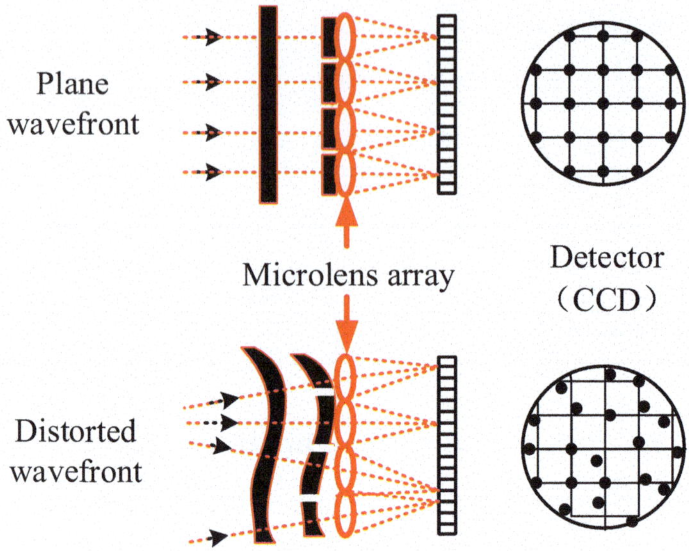

Fig. 8.19 Schematic of the Shack-Hartmann wavefront sensor principle [45]

wavefront information. The experimental system is highly dependent on the optical path and environmental conditions and requires image acquisition and calculation processing. Therefore, it is generally difficult to perform real-time measurements of the wavefront. However, the advantage of this system is its low cost. The calculation program used in the S-H wavefront sensor can efficiently output the measured wavefront information in real time. Its dependence on the experimental environment is lower than that of the shear interferometer measuring wavefront system; however, the cost is high. Therefore, to better realize the wavefront measurement of the spatially coherent optical communication system and output reliable wavefront information, the S-H wavefront sensor is used to measure the wavefront.

8.6 Wavefront Reconstruction

The conjugate method of the LC-SLM-R is used to perform the wavefront correction experiment for spatial coherent optical communication. First of all, we must measure and reconstruct the distorted wavefront caused by atmospheric turbulence. Because each term in the Zernike polynomial corresponds to a different order of aberration, a linear combination of the Zernike polynomial can easily reconstruct the measured wavefront. This is currently one of the most commonly used methods.

8.6.1 Zernike Polynomial

The Zernike polynomial is a two-dimensional polynomial of continuous functions of radial and angular variables in the unit circle domain. It can be expressed in two forms: the polar and rectangular coordinate systems. It can not only describe the phase information of the distorted wavefront but also achieve sufficient accuracy [40–42]. Therefore, Zernike polynomials are widely used in correcting the wavefront distortion caused by atmospheric turbulence. The two-dimensional Zernike polynomial can be expressed as follows [40]:

$$Z_n^m(\rho, \theta) = R_n^m(\rho)\exp(im\theta) \tag{8.48}$$

where n represents the radial sequence number, m represents the angular sequence number, and $n \geq 0$, $m \geq 0$, and $n \geq m$. $R_n^m(\rho)$ is the radial function, which can be expressed as [41]

$$R_n^m(\rho) = \begin{cases} \sum\limits_{s=0}^{(n+m)/2} (-1)^s \frac{(n-s)!}{s![(n+m)/2-s]!+[(n-m)/2-s]!}\rho^{n-2s} & (n-m=\text{even}) \\ 0 & (n-m=\text{odd}) \end{cases} \tag{8.49}$$

For convenience, orders n and m are converted to order i, and the orthogonality of the Zernike polynomial does not depend on the order of the polynomial. Formula (8.49) can be converted to a Zernike with only one order i polynomial. Its expression is [42]

$$Z_i(\rho, \theta) = \sqrt{n+1}\begin{cases} R_n^m(\rho)\sqrt{2}\cos(m\theta) & (\text{even}, i, m \neq 0) \\ R_n^m(\rho)\sqrt{2}\sin(m\theta) & (\text{odd}, i, m \neq 0) \\ R_n^m(\rho) & (m=0) \end{cases} \tag{8.50}$$

The different values of n and m in Eq. (8.50) correspond to different Zernike polynomials. Various types of Zernike polynomials can be reconstructed by linear combination of various polynomials. This combination can represent the surface of various measured wave surfaces, shapes, and aberrations. The low-order terms of the Zernike polynomial can represent various primary aberrations. Through mathematical derivation, a continuous function can be used to represent various wavefronts that are actually desired to be generated. Table 8.2 lists the polar and rectangular coordinate expressions of the first 20 Zernike polynomials and the corresponding types of wave aberrations.

The coefficients of the Zernike polynomial are calculated using the phase information of the measured distortion wavefront. The Zernike polynomials of various levels are used for linear combination to reconstruct the distorted wavefront. The conjugate wavefront of the distorted wavefront is obtained, and the corresponding

Table 8.2 First 20 Zernike polynomials

Order	Polar coordinate expression	Cartesian coordinate expression	Type of aberration
1	$2\rho\cos\theta$	$2x$	Tilt in x direction
2	$2\rho\sin\theta$	$2y$	Tilt in y direction
3	$2\rho^2 - 1$	$2(x^2 + y^2) - 1$	Defocused
4	$\rho^2\cos2\theta$	$x^2 - y^2$	0 or 90°direction astigmatism
5	$\rho^2\sin2\theta$	$2 \times y$	± 45° direction astigmatism
6	$(3\rho^3 - 2\rho)\cos\theta$	$3\,(x^3 + x\,y^2) - 2\,x$	x direction coma
7	$(3\rho^3 - 2\rho)\sin\theta$	$3\,(x^2\,y + y^3) - 2\,y$	Tilt in y direction
8	$\rho^3\cos\theta$	$x^3 + 3 \times y^2$	Trefoil astigmatism
9	$\rho^3\sin\theta$	$3x^2\,y - y^3$	Trefoil astigmatism
10	$6\rho^4 - 6\rho^2 + 1$	$6(x^4 + y^4 + 2x^2\,y^2 - x^2 - y^2) + 1$	Primary ball aberration
11	$(4\rho^4 - 3\rho^2)\cos2\theta$	$4x^4 - 4\,y^4 - 3x^2 + 3y^2$	0° or 90°primary astigmatism
12	$(4\rho^4 - 3\rho^2)\sin2\theta$	$2\,xy\,(4x^2 + 4y^2 - 3)$	± 45° primary astigmatism
13	$\rho^4\cos4\theta$	$x^4 + y^4 - 6\,x^2\,y^2$	Four-leaf astigmatism
14	$\rho^4\sin4\theta$	$4x^3\,y - 4 \times y^3$	Four-leaf astigmatism
15	$(10\rho^5 - 12\rho^3 + 3\rho)\cos\theta$	$10x^5 + 20\,x^3y^2 + 10\,xy^4 - 12\,x^3 - 12xy^2 + 3x$	Secondary coma
16	$(10\rho^5 - 12\rho^3 + 3\rho)\sin\theta$	$10y^5 + 20\,x^2y^3 + 10\,x^4y - 12\,y^3 - 12x^2y + 3y$	Secondary coma
17	$(5\rho^5 - 4\rho^3)\cos3\theta$	$5x^5 + 10\,x^3\,y^2 - 4\,x^3 - 15 \times y^4 + 12 \times y^2$	Second-order trilobal astigmatism
18	$(5\rho^5 - 4\rho^3)\sin3\theta$	$15\,x^4y - 5\,y^5 + 10\,x^2\,y^3 - 4\,y^3 - 12\,x^2\,y$	Second-order trilobal astigmatism
19	$\rho^5\cos5\theta$	$x^5 - 10\,x^3\,y^2 + 5 \times y^4$	Five-leaf astigmatism
20	$\rho^5\sin5\theta$	$5x^4y - 10\,x^2\,y^3 + y^5$	Five-leaf astigmatism

grayscale image is generated by the program and loaded on the LC-SLM-R to realize the correction of the distorted wavefront.

8.6.2 *Wavefront Reconstruction Based on Zernike Polynomial*

To obtain the conjugate wavefront required by the LC-SLM-R to correct the wavefront, it is also necessary to reconstruct the wavefront of the measured distortion wavefront. In spatial coherent optical communication, the first 20 low-order Zernike polynomials can be used to accurately reconstruct the distorted wavefront [43]. Furthermore, the algorithm is simplified, which significantly reduces the calculation time.

The main objective of using the Zernike polynomials to reconstruct the wavefront is to obtain the Zernike polynomial coefficients of the measured differential wavefront through an optimization algorithm. The measured distortion wavefront ϕ (x, y) can be expressed as a series of linear combinations of Zernike polynomials [44]:

$$\phi(x, y) = a_0 + \sum_{i=1}^{n} a_i Z_i(x, y) + \xi \tag{8.51}$$

where a_0 is the average value of the distorted wavefront, a_i is the Zernike polynomial coefficient of i term, Z_i (x, y) is the Zernike polynomial at the discrete point, and ξ is the measurement error of the distorted wavefront.

The first-order partial derivative of the Zernike polynomial represents the slope of the distorted wavefront. The wavefront slopes in the x and y directions at discrete points can be expressed as follows:

$$\begin{cases} B_x(j) = \sum_{i=1}^{n} a_i \dfrac{\iint_{S_j} \frac{\partial Z_i(x,y)}{\partial x}}{S_j} + u_j = \sum_{i=1}^{n} a_i Z_{xi}(j) + u_j \\ B_y(j) = \sum_{i=1}^{n} a_i \dfrac{\iint_{S_j} \frac{\partial Z_i(x,y)}{\partial y}}{S_j} + v_j = \sum_{i=1}^{n} a_i Z_{yi}(j) + v_j \end{cases} \tag{8.52}$$

where B_x (j) and B_y (j) are the coefficients of the Zernike polynomial; (x, y) represents the barycentric coordinates of the i th discrete point; S_j is the area of the discrete point; u_j and v_j are sampling noise. Then the Zernike polynomial coefficients of the reconstructed distorted wavefront can be represented by the following matrix:

$$\begin{bmatrix} B_x(1) \\ B_y(1) \\ B_x(2) \\ B_y(2) \\ \vdots \\ B_x(m) \\ B_y(m) \end{bmatrix} = \begin{bmatrix} Z_{x_1}(1) & Z_{x_2}(1) & \cdots & Z_{x_n}(1) \\ Z_{y_1}(1) & Z_{y_2}(1) & \cdots & Z_{y_n}(1) \\ Z_{x_1}(2) & Z_{x_2}(2) & \cdots & Z_{x_n}(2) \\ Z_{y_1}(2) & Z_{y_2}(2) & \cdots & Z_{y_n}(2) \\ \vdots & \vdots & \cdots & \vdots \\ Z_{x_1}(m) & Z_{x_2}(m) & \cdots & Z_{x_n}(m) \\ Z_{y_1}(m) & Z_{y_2}(m) & \cdots & Z_{y_n}(m) \end{bmatrix} \cdot \begin{bmatrix} a_1 \\ a_2 \\ \vdots \\ a_n \end{bmatrix} + \begin{bmatrix} \eta_1 \\ \eta_2 \\ \eta_3 \\ \eta_4 \\ \vdots \\ \eta_{2m-1} \\ \eta_{2m} \end{bmatrix} \tag{8.53}$$

Equation can be written as $B = Z \cdot A + \eta$, where A is the coefficient matrix of the Zernike polynomial, η is the measurement error matrix of the distortion wavefront, and Z is the matrix formed by the first-order partial derivatives of the Zernike polynomials. Next, the first 20 Zernike polynomials are used to obtain the partial derivative matrix, as listed in Table 8.2. According to the singular matrix decomposition method, the generalized inverse matrix Z^{-1} of Z can be calculated, and the least square solution is $A = Z^{-1} \cdot B$ For each discrete point of distorted wavefront, Z^{-1} is determined, so the distorted wavefront can be reconstructed.

8.7 LC-SLM-R Wavefront Correction Experiment

8.7.1 Static Wavefront Correction Experiment

(1) LC-SLM

Two LC-SLMs were used in this experiment: LC-SLM-R and LC-SLM-T. The LC-SLM-R is used to correct the wavefront and act as a wavefront corrector. The LC-SLM-T is used to simulate atmospheric turbulence and act as a random phase screen. The performance parameters of the two are listed in Table 8.3.

(2) Wavefront measurement experiment

In the experiment, the shearing amount of the transverse shearing interferometer was 1 mm, and the sampling area length was 400 mm. Because the phase of the distorted wavefront was unpacked using the four-step phase shift method, the phases of the x and y directions were cut out. Four interference fringes with a phase difference of 90° are depicted in Figs. 8.20 and 8.21.

The four interferograms each in Figs. 8.20 and 8.21 were extracted using the four-step phase shift method, and the respective wrapped phases were obtained as depicted in Figs. 8.22 and 8.23. The wrapped phase is limited to the range of $[-\pi, \pi]$. To eliminate the impact of the phase jump, the above-mentioned wrapped phase must be unwrapped. Finally, it must be transformed into the range of $[0, 2\pi]$. After unwrapping, the three-dimensional coordinate values of the differential wavefront in the x and y directions are depicted in Figs. 8.24 and 8.25, respectively.

Table 8.3 SLM parameters

Performance parameter	LC-SLM-R	LC-SLM-T
LCD type	Reflective	Transmissive
Pixel size	9 μm	26 μm
Resolution	1024 × 768	1024 × 768
Phase modulation capability	> 2π @532 nm	> 2π @532 nm
Refresh rate	60 Hz	60 Hz
Calibration wavelength	400–700 nm	400–700 nm
Grayscale	8 bit, 256 levels	8 bit, 256 levels
Data interface	VGA	VGA

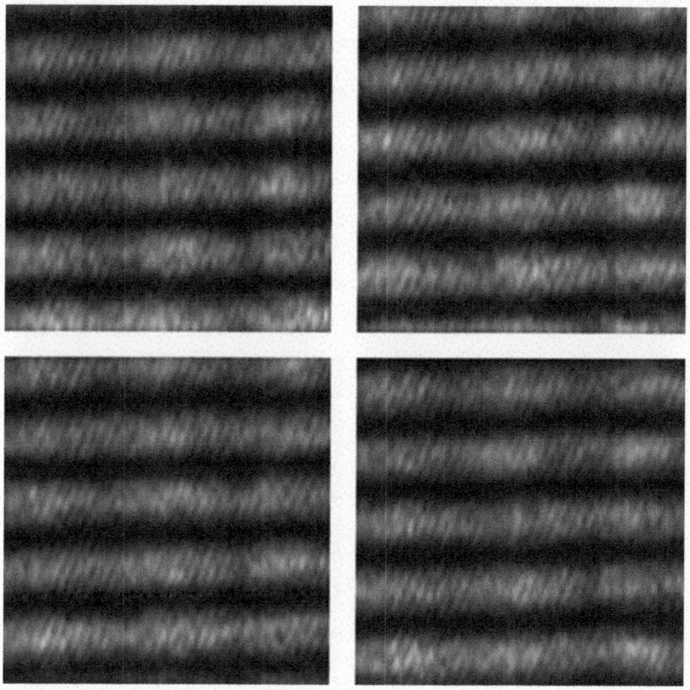

Fig. 8.20 Shearing interferogram in x direction [45]

(3) Wavefront reconstruction

Using the Zernike polynomials to reconstruct the wavefront is a complex process. After unwrapping, the three-dimensional coordinate values of the differential wavefront in the x and y directions are fitted with Zernike polynomials, and the corresponding Zernike polynomial coefficients are calculated. The coefficients obtained as such and the Zernike polynomial are combined based on certain rules to reconstruct the measured wavefront. The reconstructed wavefront is depicted in Fig. 8.26.

(4) Static wavefront correction experiment

In spatial coherent optical communication systems, the signal light wavefront phase distortion caused by atmospheric turbulence seriously affects the mixing efficiency when the signal light and the local oscillator light are coherently mixed, thereby reducing the output voltage of the difference frequency signal. One of the advantages of heterodyne detection is that it can detect weak light signals. The objective is to amplify and detect the signal light using the optical power of the local oscillator. However, when the optical power of the local oscillator is significantly large, the additional shot noise becomes overwhelming. This makes it impossible to detect the beat frequency signal, and thus cannot realize the demodulation of the beat frequency

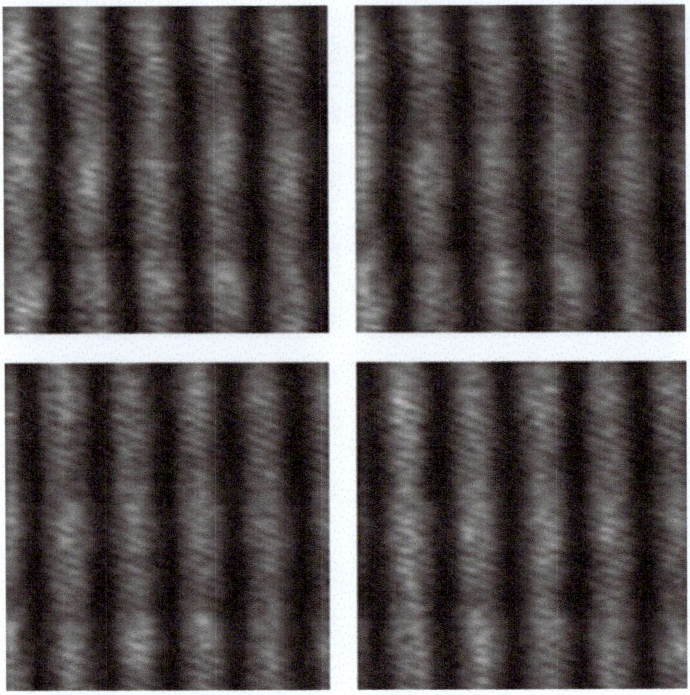

Fig. 8.21 Shearing interferogram in *y* direction [45]

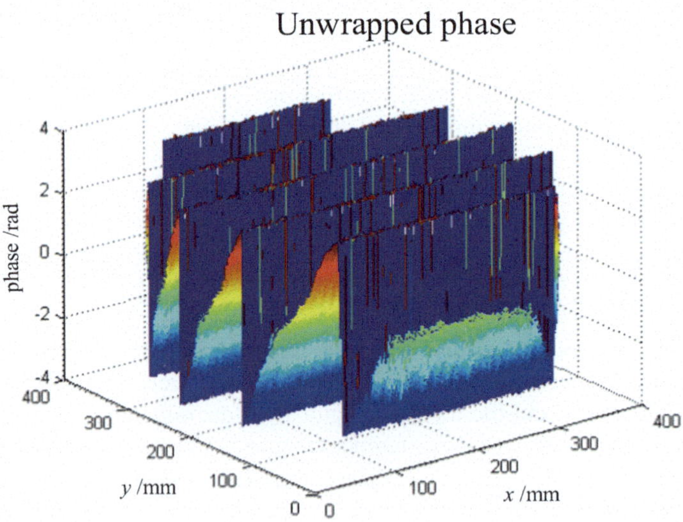

Fig. 8.22 Wrapping phase in *x* direction [45]

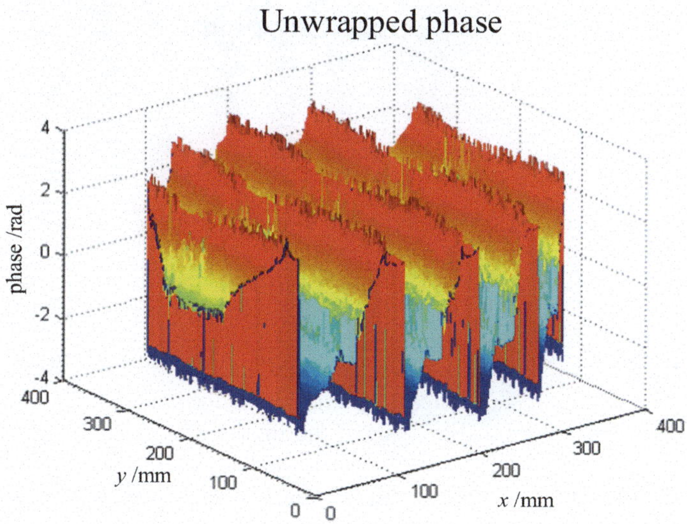

Fig. 8.23 Wrapping phase *y* direction [45]

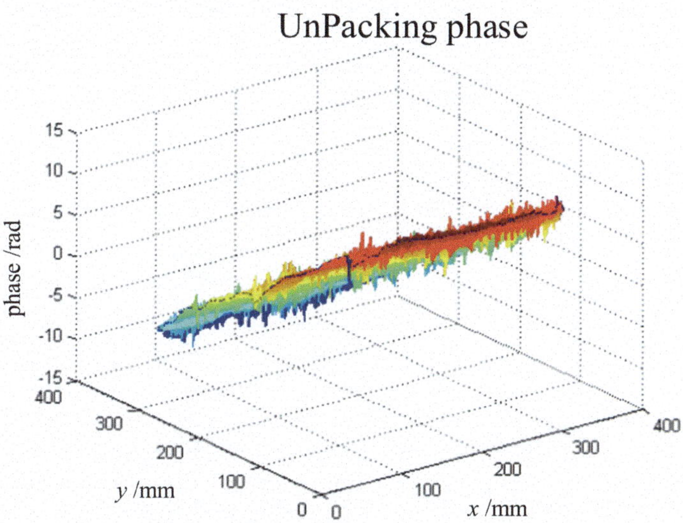

Fig. 8.24 Differential wavefront in *x* direction [45]

signal. Therefore, the measured wavefront must be used to correct the distorted wavefront, and the conjugate wavefront must be used for wavefront correction.

The working principle of wavefront correction is depicted in Fig. 8.27. The distorted wavefront measured using the transverse shearing interferometry is a beacon beam with a wavelength of 632.8 nm; the beam for transmitting information in spatial

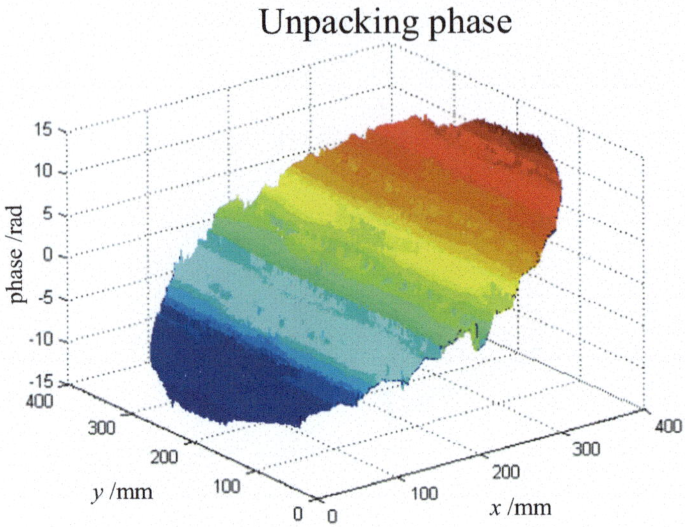

Fig. 8.25 Differential wavefront in *y* direction [45]

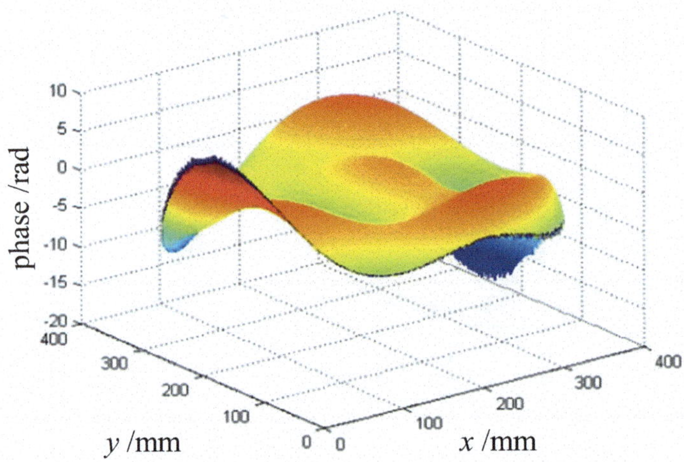

Fig. 8.26 Zernike polynomial reconstruction of 632.8 nm beam wavefront [45]

coherent optical communication is a beam with a wavelength of 1550 nm. The wavefront distortion of the 1550 nm wavelength beam must be corrected using the conjugate wavefront of the 632.8 nm wavelength beam distortion wavefront. Based on the phase calibration experiment of the LC-SLM-R in Sect. 8.3, it is evident that the LC-SLM-R has strong phase modulation ability for a beam with a wavelength of 632.8 nm and poor phase modulation ability for a beam with a wavelength of 1550 nm. In other words, the same turbulence phase screen causes less wavefront

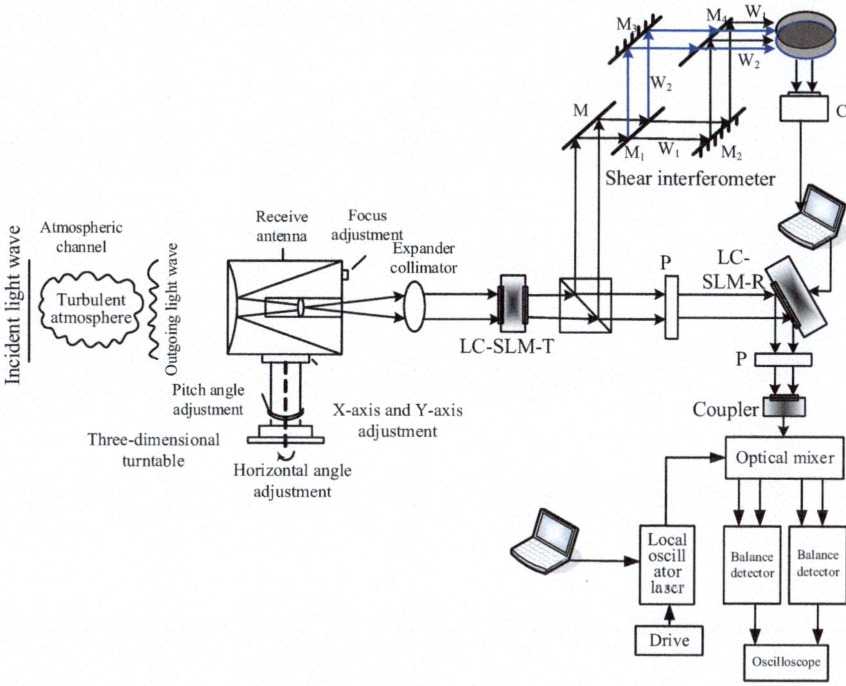

Fig. 8.27 Schematic of spatially coherent light wavefront correction

distortion for light beams with short wavelengths and larger wavefront distortion for light beams with long wavelengths. After calculating the wavefront distortion of the 632.8 nm wavelength beam of the beacon light, based on the phase modulation characteristic curve of the LC-SLM-T to the 632.8 nm wavelength beam and 1550 nm beam depicted in Fig. 8.14, the simulated turbulence phase screen pair is calculated. The distorted wavefront caused by the 1550 nm beam is depicted in Fig. 8.28.

The objective of the conjugate method for correcting the wavefront is to obtain the conjugate wavefront of the distorted wavefront reconstructed at 1550 nm, as depicted in Fig. 8.29, and convert the conjugate wavefront to that needed for the LC-SLM-R driving gray value. The driving gray values required for the 1550 nm beam under various phase delays are listed in Table 8.1. The driving gray value is generated as depicted in Fig. 8.30 and loaded into the LC-SLM-R. The LC-SLM-R loaded with the conjugate wavefront grayscale image is used to phase modulate the distorted wavefront of the signal beam to achieve the objective of wavefront correction.

Next, the grayscale image of the conjugate wavefront, depicted in Fig. 8.30, is loaded on the LC-SLM-R. In the experiment, the laser at the transmitter has an output power of 86 mW. Because the coupling efficiency and LC-SLM-T transmittance of the simulated turbulence phase screen are not high, the optical power received by the receiving end telescope is not high, and then passes through the LC-SLM-R, polarization device, coupling lens, and fiber loss, the optical power output by the

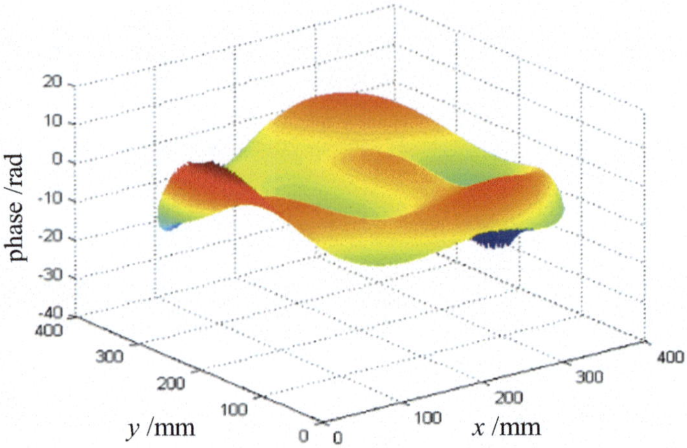

Fig. 8.28 Zernike polynomial reconstruction of 1550 nm beam wavefront [45]

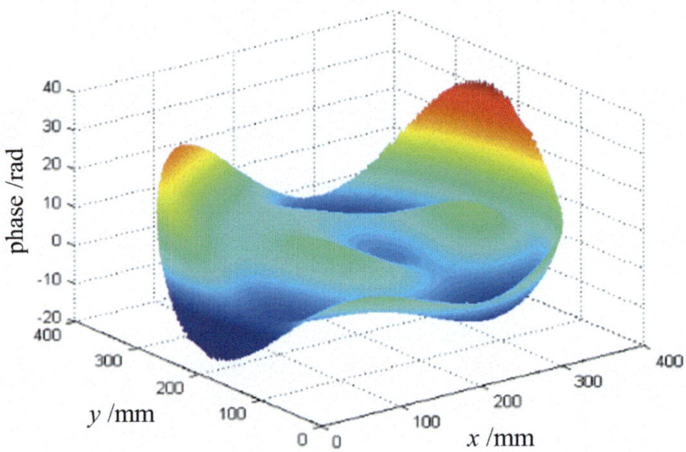

Fig. 8.29 Conjugate wavefront of 1550 nm beam wavefront distortion [45]

coupling fiber is only a few hundred nW. When the direct detection technology is used, the detector cannot respond to such a small optical power. The heterodyne detection method can effectively amplify the signal optical power using the local oscillator optical power, detect the difference frequency signal after the two beams are coherently mixed, and then demodulate the useful information using the external difference step demodulation technology. In the experiment, the signal light input through the coupling lens is a polarization-maintaining fiber. It is input to the mixer after passing through the polarization controller. By adjusting the coupling lens and polarization controller, the signal light power input to the mixer is increased to more than 1 μW. Next, the local oscillator laser is adjusted in a manner that the output power of the local oscillator laser does not exceed 10 mW. This prevents the balanced detector from entering a saturated working state owing to the excessively high input optical power.

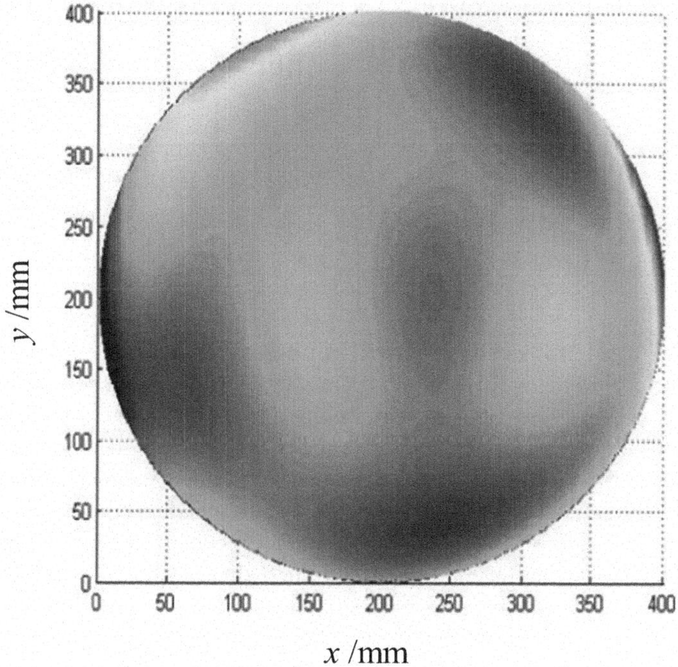

Fig. 8.30 Grayscale image of conjugate wavefront [45]

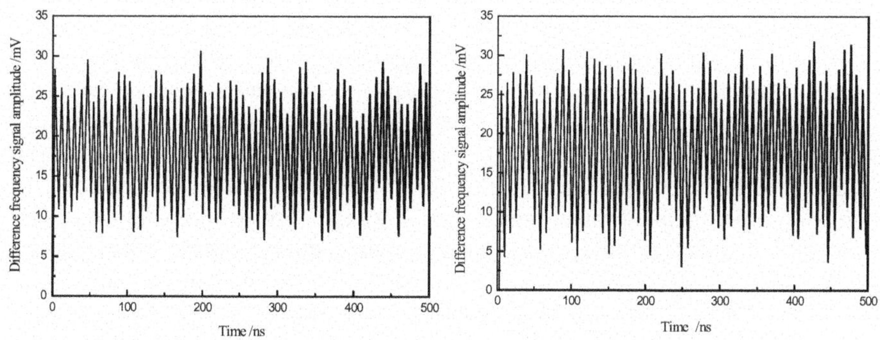

Fig. 8.31 Amplitude of difference frequency signal before correction [45]

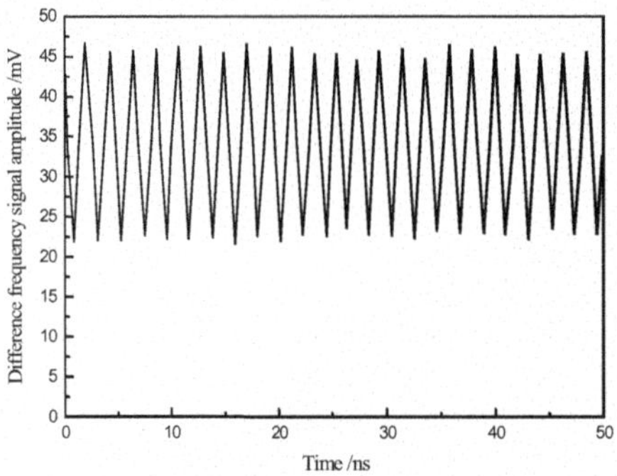

(a) In-phase channel

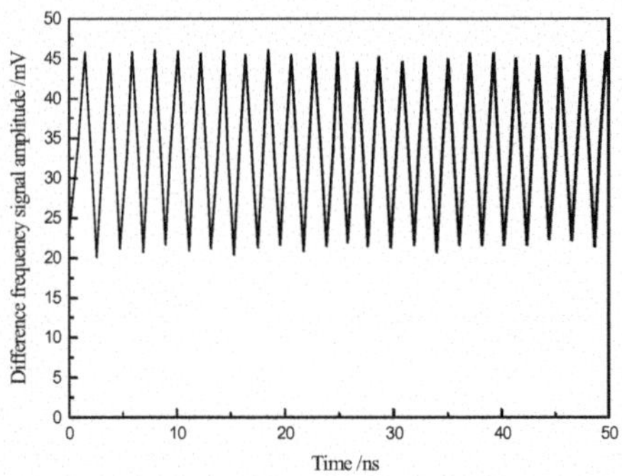

(b) Quadrature channel

Fig. 8.32 Amplitude of difference frequency signal after correction [45]

Figure 8.31 depicts the voltage value of the difference frequency signal output by the oscilloscope without wavefront correction. In this Figure (*a*) is the voltage amplitude of the difference frequency signal in the in-phase direction (In-phase channel), and (*b*) is the voltage amplitude of the difference frequency signal in the quadrature direction (Quadrature channel). There are two figures that show that there is a slight difference in voltage amplitudes between the in-phase and orthogonal directions. It is caused by the slight difference in the polarization state of the optical fiber transmitting the optical signal in the optical path and the difference in the cleanliness of the end face of the pigtail passing through the flange. In addition, because the wavefronts of the two beams differ, the sine value of the difference frequency signal has a certain distortion.

Next, the conjugate wavefront gray value, as depicted in Fig. 8.30, is loaded on the LC-SLM-R to correct the distorted wavefront. In the experiment, owing to the refresh frequency of the LC-SLM-R being low (only 60 Hz) and the constraint of the operating speed of the computer, the LC-SLM-R could not be used for dynamic wavefront correction. In addition, experiments have confirmed that the difference frequency signal fluctuation caused by atmospheric turbulence near the ground is significant before 10:30 a.m. and after 4:30 p.m. Between 10:30 a.m. and 4:30 p.m., the difference frequency signal caused by atmospheric turbulence near the ground is relatively stable, and there is no wide fluctuations. Therefore, in this experiment, the simulation experiment on wavefront correction was performed at around 1 p.m. The voltage amplitude of the difference frequency signal output by the balanced detector after correction is depicted in Fig. 8.32. In the figure, (*a*) and (b) are the voltage amplitudes of the difference frequency signals of channels I and Q, respectively.

Figure 8.32 depicts that the voltage amplitude of the difference frequency signal output by the balanced detector tends to a stable value after correction. Although this value fluctuates slightly with the time of the day in the experiment, there is no significant difference between the values of the output sine wave amplitudes, which basically remain stable at approximately 46 mV. Compared with the voltage

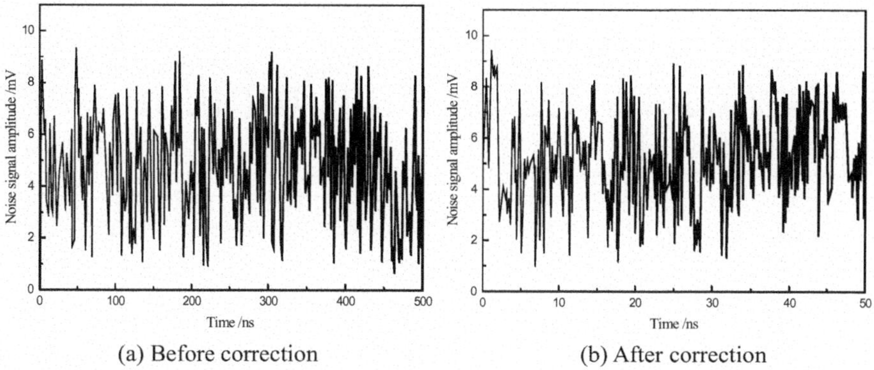

Fig. 8.33 Noise signal amplitude [45]

amplitude of the difference frequency signal before correction in Fig. 8.31, the signal voltage amplitude of the difference frequency after wavefront correction is performed by LC-SLM-R has increased by approximately 1.7 times, and the corrected signal-to-noise ratio has increased almost threefold.

Figure 8.33 depicts the noise signal output before and after correction. It is evident that the noise output by the balanced detector before and after correction has not changed significantly. This is because the detection method of heterodyne detection filters out the effect of the background light noise to a significant extent, and the effect of the dark current is ignored. The output noise signal mainly comes from the thermal noise of the detector and the shot noise generated by the optical power of the local oscillator. However, double-balanced heterodyne detection adds two beams of light with a phase aberration of 180° in the same direction and outputs them from the balanced detector, which effectively eliminates the impact of shot noise. Therefore, the main effect of the output noise signal in the mid-term is the thermal noise of the detector, and the magnitude of this noise does not change with the change in the wavefront correction.

8.7.2 Field Experiment

In a 1.3 km experiment, we used the Zernike polynomial coefficients given in the fitting process to obtain the conjugate wave of the distorted wavefront using the principle of the conjugate method. The conjugate wave was calculated based on the modulation characteristics of the LC-SLM-R on the signal light and beacon light. The wavefront was converted to the driving grayscale of the LC-SLM-R and loaded on the LC-SLM-R to realize the correction of the wavefront of the signal light distortion. Figures 8.34, 8.35, 8.36, 8.37 and 8.38 depict the results of the field experiment from 0 a.m. to 4 a.m. on May 7, 2018.

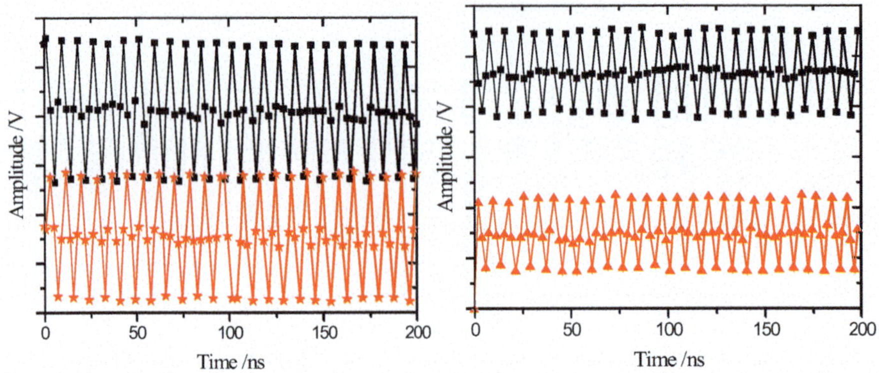

(a) Before correction (b) After correction

Fig. 8.34 Amplitude of difference frequency signal before and after correction [45]

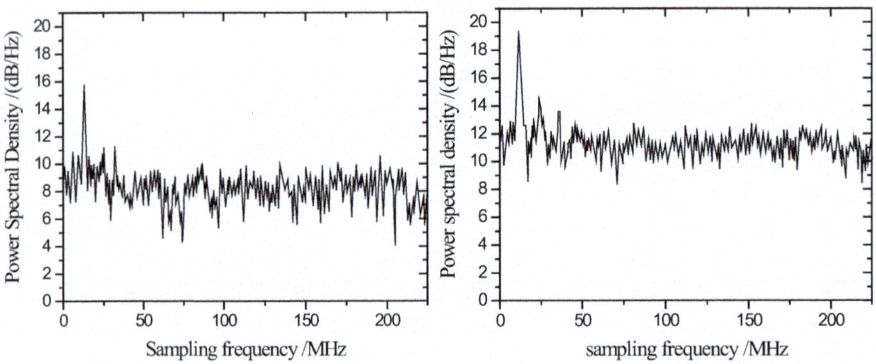

(a) Before correction (b) After correction

Fig. 8.35 Power spectrum of difference frequency signal before and after correction [45]

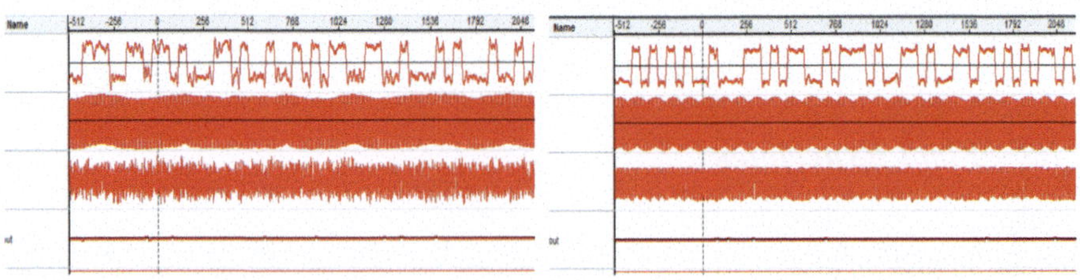

(a) Before correction (b) After correction

Fig. 8.36 Baseband signal before and after AO correction [45]

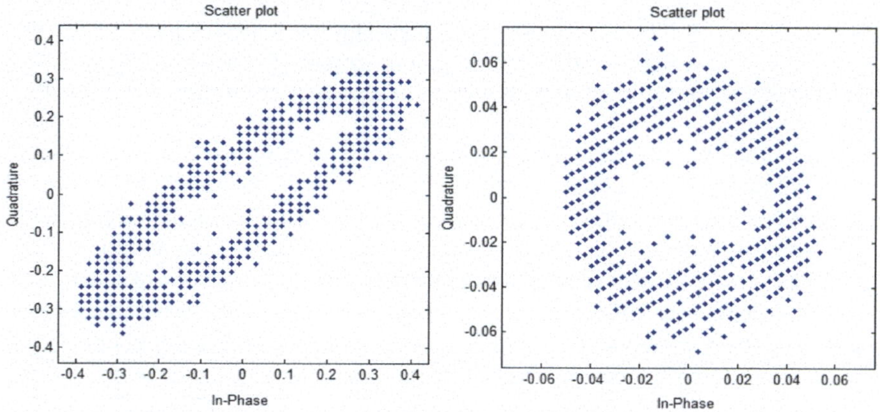

Fig. 8.37 Constellation diagram of received signal before AO correction [45]

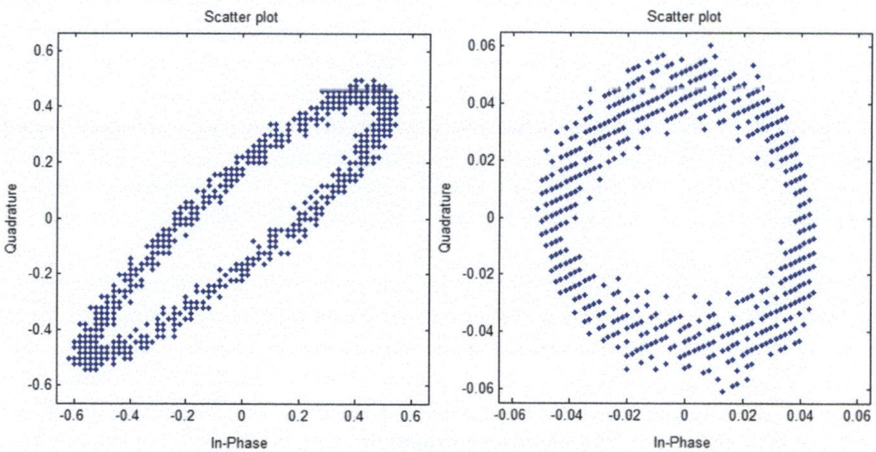

Fig. 8.38 Constellation diagrams of received signal after AO correction [45]

Figure 8.34 depicts the amplitudes of the difference frequency signal before and after the correction. The black and red sine waves in the figure depict the I- and Q-channel difference frequency signals, respectively. It is evident from the figure that the amplitude of the two signals before the correction is approximately 0.55 V, and this value changes with time to produce certain fluctuations. The amplitude of the two-channel difference frequency signal output after correction is about twice as high as that before correction, and the amplitude of each sinusoidal signal tends to a constant value.

Figure 8.35 depicts the power spectrum curves of the difference frequency signal before and after correction. It is evident from the figure that the main spectral peak of the difference frequency signal power spectrum after correction becomes sharper, and the power of the difference frequency signal at the main spectral peak is increased from 15.7 to 19.3 dB/Hz. Compared with before correction, the corrected difference frequency signal noise is filtered out and the signal-to-noise ratio of the system is improved.

After using the AO technology to correct the wavefront of spatial coherent optical communication, the optical power value of the single-mode fiber coupled into the receiving end of the coherent optical communication system is significantly improved. This indicates that the mixing efficiency of signal light and local oscillator light is greatly improved after correcting the wavefront distortion of signal light. Figure 8.36 depicts the before- and after-correction waveforms of the baseband signal obtained by the demodulation system. It is evident that the before-correction amplitude of the difference frequency signal of the system is extremely small, and the signal amplitude fluctuates widely in adjacent periods of the difference frequency signal, this does not meet the minimum amplitude decoding requirements of the receiving end, and the shape of the baseband signal output by the decoder is irregular, more burrs around. After the signal light wavefront distortion is corrected using the AO technology, the amplitude of the difference frequency signal satisfies the decoding requirements of the receiving end. The baseband signal shape and square wave obtained by the decoder can decode the transmitted video signal.

Figures 8.37 and 8.38 depict the constellation diagrams of the received signal before and after correction, respectively. Through comparison, it is evident that the dispersion between the points in the scatter diagram of the received signal before correction is relatively wide. After the AO system corrects the distortion wavefront caused by atmospheric turbulence, the distribution of the scatter points in the constellation diagram of the received signal is relatively concentrated, and the error rate of the system after the correction is reduced to 1/10th of that before correction. There is a certain difference between the constellation diagrams of the I- and Q-channel signals, which is attributed to the difference in the polarization degrees of the two fibers and the unevenness of the cleanliness of the fiber end face when the signal light is connected to the photodetector through the optical fiber.

References

1. Mcmanamon PF, Dorschner TA, Corkum DL et al (1996) Optical phased array technolgy. Proc IEEE 84(2):268–298
2. Li C, Xia M, Mu Q et al (2009) High-precision open-loop adaptive optics system based on LC-SLM. Opt Epress 17(13):10744–10781
3. Arines J, Duran V, Jaroszewicz Z et al (2007) Measurement and compensation of optical aberrations using a single spatial light modulator. Opt Express 15(23):15287–15292
4. Weiner AM (2000) Femtosecond pulse shaping using spatial light modulators. Rev Sci Instrum 71(5):1929–1960
5. Sinclair G, Leach J, Jordan P et al (2004) Interactive application in holographic optical weezers of a multi-pline Gerchberg-Saxton algorithm for three-dimensional light shaping. Opt Express 12(8):1665–1670

6. Briendc O, Faulknerge, Wilkinson TD, et al (2004) Design and analysis of an adaptive board-to-board dynamic holographic interconnect. Appl Opt 43(16):3297–3305
7. Lic LX, Muq X et al (2009) High-precision open-loop adaptive optics system based on LC-SLM. Opt Express 17(13):10774–10781
8. David E, Martin P, Jorgen B et al (2013) Calibration of spatial light modulators suffering from spatially varying phase response. Opt Express 21(13):16086–16103
9. Chen HX, Sui Z, Chen ZP et al (2000) Optical modulation characteristics of liquid crystal television (LCTV) and its application in optics information processing. Chin J Lasers 27(8):741–745
10. Li J, Tang F, Wang XC et al (2014) System errors analysis of grating lateral shearing interferometer. Chinese J Lasers 41(5):0508006
11. Yamauchi M, Marquez A, Davis JA et al (2000) Interferometric phase measurements for polarization eigenvectors in twisted nematic liquid crystal spatial light modulators. Opt Commun 181(1–3):1–6
12. Wang GJ, Zhang CL, Wang R et al (2013) Phase difference surface plasmon resonance sensor based on mach-zehnder configuration. Opt Commun 40(7):0714003
13. Ge AM, Sui Z, Xu KS (2003) Characteristics of phase-only modulation using a reflective liquid crystal on silicon device. Acta Physica Sinica 52(10):2481–2485
14. Zhang H, Zhang J, Wu L (2007) Evaluation of phase-only liquid crystal spatial light modulator for phase modulation performance using a Twyman-Green interferometer. Measur Sci Technol 18(6):1724–1728
15. Liu H, Lu ZW, Li FY (2006) CGH testing system for large aspheric surface. Infrared Laser Eng 35(2):177–182
16. Burge JH, Allderson DS (1994) Full-aperture interferometry test of convex secondary mirrors using holography test plate. Proc SPIE 2199:181–191
17. Zhang HX (2009) Research of the performance measurement and wavefront correction application of phase-only liquid crystal spatial light modulator. Harbin, Harbin Institute of Technology
18. Zhang HX, Zhang J, Wu LY (2008) Phase modulation of liquid crystal spatial light modulator measured by a Twyman Green interferometer. Chin J Lasers 35(9):1360–1364
19. Cao ZL, Xuan L, Hu LF et al (2005) Investigation of optical testing with a phase-only liquid crystal spatial light modulator. Opt Express 13(4):1059–1065
20. Zhao XF, Li DH, Chen ZP (2002) Measuring phase modulation characteristics of liquid crystal spatial modulators by using cyclic radial shearing interference. J Sichuan Univer 39(4):675–679
21. Ren XY, Cai CW, Wang Z et al (2007) The phase modulation properties measurement of LC-SLM by common path lateral shearing interferometry. Acta Photonica Sinica 36(5):899–904
22. Dou R, Giles MK (1995) Closed-loop adaptive optics system with a liquid crystal television as a phase retarder. Opt Lett 20(14):1583–1585
23. Kanghua L, Saleh BEA (1990) Theory and design of the liquid crystal TV as an optical spatial phase modulator. Opt Eng 29(3):240–246
24. Neto LG, Rogerge D, Sheng Y (1996) Full-range, continuous, complex modulation by the use of two coupled-mode liquid-crystal televisions. Appl Op 35(23):4567–4573
25. Chen HX, Sui Z, Chen ZP et al (2000) Optical modulation characteristics of liquid crystal television (LCTV)and its application in optics information processing. Chin J Lasers 27(8):741–745
26. Zhang H, Zhang J, Wu L (2007) Evaluation of phase only LC SLM for phase modulation performance. Meas Sci Tech 18(6):1724–1728
27. Makoto Y, Andrés M, Davis JA et al (2000) Interfermetric phase measurements for polarization eigenvectors in twisted nematic liquid crystal spatial light modulators. Opt Commun 181(1–3):1–6
28. Liang QY (2012) physical optics. Electronic Industry Press, Beijing
29. Davis, Jeffrey A (1999) Transmission variations in liquid crystal spatial light modulators caused by interference and diffraction effects. Opt Eng 38(6):1051–1057

30. Moreno I, Davis JA, D'Nelly K et al (1998) Transmission and phase measurement for polarization eigenvectors in twisted-nematic liquid crystal spatial light modulators. Opt Eng 37(11):3048–3052
31. Gibson JS, Chang CC, Ellerbroek BL (1999) Adaptive optics wavefront reconstruction by adaptive filtering and control. In: The 38th IEEE conference on decision and control, phoenix, AZ, USA, pp 761–766
32. Kudryashov A, Samarkin V, Aleksandrov A (2001) Adaptive optical elements for laser beam control. Proc SPIE 4457:170–178
33. Zhou RZ, Yan JX (1996) Adaptive optics theory. Beijing Institute of Technology Press, Beijing
34. Ragazzoni R, Ghedina A, Baruffolo A, et al (2000) Testing the pyramid wavefront sensor on the sky. Proc SPIE, 4007
35. Fu SQ (2010) Research on key technologies of phase unwrapping for 3D optical measurement. Nanjing University of Aeronautics and Astronautics, Nanjing
36. Sun L, Hu ML, Zhang ZY (2005) Research of the least-squares algorithms for In SAR 2D phase unwrapping. Radar Sci Technol 3(4):216–220
37. Ghiglia DC, Romero LA (1994) Robust two dimensional weighted and unweighted phase unwrapping that uses fast transforms and iterative methods. Opt Soe Am A 11(1):107–117
38. Huntley JM (1989) Noise-immune phase unwrapping algorithm. Appl Opt 28(15):3268–3270
39. Liu YT, Chen N, Gibson Adaptive JS (2005) Filtering and control for wavefront reconstruction and jitter control in adaptive optics. In: Proceedings of the 2005 American control conference. United states, Portland, OR, pp 2608–2612
40. Love DG (1997) Wave-front correction and production of Zernike modes with a liquid crystal spatial light modulator. Appl Opt 36:1517–1524
41. Wang JY, Silva DE (1980) Wave-front interpretation with Zernike polynomials. Appl Opt 19(9):1510–1518
42. Marcus B, Emile W (2006) Principles of optics. Electronic Industry Press, Beijing
43. Noll RJ (1976) Zernike polynomials and atmospheric turbulence. Opt J Soc Amer 66:207–211
44. Fred DL (1977) Adaptive optics. Opt J Soc Amer 67(3):1–422
45. Kong YX (2018) Wavefront correction technology of LC-SLM spatial coherent optical communication. Xi'an University of Technology, Xi'an

Chapter 9
Effect of Beam Mode on Coherent Detection System

Contents

We have developed a mathematical model of a partially coherent beam detection system under atmospheric turbulence. The model is based on the modified von Karman atmospheric turbulence power spectrum model, with the partially coherent Gaussian–Schell model (GSM) beam as the object of study and considering the partially coherent light transmission problem. The expressions for the weight factor of the partially coherent GSM beam at the receiver are derived based on the generalized

X. Ke and J. Wu, *Coherent Optical Wireless Communication Principle and Application*, Optical Wireless Communication Theory and Technology,
https://doi.org/10.1007/978-981-19-4823-7_9

Huygens–Fresnel principle and incoherent mode decomposition theory. Combining these two parts, a mathematical model is established to represent the effect of the beam mode on the coherent detection system. Further, the effect of the beam mode on the performance of the partially coherent beam detection system under atmospheric turbulence is theoretically analyzed. Finally, experimental verification is carried out using a 1.3 km out-of-field coherent communication link.

9.1 Basic Theory of Pattern Decomposition

Light waves are transmitted through a turbulent random medium in the atmosphere, and the beam mode is affected by atmospheric turbulence. In addition, owing to the thermal effects of the laser, optical components, and other factors, the emitted beam is no longer a fundamental-mode coherent light beam, but, in general, has a complex multimode structure [1]. In practice, multimode beams can be generated by the incoherent superposition of Hermite–Gaussian (H-G) beams or the coherent superposition of Laguerre–Gaussian beams.

9.1.1 Mathematical Model of Incoherent Mode Decomposition

The one-dimensional (1D) light intensity of a partially coherent GSM beam can be expressed as a superposition of the 1D light intensities of H-G beams that are independent of each other between modes [2], that is,

$$I(x,0) = \sum_{n=0}^{\infty} c_n \cdot |\varphi_n(x, w_0)|^2 \tag{9.1}$$

where ω_0 is the radius of the beam waist of the fundamental-mode Gaussian beam, c_n is the mode factor of the nth-order H-G beam, and φ_n is the amplitude of the nth-order H-G beam in direct coordinates, which can be expressed as follows [3]:

$$\varphi_n = \left(\frac{2}{\pi w_0^2}\right)^{1/4} \frac{1}{(2^n \cdot n!)^{1/2}} H_n\left(\frac{x\sqrt{2}}{w_0}\right) \exp\left(-\frac{x^2}{w_0^2}\right) \tag{9.2}$$

where $H_{\mathrm{n}}(x)$ is the nth-order Herrmite function, the Fourier transform of both sides of Eq. (9.1) can be processed separately to obtain Eq. (9.3) [2], and the modulus coefficients are obtained in the frequency domain using the orthogonal mode coefficient expansion method as follows:

$$I(f) = \sum_{n=0}^{\infty} c_n \psi_n\left(\pi^2 w_0^2 f^2\right) \tag{9.3}$$

where f denotes the frequency and ψ_n is the Fourier transform of $|\varphi_n(x,\omega_0)|^2$, which can be expressed as follows:

$$\psi_n(\pi^2 w_0^2 f^2) = L_n(\pi^2 w_0^2 f^2)\exp(-\pi^2 w_0^2 f^2/2) \tag{9.4}$$

Here, L_n is an nth-order Laguerre polynomial, and, using the orthogonal normalization property of the Laguerre–Gaussian function, the modulus coefficients can be expressed as follows [3]:

$$c_n = 2\pi^2 w_0^2 \int_0^{+\infty} I(f)\psi_n\left(\pi^2 w_0^2 f^2\right) f df \tag{9.5}$$

The weight factor can be expressed as follows:

$$\lambda_n = \frac{c_n}{\sum\limits_{n=0}^{\infty} c_n} \tag{9.6}$$

The M^2 factor can be expressed using the weight factor as follows [4]:

$$M^2 = \sum_{n=0}^{+\infty} (2n+1)\lambda_n \tag{9.7}$$

Any type of beam can be constituted by the incoherent superposition of H-G beams of different orders. For example, Fig. 9.1 exhibits the light intensity distribution of H-G beams of different orders, where HG_{AB} indicates the light intensity of the H-G beams ($A, B = 0, 1, 2, \ldots$). The synthetic beam is obtained from the equal proportional superposition of the H_{00} and H_{01} beams.

The expression for the light field of the H-G beam has a Hermite function term defined as follows:

$$H_n(x) = \begin{cases} 1 & n \le 0 \\ 2x & n = 1 \\ 2xH_{n-1}(x) - 2(n-1)H_{n-2}(x) & n \ge 2 \end{cases} \tag{9.8}$$

The common Hermite polynomials are as follows:

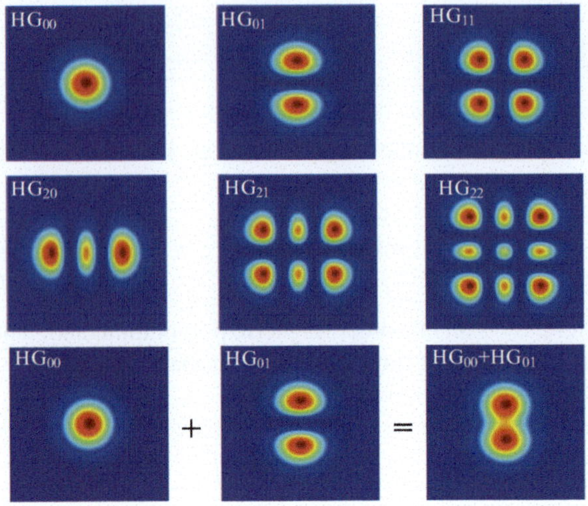

Fig. 9.1 Light intensity distribution of H-G beams of different orders and their synthetic beams [5]

$$
\begin{aligned}
H_0(x) &= 1 \\
H_1(x) &= 2x \\
H_2(x) &= 4x^2 - 1 \\
H_3(x) &= 8x^3 - 12x \\
H_4(x) &= 16x^4 - 48x^2 + 12 \\
H_5(x) &= 32x^5 - 160x^3 + 120x
\end{aligned}
\tag{9.9}
$$

9.1.2 *Coherent Module Decomposition*

At the source field, any beam field distribution can be represented by a normalized orthogonal H-G beam [6], that is,

$$E(x, 0) = \sum_{n=0}^{\infty} c_n \times \varphi_n(x) \tag{9.10}$$

Let the monochromatic field be approximated and the cross-spectral density function of the partially coherent GSM beam be expressed by the coherent mode superposition as follows [7]:

$$W(x_1, x_2) = \sum_{m=0}^{+\infty} \sum_{n=0}^{+\infty} \lambda_{mn} \varphi_m(x_1) \varphi_n^*(x_2) \tag{9.11}$$

where λ_{mn} is the mode correlation coefficient, m and n represent the mode orders, and * denotes the complex conjugate. Therefore,

$$\lambda_{mn} = \int_{-\infty}^{+\infty}\int_{-\infty}^{+\infty} \varphi_m^*(x_1)W(x_1, x_2)\varphi_n(x_2)dx_1dx_2 \tag{9.12}$$

The cross-spectral density function of the partially coherent GSM beam can be expressed as follows [8]:

$$W(x_1, x_2) = \exp\left(\frac{x_1^2 + x_2^2}{w_0^2}\right)\exp\left[-\frac{(x_1 - x_2)^2}{2\sigma_0^2}\right] \tag{9.13}$$

where σ_0 is the correlation length of the GSM beam.

Substituting Eqs. (9.2) and (9.13) in Eq. (9.12) and performing complex integration, we obtain the following expressions [7]:

$$\int_{-\infty}^{+\infty} H_m(t)e^{-(x-t)^2/2u}dt = \sqrt{2\pi u}(1 - 2u)^{m/2}H_m\left(\frac{x}{\sqrt{1-2u}}\right) \tag{9.14}$$

$$\begin{aligned}&\int_{-\infty}^{+\infty} H_m\left(\frac{x+t}{\sqrt{2}v}\right)H_n\left(\frac{y+t}{\sqrt{2}\mu}\right)e^{-pt^2}dt\\ &= \frac{\sqrt{\pi}}{v^m\mu^n p^{(m+n+1)/2}}\sum_{k=0}^{\min(m,n)}\left[\binom{m}{k}\binom{n}{k}k!(2v^2p-1)^{\frac{m-k}{2}}(2\mu^2p-1)^{\frac{n-k}{2}}\right.\\ &\quad\left.\times 2^{-\frac{m-k}{2}-\frac{n-k}{2}}H_{m-k}\left(\frac{x}{\sqrt{2v^2-1/p}}\right)H_{n-k}\left(\frac{y}{\sqrt{2\mu^2-1/p}}\right)\right]\end{aligned} \tag{9.15}$$

where

$$\binom{i}{k} = \frac{i!}{k!(i-k)!} \tag{9.16}$$

based on

$$H_{2m}(0) = (-1)^m\frac{(2m)!}{m!} \tag{9.17}$$

$$H_{2m+1}(0) = 0 \tag{9.18}$$

After integrating and simplifying, the modal correlation coefficient can be expressed as follows [7]:

$$\lambda_{mn} = \lambda_0 \left(\frac{\sqrt{\left(\frac{1}{w_0^2} + \frac{1}{2\sigma_0^2}\right)^2 - \frac{\alpha^4}{4}}}{\frac{1}{w_0^2} + \frac{1}{2\sigma_0^2} + \frac{\alpha^2}{2}} \right)^m \frac{\sqrt{m!n!}}{2^{m+n} v^m \mu^n p^{\frac{m+n}{2}}}$$
$$\times \sum_{k=0}^{\min(m,n)} \left[\frac{(-1)^{\frac{m-k}{2}+\frac{n-k}{2}} 2^k}{k!\left(\frac{m-k}{2}\right)!\left(\frac{n-k}{2}\right)!} \left(2v^2 p - 1\right)^{\frac{m-k}{2}} \left(2\mu^2 p - 1\right)^{\frac{n-k}{2}} \right] \tag{9.19}$$

When the parities of m and n are different, $\lambda_{mn} = 0$.

$$\lambda_0 = \sqrt{\frac{\pi}{\frac{1}{w_0^2} + \frac{1}{2\sigma_0^2} + \frac{\alpha^2}{2}} \frac{\alpha}{\sqrt{p}}} \tag{9.20}$$

$$p = \frac{4\sigma_0^4 \left(\frac{1}{w_0^2} + \frac{1}{2\sigma_0^2} + \frac{\alpha^2}{2}\right)^2 - 1}{4\sigma_0^4 \left(\frac{1}{w_0^2} + \frac{1}{2\sigma_0^2} + \frac{\alpha^2}{2}\right)} \tag{9.21}$$

$$v = \frac{2\sigma_0^2 \sqrt{\left(\frac{1}{w_0^2} + \frac{1}{2\sigma_0^2}\right)^2 - \frac{\alpha^4}{4}}}{\sqrt{2}\alpha} \tag{9.22}$$

$$\mu = \frac{1}{\sqrt{2}\alpha} \tag{9.23}$$

$$\alpha = \frac{\sqrt{2}}{w_0} \tag{9.24}$$

Therefore, the normalized mode correlation coefficients are expressed as follows:

$$c_{mm} = \frac{\lambda_{mm}}{\sum\limits_{m=0}^{\infty} \lambda_{mm}}, \quad c_{m,m+2} = \frac{\lambda_{m,m+2}}{\sum\limits_{m=0}^{\infty} \lambda_{mm}} \tag{9.25}$$

9.2 Effect of Beam Pattern on Performance of Coherent Detection Systems

Owing to the effect of atmospheric turbulence, the coherent mixing of the signal and local oscillation light beams leads to a mismatch between the modes and polarization states of the signal and local oscillation light beams, which reduces the heterodyne efficiency. A partially coherent GSM model is used to establish a mathematical model of the effect of the beam pattern on the partially coherent GSM detection system and analyze the varation in the beam pattern and its effect on the performance of the coherent detection system under atmospheric turbulence.

9.2.1 *Mathematical Modeling of Effect of Beam Patterns on Coherent Detection Systems Under Atmospheric Turbulence*

The large number of atoms that constitute a light source accord it the nature of random radiation; therefore, it is possible to explain the random radiation emitted by a light source at the physical level [9]. A mathematical model of a partially coherent GSM beam detection system is developed using a partially coherent GSM beam as a research model to analyze the effect of the beam pattern on the coherent detection system.

Let both the signal and local oscillation light beams be incident parallel to the optical axis and perpendicular to the detector surface. The two light fields at the surface of the detector can then be expressed as follows [10]:

$$U_L(\rho, t) = U_{L0}(\rho, t)\exp(j\omega_L t) \tag{9.26}$$

$$U_S(\rho, t) = U_{S0}(\rho, t)\exp(j\omega_s t) \tag{9.27}$$

where U_{L0} and U_{S0} and ω_{L0} and ω_{S0} represent the initial optical field distributions and frequencies of the local oscillation and signal light beams, respectively. Based on the principle of coherent detection, the intermediate frequency (IF) signal power output from the detector can be calculated based on the square rate characteristics of the detector. Let us assume that the polarizations and wavefronts of the signal and local oscillation light beams are the same. Therefore, the mixed light field on the surface of the detector can be expressed as follows:

$$U(\rho, t) = U_{L0}(\rho, t)\exp(j\omega_L t) + U_S(\rho, t)\exp(j\omega_s t) \tag{9.28}$$

The coherence process between the two optical fields of the signal and local oscillation light beams generates a mixing current, and the value of this current at any point r on the surface of the detector can be expressed as follows [11]:

$$di(\rho, t) = \Re(\rho)\left[U(\rho, t) \times U^*(\rho, t)\right] \tag{9.29}$$

where $\Re(\rho) = e\eta(\rho)/h\nu$ denotes the detection sensitivity at ρ on the surface of the detector, $\eta(\rho)$ is the corresponding quantum efficiency, e is the electronic charge, $h\nu$ denotes the energy of the photon, and * denotes the complex conjugate. Substituting Eq. (9.28) in Eq. (9.29) and filtering out the DC term of the IF current, the total IF current output from the detector can be expressed as follows:

$$
\begin{aligned}
i_{IF}(t) = \iint & \Re(\rho)\big\{U_{LO}(\rho)U_S^*(\rho)\exp[j(\omega_{LO} - \omega_S)t] + \\
& U_S(\rho)U_{LO}^*(\rho)\exp[-j(\omega_{LO} - \omega_S)t]\big\}d^2\rho
\end{aligned}
\tag{9.30}
$$

The power of the IF signal can be expressed as follows:

$$
\mathrm{p}_{IF} = \overline{i_{IF}(t)i_{IF}^*(t)R_L} \tag{9.31}
$$

where the horizontal line indicates the time averaging of the IF signal sought and R_L indicates the load resistance of the detector. Although only the time averaging of the optical field is performed in Eq. (9.31), the signal light is affected by random factors such as atmospheric turbulence and the mixed optical field is random in space. Therefore, the IF signal power is the coefficient average of the IF signals throughout the turbulent conditions, that is,

$$
P_{IF} = \langle \mathrm{p}_{IF} \rangle \tag{9.32}
$$

Substituting Eqs. (9.30) and (9.31) in Eq. (9.32), P_{IF} can finally be expressed as follows:

$$
\begin{aligned}
P_{IF} = \int\int\int\int & \Re(\rho_1)\Re(\rho_2) \times \big[\langle U_{LO}^*(\rho_1)U_S(\rho_1)U_S^*(\rho_2)U_{LO}(\rho_2)\rangle + \\
& \langle U_{LO}(\rho_1)U_{LO}^*(\rho_2)U_S(\rho_2)U_S^*(\rho_1)\rangle\big]d^2r_1d^2r_2
\end{aligned}
\tag{9.33}
$$

The signal light beam is transmitted from the transmitter to the receiver through a random turbulent medium, whereas the local oscillation light beam at the receiver is not affected by the turbulence; therefore, the two light fields can be considered to be independent of each other, and Eq. (9.33) can be further reduced to the following expression:

$$
P_{IF} = \int\int\int\int \Re(\rho_1)\Re(\rho_2) \times \mathrm{Re}\big[\Gamma_{LO}(\rho_1, \rho_2) \times \Gamma_S^*(\rho_1, \rho_2, L)\big]d^2\rho_1 d^2\rho_2 \tag{9.34}
$$

where $\Gamma_{LO}(\rho_1, \rho_2)$ and $\Gamma_S(\rho_1, \rho_2)$ represent the correlation functions of the local oscillation and signal light beams, respectively. The main source of noise in the coherent detection system is the scattered particle noise generated by the local oscillation light beam, which can be expressed in the form of the correlation function as follows [10]:

$$
N = 2eB \iint \Re(r)\Gamma_{LO}(\rho, \rho)d\rho^2 \tag{9.35}
$$

Based on the effective optical power of the IF signal expressed in Eq. (9.34) and the scattered particle noise represented by Eq. (9.35), the signal-to-noise ratio (SNR) of a coherent detection system with a partially coherent GSM beam can be defined as follows:

$$SNR_{GSM} = \frac{\int\int\int\int \Re(\rho_1)\Re(\rho_2) \times \mathrm{Re}\left[\Gamma_{LO}(\rho_1, \rho_2) \cdot \Gamma_S^*(\rho_1, \rho_2, L)\right] d^2\rho_1 d^2\rho_2}{2\mathrm{eB} \iint \Re(\rho)\Gamma_{LO}(\rho, \rho) dr^2} \tag{9.36}$$

Let the detection responsivity $\Re$ on the photosensitive surface of the detector be a constant; it can be seen that the constant in Eq. (9.36) does not affect the trend of SNR_{GSM}. Therefore, the equation can be further reduced to

$$SNR'_{GSM} = \frac{\int\int\int\int \mathrm{Re}\left[\Gamma_{LO}(\rho_1, \rho_2) \cdot \Gamma_S^*(\rho_1, \rho_2, L)\right] d^2 r_1 d^2 r_2}{\iint \Gamma_{LO}(\rho, \rho) dr^2} \tag{9.37}$$

Comparing Eqs. (9.36) and (9.37), it can be seen that SNR' GSM $=$ $2eB\mathrm{SNR}_{GSM}/\Re$. Based on the definition of the heterodyne efficiency, an expression for the heterodyne efficiency of a partially coherent GSM beam can be written as follows:

$$\eta_{het\_GSM} = \frac{\int\int\int\int \Re(\rho_1)\Re(\rho_2) \times \mathrm{Re}\left[\Gamma_{LO}(\rho_1, \rho_2) \cdot \Gamma_S^*(\rho_1, \rho_2, L)\right] d^2\rho_1 d^2\rho_2}{\iint \Re(\rho)\Gamma_{LO}(\rho, \rho) dr^2 \iint \Re(\rho)\Gamma_{SO}(\rho, \rho) dr^2} \tag{9.38}$$

For the sake of analysis, let the quantum efficiency η on the photosensitive surface of the detector be a constant. Thus, Eq. (9.38) can be further reduced to the following expression:

$$\eta'_{het\_GSM} = \frac{\int\int\int\int \mathrm{Re}\left[\Gamma_{LO}(\rho_1, \rho_2) \cdot \Gamma_S^*(\rho_1, \rho_2, L)\right] d^2\rho_1 d^2\rho_2}{\iint \Gamma_{LO}(\rho, \rho) dr^2 \iint \Gamma_{SO}(\rho, \rho) dr^2} \tag{9.39}$$

Both the signal and local oscillation light beams are partially coherent GSM beams in the source field, and the corresponding mutual correlation functions can be expressed as follows [12]:

$$\Gamma_\alpha(r_1, r_2, 0) = I_\alpha \exp\left[-\frac{\left(r_1^2 + r_2^2\right)}{4\sigma_\alpha^2} - \frac{(r_1 - r_2)^2}{2\delta_\alpha^2}\right] \tag{9.40}$$

where r_1 and r_2 denote the position vectors at any two points in the source field $L = 0$; I_α is the light intensity, where $I_\alpha = 1$ denotes the ideal light intensity at the transmitting end; σ_α is the waist radius of the beam; and δ_α is the spatial coherence length, describing the partial coherence characteristics of the light wave. The subscripts $\alpha = L_0$ and S_0 correspond to the principal vibration and signal light beams, respectively. Let the signal light be transmitted horizontally along the Z-axis. Therefore, the correlation function of the partially coherent GSM beam transmitted through an atmospheric turbulent medium is easily obtained based on the generalized Huygens–Fresnel diffraction principle and can be expressed as follows [13]:

$$\Gamma_S(\rho_1, \rho_2, L) = \left(\frac{k}{2\pi L}\right)^2 \iint d^2\rho_1 \iint d^2\rho_2 \Gamma_{S_0}(r_1, r_2, 0) \cdot \exp\left\{-\frac{ik}{2L}\left[(r_1 - \rho_1)^2 - (r_2 - \rho_2)^2\right]\right\} \cdot \left\langle \exp\left[\psi^*(r_1, \rho_1, L) + \psi(r_2, \rho_2, L)\right]\right\rangle \tag{9.41}$$

where ρ_1 and ρ_2 denote the position vectors at any two points on the face of the receiver. The correlation function of the complex phase of the spherical wave can be expressed as follows [14]:

$$\left\langle \exp\left[\psi^*(r_1, \rho_1, L) + \psi(r_2, \rho_2, L)\right]\right\rangle = \exp\left[-\frac{1}{2} D_\phi(\rho_1 - \rho_2, r_1 - r_2)\right] \tag{9.42}$$

where $D_\phi(\rho_1\text{-}\rho_2, r_1\text{-}r_2)$ is the wave structure function.

Using the modified Hill spectral model, the signal optical ramp is transmitted to the receiver under atmospheric turbulence. A complex integral reduction reduces the final wave structure function to the following expression:

$$\left\langle \exp\left[\psi^*(r_1, \rho_1, L) + \psi(r_2, \rho_2, L)\right]\right\rangle \cong \exp\left[-\frac{1}{2}\left(Q_1 r_d^2 + Q_2 r_d \rho_d + Q_3 \rho_d^2\right)\right] \tag{9.43}$$

where $\rho_d = \rho_1 - \rho_2$, $r_d = r_1 - r_2$, and Q_1, Q_2, and Q_3 are:

$$\begin{cases} Q_1 = 2\pi^2 k^2 \sec\varphi \int_{h_0}^{H} \int_0^\infty \Phi_n(\kappa, h)(1 - \xi)^2 \kappa^3 d\kappa dh \\ Q_2 = 4\pi^2 k^2 \sec\varphi \int_{h_0}^{H} \int_0^\infty \Phi_n(\kappa, h)(1 - \xi)\xi \kappa^3 d\kappa dh \\ Q_3 = 2\pi^2 k^2 \sec\varphi \int_{h_0}^{H} \int_0^\infty \Phi_n(\kappa, h)\xi^2 \kappa^3 d\kappa dh \end{cases} \tag{9.44}$$

where $\xi = 1 - (h - h_0)/(H - h_0)$ in the uplink and $\xi = (h - h_0)/(H - h_0)$ in the downlink. When the transmission is horizontal, in Eq. (9.44), $Q_1 = Q_2 = 0$, $Q_3 = (\pi^2 k^2 L/3)\int_\infty^0 \kappa^3 \Phi_n(\kappa) d\kappa$, and the atmospheric refractive index structure constant is denoted by C_0.

By substituting Eqs. (9.40) and (9.43) in Eq. (9.41), it can be rewritten as follows:

$$\Gamma_S(\rho_1, \rho_2, L) = \left(\frac{k}{2\pi L}\right)^2 \iint d^2\rho_1 \iint d^2\rho_2 I_{S0} \exp\left[-\frac{(r_1^2 + r_2^2)}{4\sigma_{S0}^2} - \frac{(r_1 - r_2)^2}{2\delta_{S0}^2}\right]$$
$$\cdot \exp\left\{-\frac{ik}{2L} \times [(r_1 - \rho_1)^2 - (r_2 - \rho_2)^2]\right\}$$
$$\cdot \exp\left[-\frac{1}{2}(Q_1(r_1 - r_2)^2 + Q_2(r_1 - r_2)(\rho_1 - \rho_2) + Q_3(\rho_1 - \rho_2)^2)\right] \quad (9.45)$$

Let $u = (\rho_1 - \rho_2)/2$, $v = \rho_1 - \rho_2$, $p = r_1 - r_2$, and $q = r_1 + r_2$, that is, $\rho_1 = u + v/2$ and $\rho_1 = u - v/2$. Therefore, Eq. (9.45) can be simplified as follows:

$$\Gamma_S(\rho_1, \rho_2, L) = I_{S_0}\left(\frac{k}{2\pi L}\right)^2 \iint d^2u \exp\left(-\frac{u^2}{2\sigma_{S_0}^2}\right) \exp\left[-ik\frac{(v-p)}{L}u\right]$$
$$\times \iint d^2v \exp\left[-\left(\frac{Q_3}{2} + \frac{1}{8\sigma_{S_0}^2} + \frac{1}{2\delta_{sg}^2}\right)v^2 - \frac{Q_2}{2}pv\right] \quad (9.46)$$
$$\times \exp\left(ik\frac{qv}{2L}\right) \exp\left[-\frac{Q_1}{2}p^2 - ik\frac{pq}{2L}\right]$$

$$\int_{-\infty}^{\infty}\int_{-\infty}^{\infty} \exp[-(a\rho^2 + 2b\rho + c)] \exp[\pm i(d\rho^2 + 2e\rho + f)] d^2\rho$$
$$= \frac{\pi}{\sqrt{a^2 + d^2}} \exp\left[\frac{a(b^2 - e^2) + 2dbe}{a^2 + d^2} - c\right] \quad (9.47)$$
$$\times \exp\left\{\pm i\left[\arctan\frac{d}{a} - \frac{d(e^2 - b^2) + 2abe}{a^2 + d^2} + f\right]\right\} \quad (a > 0)$$

Carrying out the integration in Eq. (9.46) with respect to u based on Eq. (9.47), the result of the collation is as follows:

$$\Gamma_S(\rho_1, \rho_2, L) = \frac{\sigma_{S_0}^2 I_{S_0} k^2}{2\pi L^2} \exp\left(-\frac{\sigma_{S_0}^2 k^2}{2L^2}p^2 - \frac{Q_1}{2}p^2\right) \exp\left(-ik\frac{pq}{2L}\right)$$
$$\times \iint d^2v \exp\left[-\alpha v^2 - \left(\frac{Q_2}{2} - \frac{\sigma_{S_0}^2 k^2}{L^2}\right)pv\right] \exp\left[ik\frac{qv}{2L}\right] \quad (9.48)$$

where $\alpha = Q_3/2 + 1/8\sigma_{S_0}^2 + 1/2\delta_{S_0}^2 + k^2\sigma_{S_0}^2/2L^2$. Again, the integration with respect to v in Eq. (9.48) is simplified to obtain the following expression:

$$\Gamma_S(\rho_1, \rho_2, L) = \frac{I_{S_0}}{Q_s} \exp\left[-H_s(\rho_1 - \rho_2)^2 - \frac{1}{8\sigma_{S_0}^2 Q_s}(\rho_1 + \rho_2)^2 + ikT_s(\rho_2^2 - \rho_1^2)\right] \quad (9.49)$$

$$\begin{cases} Q_s = 1 + \left(\frac{L}{k\sigma_{S_0}\theta_s}\right)^2 + \frac{Q_3 L^2}{k^2\sigma_{S_0}^2} \\ \frac{1}{\theta_s^2} = \frac{1}{4\sigma_{S_0}^2} + \frac{1}{\delta_{S_0}^2} \\ T_s = \frac{k^2\sigma_{S_0}^2 Q_s + \frac{Q_2}{2}L^2 - k^2\sigma_{S_0}^2}{2Lk^2\sigma_{S_0}^2 Q_s} \\ H_s = \frac{1}{2\theta_s^2 Q_s} + \frac{1}{Q_s}\left(\frac{Q_3}{2} + \frac{Q_2}{2}\right) + \frac{Q_1}{2} - \frac{\left(\frac{Q_2}{2}\right)^2 L^2}{2k^2\sigma_{S_0}^2 Q_s} \end{cases} \tag{9.50}$$

Substituting Eqs. (9.40) and (9.49) in Eq. (9.34) and translating the operation to the polar angle coordinate system, the IF signal output from the detector can be expressed as follows:

$$P_{IF} = \int_0^{2\pi}\int_0^{2\pi}\int_0^{D/2}\int_0^{D/2} \frac{I_{S_0} I_{L_0}}{Q_s} \mathrm{Re}\left\{ \begin{array}{l} \exp\left\{-\left[\frac{1}{2\delta_{L_0}^2} + H_s\right](\rho_1 - \rho_2)^2\right\} \\ \exp\left[-\frac{1}{8\sigma_{S_0}^2 Q_s}(\rho_1 + \rho_2)^2\right] \end{array} \right. \times \exp\left(-\frac{\rho_1^2 + \rho_2^2}{4\sigma_{L_0}^2}\right) \times \exp\left[ikT_s\left(\rho_1^2 - \rho_2^2\right)\right]\Bigg\} \rho_1\rho_2 \mathrm{d}\rho_1 d\rho_2 \mathrm{d}\varphi_1 \mathrm{d}\varphi_2 \tag{9.51}$$

where D is the diameter of the effective detection area of the detector. To facilitate the calculation, the Gaussian or soft aperture is considered instead of the hard aperture, and the relationship between the soft aperture W and hard aperture D is $D^2 = 8W^2$ [13]. Based on this relationship as well as the relation $\rho_1\rho_2 = \rho_1\rho_2\cos(\varphi_1 - \varphi_2)$, Eq. (9.51) can be rearranged as follows:

$$P_{IF} = \frac{I_{S_0} I_{L_0}}{Q_s}\int_0^{2\pi}\int_0^{2\pi}\int_0^{\infty}\int_0^{\infty} \mathrm{Re}\left\{\exp\left[-g\left(\rho_1^2 + \rho_2^2\right)\right] \times \exp[q2\rho_1\rho_2\cos(\varphi_1 - \varphi_2)] \times \exp\left[is\left(\rho_1^2 - \rho_2^2\right)\right]\right\}\rho_1\rho_2 \mathrm{d}\rho_1 d\rho_2 \mathrm{d}\varphi_1 \mathrm{d}\varphi_2 \tag{9.52}$$

where

$$\begin{cases} g = \frac{1}{2\delta_{L_0}^2} + H_s + \frac{1}{8\sigma_{S_0}^2 Q_s} + \frac{1}{4\sigma_{L_0}^2} + \frac{1}{W^2} \\ q = \frac{1}{2\delta_{L_0}^2} + H_s - \frac{1}{8\sigma_{S_0}^2 Q_s} \\ s = kT_s \end{cases} \tag{9.53}$$

Performing the integration in Eq. (9.52) with respect to φ_1, Eq. (9.52) can be rewritten as follows:

$$P_{IF}=\frac{I_{S_0}I_{L_0}}{Q_s}\int_0^{2\pi}\int_0^{\infty}\int_0^{\infty}\mathrm{Re}\{\exp[-g(\rho_1^2+\rho_2^2)]\times\exp[is(\rho_1^2-\rho_2^2)]\,\rho_1\rho_2 d\rho_1 d\rho_2 d\varphi_2 \\ \times\int_0^{2\pi}\exp[2q\rho_1\rho_2(\cos\varphi_1\cos\varphi_2 - \sin\varphi_1\sin\varphi_2)]d\varphi_1\Big\} \tag{9.54}$$

$$\int_0^{2\pi}\exp(m\cos x+n\sin x)\exp(ir\cos x)dx=2\pi I_0\left(\sqrt{E+iF}\right),\ m^2+(r+n)^2>0 \tag{9.55}$$

where $E=m^2+n^2-r^2$ and $F=2mr$. Using Eq. (9.55), we obtain the following [15]:

$$P_{IF}=\frac{(2\pi)^2 I_{S_0}I_{L_0}}{Q_s}\int_0^{\infty}\int_0^{\infty}\mathrm{Re}\left\{\begin{array}{l}\exp[-(g+is)\rho_2^2+(is-g)\rho_1^2]\\ \times I_0(2q\rho_1\rho_2)\rho_1\rho_2 d\rho_1 d\rho_2\end{array}\right\} \tag{9.56}$$

Performing the integration in Eq. (9.56) with respect to ρ_1 again, we obtain the integral Eq. (9.57) as follows:

$$\int_0^{\infty}xe^{-\alpha x^2}I_v(\beta x)J_v(\gamma x)dx=\frac{1}{2\alpha}\exp\left(\frac{\beta^2-\gamma^2}{4\alpha}\right)J_v\left(\frac{\beta\gamma}{2\beta}\right),\ [\mathrm{Re}\alpha>0,\ \mathrm{Re}v>-1] \tag{9.57}$$

Equation (9.56) can finally be rewritten as follows:

$$\begin{aligned}P_{IF}&=\frac{(2\pi)^2 I_{S_0}I_{L_0}}{Q_s}\int_0^{\infty}\mathrm{Re}\left\{\exp[-(g+is)\rho_2^2]\rho_2 d\rho_2\left[\frac{1}{2(g-is)}\exp\left(\frac{q^2\rho_2^2}{g-is}\right)\right]\right\}\\ &=\frac{(2\pi)^2 I_{S_0}I_{L_0}}{Q_s}\frac{1}{2(g-is)}\int_0^{\infty}\mathrm{Re}\left\{\rho_2\exp\left[-\left(g+is-\frac{q^2}{g-is}\right)\rho_2^2\right]d\rho_2\right\}\end{aligned} \tag{9.58}$$

$$\int_0^{\infty}x^m\exp(-\beta x^n)dx=\frac{\Gamma(\gamma)}{n\beta\gamma},\ \gamma=\frac{m+1}{n}\ [\mathrm{Re}\beta>0,\ \mathrm{Re}m>0,\ \mathrm{Re}n>0] \tag{9.59}$$

An integral simplification with respect to ρ_2 of Eq. (9.58) based on Eq. (9.58) can be expressed as follows:

$$P_{IF}=\frac{(2\pi)^2 I_{S_0}I_{L_0}}{Q_s}\frac{1}{2(g-is)}\times\frac{1}{2\left(\frac{g^2+s^2-q^2}{g-is}\right)}=\frac{\pi^2 I_{S_0}I_{L_0}}{Q_s}\frac{1}{(g^2+s^2-q^2)} \tag{9.60}$$

Based on Eq. (9.40), the magnitudes of the optical power of the signal and local oscillation light beams in the source field can be calculated as follows:

$$P_{L_0}=2\pi I_{L_0}\int_0^{\infty}\rho\exp\left[-\left(\frac{1}{2\sigma_{L_0}^2}+\frac{1}{W^2}\right)\rho^2\right]d\rho=\frac{2\sigma_{L_0}^2W^2\pi I_{L_0}}{W^2+2\sigma_{L_0}^2} \tag{9.61}$$

$$P_{S_0}=\frac{2\sigma_{S_0}^2W^2\pi I_{S_0}}{W^2+2\sigma_{S_0}^2} \tag{9.62}$$

Substituting Eqs. (9.60, 9.61 and 9.62) in Eq. (9.37), the SNR of the partially GSM beam coherent detection system can be easily obtained, that is,

$$SNR'_{GSM}=\frac{\frac{\pi^2 I_{S_0}I_{L_0}}{Q_s}\times\frac{1}{(g^2+s^2-q^2)}}{\frac{2\sigma_{L_0}^2W^2\pi I_{L_0}}{W^2+2\sigma_{L_0}^2}}=\frac{\pi I_{S_0}(W^2+2\sigma_{L_0}^2)}{2\sigma_{L_0}^2W^2Q_s(g^2+s^2-q^2)} \tag{9.63}$$

Similarly, the heterodyne efficiency of the partially coherent GSM beam detection system is expressed as follows:

$$\eta'_{\text{het\_}GSM}=\frac{(W^2+2\sigma_{L_0}^2)(W^2+2\sigma_{S_0}^2)}{4\sigma_{L_0}^2\sigma_{S_0}^2W^4Q_s(g^2+s^2-q^2)} \tag{9.64}$$

For a continuous narrow-linewidth light source, the cross-spectral density of the beam and the correlation function can be treated as approximately equivalent to each other [16]. When the signal light beam is transmitted upstream to the receiver, based on Eq. (9.40), let $\rho_1=\rho_2=\rho$. Following a complex integration, the cross-spectral density function is expressed as follows:

$$W_S(\rho,\rho,L)=I_S(\rho,L)=\frac{I_{S_0}}{Q_s}\exp\left(-\frac{\rho^2}{2\sigma_{S_0}^2Q_s}\right) \tag{9.65}$$

Applying the Fourier transform to the expression (9.65) for light intensity after passing through atmospheric turbulence, we obtain the following expression:

$$I(f,L)=\sqrt{2\pi}\sigma_{S_0}\frac{1+\left(\frac{L}{2k\sigma_{S_0}^2}\right)^2}{Q_s}\exp[-2\pi^2f^2\sigma_{S_0}^2Q_s^2] \tag{9.66}$$

The 1D light intensity of a partially coherent GSM beam can be obtained through the superposition of H-G beams with mutually orthogonal modes, whose corresponding mode coefficients can be expressed as follows [2]:

$$c_n=2\pi v_0^2\int_0^{\infty}I(f,z)\Phi_n(\pi^2v_0^2f^2)f\,df \tag{9.67}$$

where $\Phi_n(\pi^2 v_0^2 f^2)$ is the Fourier transform of the normalized optical intensity of the 1D H-G beam and v_0 is the waist width of the expanded H-G beam; these two variables are expressed as follows [2, 3]:

$$\Phi_n(\pi^2 v_0^2 f^2) = L_n(\pi^2 v_0^2 f^2)\exp\left(-\frac{\pi^2 v_0^2 f^2}{2}\right) \tag{9.68}$$

$$v_0 = \frac{2\sigma_{S_0}}{\left(1+\frac{4\sigma_{S_0}^2}{\delta_{S_0}^2}\right)^{1/4}} \tag{9.69}$$

Substituting Eqs. (9.68) and (9.69) in Eq. (9.67) and simplifying the integral yields the following expression:

$$c_n = \sqrt{\frac{\pi}{2}}\left[1+\left(\frac{L}{2k\sigma_{S_0}^2}\right)^2\right]\frac{\frac{v_0^2}{\sigma_{S_0} Q_s^3}}{1+\left(\frac{v_0}{2\sigma_{S_0} Q_s}\right)^2}\left[\frac{1-\left(\frac{v_0}{2\sigma_{S_0} Q_s}\right)^2}{1+\left(\frac{v_0}{2\sigma_{S_0} Q_s}\right)^2}\right]^n \tag{9.70}$$

The corresponding weight factors are as follows:

$$\lambda_n = \frac{c_n}{\sum\limits_{n=0}^{\infty} c_n} \tag{9.71}$$

The M^2 factor can be defined as follows:

$$M^2 = \sum_{n=0}^{\infty}(2n+1)\lambda_n \tag{9.72}$$

9.2.2 *Effect of Beam Pattern on Performance of Coherent Detection Systems*

Based on Eqs. (9.69, 9.70, 9.71 and 9.72), the effects of the light source parameters, turbulence factor, and transmission distance on the weight factor and the normalized M^2 factor on the performance of the partially coherent GSM beam detection system after the GSM beam passes through atmospheric turbulence are analyzed. Based on the typical values of the system parameters, the numerical simulation parameters are set to $\lambda = 1550$ nm, $l_0 = 1$ cm, $L_0 = 10$ m, $C_0 = 10^{-14}$ m$^{-2/3}$, $L = 5$ km, $\sigma_{s0} = 5$ cm, $\varphi = \pi/6$, and $M^2 = 1.2$ without any special considerations. Based on the formula of the M^2 factor for the GSM beam, $M^2 = [1 + (2\sigma_{s0}/\delta_{s0})^2]^{1/2}$, we consider $\delta_{s0} = 0.15$ m.

To compare with the weight factor distribution of the GSM beam after passing through atmospheric turbulence, the weight factor distribution of the GSM beam at $L = 0$ following incoherent mode decomposition and the superposition of the decomposed H-G beam are illustrated first.

Figure 9.2 depicts the mode decomposition of the partially coherent GSM beam at the source field. Part (a) indicates the variation in the weight factor with respect to the mode order, and it can be seen that there are two primary modes at the source field, which are the fundamental mode with 91% content and the first-order mode with 8.3% content, and the proportion of the other higher-order modes is basically negligible. Figure 9.2b depicts the 1D intensity distribution of the partially coherent GSM beam and its corresponding decomposed 1D intensity distribution for each order of the H-G beam. $\varphi_0(x)$ denotes the zeroth-order H-G beam (that is, the fundamental-mode Gaussian beam), $\varphi_1(x)$ denotes the first-order H-G beam, $\varphi_2(x)$ denotes the second-order H-G beam, and the star-dotted line depicts the intensity of the GSM beam after the superposition of the three modes. The figure indicates that the $\varphi_0(x)$ light intensity curve almost coincides with the star-dotted line, which, in turn, indicates that the base film accounts for the largest proportion of the partially coherent GSM beam at the source field.

Figure 9.3 depicts the curve of the heterodyne efficiency of the coherent detection system with an effective detection aperture when the light field distributions of the signal and local oscillation light beams are the same, that is, both are fully coherent light beams or partially coherent GSM beams. The figure indicates that when the source field parameters of the signal and local oscillation light beams are the same, the signal light beam is transmitted to the receiver through the atmospheric turbulent random medium for coherent mixing with the local oscillation light beam. The heterodyne efficiency decreases with an increase in the effective detection aperture, and the maximum efficiency can reach more than 98%. The heterodyne efficiency of the partially coherent GSM beam for both the signal and local oscillation light beams is slightly lower than the corresponding heterodyne efficiency at the receiver when both the beams are fully coherent, because the GSM beam contains some other higher-order mode content at the source plane in addition to the more proportional fundamental-mode content, which also verifies the correctness of the analysis described above.

Figure 9.4 depicts the effects of the source coherence length and radius of the beam waist on the weight factor when the GSM beam is transmitted under horizontal atmospheric turbulence. Figure 9.4a indicates that the mode content at different orders varies with respect to the coherence length, and the larger the coherence length of the light source, the higher the amount of fundamental-mode content it contains. Figure 9.4b indicates that the value of the fundamental-mode content of the light source can be the maximum only when the radius of the bundle waist has a suitable value. Comparing Figs. 9.4a and 9.2a, it can be seen that the higher-order-mode content in the partially coherent GSM beam increases and the fundamental-mode content decreases from 91 to 40.37% at the source field after 5 km of transmission under atmospheric turbulence. This is attributed primarily to the random diffraction

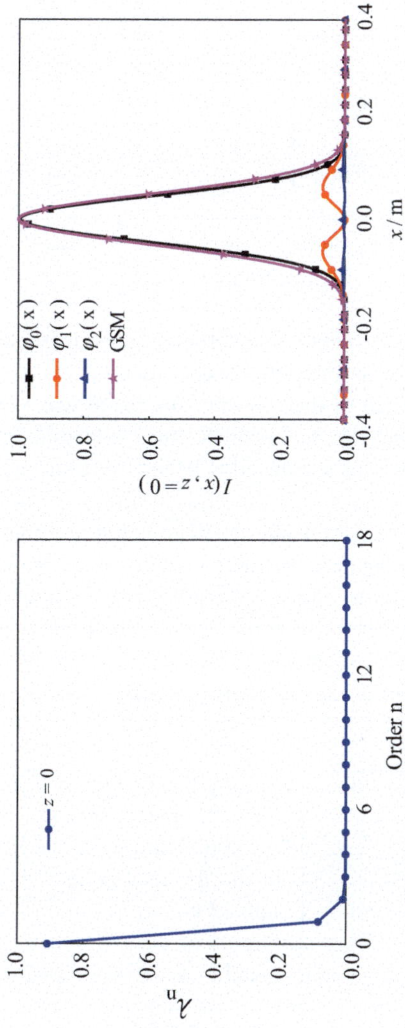

(a) Variation of weighting factors with respect to the order

(b) Superposition of H-G beams of various orders

Fig. 9.2 Mode decomposition of the GSM beam at the source field [5]

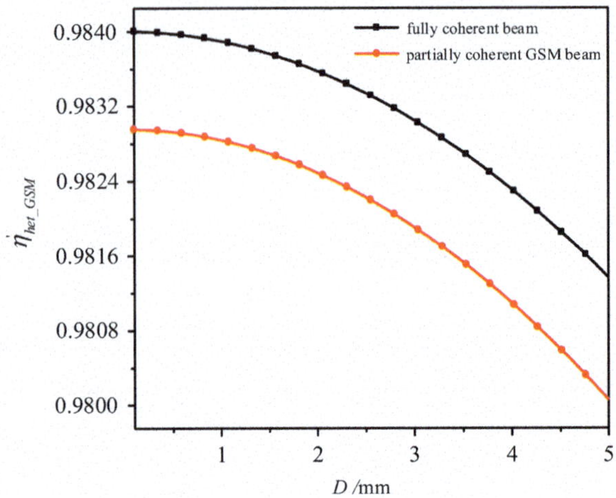

Fig. 9.3 Variation of heterodyne efficiency of coherent detection system with effective detection aperture [5]

of the beam under atmospheric turbulence, which generates diffraction flaps. Thus, the low-order modes are transformed into high-order modes.

Figures 9.5a–c depict the effect of the radius of the beam waist on the base-film content when the coherence length, transmission distance, and atmospheric refractive index structure constant, respectively, are varied. All the three plots indicate that when the proportion of the fundamental mode is the highest, the corresponding radius of the beam waist increases with an increase in the coherence length of the beam, transmission distance, and atmospheric refractive index structure constant. When the coherence length $\sigma_g = 0.15$ m and the transmission distance $z = 5$ km, the fundamental-mode content reaches the extreme value corresponding to the radius of the beam waist (σ_s) = 0.1 m. Moreover, Fig. 9.5a indicates that when $\sigma_s = 0.01$ m, the number of transmitted beam modes increases, whereas the trend in the weight

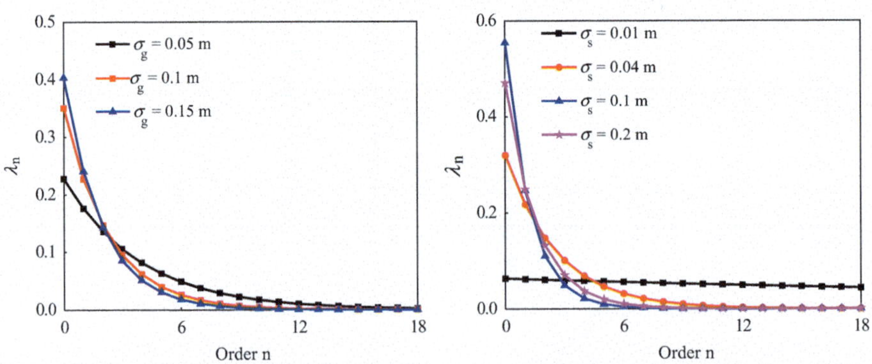

(a) Various coherence lengths (b) Various girdle radii

Fig. 9.4 Effect of light source parameters on the weight factor [5]

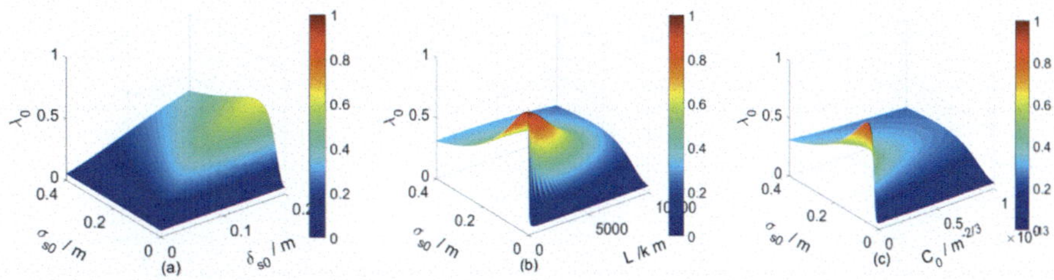

Fig. 9.5 Effect of waist radius on base-mode content [5]: **a** Coherence length, **b** Ttransmission distance, and **c** Structural constant of the atmospheric refractive index

factor varies. Therefore, the selection of the appropriate light source parameters is critical when the beam is transmitted under different environments.

Figure 9.6 depicts the variations in the beam-mode order and the corresponding mode coefficients with respect to the transmission distance. Figure 9.6a indicates that when the beam is transmitted horizontally, the mode order increases continuously and the fundamental-mode content decreases as the transmission distance increases. Figure 9.6b depicts the variation in the mode order for mode content higher than 5% on the horizontal and oblique range transmission paths for various transmission distances. A comparison of Figs. 9.5a, b, and 9.2a indicates that, when the beam is transmitted horizontally, the mode order remains almost constant for transmission distances up to 1 km. The effect of turbulence can be ignored temporarily when the beam is transmitted along a short distance under atmospheric turbulence. Further, when the transmission distance continues to increase, the turbulence effect gradually increases, beam-mode order increases, and coherence length of the beam decreases. Figure 9.6b indicates that the beam contains the most mode orders when it is transmitted horizontally and the least mode orders when it is transmitted obliquely downstream, which is attributed to the various turbulence terms Q_3 in Eq. (9.44). As the transmission height increases, the intensity of turbulence becomes weaker, causing the horizontal transmission to be more severely affected by atmospheric turbulence when compared with the oblique transmission path. For oblique beam transmission, the ξ-value of the downlink is lower than that of the uplink. Therefore, the beam is least affected by turbulence for oblique downlink transmission. In the following, we will analyze the transmission of the oblique-range uplink beam.

Figure 9.7 depicts the effect of the fundamental-mode content on the heterodyne efficiency for the transmission of the GSM beam in the oblique range under atmospheric turbulence at various coherence lengths and radii of the beam waist of the source. The figure indicates that the heterodyne efficiency increases with an increase in the fundamental-mode content. In addition, as the partial coherence length increases, the fundamental-mode content and the corresponding heterodyne efficiency increase. Figure 9.7b indicates that the fundamental-mode content increases and then decreases as the radius of the waist of the signal light beam increases. This variation can be attributed to the fact that when the radius of the waist of the signal

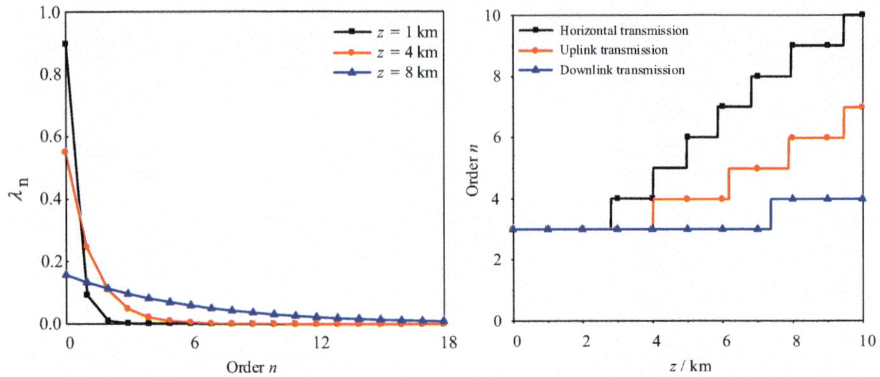

(a) Variation of weight factor with mode order

(b) Variation of mode order with transmission distance

Fig. 9.6 Variation curve of beam mode order and corresponding mode coefficient with transmission distance

light beam increases to match that of the local oscillation light beam, the heterodyne efficiency is the highest and the corresponding fundamental-mode content is the largest. Further, if the radius of the waist of the signal light beam continues to increase, the fundamental-mode content and heterodyneefficiency decrease.

Figure 9.8 represents the variation in the heterodyne efficiency with respect to the fundamental-mode content when the intraturbulent scale, extraturbulent scale, zenith angle, and intensity of turbulence are different for the signal optical ramp transmission to the receiver. Figure 9.8a indicates that the smaller the intraturbulent scale, the lower the fundamental-mode content and the corresponding heterodyne efficiency. This is because the lower-order modes are impacted more by the atmospheric turbulence effect, and as the inner scale decreases, the corresponding atmospheric turbulence intensity gradually increases. A significant diffraction phenomenon occurs during beam transmission, causing increasingly more lower-order modes in the beam to

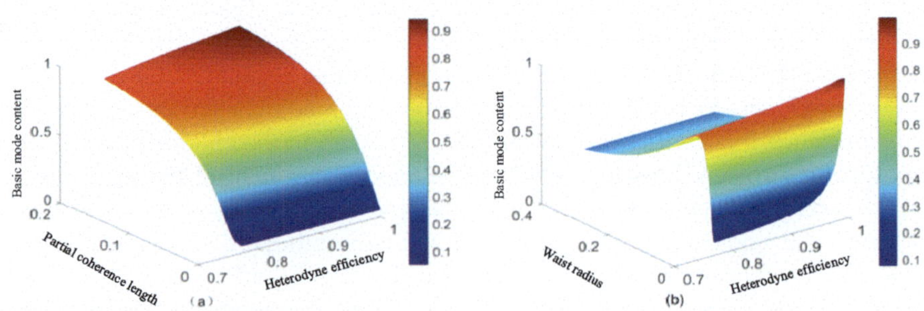

(a) Coherence length

(b) Girdle radius

Fig. 9.7 Effect of base-mode content on heterodyne efficiency [5]

transform into higher-order modes. The more uniform the beam-mode distribution becomes, the less the mode content matches with the fundamental light.

Figure 9.8b indicates that the extraturbulent scale has less effect on the beam pattern and heterodyne efficiency when compared with the intraturbulent scale. Figure 9.8c depicts the variation in the fundamental-mode content and heterodyne

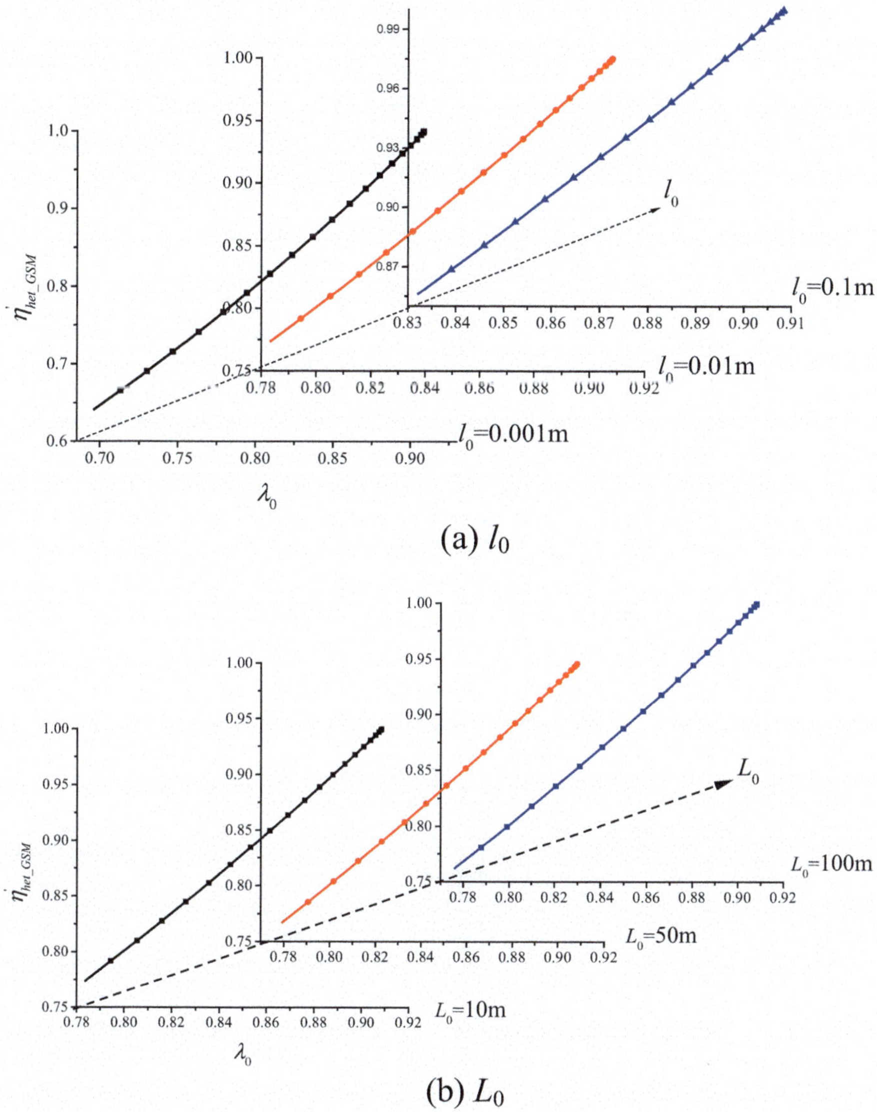

Fig. 9.8 Effect of turbulence parameters on weighting factor and heterodyne efficiency at $L = 5$ km [5]

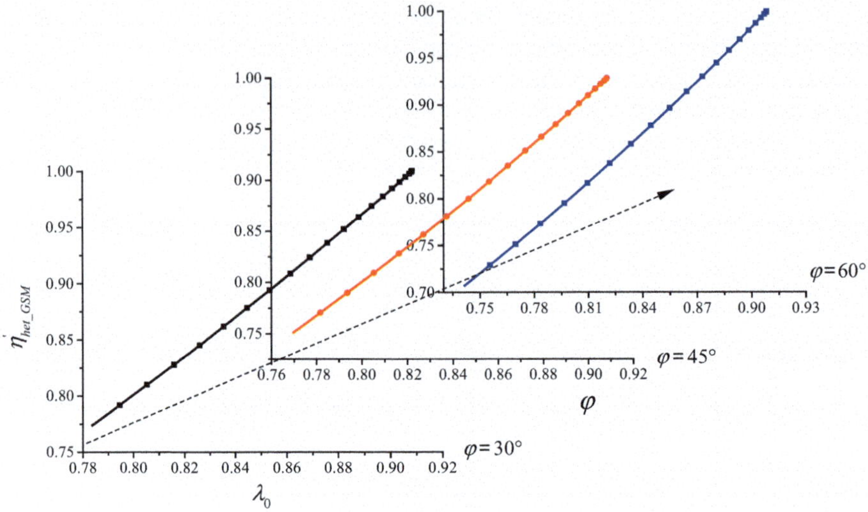

(c) φ

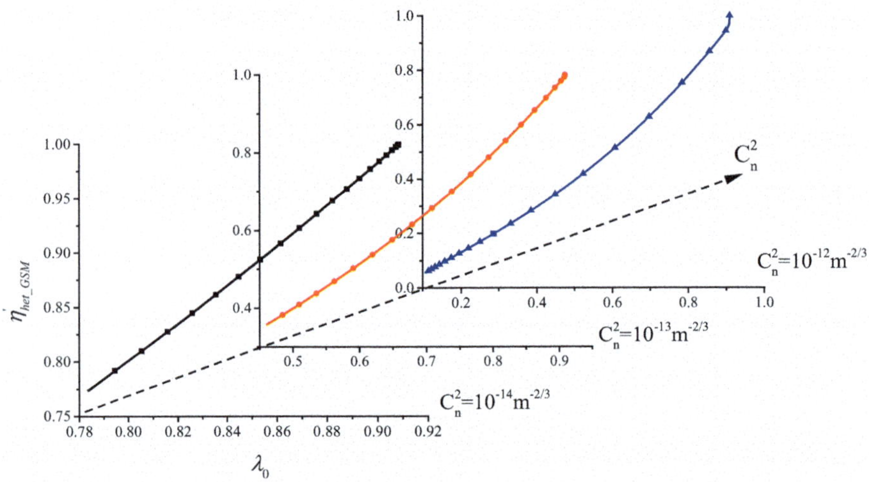

(d) C_n^2

Fig. 9.8 (continued)

efficiency when the zenith angle is different. The figure indicates that as the zenith angle increases, based on how close the transmission path is to horizontal transmission, the path of signal light transmission in the region of strong turbulence increases, higher-order-mode component increases, and fundamental-mode component decreases, which in turn leads to a decrease in the heterodyne efficiency.

Figure 9.8d depicts the distribution of the fundamental-mode content and the variation in the heterodyne efficiency under various intensities of atmospheric turbulence, which indicate that the beam-mode order increases rapidly as the intensity of turbulence increases. For the coherent detection communication system with a partially coherent beam, the impact of the intensity of atmospheric turbulence increases the beam-mode order and decreases the proportion of corresponding fundamental-mode content, which causes a sharp decrease in the heterodyne efficiency.

Figure 9.9 depicts the effects of light source parameters on the normalized M^2 factor and heterodyne efficiency. Figure 9.9a presents the variation in the heterodyne efficiency and normalized M^2 factor with respect to the transmission distance for various coherence lengths. The plots indicate that the normalized M^2 factor becomes larger and the heterodyne efficiency decreases as the transmission distance increases. When the transmission distance is constant, the normalized M^2 factor increases with an increase in the coherence length, and the beam quality of the corresponding signal light and the degree of match with the fundamental light becomes worse, leading to a decrease in the heterodyne efficiency. This is consistent with the fact that the worse the spatial coherence, the less the beam is affected by transmission under atmospheric turbulence.

Figure 9.9b depicts the effects of the radius of the beam waist and transmission distance on the normalized M^2 factor and heterodyne efficiency. The figure indicates that when the radius of the beam waist is $\sigma_{s0} = 0.01$ m, the trends in the corresponding M^2 factor and heterodyne efficiency are different from the other two curves. This phenomenon is illustrated in Fig. 9.9c: when the radius of the beam waist increases, the normalized M^2 factor decreases significantly at a certain value (which is related to the transmission distance and radius of the beam waist), and, thereafter, the decreasing trend gradually tapers down. The figure indicates that the presence of this radius allows the beam quality to be optimal, which is in agreement with the results presented in Fig. 9.4b.

Figures 9.10a–d, respectively, depict the effects of the intraturbulent scale, extraturbulent scale, zenith angle, and intensity of turbulence on the normalized M^2 factor and heterodyne efficiency with respect to increasing transmission distance. It is evident from the figures that as the inner scale decreases and the outer scale, zenith angle, and intensity of turbulence increase, the normalized M^2 factor increases and the beam quality becomes worse; this conclusion is consistent with the reference [17]. In addition, the effect of the outer scale on the M^2 factor and heterodyne efficiency is much smaller when compared with the inner scale. This conclusion is consistent with the conclusion depicted in Fig. 9.8, that is, as the intensity of atmospheric turbulence increases, a significant diffraction occurs in the beam, the corresponding mode order increases, the normalized M^2 factor increases, and the beam quality decreases, which eventually leads to a decrease in the heterodyne efficiency.

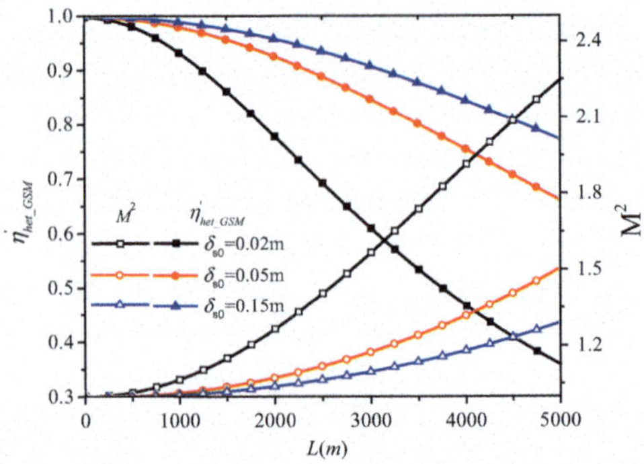

(a) Variation in normalized M^2 factor and outlier efficiency with respect to transmission distance at various coherence lengths

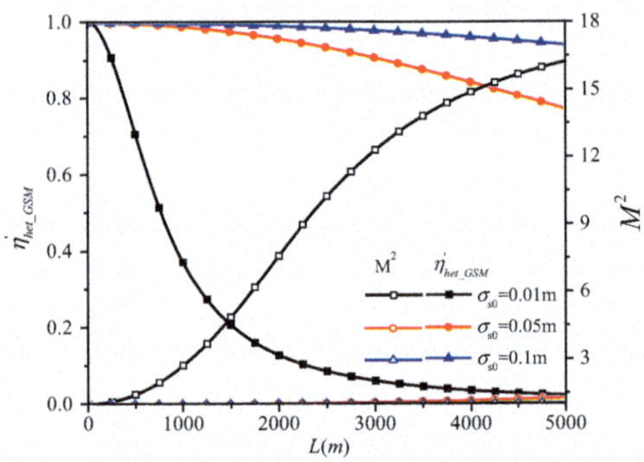

(b) Variation in normalized M^2 factor and outlier efficiency with respect to the transmission distance at various values of the radius of the signal beam waist

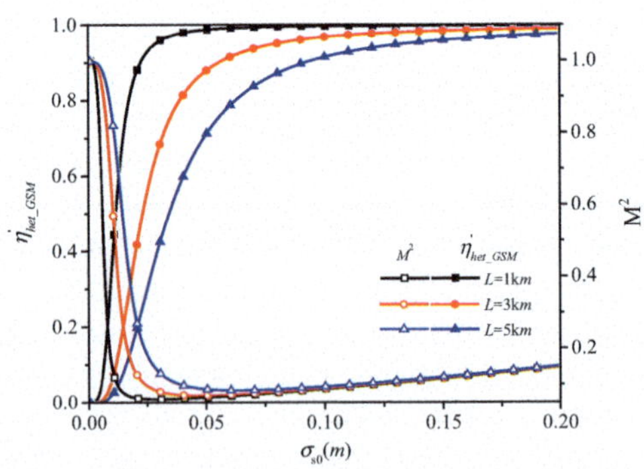

(c) Variation in normalized M^2 factor and outlier efficiency with respect to the radius of the signal beam waist at various values of the transmission distance

Fig. 9.9 Effect of light source parameters and transmission distance on the normalized M^2 factor and heterodyne efficiency [5]

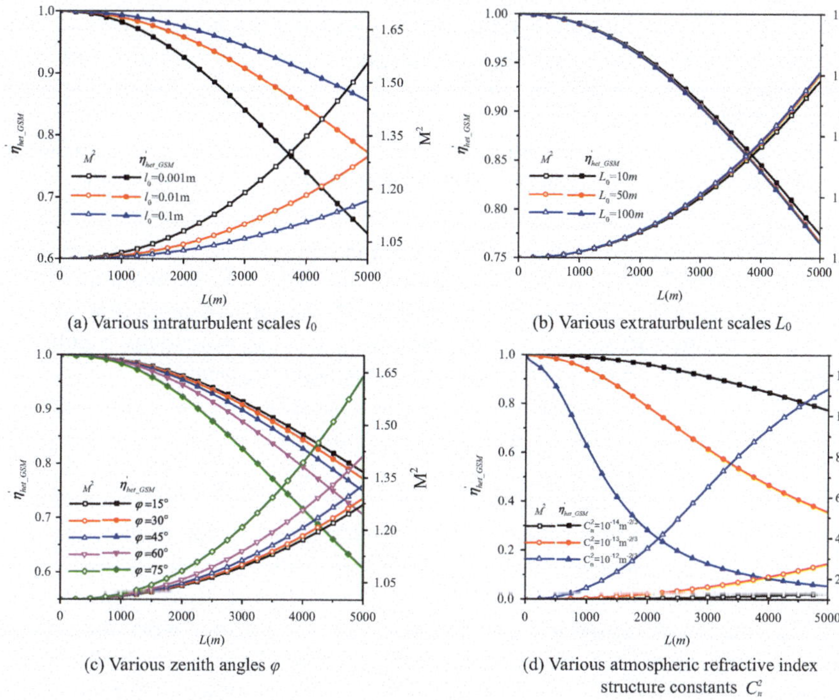

(a) Various intraturbulent scales l_0

(b) Various extraturbulent scales L_0

(c) Various zenith angles φ

(d) Various atmospheric refractive index structure constants C_n^2

Fig. 9.10 Effect of atmospheric turbulence on the normalized M^2 factor and heterodyne efficiency [5]

9.3 Experimental Results of Coherent Detection System—1.3-km Analysis

At the experimental distance of 1.3 km, the optical field distribution of the fundamental-mode beam is a standard Gaussian function, whereas for beams containing higher-order modes, the optical field distribution is no longer a Gaussian function; therefore, it is reasonable to use a Gaussian fit to reflect the fundamental-mode content. The Spiricon beam analyzer on Gaussian fit is defined as follows: an equation that fits two univariate variables through an orthogonal data line at the light scale or a least-squares fit. Using the ISO method to calculate the Gaussian fit, the maximum deviation of the theoretical fit from the measured value is referred to as the roughness of fit, which can be expressed as follows:

$$Q = \frac{\left|E_{ij} - E_{ij}^{f}\right|_{\max}}{E_{\max}} \tag{9.73}$$

where E_{ij} denotes the measured light intensity distribution and E_{ij}^{f} denotes the theoretical light intensity distribution of the Gaussian beam. This equation indicates that the value of Q falls within the range 0–1. A lower value of Q indicates a lower roughness of fit of the beam.

As depicted in Fig. 9.11, the spot distribution is measured under various weather conditions. It can be seen that the measured spot is hollow, mainly because a Cassegrain receiver optical system with a blocking ratio of 0.2 is used at the receiver terminal of the coherent detection system. A comparison of Figs. 9.11a, b indicates that the spot size in Fig. 9.11a is large, the spot is diffuse, and the beam quality is poor, whereas the spot in Fig. 9.11b is complete, light energy is concentrated, and beam quality is better. This is because the turbulence is weaker in cloudy weather as compared with sunny weather and has a lower impact on the beam.

Table 9.1 presents the Gaussian fit information measured by the beam analyzer. It can be seen that the Gaussian fit is smaller and rougher on sunny days than on cloudy days. The measured data are consistent with the theory; both sets indicate that the intensity of turbulence is higher on sunny days when compared with cloudy days and that the beam gets affected more by turbulence, which implies that the fundamental-mode content in the beam after transmission is lower on sunny days than on cloudy days.

The information presented in Figs. 9.11 and 9.12 was acquired when the experimental environment and parameters were identical. Figure 9.12 depicts the IF signals acquired using the oscilloscope under different weather conditions after the signal and local oscillation light beams were mixed coherently at the receiver; Figs. 9.12a and b correspond to a sunny day and a cloudy day, respectively. It is evident from

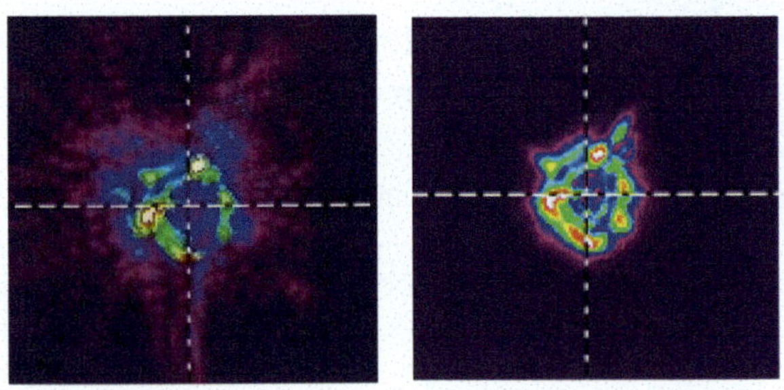

(a) Sunny day (b) Cloudy day

Fig. 9.11 Measured light spots under different weather conditions [5]

Table 9.1 Results of Gaussian fit coincidence and roughness under different weather conditions

Weather conditions	2D Gaussian coincidence (%)	*X*-axis Gaussian coincidence (%)	*Y*-axis Gaussian coincidence (%)	2D Gaussian roughness (%)	*X*-axis Gaussian roughness (%)	*Y*-axis Gaussian roughness (%)
Sunny day	73.1	58.9	57.8	76.2	61.6	66.5
Cloudy day	80	62.1	62.4	76.2	58.3	60.5

the figures that the peak-to-peak amplitude of the IF signal is only 720 mV on sunny days and 2.04 V on cloudy days, which can be attributed to the fact that the quality of the signal spot under sunny conditions is worse than that on cloudy days, thus leading to poor mixing of the signal and local oscillation light beams and, in turn, lower amplitude of the IF signal.

The results of the measurement of the light spot distribution under different weather conditions at different moments on the same day are presented in Fig. 9.13. Figures 9.13a–d present the light intensity distributions measured during moderate rain, during light rain, just after the rain has stopped, and at night after the rain has stopped, respectively. Comparing the light spot distributions depicted in the four figures, it can be seen that the light spot distribution is the most scattered and the light intensity distribution is the most uneven in the moderate rainy days, the light spot energy is gradually enhanced as the rainfall decreases, and the quality of the light spot is the best at night.

Table 9.2 presents the 2D fit coincidence and 2D fit roughness of the light intensity at different moments corresponding to Fig. 9.13. It can also be seen from the table that the fit coincidence gradually increases and the opposite fit roughness gradually decreases as the amount of rainfall decreases. This implies that the proportion of fundamental modes in the beam gradually increases as the rainfall decreases.

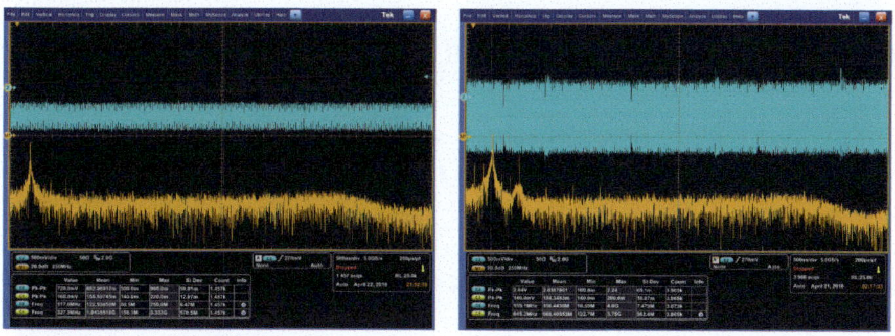

(a) Sunny day (b) Cloudy day

Fig. 9.12 IF signals under different weather conditions [5]

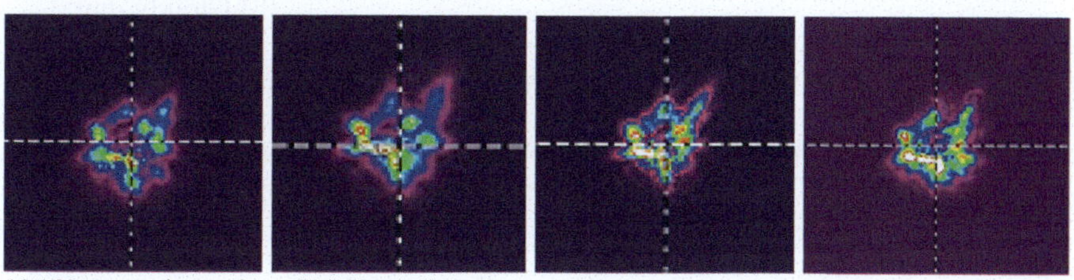

(a) 4:30 pm - moderate rain (b) 4:30 pm - light rain (c) 8:30 pm - no rain (d) 10:30 pm - cloudy at night

Fig. 9.13 Light spots at different time instants on May 2, 2018 [5]

Table 9.2 Results of Gaussian fit coincidence and roughness at different time instants on May 2, 2018

Time of experiment	4:30 pm (%)	4:30 pm (%)	8:30 pm (%)	10:30 pm at night (%)
2D Gaussian fit coincidence	42	65	72	78
2D Gaussian fit roughness	99.7	85.1	81.4	71

The light spots presented in Fig. 9.13 were measured when the experimental environment and parameters were exactly the same. Figure 9.14 depicts the IF signal acquisition on the same day at different time instants and under different weather conditions. The graphs presented in Fig. 9.14a–d indicate the IF signal data for moderate rain, light rain, just after the rain has stopped, and two hours after the end of rain, respectively. The four plots indicate that the peak value of the IF signal is the smallest for moderate rain; it gradually increases as the rainfall decreases, at which time the quality of the light spot corresponding to Fig. 9.14 gradually becomes better. When the rain ends, owing to the heavy moisture in the air, there is a certain attenuation of the light energy, and the peak value of the IF signal depicted in Fig. 9.15d is far less than that corresponding to cloudy conditions. However, in comparison with the data for the same day, the light spot quality is the best at the end of the rain; the peak-to-peak value of the IF signal is also the highest at this time.

9.4 Light Intensity Distribution of Partially Coherent GSM Beams Under Turbulence

Using the partially coherent GSM model beam as the light source, the cross-spectral density function in the $z = 0$ plane can be expressed as follows [7]:

$$W^{(0)}(\rho_1, \rho_2, 0) = I_0 \exp(-\frac{|\rho_1|^2 + |\rho_2|^2}{4\sigma_s^2}) \times \exp(-\frac{|\rho_1 - \rho_2|^2}{2\sigma_g^2}) \tag{9.74}$$

where ρ_1, ρ_2 are the position vectors of any two points at $z = 0$, $I_0 = 1$ is the light intensity at the source plane, σ_s is the beam radius, and σ_g is the spatial coherence length.

The beam is transmitted under atmospheric turbulence, and based on the formula for the Huygens–Fresnel diffraction integral, the cross-spectral density at the z-plane can be expressed as follows [18]:

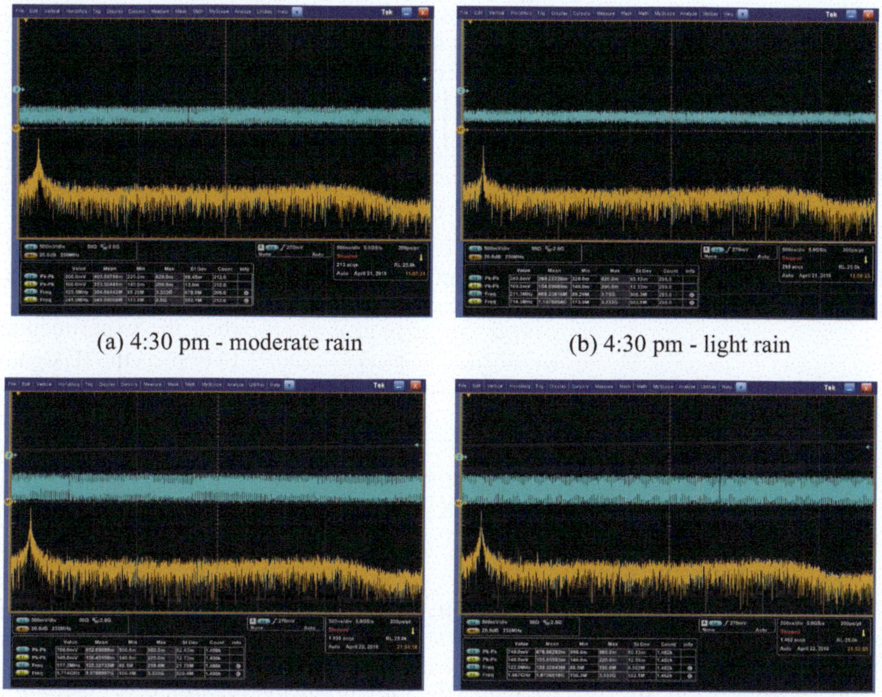

(a) 4:30 pm - moderate rain (b) 4:30 pm - light rain

(c) 8:30 pm - no rain (d) 10:30 pm - cloudy at night

Fig. 9.14 IF currents at different time instants on May 2, 2018 [5]

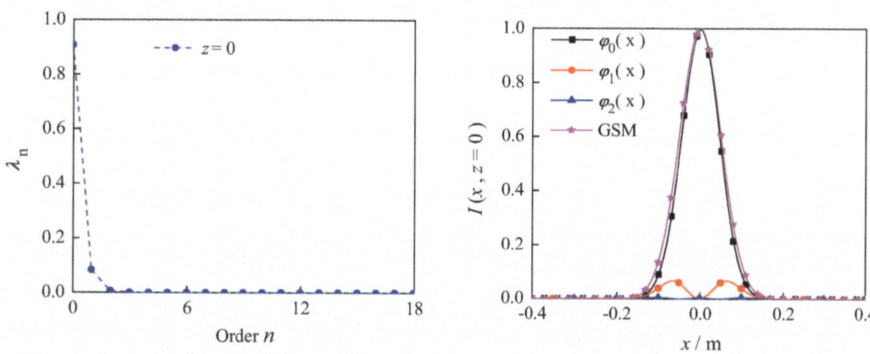

(a) Weighting factors (b) Superposition of H-G beams

Fig. 9.15 Mode decomposition of the GSM beam at the source plane [22]

$$W(r_1, r_2, z) = \left(\frac{k}{2\pi z}\right)^2 \iint d^2\rho_1 \iint d^2\rho_2 W^{(0)}(\rho_1, \rho_2, 0)$$
$$\cdot \exp\left\{-\frac{ik}{2z}\left[(r_1 - \rho_1)^2 - (r_2 - \rho_2)^2\right]\right\} \cdot \left\langle \exp\left[\psi^*(r_1, \rho_1, z) + \psi(r_2, \rho_2, z)\right]\right\rangle \tag{9.75}$$

where r_1, r_2 are the position vectors of any two points on the z-plane and $\langle \bullet \rangle$ represents the coefficient average of the turbulent medium, which can be expressed as follows [17]:

$$\left\langle \exp\left[\psi^*(r_1, \rho_1, z) + \psi(r_2, \rho_2, z)\right]\right\rangle = \exp\left[-\frac{1}{2} D_\phi(\rho_1 - \rho_2, r_1 - r_2)\right] \tag{9.76}$$

where $D_\phi(\rho_1 - \rho_2, r_1 - r_2)$ is the wave structure function.

In this study, we use the modified von Karman spectrum [19], which is expressed as follows:

$$\phi_n(\kappa, h) = 0.033 C_n^2(h) \exp(-\kappa^2 / \kappa_m^2)(\kappa^2 + \kappa_0^2)^{-11/6} \tag{9.77}$$

where $\kappa_m = 5.92/l_0$, $\kappa_0 \approx 2\pi / L_0$, l_0 is the intraturbulent scale, L_0 is the extraturbulent scale, and $C_n^2(h)$ represents the structural model of the atmospheric refractive index, which is adopted in this study based on the height-dependent model proposed in Ref. [20] and can be expressed as follows:

$$C_n^2(h) = 8.148 \times 10^{-56} v_{RMS}^2 h^{10} e^{-h/1000} + 2.7 \times 10^{-16} e^{-h/1500} + C_0 e^{-h/100} \tag{9.78}$$

where $v_{RMS}^2 = \sqrt{v_g^2 + 30.69 v_g + 348.91}$ is the vertical path wind speed, v_g is the near-surface wind speed, C_0 is the near-surface atmospheric refractive index structure constant (with the range $10^{-17}\text{m}^{-2/3}$–$10^{-13}\text{m}^{-2/3}$), and h is the vertical distance between the transmitter and the receiver.

After a complex integration for the simplification of the oblique-range transport under atmospheric turbulence, the resulting wave structure function can be expressed as follows [21]:

$$D_\phi(Q, p, z) = B_0(B_1 p^2 + B_2 p Q + B_3 Q^2) \tag{9.79}$$

$$B_0 = 2\pi^2 k^2 \sec\theta \int_0^\infty 0.033 \kappa^3 (-\kappa^2 / \kappa_m^2)(\kappa^2 + \kappa_0^2)^{-11/6} d\kappa \tag{9.80}$$

$$B_1 = \int_{h_0}^{H} C_n^2(h)(1 - \xi)^2 dh \tag{9.81}$$

$$B_2 = 2\int_{h_0}^{H} C_n^2(h)(1-\xi)\xi dh \tag{9.82}$$

$$B_3 = \int_{h_0}^{H} C_n^2(h)\xi^2 dh \tag{9.83}$$

where $p = r_1 - r_2$, $Q = \rho_1 - \rho_2$, and θ is the zenith angle. For the uplink, $\xi = 1 - (h-h_0)/(H-h_0)$, where H is the distance of the receiver from the ground and h_0 is the distance of the transmitter from the ground. For the downlink, $\xi = (h-h_0)/(H-h_0)$, where h_0 is the distance of the receiver from the ground and H is the distance of the transmitter from the ground.

For slant-range transmission, we have

$$M = \frac{1}{2}B_0 B_3 \tag{9.84}$$

For horizontal transport, the atmospheric refractive index structure constant $C_n^2(h)$ being constant, we have

$$M = \frac{1}{3}\pi^2 k^2 z \int_0^{\infty} \kappa^3 \phi_n(\kappa) d\kappa \tag{9.85}$$

At the receiver, let $r_1 = r_2 = r$. Following complex integration, the cross-spectral density function is obtained as follows [18]:

$$W(r, r, z) = I(r, z) = \frac{1}{\Delta^2(z)} \exp\left[-\frac{|r|^2}{2\sigma_s^2 \Delta^2(z)}\right] \tag{9.86}$$

where

$$\Delta^2(z) = 1 + \left[\frac{z}{k\sigma_s \delta}\right]^2 + \frac{2Mz^2}{k^2\sigma_s^2} \tag{9.87}$$

$$\frac{1}{\delta^2} = \frac{1}{4\sigma_s^2} + \frac{1}{\sigma_g^2} \tag{9.88}$$

9.5 Modal Coefficient Distribution of Partially Coherent GSM Beam Under Turbulence

Applying the Fourier transform to the light intensity expression (9.86) after the beam passes through atmospheric turbulence, we have

$$I(f,z)=\sqrt{2\pi}\sigma_s\frac{1+\left(\frac{z}{2k\sigma_s^2}\right)^2}{\Delta(z)}\exp\left[-2\pi^2f^2\sigma_s^2\Delta^2(z)\right] \tag{9.89}$$

where f is the spatial frequency.

The 1D light intensity of a partially coherent GSM beam can be expressed as a superposition of H-G beams with mutual orthogonality between modes with the coefficients [3]

$$c_n=2\pi v_0^2\int_0^{\infty}I(f,z)\Phi_n(\pi^2v_0^2f^2)fdf \tag{9.90}$$

where $\Phi_n(\pi^2v_0^2f^2)$ is the Fourier transform of the normalized optical intensity of the 1D H-G beam and v_0 is the width of the expanded H-G beam waist, expressed as follows [2, 3]:

$$\Phi_n(\pi^2v_0^2f^2)=L_n(\pi^2v_0^2f^2)\exp(-\frac{\pi^2v_0^2f^2}{2}) \tag{9.91}$$

$$v_0=\frac{2\sigma_g}{\left(1+\frac{4\sigma_s^2}{\sigma_g^2}\right)^{1/4}} \tag{9.92}$$

Substituting Eqs. (9.89) and (9.91) in Eq. (9.90) to simplify the integral, we have the following expression:

$$c_n=\sqrt{\frac{\pi}{2}}\left[1+\left(\frac{z}{2k\sigma_s^2}\right)^2\right]\frac{\frac{v_0^2}{\sigma_s\Delta^3(z)}}{1+\left(\frac{v_0}{2\sigma_s\Delta(z)}\right)^2}\left[\frac{1-\left(\frac{v_0}{2\sigma_s\Delta(z)}\right)^2}{1+\left(\frac{v_0}{2\sigma_s\Delta(z)}\right)^2}\right]^n \tag{9.93}$$

9.6 Effect of Turbulence on Beam Pattern

Based on Eqs. (9.93), (9.71), and (9.72), respectively, the effects of the source parameters, turbulence factor, and transmission distance on the weighting and normalized M^2 factors of the GSM beam after passing through atmospheric turbulence are simulated and analyzed. Based on the typical values of the system parameters [18], the numerical simulation is performed with the following specifications: $\lambda=1550nm$, $l_0=1$ cm, $L_0=10$ m, $C_0=10^{-14}$ m$^{-2/3}$, $z=5$ km, $\sigma_s=5$ cm, $\theta=\pi/4$. The laser output beam is no longer an ideal fundamental-mode Gaussian beam owing to the thermal effects, the resonant cavity, and optical elements [1], that is, $M^2 \neq 1$. Therefore, $M^2 = 1.2$ is set in this study. Based on the formula $M^2 = (1 + (2\sigma_s/\sigma_g)^2)^{1/2}$ for the M^2 factor of the GSM beam, we set $\sigma_g = 0.15$ m.

9.6.1 *Effect of Light Source Parameters on Weighting Factors*

To compare with the weighting factor distribution of the GSM beam under atmospheric turbulence, the weighting factor distribution of the GSM beam after incoherent mode decomposition and the superposition of the decomposed H-G beam are considered.

Figure 9.15 depicts the mode decomposition of the partially coherent GSM beam at the source plane. It can be seen that the partially coherent GSM beam consists of a superposition of multiple orders of H-G beams. When the coherence length is different, the mode content of the different orders will be different, and the fundamental mode is the main component at the source plane. Figure 9.15a depicts the variation in the weighting factor with respect to the mode order: it can be seen that for the values of the parameters considered in this chapter, two modes exist at the source plane: the fundamental-mode content constitutes 91% and the first-order modes constitute only 8.3%; the remaining higher-order mode content is basically negligible. Figure 9.15b presents the 1D light intensity distribution of the corresponding H-G beam after the decomposition of the GSM beam. $\varphi_0(x)$ represents the zeroth-order H-G beam (that is, the fundamental-mode Gaussian beam), $\varphi_1(x)$ represents the first-order H-G beam, $\varphi_2(x)$ represents the second-order H-G beam, and the star-dotted line indicates the light intensity of the GSM beam after the superposition of several modes.

Figure 9.16 depicts the GSM beam transmission under horizontal atmospheric turbulence along with the effect of the source coherence length and radius of the beam waist on the weighting factor. A comparison of Figs. 9.15 and 9.16 indicates that the number of modes increases and the fundamental-mode content decreases from 91 to 40.37% at the light source plane after 5 km of GSM beam transmission under atmospheric turbulence, which is mainly caused by the variation in the shape of the wavefront owing to the diffraction of the beam under turbulence. The diffraction produces diffraction flaps and transforms the low-order modes into high-order modes. Figure 9.16a indicates that the greater the coherence length of the source, the higher the amount of fundamental-mode content. Figure 9.16b indicates that the value of the fundamental-mode content is maximum only when a suitable value is considered for the bundle radius.

Figures 9.17a–c provide further details on Fig. 9.16b. Figure 9.17b indicates that when the fundamental-mode content is the maximum, the radius of the bundle waist is considered to increase with an increase in the coherence length, transmission distance, and intensity of turbulence. The fundamental-mode content near $\sigma_s = 0.1$ m is the maximum when $\sigma_g = 0.15$ m and z = 5 km. The trend of the weighting factors changes at $\sigma_s = 0.01$ m, where the number of modes increases

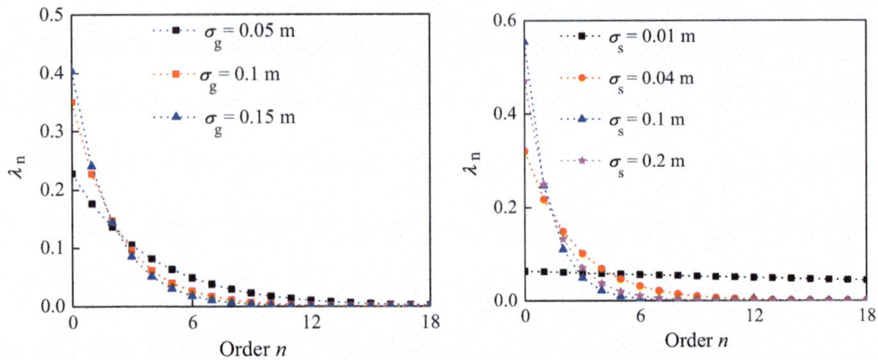

(a) Coherence length (b) Girdle radius

Fig. 9.16 Effect of light source parameters on weighting factors [22]

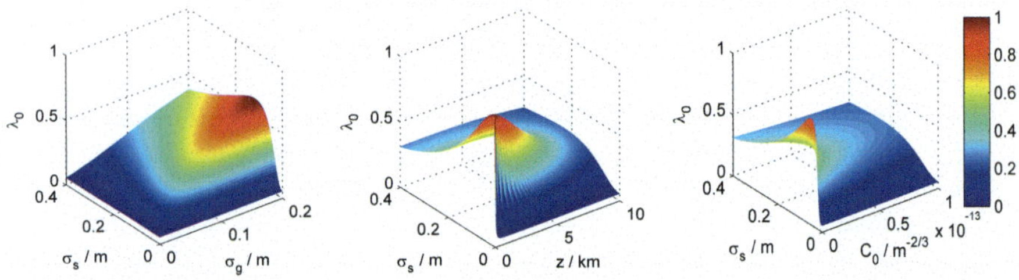

(a) Coherence length (b) Transmission distance (c) Atmospheric refractive index structure constant

Fig. 9.17 Effect of girdle radius on base-mode content [22]

and the weighting factors are evenly distributed. Therefore, for the transmission of beams under different environments, it is crucial to select the appropriate light source parameters.

9.6.2 *Effect of Turbulence Parameters on Weighting Factors*

Figure 9.18a indicates that the smaller the intraturbulent scale, the lower the fundamental-mode content. Comparing the cases $l_0 = 0.001$, 0.01, and 0.1 m, we see that the three curves intersect. The content of the low-order modes to the left of the intersection point decreases with a decrease in the inner scale, and the content of the modes to the right of the intersection point increases with a decrease in the inner scale. Based on an analysis of the physical mechanism of turbulence, because the lower-order modes are more impacted by turbulence, a decrease in the inner scale implies an increase in the intensity of turbulence; the stronger the turbulence,

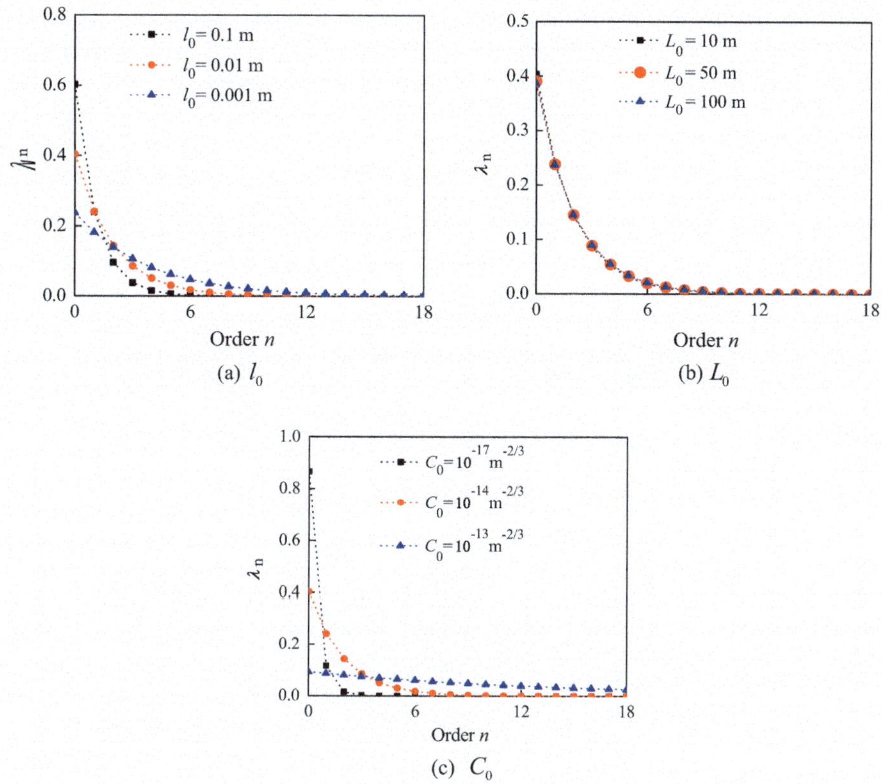

Fig. 9.18 Effect of turbulence on weighting factor at $z = 5$ km [22]

the more significant the diffraction phenomenon of the beam, the more components of the lower-order modes are transformed into higher-order modes, and the more uniform the mode distribution. Figure 9.18b indicates that the extraturbulent scale has little effect on the beam pattern. Figure 9.18c depicts the distribution of weighting factors under weak, moderate, and strong turbulence, indicating that the stronger the turbulence, the more significant the increase in the number of modes. For the fundamental-mode laser communication system, the efficiency of outlier mixing at the receiver side decreases sharply under strong turbulence owing to the increased number of mode and reduced fundamental-mode content.

9.6.3 Effect of Transmission Distance on Weighting Factor

Figure 9.19 depicts the variation in the modes with respect to the transmission distance. Figure 9.19a indicates that the longer the transmission distance under horizontal atmospheric turbulence, the larger the number of modes and the lower

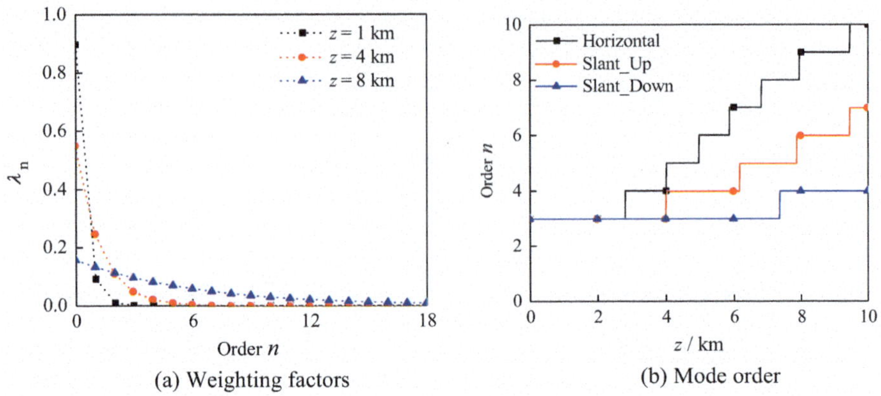

Fig. 9.19 Curve depicting variation in the mode with respect to transmission distance [22]

the fundamental-mode content. Figure 9.19b depicts the number of modes with a weighting factor greater than 5% for horizontal, oblique-range transport with respect to variation in the transport distance. A comparison of Figs. 9.19a, b, and 9.16a indicates that the mode of the beam is almost unchanged within 1 km for horizontal transmission, which is attributed to the fact that the variation in mode is determined by both the light source parameters and the intensity of turbulence. The turbulence effect is relatively small when the beam is transmitted along a short distance under turbulence, which mainly manifested as free diffraction. As the distance increases, the turbulence effect is gradually enhanced, number of modes increases, and coherence length decreases. Figure 9.19b indicates that the number of modes in the horizontal transmission beam is the largest and the number of modes in the downlink varies less in the case of oblique transmission. This is because the turbulence term M varies for horizontal and oblique-range transmission, and it becomes smaller as the atmospheric structure constant gradually decreases with increasing altitude. Therefore, horizontal transmission is more affected by turbulence than by oblique-range transmission. Because the downward GSM beam has a smaller ξ-value than the upward one during oblique transmission, the downward transmission is the least affected by turbulence. Thus, the analysis indicates that oblique transmission is recommended in base-mode laser communication.

9.7 Effect of Turbulence on M^2 Factor

The M^2 factor is obtained at the receiver using mode decomposition after the partially coherent GSM beam passes through atmospheric turbulence. Based on Eq. (9.23), the effect of the optical parameters and turbulence factor on the normalized M^2 factor is analyzed through numerical simulation.

9.7.1 Effect of Light Source Parameters and Transmission Distance on M^2 Factor

Figure 9.20 depicts the effect of light source parameters on the normalized M^2 factor; parts (a) and (c) depict the effect of coherence length and transmission distance, respectively. This is consistent with the fact that "the worse the spatial coherence, the less the beam is affected by the transmission in turbulence [23]." Fig. 9.20c indicates that the normalized M^2 factor remains essentially constant at $z = 1$ km, that is, the beam quality remains constant at shorter transmission distances owing to the combined effect of turbulence and light source parameters. At the same transmission distance, the normalized M^2 factor first increases and gradually tends to a stable value with an increase in the coherence length. Figures 9.20b, d depict the effects of the bundle radius and transmission distance, respectively, on the normalized M^2 factor. Referring to Fig. 9.20b, a comparison of the cases $\sigma_s = 0.01$ m, $\sigma_s = 0.05$ m, and $\sigma_s = 0.05$ m indicates that the trend of the curve at $\sigma_s = 0.01$ m is different from that in the other two cases. This phenomenon can be explained through Fig. 9.20d: the normalized M^2 factor increases with the bundle radius, decreases sharply when it is less than a certain value (the value is related to the transmission distance and coherence length), and then gradually decreases with an increasing trend. This can also be explained by the variation in the trend of the weighting factor at $\sigma_s = 0.01$ m in Fig. 9.17b, which explains the variation in the beam quality. It can also be seen from the trend of the curve presented in Fig. 9.20d that the presence of the radius of the beam waist can optimize the beam quality, which is consistent with Fig. 9.17b.

9.7.2 Effect of Atmospheric Turbulence on M^2 Factor

Figure 9.21 depicts the effect of turbulence on the normalized M^2 factor. Figure 9.21a, c indicates that the smaller l_0, the larger C_0 and the normalized M^2 factor are. Figure 9.21b indicates that the extraturbulence scale has little effect on the normalized M^2 factor. This conclusion is consistent with that represented in Fig. 9.19. As the effect of turbulence increases, the diffraction becomes more severe, number of modes increases, beam wavefront is affected, correlation of the wavefront phase decreases, normalized M^2 factor increases, and beam quality decreases. These conclusions are in agreement with those presented in the Ref. [21].

9.8 Numerical Simulation

As described in this section, the phase screen is mostly used to simulate the atmospheric turbulence, and the light intensity distribution of the beam is numerically simulated when it is transmitted along various distances under moderate turbulence.

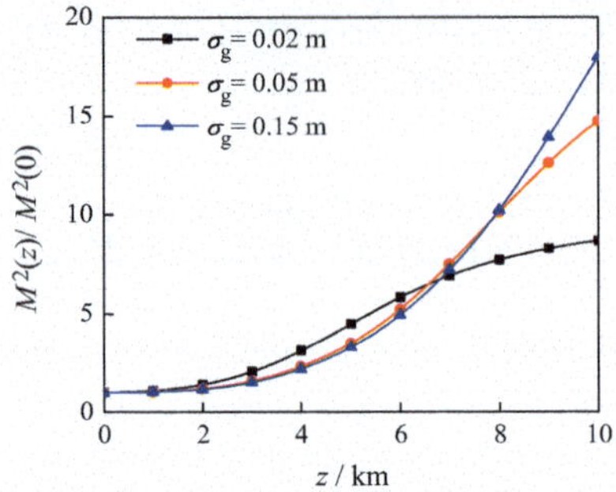

(a) Variation in normalized M^2 factor with respect to the transmission distance for various coherence lengths

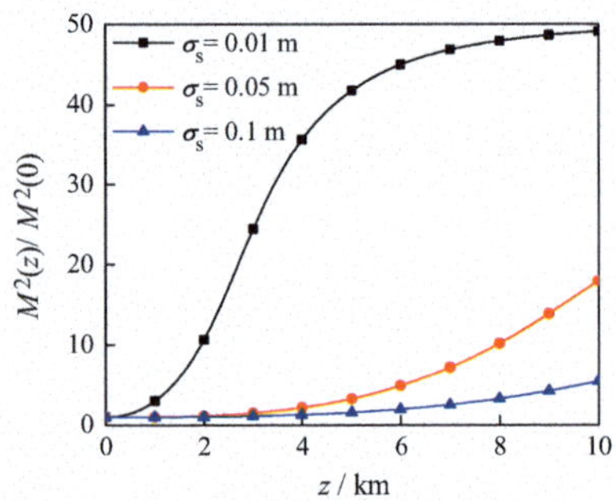

(b) Variation in normalized M^2 factor with respect to the transmission distance for various bunching radii

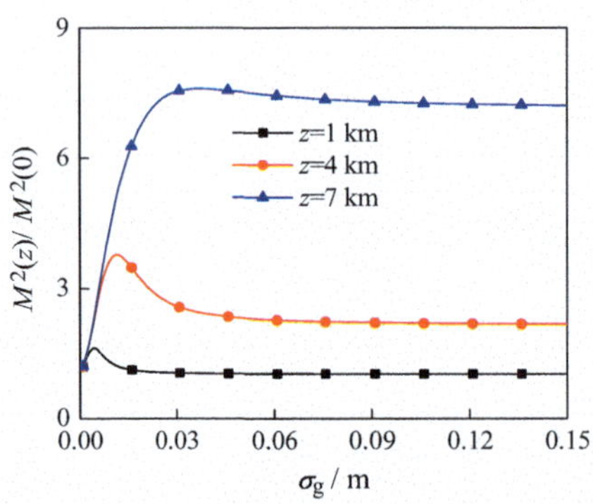

(c) Coherence length

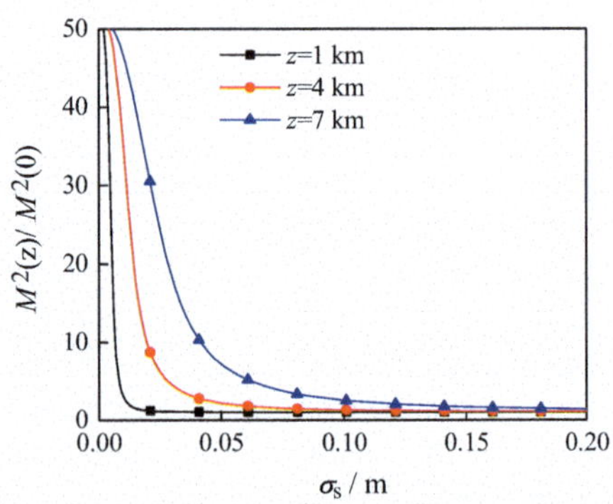

(d) Girdle radius

Fig. 9.20 Effect of light source parameters on the normalized M^2 factor [22]

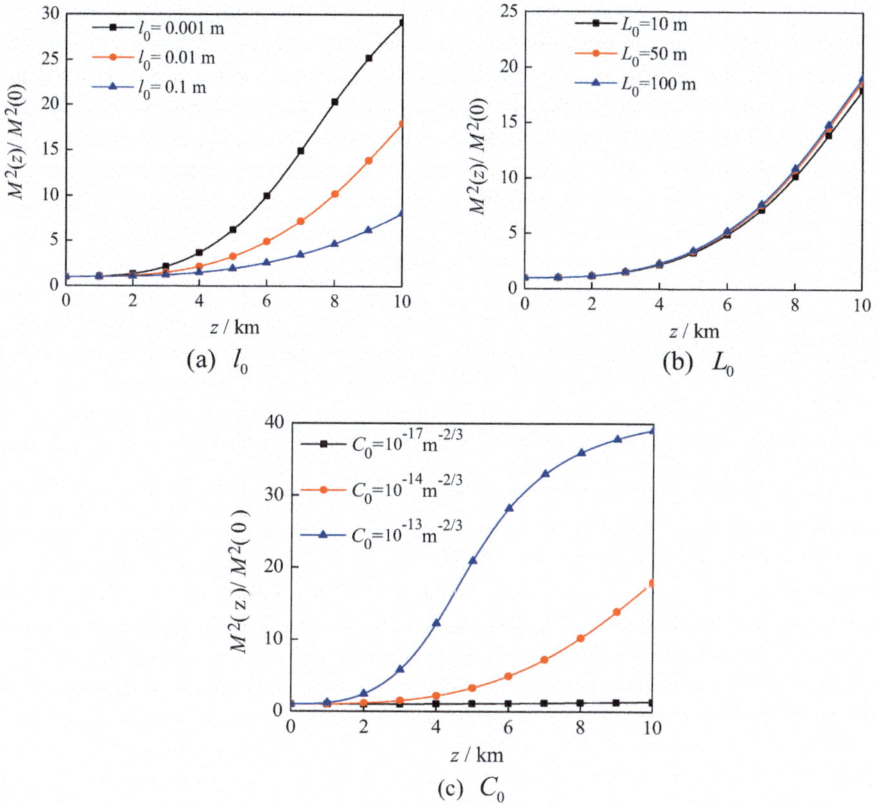

Fig. 9.21 Effect of turbulence on the normalized M^2 factor [22]

The distribution of the mode-weighting factor is calculated based on the light intensity distribution to verify the theoretical analysis presented in the preceding sections of this chapter.

9.8.1 Calculation Method of Numerical Simulation Beam Pattern

The theoretical derivation and simulation of the beam-mode variation under turbulence were described in the preceding sections, and the results indicate that the mode of the beam is related to the light source parameters, turbulence parameters, and transmission distance. In this section, the numerical simulation of the variation in the mode of beam transmission under turbulence using the phase screen theory is described. The transmission of beams was numerically simulated under turbulence to obtain the light intensity information for various turbulent intensities, extract the

1D light intensity information contained in the light spot, and solve for the mode coefficients. Because the light spot drifts and splits after the beam passes through the random phase screen, the light intensity distribution is no longer an ideal Gaussian function and there will be multiple extreme points. When we want to solve for the mode coefficients based on the directly extracted light intensity information, a large error may occur; however, based on the phase screen theory, it is known that the light intensity distribution of the beam transmitted under turbulence is always a Gaussian function no matter how large the turbulence and transmission distance are. Therefore, we adopt the curve-fitting method to fit the Gaussian function to the extracted light intensity information, determine the parameters of the Gaussian function to obtain the Gaussian curve of the light intensity information after the beam passes through turbulence, and carry out mode decomposition on the curve to obtain the mode coefficients.

The 1D light intensity, after passing through the phase screen, is represented by I. When $N = 256$, I is a 1D vector (1×256), the orthogonal mode system uses the H-G mode system with the same number of points, the order N is the simulation parameter, and $n = 18$. Therefore, the orthogonal mode matrix is an $N \times n$ matrix and the coefficient matrix is a $1 \times n$ matrix, which can be expressed as follows:

$$[I_0 \quad I_1 \quad \cdots\cdots\cdots I_{N=255}] = [C_0 \quad C_1 \quad \cdots\cdots\cdots C_{n=17}] \begin{bmatrix} H_{00} & H_{01} & \cdots\cdots\cdots & H_{0N=255} \\ \vdots\ \vdots & & \cdots\cdots & \vdots\ \vdots \\ \vdots\ \vdots & & \cdots\cdots & \vdots\ \vdots \\ H_{n=17,0} & H_{n=17,1} & \cdots & H_{n=17N=255} \end{bmatrix} \tag{9.94}$$

Thus, the coefficient matrix is expressed as follows:

$$C = I \cdot H^{-1} \tag{9.95}$$

9.8.2 *Effect of Transmission Distance on Mode Coefficients*

For beam transmission under turbulent flow, with an increase in the transmission distance spot energy decay, the beam size becomes larger and, under strong turbulence, even spot breakup can occur, affecting the beam pattern. In this section, an analysis of the variation in the beam pattern at various distances through numerical simulation using the phase screen theory is described. The following parameters are used [19, 23]: beam waist radius $w_0 = 0.05$ m, $\lambda = 1550$ nm, $C_n^2 = 10^{-14}$ m$^{-2/3}$, grid number $N = 256$, turbulence window width $L = 0.4$ m, intraturbulence scale $l_0 = 0.01$ m, extraturbulence scale $L_0 = 10$ m, $c = 3 \times 10^8$ m/s. The mode-coefficient distributions are analyzed for $z = 1$, 3, and 5 km.

As depicted in Fig. 9.22, the spot at $z = 1$ km is a complete spot, which still conforms to the Gaussian function distribution, except that its energy is attenuated and the light intensity is decreased. Therefore, the beam is transmitted under close weak and moderate turbulences, the turbulence has a lower impact on the beam, and the spot information is more complete.

Figure 9.23 depicts the extracted 1D light intensity information using the Curve Fitting toolbox in the Matlab application. The Gaussian function was fitted to the first order, and it can be seen from the fitted plot that the light intensity distribution after the phase screen is very similar to the standard Gaussian function. The values of the parameters for the fitted Gaussian function are as follows: $a_1 = 0.8556$,$b_1 = 0.00551$, $c_1 = 0.03564$; the standard expression for the Gaussian function is $f(x) = a_1 * \exp\left(-\left((x - b_1)/c_1\right)^2\right)$, the fitted valuation parameters variance (SSE) = 0.03858, coefficient of determination (R-square) = 0.9976, and root mean square error (RMSE) = 0.01235.

The fitted expression is used as the target light intensity and the H-G beam is used as the orthogonal mode system for mode decomposition; the results of the decomposition are presented in Fig. 9.24. The figure presents the variation in the weighting factor with the mode order at $z = 1$ km, from which it can be seen that the fundamental mode is mostly dominant at $z = 1$ km and the content of other higher-order modes is relatively low.

Figure 9.25 depicts the light intensity distribution when the beam is transmitted under turbulence for 3 km. It is evident from the figure that the intensity of the beam decreases owing to the impact of turbulence, the spot is no longer a complete circle, and the radius of the spot increases. Therefore, with an increase in the transmission distance, the turbulence effect becomes increasingly evident, the beam transmission

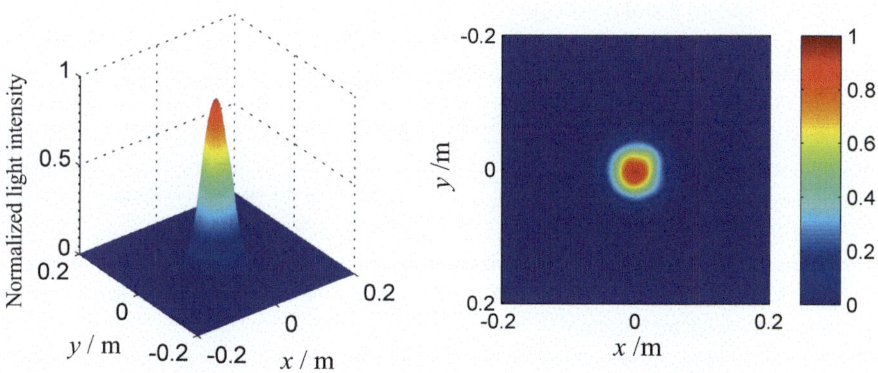

(a) 3D map (b) 3D diagram

Fig. 9.22 Light intensity distribution for $z = 1$ km [22]

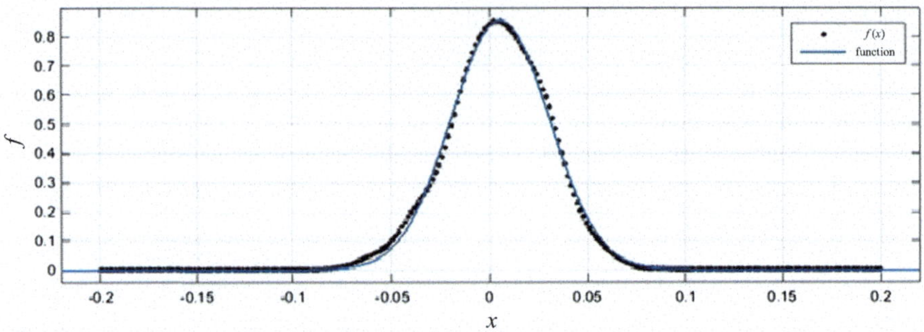

Fig. 9.23 Curve depicting 1D light intensity Gaussian fitting at $z = 1$ km [22]

Fig. 9.24 Distribution of weighting coefficients at $z = 1$ km [22]

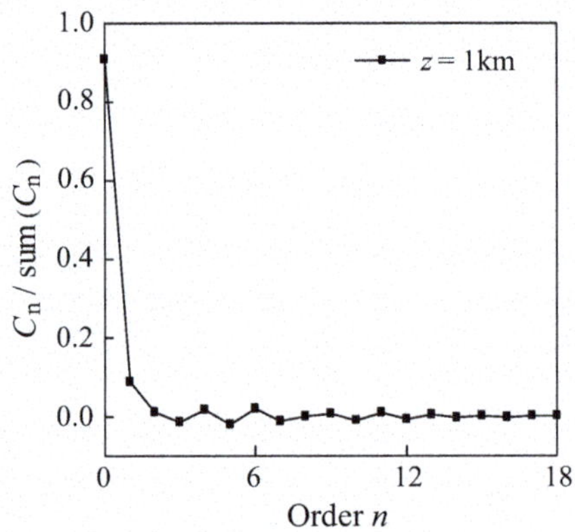

under turbulence is increasingly affected, and the distribution of the spot energy no longer conforms to a standard Gaussian function.

As depicted in Fig. 9.26, the distribution of the light intensity is no longer a circular curve, but all points are distributed in an undulating form resembling the standard Gaussian curve, that is, the general trend still follows a Gaussian function. Similar to the representation in Fig. 9.23, a first-order fit using a Gaussian function is obtained, where $a_1 = 0.4874, b_1 = -0.01765, c_1 = 0.05273$ and the fitted evaluation parameters are SSE = 0.3272, R-square = 0.9473, and RMSE = 0.03596.

Figure 9.27 depicts the proportion of mode coefficients of the beam after 3 km of beam transmission under turbulence, from which it can be seen that the fundamental-mode content decreases and the number of beam modes increases after 3 km of beam transmission under turbulence.

Figure 9.28 depicts the light intensity distribution of the beam after 5 km of transmission under moderate turbulence, from which it can be seen that the light intensity decays very severely and the spot also drifts and widens.

The discrete points depicted in Fig. 9.29 represent the extracted 1D light intensity distribution after 5 km of transmission, and the light intensity profile is fitted using

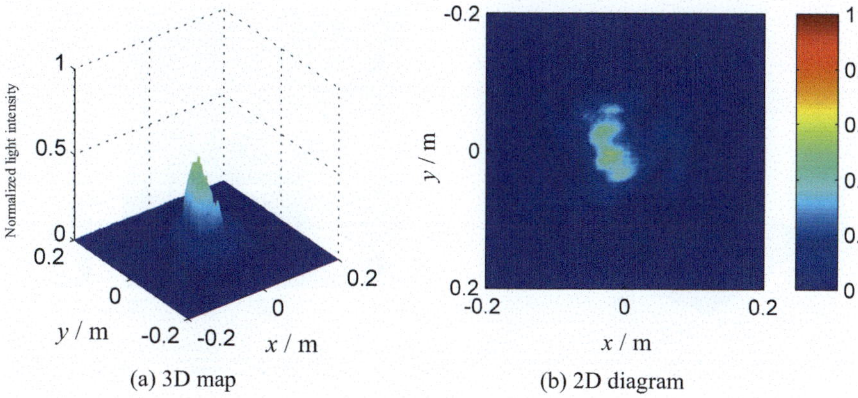

Fig. 9.25 Light intensity distribution at $z = 3$ km [22]

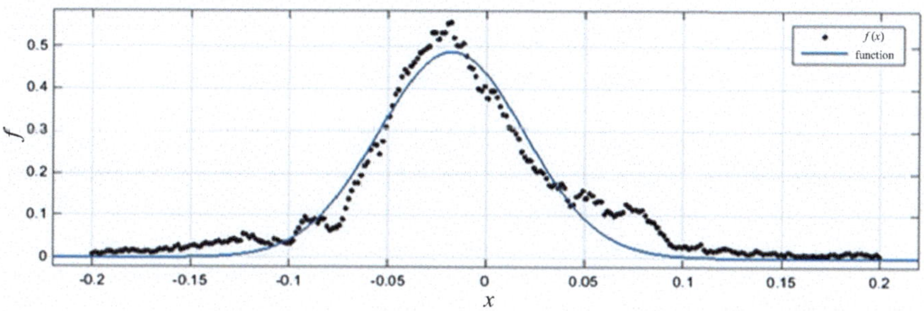

Fig. 9.26 Curve depicting 1D light intensity Gaussian fitting at $z = 3$ km [22]

Fig. 9.27 Distribution of weighting coefficients at $z = 3$ km [22]

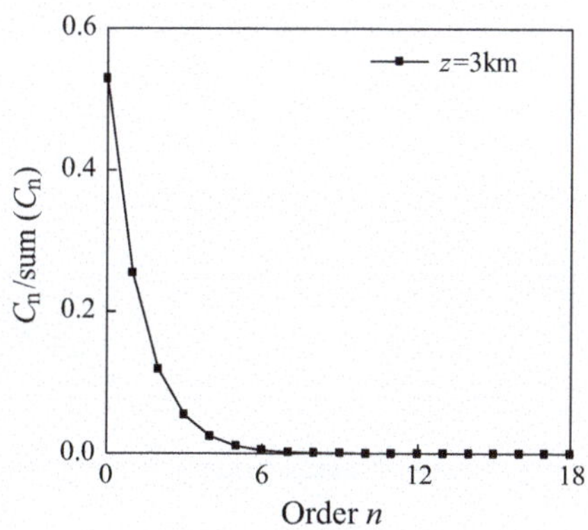

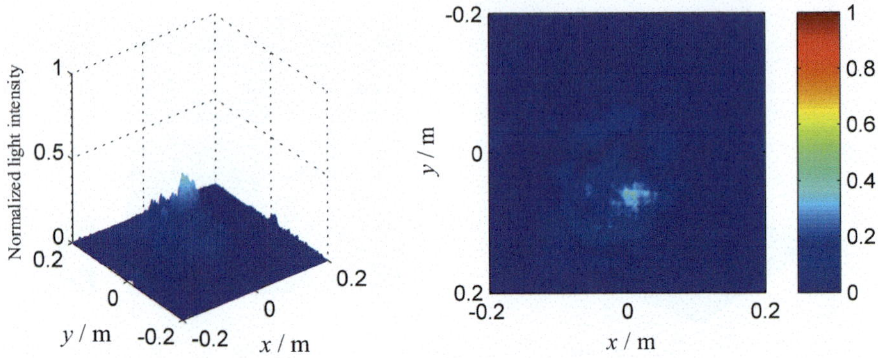

(a) 3D map (b) 2D diagram

Fig. 9.28 Light intensity distribution at $z = 5$ km [22]

the same fitting method to obtain the parameters $a_1 = 0.2715$, $b_1 = -0.003948$, $c_1 = 0.06041$; the fitted evaluation parameters are SSE = 0.1766, R-square = 0.9231, and RMSE = 0.02642.

Figure 9.30 depicts the distribution of the mode-weighting coefficients after 5 km of beam transmission. From the figure, it can be seen that when the transmission distance is relatively long, the distribution of the weighting coefficients is different from the theoretical one because of the randomness of the turbulent phase screen and the fitting error, but it is consistent with the theoretical one that is applicable to the general trend.

After analyzing the distribution of the mode-weighting coefficients at $z = 1$, 3, and 5 km, it can be seen that with an increase in the transmission distance, the beam intensity decreases, spot radius becomes larger, and spot drifts. Further, when the

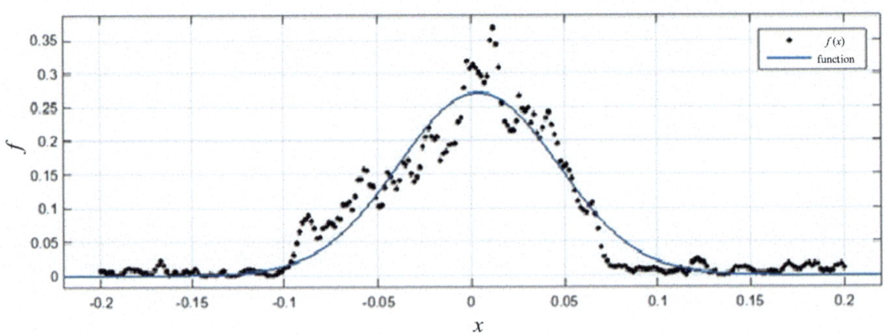

Fig. 9.29 Curve depicting 1D light intensity Gaussian fitting at $z = 5$ km [22]

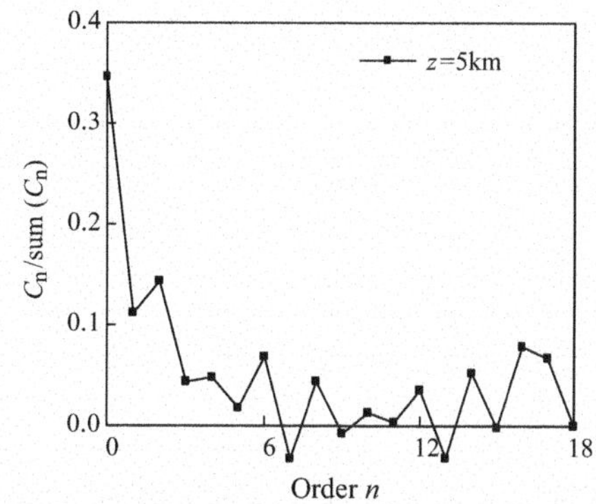

Fig. 9.30 Distribution of weighting coefficients at $z = 5$ km [22]

transmission distance is larger, the spot splits and is no longer intact; the fundamental-mode content also decreases and diffracts the higher-order components. In atmospheric laser communication, the mixing efficiency at the receiver decreases and the communication performance deteriorates owing to the change in mode.

9.9 Experimental Studies

9.9.1 Experimental Principle

The experiments focus on measuring the spot, Gaussian fit, and Gaussian roughness and, thereafter, using the Gaussian fit and roughness to reflect the fundamental-mode content of the beam following transmission. The Spiricon beam analyzer defines the Gaussian fit as either a least-squares fit for all the data or a fit to two univariate equations performed on orthogonal data lines defined at the location of the light scale. Here, the fit is calculated using the ISO method, and the roughness of fit is equal to the maximum deviation of the theoretical fit from the measured distribution as follows:

$$R = \frac{\left| E_{ij} - E_{ij}^{f} \right|_{\max}}{E_{\max}} \tag{9.96}$$

where E_{ij} is the light intensity distribution of the input beam analyzer, E_{ij}^{f} is the theoretical Gaussian light intensity distribution, and $0 \leq R < 1$; the smaller the value of R, the better. The distribution of the light field of the fundamental-mode Gaussian beam is a standard Gaussian function, whereas for the higher-order modes, the distribution of the light field is no longer a Gaussian function. Therefore, it is reasonable to measure the amount of fundamental-mode content using the Gaussian fit; the larger the amount of fundamental mode, the better the Gaussian fit.

9.9.2 Experimental Results

The beam parameters were measured under various weather conditions, as depicted in Fig. 9.31; parts (a) and (b) depict the spot diagram measured in clear and cloudy weather, respectively. The crossed lines in the diagram are used to determine the location of the spot center of mass, and it can be seen that the spot is hollow, which is primarily owing to the fact that the receiving telescope in the experiment is a Cassegrain telescope with a 20% blocking ratio. A comparison of Figs. 9.31a, b indicates that the spot light intensity depicted in (a) is significantly weaker, spot radius is larger, and beam quality is worse; the spot in (b) is more complete, energy is concentrated, and beam quality is better. This is attributed mainly to the fact that the turbulence is higher in clear weather and, therefore, has a more significant effect on the beam.

Table 9.3 presents the Gaussian fit information measured by the beam analyzer. It can be seen that the Gaussian fit is smaller and rougher on sunny days when compared with cloudy days. The measured data are consistent with the theory; both indicate that the intensity of turbulence is higher on sunny days than on cloudy days, and the beam is affected more by turbulence, which implies that the fundamental-mode content of the beam after transmission is lower on sunny days when compared with cloudy days.

Figure 9.32 depicts the measurement of light intensity information at different moments on the same day and under different weather conditions. Parts (a), (b), (c), and (d) depict the light intensity distribution measured during moderate rainfall,

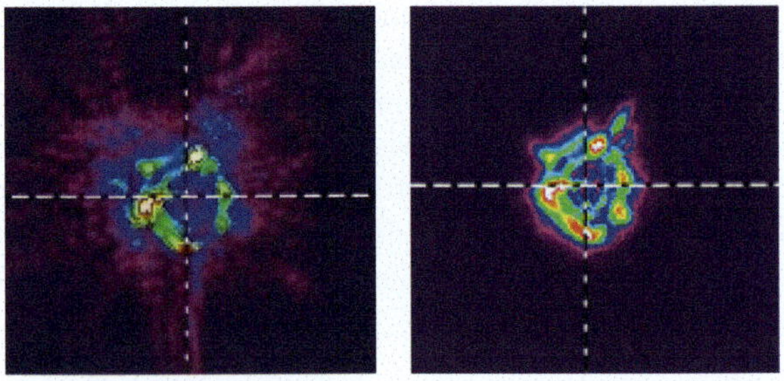

(a) Sunny day (b) Cloudy day

Fig. 9.31 Light spot distribution under different weather conditions [22]

Table 9.3 Gaussian fitting parameters under different weather conditions

Measurement/ Weather	2D fit coincidence	*X*-axis fit coincidence	*Y*-axis fit coincidence	2D fit roughness	*X*-axis fit roughness	*Y*-axis fit roughness
Sunny day	0.731	0.589	0.578	0.762	0.616	0.665
Cloudy day	0.8	0.621	0.624	0.762	0.583	0.605

during light rainfall, after the rain stops, and at night after the rain stops, respectively. The four graphs indicate that on days with moderate rain, the light spot is more scattered and slightly larger and the light intensity distribution is not uniform. With a decrease in the rainfall, the light spot energy increases, the light intensity distribution becomes increasingly uniform, and the quality of the light spot becomes better towards the night.

Table 9.4 presents the distribution of the 2D fit agreement and fit roughness of the light intensity at moments corresponding to those selected for Fig. 9.32. The table indicates that the fit agreement increases and roughness decreases with decreasing rainfall. Therefore, the fundamental-mode content of the beam increases as the rainfall decreases.

The partially coherent GSM model beam is used as the object of study, and the expression for the weighting factor of the GSM beam at the receiver is derived using the generalized Huygens–Fresnel principle and incoherent mode decomposition theory. The effects of the optical parameters and turbulence factor on the weighting factor and normalized M^2 factor are analyzed through numerical simulation using

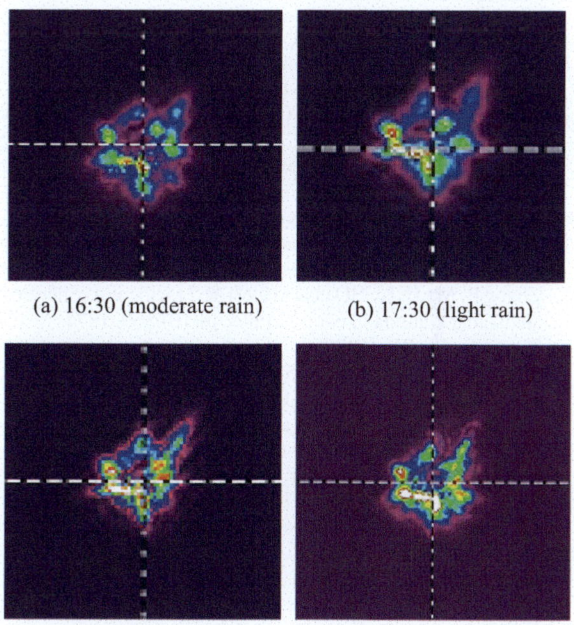

(a) 16:30 (moderate rain) (b) 17:30 (light rain)

(c) 20:30 (no rain) (d) 22:30 (no rain)

Fig. 9.32 Light intensity distribution at different moments on the same day [22]

Table 9.4 Gaussian fitting parameters at different moments

Measurement/time	16:30	17:30	20:30	22:30
2D fit	0.42	0.65	0.72	0.78
2D fitted roughness	0.997	0.851	0.814	0.71

the equation method, and the effect of the transmission distance on the beam-mode-weighting factor is numerically simulated by modeling the atmospheric turbulence with a phase screen. Finally, the fundamental-mode content under different turbulence intensities is verified based on the coherent communication system. The results indicate that there exists a maximum value of the fundamental-mode content with respect to the variation in the radius of the beam waist. This maximum value increases with an increase in the coherence length, transmission distance, and intensity of turbulence. The turbulence has the least impact on the beam mode and quality when the GSM beam is transmitted downstream: the smaller the intraturbulence scale, the larger the atmospheric refractive index structure constant and lower the fundamental-mode content; the extraturbulent scale does not affect the beam mode at the receiver. The M^2 factor is calculated using the mode decomposition method, and the simulation results are consistent with the results obtained from the calculation of the M^2 factor based on the higher-order moments.

References

1. Wu HS, Yan P, Gong ML et al (2002) $M^2 \leq 1.14$ diode pumped Yb: YAG microchip laser. China Laser 29(11):961–964
2. Starikov A, Wolf E (1982) Coherent-mode representation of Gaussian Schell-model sources and of their radiation fields. J Opt Soc Am A 72(7):923–928
3. Gori F, Santarsiero M, Borghi R et al (1998) Intensity-based modal analysis of partially coherent beams with Hermite-Gaussian modes. Opt Lett 23(13):989–991
4. Siegman AE (1990) New developments in laser resonators. Optical Resonators 1224:2–14
5. Tan ZK (2019) Influence factors and experimental study on the performance of differential coherent detection in wireless optical communication. Xi'an University of Technology, Xi'an
6. Zhang B, Ma H, Lv BD et al (1999) M 2 factor and coherent mode decomposition of hyperbolic sinusoidal Gaussian beams. J Phys 48(10):869–1874
7. Zhang B, Lv BD (1999) Mode correlation and coherent mode representation of multimode lasers beams. J Phys 48(1):58–64
8. Shirai T, Dogariu A, Wolf E et al (2003) Mode analysis of spreading of partially coherent beams propagating through atmospheric turbulence. J Opt Soc Am A 20(6):1094–102
9. Born M, Wolf E (2000) Principles of optics. Cambridge university press
10. Li QC, Wang TF, Zhang HY et al (2015) The performance of heterodyne detection system for partially coherent beams in turbulent atmosphere. Optics Commun 356:620–627
11. Osche GR (2002) Optical detection theory. Wiley, New York
12. Salem M, Rolland JP (2010) Heterodyne efficiency of a detection system for partially coherent beams. J Optical Soc Am A 27(5):1111–1119
13. Andrews LC, Phillips RL (2001) Laser beam scintillation with applications. SPIE Optical Engineering Press, Bellingham, Washington USA
14. Alavinejad M, Ashiri F, Ghafary B (2008) Propagation properties of partially coherent beams through turbulent media with coherent modes representation. Appl Optics 38(3): 549–558
15. Jeffrey A, Zwillinger D (2007) Table of integrals, series, and products. Elsevier, USA
16. Andrews LC, Phillips RL (2005) Laser beam propagation through random media. SPIE Optical Engineering Press, Bellingham, Washington USA
17. Tanaka K, Ohta N (1987) Effects of tilt and offset of signal field on heterodyne efficiency. Appl Optics 26(4): 627–632
18. Li CQ, Zhang HY, Wang TF et al (2013) Coherent properties of Gauss-Schell mode beam transport in atmospheric turbulence. J Phys 62(22):191–197

19. Kaimal JC (1972) Spectral characteristics of surface-layer turbulence. Q J Roy Meteor Soc 98(2):563–589
20. Wei HY, Wu ZS, Peng H (2008) Scattering from a diffuse target in the slant atmospheric turbulence. Acta Physica Sinica 57(10):6666-6671
21. Ke XZ, Wang WT (2015) Light intensity and expansion of partially coherent light in oblique range and horizontal atmospheric turbulence. J Appl Sci 33(2):142–154
22. Mao YZ (2018) Study of mode degeneracy and polarization characteristics in coherent optical communication. Xi'an University of Technology, Xi'an
23. Wu J (1990) Propagation of a Gaussian-Schell beam through turbulent media. Optica Acta Internation Journal of Optics 37(4):671-684

The manufacturer's authorised representative in the EU is Springer Nature Customer Service Centre GmbH, Europaplatz 3, 69115 Heidelberg, Germany. If you have any concerns regarding our products, please contact ProductSafety@springernature.com

Printed and bound by CPI Group (UK) Ltd, Croydon, CR0 4YY
30/09/2026
02225315-0015